AF769643

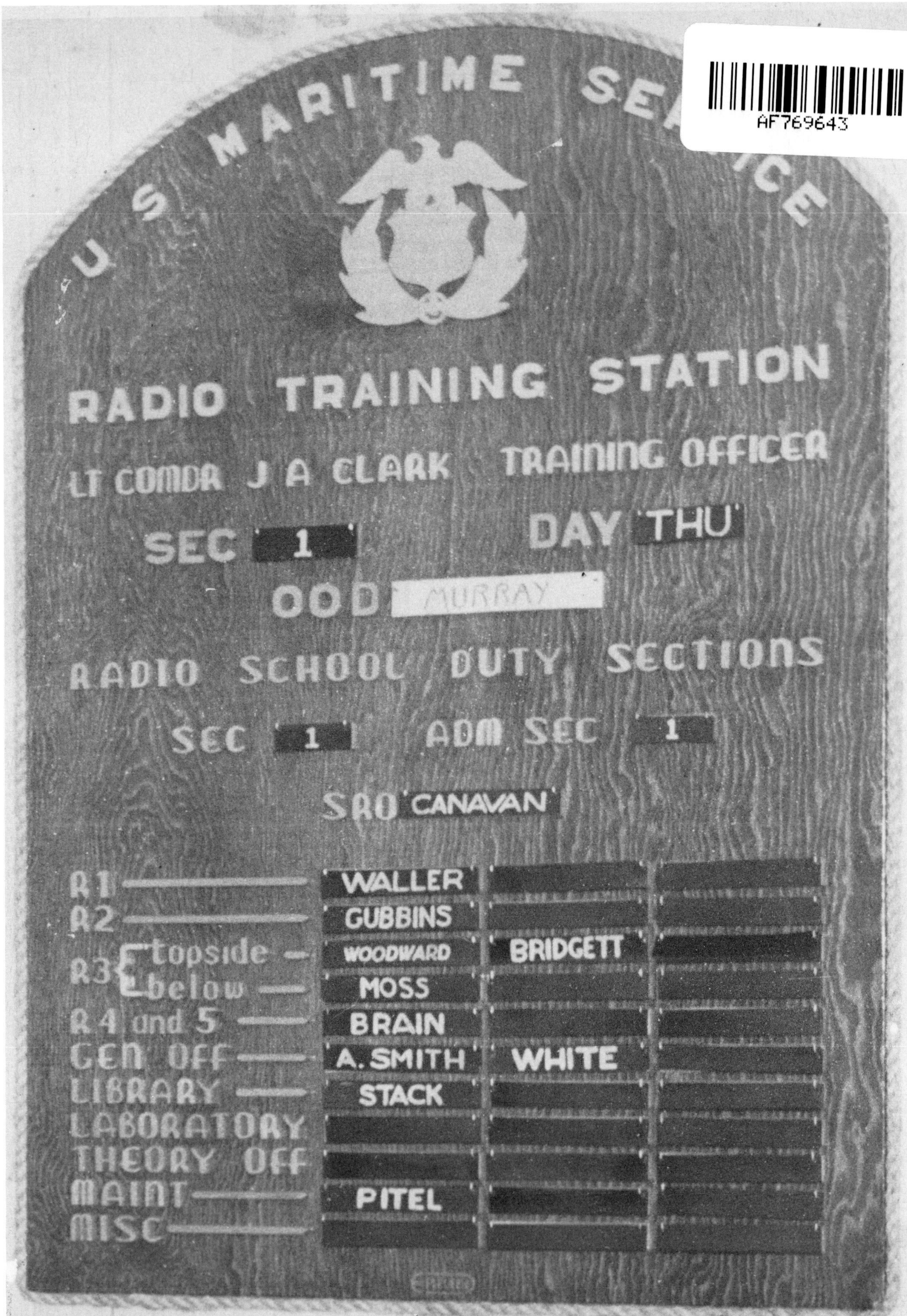
U S MARITIME SERVICE
RADIO TRAINING STATION
LT COMDR J A CLARK TRAINING OFFICER
SEC 1 DAY THU
OOD MURRAY
RADIO SCHOOL DUTY SECTIONS
SEC 1 ADM SEC 1
SRO CANAVAN
R1 WALLER
R2 GUBBINS
R3 topside WOODWARD BRIDGETT
below MOSS
R4 and 5 BRAIN
GEN OFF A. SMITH WHITE
LIBRARY STACK
LABORATORY
THEORY OFF
MAINT PITEL
MISC

WE CAME FROM ALL OVER, WE WENT EVERYWHERE

This Was Gallups Island

"Going Home", an original watercolor of a war weary Liberty Ship returning victorious after the war by Jerry McClish, A-2.

Gallups Island Radio Association

A History of Gallups Island, Its Veterans and Their Role in Our Freedom

Turner Publishing Company
Paducah, Kentucky

Turner Publishing Company
P.O. Box 3101
412 Broadway
Paducah, KY 42002-3101

Copyright © 1994 Gallups Island Radio Association
All Rights Reserved

Coordinated and Designed by:
David A. Hurst, Publishing Consultant

This book or any part thereof may not be reproduced without the written consent of the Publisher.

This book was compiled and produced using available and submitted information; the Publisher and Gallups Island Radio Association regret they cannot assume liability for errors or omissions.

Library of Congress Catalog Card No. 93-061020
ISBN: 1-56311-115-2
Printed in the USA

Additional books may be available directly from the Publisher.

CONTENTS

(Sections in light-face Italics are Pictorials)

WE CAME FROM ALL OVER . . . WE WENT EVERYWHERE

This was Gallups Island

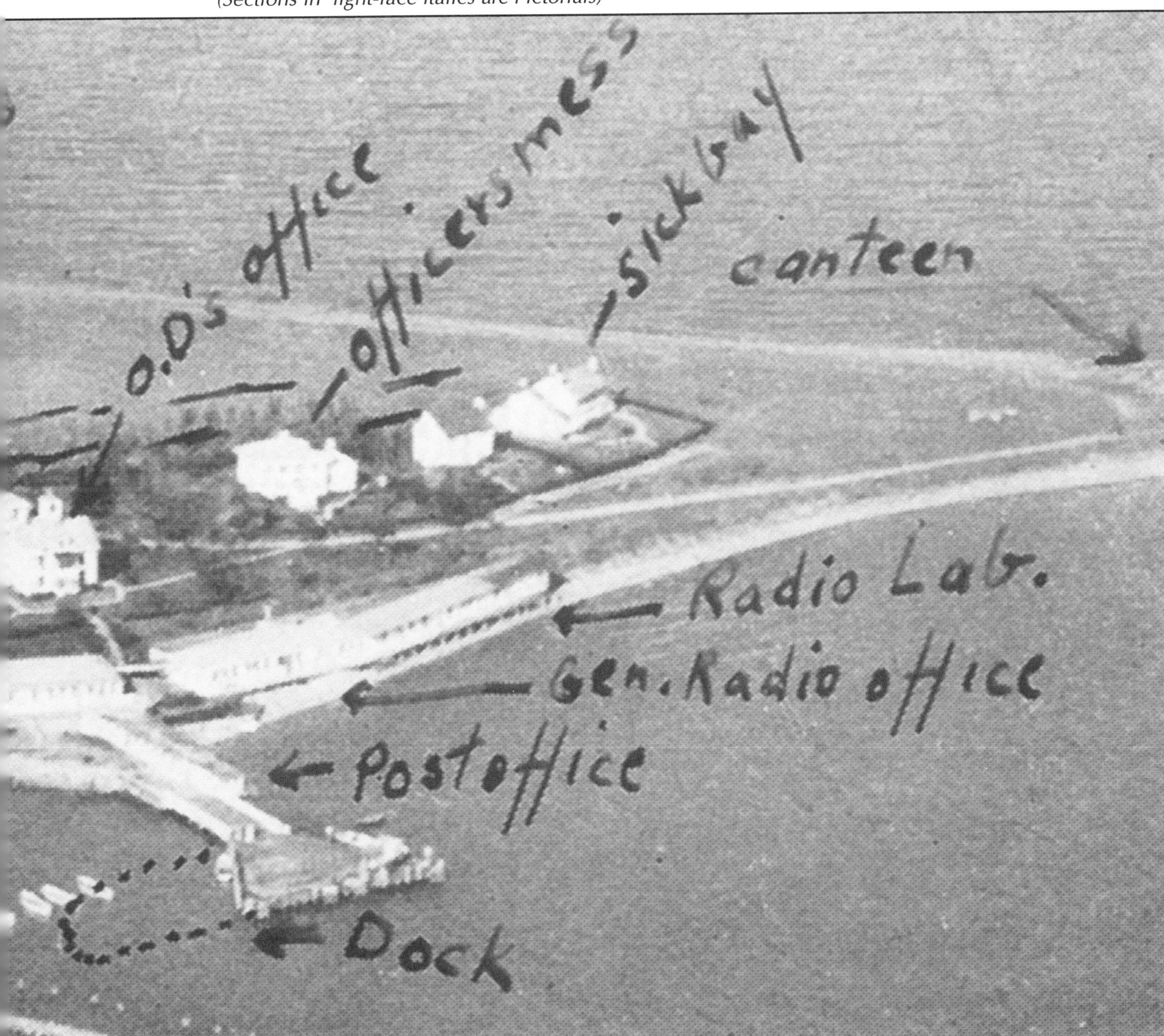

*This book is dedicated to **Woodward Rion Dixon**, R-17, President of GIRA, 1992-1993, who believed these stories were worth preserving and whose efforts made this book possible.*

Gallups Island Radio Association

Dedicated to those who served as Radio Officers in the U.S. Merchant Marine during World War II.

Gallups Island Radio Association
7151 E. Speedway Blvd., Box 108
Tucson, AZ 85710-1317

Publication:	The Gallups Islander
Editor:	Stan Jennings, R-58
Production Editor:	Jane Jennings, XYL

Publication:	The Spark Gap
Editor:	Tom Cruse, R-50

Officers 1992-1993

President:	Woodward Rion Dixon
Vice President:	Robert Lanphere
Sec/Treasurer:	Raymond Anderson

Officers 1994-1995

President:	Samuel Hucke
Vice President:	Lester Rauber
Sec/Treasurer:	William Corcoran

Directors

Region 1: Raymond E. King
Region 2: Stephen C. Garcia
Region 3: Frank Conwell
Region 4: William Yount
Region 5: Joseph Wagner
Region 6: Robert Mitchell
Region 7: Kenneth Long
Region 8: Arnold O. Jangord
Region 9: James A. Jolly
Region 10: Don Runmark

Photo on pages 2-3: Gallups Island before there was a radio school. Dotted line by the dock is present size. Square lines back of OD's office represent two long buildings, one housing trainees and the other having four lecture rooms, one seaman lecture room and five code rooms. Naturally, X marks my barracks. The terrain looks flat, but it is plenty hilly. It's a steep trip from the dock to the OD's office. Tide was out at the time of the picture.—Ed Frank (R-30)

Introduction

When I was in Boston at Gallups Island, my brother back in Indiana sent me a post card. He had pasted thereon a news clipping from Time magazine, which read: A man in Boston jumped into a taxi which in wartime practice he was sharing with a complete stranger. As the cab pulled out he said to his new companion, "My name is Jennings." The other more proper Bostonian replied, "Mine isn't."

My brother wrote one line beneath the clipping, for me: "Don't do this again."

And this is good old Boston,
The home of the bean and the cod,
Where the Lowells talk only to the Cabots
And the Cabots talk only to God.

But Boston was changing. The famous lines above were written by an Irishman. Another Irishman soon would represent the Harvard constituency in Congress and defeat a Cabot on the way to the U.S. presidency.

Gallups Island radio operators contributed to the change. We came from all over . . . we went everywhere. But while at Gallups we gathered occasionally on the Boston Common. We dated delightful Boston girls. We did sleep in the Boston jail. We did go to Scollay Square. This outlander even went to Harvard! For one night. As guest of a Navy J.G. quartered there, whom he had known in high school.

Says Bob Mitchell (R-34), ". . . many other things we remember about Boston. Sure, I'll bet some of us went to the Old Howard in Scollay Square, but how about the Boston Common, or Boston Garden, or Symphony Hall, or Merchants Building, the Museum of Modern Art, the Natural History Museum, or the Old North Church, or the *USS Constitution,* Bunker Hill, Faneuil Hall, Lexington and Concord, and so much more."

Yes, yes, Bob, but your list, too, is only a part of the story. For GIRA members, WWII radio operators, Gallups Island and Boston were just our first port of call. We all saw many things, different things, there. Like the courses at Gallups, even when you go through them, no one can comprehend it all. Fats Waller put it, "One never knows, do one?"

But we can try, with the help of all these people who were there.

—Stan Jennings, Editor, The Gallups Islander

BOUILLABAISSE

A HISTORY OF GALLUPS ISLAND BY SPARKS HIMSELF.... SPARKS IS A PLURAL PERSONALITY

ART BRADFORD • JAMES SMART • JERRY McCLISH • TOM CRUSE
ED HAYDEN • JIM KINKEL • RAY C. MOORE • AL HADAD
RION DIXON • BILL ANDERSON • MRS. SHERMAN REED • ED FRANK
RAY ANDERSON • ELLENOR YAHRMARKT • KEVIN GALVIN
KEN WALLER • CHET SZYMANSKI • HARLEY A. OLSON • VINCENT KIRK
STEPHEN C. GARCIA • DAVE BULKLEY • RALPH ALBERS • BOB CLOUGH

. . . All had a hand in preparing these histories of the WWII Radio Operators aboard U.S. merchant ships. So did the 5000 Operators who sailed from Gallups Island.

Boston Harbor Islands

Boston Harbor, part of the Boston Basin, is a geologic phenomenon created millions of years ago by a shift in the earth's crust. In a later geologic period, glaciers covered the Boston area, grinding down rock ridges and leaving behind new hill formations called drumlins. The subsequent gradual sinking of the land and the rising of the sea level isolated the easternmost drumlins and rocky ridges, forming the Boston Harbor Islands. The more than 30 Islands are spread over 50 square miles of the Harbor, and include 1200 acres of land.

Remains of campsites and other signs provide evidence that Indians frequented the Harbor Islands as far back as 6,000 years ago. Captain John Smith visited in 1614 and described Boston Harbor as a paradise, ". . . for here are many isles all planted with corn, groves, mulberries, savage gardens, and good harbors. The seacoast . . . shows all along large cornfields, and great troops of well-proportioned people."

Colonial farms, eighteenth century resorts, and numerous fortifications dotted the islands. Remains of these previous uses are still visible today.

The islands' unique geology, history, and abundant natural resources combine to make the Boston Harbor Islands one of the most exciting Parks in the Nation.

1990—Two aging Sparks revisit where it all began.

Photo by Stan Jennings

Chronology of Gallups Island

1630's Captain John Gallop acquired Gallops Island.
1636 Nearby island of Nix's Mate granted to John Gallops for use as pasture. Since eroded, is now site of lighthouse.
1649 Gallops Island listed in John Gallop's will as worth 75 pounds.
1650 The "MUNT" on Nix's Mate used as hayfield. John Gallop dies. Value of Nix's Mate, 12 pounds. Cites island as 16 acres.
1669 Joseph Rock paid Edward Tying 20 pounds for parts of Gallops from Henry Kemple, a Boston blacksmith who obtained the land from Gallops heirs around 1650.
1683 Rock sold his share to Edward Bromfield.
1689 Hung on Nix' Mate: Tom Pound, Thomas Hankins. These were criminals, possibly pirates, as were the ones listed below.
1704 John Quelch and five men executed. (Where?)
1726 Executed pirates buried on Nix's Mate.
1750's Elisha Leavitt and James Brackett purchased island. James sold his parcel to Elisha.
1790 Leavitt left the island to Caleb Rice, his grandson. Lemuel Brackett of Quincy paid $1,630 for the island.
1819 Island purchased by Margaret and Peter Newcomb on July 1st.
1822 Estimated date of construction of beacon on Nix's Mate.
1827 $2,426.51 appropriated to cut down cliff on Gallops Island—now spelled Gallups.
1833 Widowed Margaret Newcomb established restaurant/inn on Gallups. When Mrs. Newcomb died, her son Charles sold the island to the city of Boston for $6,500.
1855 Inn purchased by Joseph Snow. Peggy's Point named after his daughter, used to be called Beachy Point.
1856 State bought island for use as pauper's home for Boston.
1860 Purchased by City of Boston for $6,600, then dedicated to the federal government as campgrounds for the Civil War. Civil War barracks constructed. 3,000 Union troops on the island.
1865 July 6, 38th Massachusetts Regiment arrived on Gallups, discharged a week later.
1865 August 28, 54th Regiment reached Gallups to be discharged September 2.
1866 Quarantine Hospital built, operated 1867-1937.
1870 North and west sides of granite seawall built.
1881 In fifteen years since 1866, 181,765 persons placed on the island afflicted with disease, most likely smallpox or yellow fever.
1882 King's Handbook states 700 bushels of vegetables yielded annually, 10 tons of hay, milk and butter for local demand.
1908 August—309 immigrants from Portugal detained, suspected of bubonic plague.
1916 Federal government paid Boston $150,000 and built immigration station. US Public Health Service Quarantine Station from Rainsford Island to Gallups.
WWI German prisoners interned here for the duration of the war—possibly.
1920 Immigration station moved off.
1927 Dr. Sweeny, director of hospital, planted 2500 pine trees and other ornamental shrubs.
1934 Trees planted by CCC youth.
1937 Quarantine station disbanded, moved to Boston.
1940 Maritime Radio School established by Coast Guard.
1941 Gallups Island Radio School, capacity 325; had phones, electricity, water from mainland. Added school for bakers and cooks, 150 capacity. Barracks were 100' x 200' x 30', set on pilings.
1947 Gallups sold at auction; houses floated off or dismantled. One house to Weymouth, two houses to Quincy, one to Hingham. Monument moved to Fenway, later to North End.
1953 Island was sold to Charles Sassone of Dedham.
1964 Resold to Palumbo in November, 1964.
1968 Boston Harbor Islands Commission appointed an MIT study on noise level, pollution, waste disposal, as constraints on proposed recreational use.
1970 Act of 1970, Chapter 742, authorizing park.
1973 Sold to State for $63,000. Park begun as Boston Harbor Island State Park.
1975 Pier built on Gallups.
1976 Selected by Massachusetts Audubon as interns, island managers begin.
1982 Muskrats sighted on Gallups.
1990 Rabbits sighted on Gallups.
1992 July 10, 2,600 people view Tall Ships from Gallups Island.

Chronology of Gallups Island compiled by ELLENOR YAHRMARKT
Visitor Services Supervisor
Department of Environmental Management
Boston Harbor Island State Park
The story of the Boston Harbor Islands is from a photograph of a State Park sign on Gallups Island.

THE CHRISTIAN SCIENCE MONITOR, BOSTON, TUESDAY, JUNE 3, 1941

'Ghost City' of Boston Harbor Thrives as Training Base for Merchant Marine

Radio Men and Cooks Learn Their Trade at Gallop's Isle

By Leon V. Stone

Staff Writer of The Christian Science Monitor

Boston harbor's "ghost city" has been restored to an active community of more than 600 inhabitants.

And four of its new citizens have already

nine-month course, with about 325 studying to become operators on freighters having but a single radio man. The aim is to turn out more than "key-pounders."

BOSTON SUNDAY POST, MAY 25, 1941

Landlubbers + 40 Weeks = Sea Dogs

From Gallup's Island in Boston Harbor, U. S. Coast Guard Is Turning Out Seamen, Radio Experts Who Earn Big Wages

Ken Waller, Jerry McClish, A1 and A2, Recall How It Was When Radio School Started

by Gerald McClish, R-A2
Bradenton, Florida

Okay. My friend Art Bradford wanted me to give a few details, as far as I know, on the start of Gallups Island. So, if my memory doesn't fade, I guess I can figure it out.

I got on a troop train in Anderson, Indiana, about the middle of June. The troop train started in Oregon, I believe, and came all the way through the midwest and picked up fellows from CCC camps, and I was told I was going to be in the Maritime Service and probably be an AB on a ship, or something like that. They didn't really tell us.

The troop train finally got into Fort Dix, NJ. They unloaded us, put us in tents and then they inducted us into the Maritime service. After about four or five days of struggling through the mud, they brought in some buses. And in one day all these farm boys, many of whom had never seen the ocean, were put on a ship at the Brooklyn Navy Yard and headed out to sea.

We went right downtown in New York, through the Holland Tunnel, which was great, across the Brooklyn Bridge into Brooklyn, and out to sea. About three or four days out, with everybody getting seasick, and so forth, they had a muster on deck. As I recall, there were around 1,000 to 1,100 recruits on board. And they had us all muster on deck in our gob's uniforms which they had issued us on the ship and then they requested 500 recruits to step forward for Radio School in Boston.

Well, I didn't want any more seasickness or fiddling around or polishing brass on a boat, or chipping paint, so I stepped forward. So then the ship took off for Boston and got up there on July 3, and on July 4 they brought in, I think it was, an old wooden mine sweeper. They started carrying us out to the island. Radio school had not been built; it was being built. There was really no radio school there. The barracks that we lived in were also not quite all finished. But then they gave us a full course in everything from pyrotechnics to whaleboat racing, and everything concerning seamanship, and as I recall, it took almost a month before we finally got into radio school. We had A1, A2, A3, B1, B2, B3, and we had C1, C2, and C3. That would be nine classes and about 50 in a class—would be about 450 people. So that meant about 500 and that, to my knowledge, is how school started and I graduated in August or September of 1941, and I was going to sea when the war broke out. So that's about the story, without going into all the details.

And now, after that long story of how I got to Gallups Island, along with a whole group of other people, 50 years later in 1990, Art Bradford, and a friend and I got into a little boat and out we went. And as near as I can tell it was the same day 50 years later and it was probably close to the same hour. I got there around noon on that day. So that was my reunion on Gallups Island, as near as I can remember, and the accuracy of the dates can't be very far off, because all the papers I have indicate what I've told you is right. The date that we got there in 1940, July 4, to my graduation is a little over a year later and going to sea with the Great White Fleet—that pretty well wraps it up.

By Ken Waller, RA-1, CM-70
Scottsdale, Arizona

I was in the first wave ashore on Gallups in June 1940, one of the early "selectees," from the CCC. We assembled, some 60 of us, at Camp Dix, NJ, where we got a week or so of indoctrination and close-order drill from the U.S. Army. We were then hustled aboard the *SS Seaman* in New York and got our first little taste of life at sea en route to Boston—and the "Rock." We became classes A1 and A2, our tenure stretching out for about 15 months before finally getting our license and transfer to Hoffman Island for shipping assignment. My first assignment was as a "radio cadet" on the *SS Memphis City*, with my chief being none other than Elza (Al) Blan from Class A1. We were in Hawaiian waters on Sunday morning, Dec. 7, 1941. Our only inkling of what was happening was the 500 kc was so jammed as to be unreadable.

My next contact with a G.I. grad was on the transport *SS African Comet*, then chartered by the Navy as the *USNTS Arthur Middleton*. We became part of "Operation Bobcat," a SE Pacific convoy that established a refueling base at Bora Bora, a most memorable experience. One of the soldiers aboard turned out to be a fellow from Class A3, who I remember only as "Boots." He dropped in on the radio shack one day when I was on watch and we had quite a ragchew. He explained that he had goofed around a little bit too long after leaving Gallups and got himself drafted.

"Boots, if you read this, please drop me a line and let me know what happened to you fellows after we abandoned you on Bora Bora."

After another 9-month stint on a new Liberty, the *SS Robert Treat Paine*, I was recalled to active duty as an instructor in the shipboard lab at Gallups. As a long-term staff member there, teaching the mysteries of the RMCA and the Mackay "coke machines," I was always impressed with our sharp, clean-cut young trainees, many of whom were overflow from the academy at Kings Point, and am gratified to hear that most of our grads have pleasant memories of their G.I. experience.

From CCC to SEA, SEA, SEA. Also, Jerry McClish went from chipping paint on deck to brushing paint on canvas. Photo by Stan Jennings

A small island off of Boston's harbor has a special nostalgia for many veterans, for here the Coast Guard built a school in WWII which trained thousands of Merchant Marine seamen in the art of radio communication.

THE LEGEND OF GALLUPS ISLAND

by Edward L. Hayden

... is reprinted here by courtesy of Ed Hayden, R-20, K30KL, and Sea Classics, September 1992, where it originally appeared.

On 18 May 1942—during World War II—a British trawler on anti-submarine patrol near Key West, Florida, spotted something in the water. It was Harley A. Olson on a life raft. Olson claimed he had been torpedoed and had spent the last 29 days alone at sea. But the British thought they smelled something. He looked too healthy and too comfortable. They suspected they had caught a spy and called in the FBI.

Olson had no papers and no identification. The FBI grilled him for three days trying to break his story. But Harley A. Olson was telling the truth. He had indeed spent the last 29 days in the North Atlantic on a life raft. Olson was not a spy but a graduate of the radio school established by the US Coast Guard on Gallups Island in Boston Harbor to train radio officers for the wartime Merchant Marine.

Olson was on the *SS Steel Maker*, en route from New York to Iran with 7,000 tons of war supplies, when on the night of 19 April she was torpedoed by the German submarine, *U-654*. Olson, the last man on the ship —trying to send an SOS—realized he had stayed too long when the vessel started to go down. He jumped into a dark void toward the sea and the suction pulled him under.

With a mighty effort he struggled to the surface. And from the debris that covered the area, he grabbed a hatch cover and held on.

At light, he found a life raft loaded with provisions, then another, from which he gathered additional supplies: graham crackers, malted milk, chocolate bars, pemmican, matches, first aid kits, and a spare flashlight.

A settee floated by and he hauled it on board. It became a comfortable bed. He constructed a sail out of signal flags, using safety pins from the first aid kits. And he devised a homemade fish net. In addition to his well stocked supply of rations, he ate raw fish for dinner. He navigated at night by the north star and by the sun during the day. Olson had started with 20 gallons of water, saved from the two rafts, and he collected rain water for insurance. When he was rescued, he had 17 gallons left.

The maritime school on Gallups Island—called by Governor Saltonstall, during his visit there in 1943, the finest of its kind anywhere—had taught Olson well.

Neither the British nor the FBI could believe the young radio officer had survived for 29 days on a raft in the Atlantic—wet, cold, and alone—and was able to climb on board the British trawler, unassisted, looking as if he had come off a month's holiday.

He had been given up for lost, and the insurance company was ready to write a check to his family—when a healthy, relaxed, and penniless Harley A. Olson called home for train fare.

The Coast Guard started the school on Gallups Island in the spring of 1940 for the US Maritime Service. In late June of that year, the first contingent, consisting of volunteers from the Civilian Conservation Corps (C.C.C.) reported on the Island to build houses, barracks, classrooms, and a water tower—and to become members of the first class.

Later radio hams were recruited with the assistance of QST magazine, the publication of the American Radio Relay League in Hartford, Connecticut. By September 1942, the school was a hive of activity and was taken over by the US Maritime Service.

The school drew students from all over the country, and due to its location, attracted many from the New England area. One was Vincent Kirk of Bailey Street, Boston.

The first person I met, when reporting on Gallups Island in 1942, was Vincent Kirk. He was soft-spoken with an easy smile. He looked at me with a grin and said, "It might be hot now, but wait 'til winter. It gets *cold*." He was right. It didn't snow vertically. It swept in horizontally across the island. To get to the mess hall, you had to go with your head down at a 45-degree angle. And it *was* cold.

That was Kirk's problem. He hated the cold. He shivered and hugged himself. And cussed the cold. He swore he was heading for the south seas when he graduated.

And he made it. He graduated in April 1943, and was assigned to a ship that sailed to the South Atlantic. Fate, thus far, had been kind to Vincent Kirk.

In January 1944, I was on a Liberty ship in Convoy ON-210—100 ships westbound out of the British Isles. The weather was brutal: 60-foot waves and hurricane force winds. The temperature, calculated in today's wind chill terms, was minus 100 degrees F. A tanker and a freighter in the column to our port disappeared during the night. And the rumor was they had collided and were lost to the weather.

Vincent Kirk was also in the convoy on a Liberty ship—the *Sumner I Kimball*. Battling to hold her position, she straggled from the convoy and became a choice target for a stalking U-boat. On 6 January at 2048GMT—at about the point of no return—she was torpedoed by *U-260*. At 2050GMT, Kirk got off a distress message. But it is unlikely that they launched any lifeboats. Nor could anyone have survived for long in that arctic North Atlantic winter if they had.

Captain Arthur Moore of Hallowell, Maine, author of **A Careless Word . . . A Needless Sinking**, reports that the *SS Sumner I. Kimball* was one of 30 US flagships that disappeared with all hands during World War II. And with the exception of the report from the attacking submarine, the ship and crew were never heard from again.

In 1978, a group of us—middle aged men—went back to Gallups Island. "In essence," one returning graduate said, " we have nowhere else to go. The ships we sailed on during the war—the Liberty ships, the Victory ships, the T-2 tankers—have gone. But the Island is here."

From circa 1630 when it was granted to John Gallop for farming, to playing a part in the Revolutionary War, to having an inn and restaurant in the 1800s, to a staging and payoff area for Yankee troops in the Civil War, a quarantine station, an internment site for World War I German soldiers, the Maritime Radio Training School during World War II, to today—the Island has endured.

It provides the place and atmosphere for the stuff of which memories are made.

The buildings are gone, the trees have grown up where once we studied, sweated out the code, and marched; and one had to stretch his imagination to remember how it was—all the eager young men, the Island alive with movement, the disciplined music of Morse code in the background, and the smell of the sea to the east.

For some, the memories were of exploding depth charges, convoy battles, air attacks, a crashing torpedo, a shout to abandon ship, and sending distress messages in a cold sweat.

For others, it was shore leave, the pubs, warm beer, and the girls. And for some, the sights and sounds of World War II Boston—Scollay Square, the Buddies' Club, the Commons, the concerts, Commercial Wharf, and the tragic Coconut Grove fire (I was there on the night of 28 November 1942).

Gallups Island, situated at the eastern end of Boston Harbor, looks toward the Atlantic. It was for the Battle of the Atlantic that most of us were trained. Everyone knows about the Battle of Britain and how it saved England. But not everyone knows the Battle of the Atlantic helped save England and Europe. And it is usually unspoken, but, for us, there is enormous pride and satisfaction in knowing, for all the right reasons, we fought the right battle at the right time.

As a result of our training on Gallups Island, most of us have an intangible asset—the ability to communicate with each other. Many of us get together each morning at 7 am over coffee. Not down at the local coffee shop—we're located all over the country, and the world—but via ham radio. Bill Anderson of East Kingston, New Hampshire, starts the roll call shortly before 7 am on a single sideband on the the 75-meter band—and brings in the New England and East Coast states.

Later in the day, when the 15-meter band "opens up," Al Newbold, a transplanted New Englander, calls the roll from California. And the transcontinental Gallups Island Radio Net is in operation. Then, a Gallups man on a supertanker off the west coast of Africa calls in.

"W6MMG, this is W0EXI. Al, this is Chet, do you copy?"

And one from a ship in the Pacific sailing to Japan to pick up 5000 Toyotas.

"Al, this is Tom, can you get a message to my wife in Baltimore?"

Another from a tanker near Alaska, and the net becomes global. Everyone gets his turn to talk. Old friends get together again, some of whom haven't talked to each other in 50 years.

Today there is a shortage of trained shipboard radio officers. When, during Desert Shield and Desert Storm a call went out for operators, many Gallups Islanders in their 60s and early 70s went back to sea to help move troops and supplies. Some liked the life and are still sailing.

One tangible thing from the Island remains—a monument dedicated during the war to radio officers lost at sea. When the school closed in 1946, it was given to the people of Boston. But over the years, the monument disappeared—until 78-year-old Gladys Reed, widow of Captain Sherman Reed, the superintendent of the school, started a search to find it. And located it in the Fenway, a park in Boston.

With the help of the Fenway Civic Association, former radio officers from the school, the Boston Arts Commission, the Propeller Club and others, she raised money to have it refurbished and moved to a new site.

It was re-dedicated on 11 November 1988, and sits on a grassy plot—Northend Park, on the Boston waterfront—near the *USS Constitution* (*"Old Ironsides"*), the old North Church, the Bunker Hill Monument and Copps Hill Burial Ground. It is now an official part of Boston's maritime history.

There are names engraved on the monument—the names of some of the graduates of the Gallups Island Radio School who died at sea during World War II. One name picked at random was Arnold Tangen.

Arnold was a radio officer on

the troopship, *Henry R. Mallory*. She was en route from New York to Reykjavik, Iceland via Halifax, Nova Scotia, when on 2 February 1943, she was torpedoed by the *U-402*. The ship was 600 miles from Iceland in frigid, bone-chilling weather, and carried men from the Army, Navy, Marine Corps, Navy Armed Guard, and Merchant Marine. The Convoy Commodore never received the message that the *Mallory* had been torpedoed. Unaware she was sinking, he called back a rescue ship that was steaming in her direction. More than half the men, in lifeboats and in the water, were poorly dressed. The arctic air and water, the action of hypothermia, confusion (and bravery), no communications, a rescue vessel mistakenly directed away, all added to the result—of 492 souls on board, 270 were lost, including Arnold Tangen.

Each name on the monument has its own story.

In August 1990, graduates of the school—now a graying group of veterans—held a 50-year reunion on Gallups Island. Their enthusiasm for the Island and its place in history has not abated. One man made a sentimental trip back on 4 July to step on board on the exact day and hour, 50 years from the time he did it in 1940.

Another, John A. Pino, an Associate Justice of the Boston Municipal Court, who was a chief petty officer on the Island during its heyday, said, "I must admit that my heart raced a bit when I saw the USMS emblem and the words 'Gallups Island' on your letter."

The next big reunion on Gallups Island will be for the 60th anniversary—to be held in the year 2000.

In the meantime, with the exception of an occasional graduate who might make the trip, possibly with grandchildren, just to walk the paths, the sea gulls and tourists will enjoy the Island—described by the Massachusetts Department of Environmental Management as, "... 16 acres in area and has a pier with gazebo, picnic grounds, and varied paths. It is served by the DEM water taxi. The Island is a high drumlin with areas of shrubs, trees, meadow, saltlands, and landscaping from previous uses. There is a wooden gazebo at the pier and an excellent lookout at the top of the main path. . . ."

A graduate of Gallups Island, radioman Chet Szymanski was still plying his trade during Operation Desert Storm in 1991. Chet is seen in the modern radio room of a supertanker.

(From BOSTON BUSINESS, September, 1944)

Making "Sparks" on Gallup's Island

This is the story of Gallup's Island, the United States Maritime Service Training School for Radiomen—and of the men who go there to train as "sparks" aboard the ships of the United States Merchant Fleet.

These men are the unsung heroes of many a disaster at sea—men noted for their alertness and ability to handle themselves under trying conditions. Such is the type of men the U.S. Maritime Service trains there to control, operate and repair the "ears" and "vocal chords" of ships.

Gallup's Island is located in Boston Harbor, about two miles off shore, nearly opposite Quincy, Mass. This training school has been acclaimed by experts one of the finest, if not the finest in the world.

It was originally a U. S. Public Health Service Station until acquired by the U.S. Maritime Service for the training of radiomen and cooks. The cook's school has long since been discontinued and the station is now exclusively one for radio students.

1000 Students

Upon approach to Gallup's Island, coming down the harbor, it is hard to visualize the place as a school in which some 1000 students are hard at work. It resembles a colony of comfortable homes rising, row on row, from the water's edge to near the summit of a hill, topped by a huge silver water tower. On closer examination, however, the illusion disappears and bustling activity is discovered.

All buildings on the island, with the exception of the powerhouse, are of frame construction. The island has telephone connections with the shore and obtains water and electric power from ashore. Regular boat schedules are maintained between India Wharf in Boston for the transfer of personnel and supplies.

The buildings devoted to the radio school have been remodeled and refinished to provide an ideal arrangement of classrooms. The health and vision of the students received first consideration in this rebuilding and the classrooms, situated just a few feet from the beach, are light and airy, with thermostatically controlled temperatures. Fluorescent lighting is used during evening classes to make the classrooms bright as day.

Originally built to accommodate 400 students, the school has increased by leaps and bounds until today it houses and instructs 1000 men, proof of the popularity of the school and its officers. Many of these new applicants are obtained as a direct result of the high reputation the school has established, applicants usually being told of the excellent opportunities available at Gallup's Island through friends or acquaintances who have attended.

Most Modern Equipment

The school has the most modern equipment available for instruction purposes. The standard marine radio installation laboratory comprises at least one of each type of radio equipment now being installed aboard vessels constructed for the U. S. Maritime Service, including crystal-controlled intermediate and high frequency transmitters, tuned radio frequency and superheterodyne receivers, radio direction finding equipment, automatic distress signal alarm equipment, stand-by equipment, lifeboat transmitting and receiving equipment, small-craft radiotelephone equipment, and accessories.

Everything possible has been done to maintain the high standards of this school and the equipment used by the students—a reflection on the initiative and ability of the officers and the faculty at Gallup's Island.

The Gallup's Island Radio School is under the very able direction of Commander Sherman W. Reed, USMS, superintendent. As his assistants are Lt. Comdr. F. C. Krastin, USMS, executive officer, and Lt. Comdr. John A. Clark, USMS, training officer.

To these officers is entrusted the training and education of the men assigned to Gallup's Island. So great is the present demand for qualified radio operators that the length of the course of study has been shortened from the original 40 weeks to 20 weeks.

Explanatory Statement

The following statement by Lt. Comdr. Clark explains it: "It usually takes about two years to turn out a good, all-around radio operator, but we are faced with the immense task of supplying a maximum number of trained men to meet the present emergency. We are getting good results with 20 weeks of intense training. One of the reasons for this is that the men are willing and eager to learn all they can in as short a time as possible. Another is that they are working with the best of equipment."

The statement by Lt. Comdr. Clark speaks for itself.

All enrollees must first complete six weeks' "boot" training at one of the U. S. Maritime Service Apprentice Seaman Training Stations. Enrollees for Apprentice Seaman training must be between the ages of 16 and 17 1/2 and 26 to 35 1/2.

Radio trainees are selected on the basis of competitive examinations. General scholastic ability, mechanical comprehension, arithmetic computation, proficiency in algebra, code learning ability, previous education, proficiency in swimming and a personal interview "make" or "break" the applicant.

Upon assignment to Gallup's Island, the student is advanced to the grade of Seaman Second Class and after successful completion of 12 weeks of the course is advanced to Seaman First Class.

Broad in Scope

The 20-week course of instruction includes typing, theory, code, laboratory, procedure, watch standing, convoy procedure and advanced seamanship. Following this, one week is spent in participating in regular FCC examinations and securing all necessary shipping papers.

All the latest standard radio and electrical textbooks are utilized in the course of instruction, together with specially prepared material compiled by the instruction staff. These latter pamphlets are thorough and well-illustrated, covering every phase of the work. All books and technical magazines are available for student reference.

Classrooms are open evenings and instructors are in attendance to assist and give individual instruction to students who may be experiencing difficulty. During this period, the student may obtain more complete and detailed explanations of troublesome problems and receive the advice and counsel of experienced instructors.

Although the present course prepares the student to pass the examination for an FCC Temporary Limited Telegraph Operators' License, extra supervised instruction and all school facilities are available for men who desire to sit for the full 2nd Class license at the end of their training period, and all men are encouraged to prepare for this higher grade of license. In the event the student satisfies the code requirements without passing the written examination, he will be issued an appropriate Maritime Service certificate which will permit assignment aboard merchant vessels as a radio watchstander with the pay of a Third Radio Operator. Students are also eligible to take, without charge, a correspondence course in radiotelegraph theory given by the Maritime Service Institute in order that they may continue to prepare themselves for a higher FCC radio license.

Students qualifying for an FCC Second Class Radiotelegrapher's License will be graduated as Warrant Radio Electricians; those obtaining and FCC Temporary Limited Radiotelegrapher's License will graduate as radiomen third-class; those satisfying only the code requirements will graduate as Seaman First Class.

A Real Opportunity

Due to the fact that a continuous radio watch must be maintained aboard American Flag vessels— three radio operators, each standing an eight-hour watch—the demand for trained radiomen has increased greatly. Young men in this branch of the service are not only offered excellent training for well-paid positions aboard ship but are given the opportunity of a lifetime to fit themselves for profitable postwar positions.

As radio operators, they are specialists aboard ship, enjoying privileges usually accorded officers and furnished quarters and mess facilities commensurate with their position. Base pay for radio operators, exclusive of bonus and overtime, ranges from $155.00 to $175.00 per month.

QST Returns to Gallups Island

How Marine Radio Operators Are Trained Under the U. S. Maritime Service

BY CLINTON B. DE SOTO,* W1CBD

It was nearly two years ago that *QST*, in the persons of Clark Rodimon, then managing editor, and Frank Beaudin, our staff photographer, first visited the Gallups Island Radio School in Boston Harbor and brought back the story appearing in the June, 1941, issue. Widely reprinted and distributed by the U. S. Maritime Service, that *QST* story became official recruiting literature for the merchant marine training program. In fact, so successful was it that, early this year, the War Shipping Administration's director of training suggested a return visit.

In the intervening two years a number of changes occasioned by the developing urgency of war needs had taken place, and it was felt that the 1941 story no longer revealed an accurate picture of the school. Last September, for example, the entire administration of the school changed hands when the Maritime Service was transferred from the Coast Guard to the War Shipping Administration. Enrollment is now more than double the original figure, and new buildings have been constructed to provide additional capacity for the increased numbers. Entrance qualifications were revised to cope with the competitive manpower situation, and the length of the course has been progressively reduced. There have been other changes, too.

Even the character of the students and the nature of the job they are training for has changed. Two years ago the United States was still looking inward, clinging doggedly to its isolation in a world at war. There was a shortage of maritime radio operators for the rapidly-growing merchant fleet, but the job itself was scarcely more than the humdrum equivalent of peacetime voyages.

Nevertheless, some members of that first 40-week class at Gallups Island had yet to step aboard their own decks when dawn broke with the ruddy glow of Hell-fire over Pearl Harbor on December 7th and the U. S. awoke from its Sunday-afternoon nap to find itself at war. For the men of the merchant marine the declaration of war meant that hardship and death had become their constant companions. Many a Sparks in his lonely radio room had scarcely finished copying the electrifying words "Pearl Harbor attacked!" when he felt the deck shiver beneath his feet, vibrating until the dull "boom" caught up with the explosion, and he turned his attention to rapping out a despairing SOS while a dying freighter or tanker sank wearily into the sea. Some, even, heard the hiss of enemy-launched torpedoes before ever they heard the news that their country was at war.

Yes, the job changed then. It became a grim and reckless one, a death-defying calling whose followers literally thumbed their noses at the enemy. We've reported the experiences of one or two of these men in *QST* — the few about which details could be released. But for the most part they are unsung heroes — hard, jesting, self-reliant men of the sea, with the kind of guts it takes to face death not once but many times out there in the lonely wastes of the North Atlantic or the deceptive sultry blue of the Caribbean or the rolling vastness of the South Pacific. Men with the kind of courage it takes for survivors of soul-searing voyages on lifeboat or raft to come ashore grinning, anxious for another crack at Jerry or the Jap.

That's the kind of men they're training at Gallups Island — men who are resolved, despite what may lie ahead, to do their part in maintaining the lifeline of supplies to our fighting forces.

By Cutter to Gallups Island

It was to see these men and the training they receive that we went to No. 22 India Wharf on Boston Harbor's motley waterfront one morning not long ago. There, in company with Ensign Ralph J. Meyers, USMS, of the local Maritime Service trainee procurement office, we boarded the cutter *Roderick* on the outgoing leg of its daily round-trip voyage to the Island.

* Executive Editor, *QST*.

Signal flags flying in conformity with strict war-time harbor regulations, the sturdy old cutter, aged but still salty, pushed through the restless water while all around the big harbor strained at its work. Vigilant Coast Guard patrol boats scurried by; overhead, planes circled and roared.

Some forty minutes later we approached Gallups Island. As the cutter swung around and approached from seaward, the island resembled a seaside colony of comfortable white-painted homes rising tier on tier from the water's edge to the summit of an enlongated hill. Then, as the soft thrum of the Diesels brought us nearer, the seashore-resort mirage metamorphosed into rows of long frame barracks and school halls.

As the *Roderick* slipped smoothly into its mooring, its passengers disembarked onto the husky new dock, recently completed as part of the school's expansion program. There were the usual formalities — the Maritime Service is as vigilant as any military post, and for the same reasons. Finally, after being duly checked and logged aboard by the officer of the day, the executive officer, Lt. Comdr. Richard J. Casilli, USNR, informed us that the superintendent, Commander Sherman W. Reed, USMS, was waiting to give us our official welcome aboard the station.

Comdr. Reed proved to be an old sea-faring man with a record of 25 years at sea, an able executive whose affability and suavity failed to conceal the alert intelligence that makes his administration of the Gallups Island establishment so successful. From him we gained a general picture of the Island's war-time development.

Expansion for War Needs

The basic story of the Island's conversion from a U. S. Public Health Service quarantine station was told in the June, 1941, *QST* story. The original school was designed to be the best-equipped and most efficient radio school ever established — and the result was just about that. The superior quality of Gallups Island alumni became famed from coast to coast, and the successful training methods developed there were adopted in military-training schools subsequently established.

But within a few months after its completion the original school, built to accommodate 400 students, bulged at the seams. Additional buildings and facilities were constructed — in itself a major feat, since every item of material down to the last nail had to be transferred to the Island by boat — until the capacity was increased to handle a full quota of 1040 men, in addition to some 280 men in the permanent company including 34 instructors.

As the school expanded to keep pace with the mushrooming merchant marine, the course of study was reduced in length from 40 to 29 weeks and the entrance requirements were lowered to ensure the needed constant flow of trainees. At the outset a candidate for Gallups Island training was required to have two years of high school math and one year of physics. He must also have passed an elementary examination in radio and be able to take code at a speed of 13 w.p.m. — in other words, the equivalent of a newly-licensed Class B amateur. Now only the first requirement — two years of high school math, including algebra — remains; no prior radio training is required.

Yet the surprising fact is that, despite the changed preliminary requirements and the condensed course, graduates of Gallups Island rate as high or higher than the product of any comparable school. They matriculate equipped to pass the FCC exam for second-class radiotelegraph license by a comfortable margin. They come off the Island ready to step on deck and take over a watch on any ship in Uncle Sam's merchant fleet.

Following the interview with Comdr. Reed, Lt. Comdr. John A. Clark, USMS, the training officer in charge of the radio school, took us in tow. A tall, laconic veteran of the Maritime Service with a reassuring proclivity for understatement, Lt. Comdr. Clark outlined the course of instruction.

The actual training period is concentrated into 28 weeks, the final week being reserved for review and final examinations — including the culminating FCC examination at the Boston inspection office. The course is organized under five basic headings — code, procedure, theory, lab and seamanship training. Code training is carried on throughout the 28 weeks, with seamanship training terminating in the 26th week. Theory classes are held for the first 21 weeks, overlapping the laboratory work, which begins with the 16th week. Lab continues until the final week of the course, as does the procedure class which begins with the 18th week.

Code and Procedure

Code training is by the original "GI" system — with modifications. In the first week the entire alphabet and all numerals are memorized in the basic EISH5, TMOJØ, etc., combinations. By the end of the second week the student is able to copy plain language at 4 w.p.m. Thereafter, if he keeps up to schedule, his speed advances at the rate of about 1 word per week until at the end of 25 weeks he must take a minimum of 23 w.p.m. plain language and 18 w.p.m. in coded groups. These minimum speeds, although 2 w.p.m. less than the previous standards, provide a satisfactory margin over FCC requirements.

Most recent among the modifications in the code-training course is the separation of initial code and touch-typing practice. Originally both were begun simultaneously, but under the shortened course basic code and typing instruction are given separately for the first four weeks — two hours of code and one of typing daily. In the following weeks the student learns to correlate the two, and after the fifth week all copying is directly on the mill.

Training in sending commences with the 13th week, the student's receiving speed then having reached a minimum of 13 w.p.m. Practice in both sending and receiving — plain language and mixed code — continues until the end of the course. After the 17th week, however, the code-class periods are reduced from 15 hours per week to 5

Reprinted courtesy

QST for May 1943

with the introduction of procedure training.

In procedure class all phases of marine operating are covered, including calling and answering, use of frequencies, Q signals and other operating abbreviations, radio laws and regulations, international treaties and agreements, traffic handling (including such details as computing charges, conversion of the gold franc, etc.), direction finding, the complicated details of distress communications and the myriad other procedural details the accomplished marine operator must know. This part of the course concludes in a series of practical drills and watch-standing sessions, closely simulating operation at sea.

Both code and procedure classes are conducted in an atmosphere calculated to give the student the true feeling and spirit of an operator. In the watch-standing classes the individual positions exactly duplicate standard shipboard operating conditions — two rows of positions on either side of a long room, one side equipped with Mackay equipment and the other with RCA.

Theory and Lab

That's the operating side of the picture. In parallel with it is the theory and lab instruction. There are some radio operating jobs in which the man at the key needs only to know which knob to turn or which switch to throw, but the marine radio operator must be the "complete radioman," familiar with every technical detail of his equipment and qualified to service and repair it.

As much training time is devoted to the technical end as to operating, therefore. Theory classes begin with the simplest electrical fundamentals and work up through a well-rounded course to complete transmitters and receivers, winding up with a week on wave propagation and antennas. Qualified instructors — most of them hams — lead the novice through the technical maze along a well-thought-out route which skirts the non-essentials while providing a well-balanced understanding not only of how radio works but why. The theory department, appropriately enough, is staffed almost entirely by hams, including W9QMG, who is in charge, W9CEE, W1NGL-ex-W4ARP, W9OMU, W9MUQ, W9KWI and W9VQE.

Texts for the theory classes are pamphlets prepared by the staff to meet the specialized requirements of the course. A comprehensive reference library of standard reference texts and handbooks is also provided.

Practical application of the theoretical knowledge thus acquired comes with the beginning of laboratory classes in the 14th week. There are three laboratory classrooms — the general lab, storage battery lab and marine radio lab. Each has its own important role.

Work in the general laboratory begins with the types and uses of tools and continues through elementary electrical experiments and assorted constructional projects involving the building of oscillators and simple receivers. This well-equipped lab has been the envy of many a service school, with its individual tool drawer complete with every needed hand tool for each student and its well-stocked supply room outfitted with ample quotas of everything from condensers and resistors to signal generators and oscilloscopes.

The storage-battery lab is just what its name implies — a place where students learn all about the maintenance, care and operation of storage cells, those indispensable elements of marine-radio installations.

It is in the third room of the laboratory section that the equipment-minded visitor lingers longest, however. Here, in the standard marine radio installation laboratory, is to be found at least one example of each type of radio equipment now being installed aboard vessels constructed for the U. S. Maritime Service, including crystal-controlled intermediate and high-frequency transmitters, t.r.f. and superhet receivers, radio direction-finding equipment, automatic distress-signal alarm equipment, stand-by equipment, lifeboat transmitting and receiving equipment, small-craft radiotelephone equipment and all the associated accessories.

This equipment is assembled in rows along the room, each unit completely installed and operating. Classes assemble on folding chairs in small groups before individual installations, several such groups assembling simultaneously in various parts of the room, while the instructors take the equipment up unit by unit. On completing this part of the course the student knows every electrical and mechanical detail of each type of equipment he may be called upon to operate. Coupled with his operating training, it makes him a thoroughly proficient radio man.

Seamanship

But a marine radioman must be more than a good radio operator and technician. He must be a seaman, as well, and two hours of each week throughout the course are devoted to teaching seamanship. It takes an astonishing amount of specialized knowledge to be a sailor, as anyone who has fitted and sailed even an 18-foot yawl can testify. Of course, a radioman doesn't need to know everything a regular seaman does — but he must learn the names and uses of the basic nautical devices from anchors to marlinspikes, the hundred and one varieties of sailor's knots, the art of getting into rubber life-saving suits and the technique of lifeboat drills.

All in all there's plenty to do in the formal classroom day from 9 A.M. to 4 P.M. — to which may be added evening periods for boning up. In this connection, classrooms are kept open evenings and instructors are on hand to assist with individual instruction and counsel where needed. For students who drop behind in their work special evening classes are made compulsory. Those who fall too far behind are dropped back a class, while for those who make exceptional progress the converse may be true. If a man fails to make the grade after several chances he is disenrolled. This never happens until after the fourth week, however; every student is given at least one full month to get straightened out and show what he can do.

Throughout the course a close relationship exists between students and instructors, and much of the training is on a personalized, individual basis.

All training activities are carried on in terms of classes — the weekly quotas of newly-arriving students. Each class is given a number; although individual students may drop behind or move ahead, the class as a whole remains a unit. For administrative purposes classes are organized as platoons, a leader and two assistants being selected for each platoon from among its members. The leader is given a temporary rating two grades higher than his fellows, and his assistants are given ratings one grade higher.

From 8 A.M. Monday mornings until Saturday noon the students are required to remain aboard the station. Over the week-end shore liberty is granted, and each Saturday noon finds the cruise decks of the converted Long Island Sound pleasure boat *Calvert* overflowing with hundreds of holidaying students going ashore.

Certain platoons are required to stay aboard each week-end for fire watch and other duties. These platoons then receive liberty on Wednesday night of the following week as compensation. In addition, the platoon which ranks highest in scholastic accomplishment for the week rates special Wednesday-night liberty the following week, and a perpetual competition is carried on between the platoons for this privilege.

Even during the week, however, the students do not lack for recreation. Excellent facilities for ping pong and other sports are provided, and a large new recreation hall is now under construction. Motion pictures are shown nightly, supplemented by occasional smokers, boxing matches, amateur nights and other entertainment supplied by Boston organizations. There is a station newspaper and a p.a. system that broadcasts music after school hours (not to mention orders of the day, announcements of disciplinary actions and other less welcome entertainment). And then, of course, there is the ocean — itself a major attraction in good weather.

The FCC Ticket

Returning now to more academic details, at the end of the 28th week the course outline has been completed and a final exam, based on the government license examination, is given. During the week the class will go ashore in groups to be given the FCC commercial license exam by the local inspector. Those whose relative standings are highest go first; the rest are given additional time for intensive brushing up on code or weak spots in theory or procedure. No student is allowed to take the FCC examination until he has displayed to the satisfaction of the assistant training officer, Lt. T. H. Grant — himself a former FCC inspector — that he is qualified.

Usually graduates of Gallups Island pass the government exam without a hitch; most classes go through intact. Now and then, however, student acquires an acute case of buck fever. At the school he may have been letter perfect, but when he faces the awesome finality of the inspector's office he becomes as nervous as the most frightened would-be ham. In fact, they tell of on apprehensive lad who went up for his ticket thr times. Each time he'd break into a nervous sweat. His muscles would tighten and his mind would go blank, and he'd become too paralyzed to write. So acute was his distress that blisters broke out on his hands, and the last time he went up they were entirely swathed in bandages. That time, though, he made it.

For most the exam is not so difficult, however — just one more formality on the road to shipping out. Once over, they return to the Island one last time to collect their personal gear. Then they receive a certificate of graduation and, ticket in hand, prepare to put their training to actual use.

After Graduation

A part of each class may be retained on active duty in the Maritime Service, some as operators and others who display special aptitude occasionally being kept on as instructors. But most are placed in inactive status in the Maritime Service and prepare to enter the merchant marine.

Perhaps a word of definition is in order here. The U. S. Maritime Service, formerly a branch of the Coast Guard, is now part of the War Shipping Administration, the federal agency which has control of U. S. shipping during the war. The Maritime Service has the specific job of training personnel for the merchant fleet. The merchant marine comprises all privately-owned U. S. shipping — the activities of which, under war-time conditions, are rigidly controlled and administered by WSA, including not only the loading and routing of vessels but the supplying of personnel.

Officers and men of the Maritime Service wear the same type of uniform with Maritime insignia and hold the same kinds of ratings as those in the Navy. Men in the merchant marine, on the other hand, may but are not required to wear regulation uniforms and do not hold permanent ratings. No officer will wear the Maritime Service insignia of a higher rank than called for by his position aboard ship, regardless of the designation of grade held by said officer, their nominal rank at any given time depending on the ship and the nature of their berth.

The first step taken by the prospective merchant marine operator after graduation from Gallups Island is to register at the WSA manning office to which he has been ordered to report. He may go home on leave for a time without pay before shipping out at the discretion of the manning officer.

When he does sign up he'll be assigned an order of precedence. As soon as his number comes up he'll be assigned a ship, all jobs being filled in rotation. If there's no immediate vacancy he'll be placed in "stand-by" status, receiving $7 per day until assigned — his time meanwhile being his own.

Once on active duty he'll find that he's a pretty important man aboard ship. He'll enjoy privileges

Reprinted courtesy

QST for May 1943

usually accorded officers and be furnished quarters and messing facilities appropriate to that status. His base pay as a radio operator in the merchant marine may range from $105 to $220, probably averaging around $165-$185, depending on his experience and the nature of the assignment — and it'll all be money in his pocket when he comes ashore. If he's required to perform clerical duties in addition, as happens in many cases, he'll be paid an extra $25 to $50 a month. On top of that, there's a 100 per cent bonus for off-shore runs in war-time, doubling his base pay.

Which adds up to a very nice figure — particularly considering that he may know he is serving his country in a critically-valuable role, and that during his training period he has been paid at the rate of $54 for the first three months and $66 thereafter. This on top of being furnished quarters and subsistence, as well as all his clothing, texts and other needs.

Enrollment in the Maritime Service

The students assigned to Gallups Island come from three sources — U. S. Maritime Service Training Stations at Sheepshead Bay, N. Y., Avalon, Calif., and St. Petersburg, Fla. Before attending radio school, they must have apprentice seaman training for five weeks and, if found to have special aptitude, will be sent to the preliminary radio school at Huntington, L. I., for two months and then to Gallups Island for four additional months.

For applicants who seek assignment, the basic qualification is two years of high-school math. Beyond that, to be eligible for enrollment as an apprentice seaman in the Maritime Service, the applicant must be "of good moral character, physically sound, well-formed and of robust constitution." He must be a U. S. citizen, between 18 and 35, and must agree to serve at least one year in the American merchant marine after completing the course.

Naturally, anyone who has a background of training and experience superior to the basic requirements will be all the more heartily welcomed. A man who holds any kind of ticket, amateur or commercial, will be given first consideration for selection. Such a man should, of course, complete the training in less than the regular time; we were told of one Class A ham who got through the entire course in just five weeks.

To attract a maximum number of qualified candidates, the enrollment procedure has been made as simple as possible. U. S. Maritime Service enrolling offices have been established in all coastal cities and in most of the principal inland cities; if you live in one of these, all you do is apply at the local office. If not, you can start the ball rolling by writing the Commandant, U. S. Maritime Service, War Shipping Administration, Training Organization, Washington, D. C.

In either case you'll be given, locally, a simple entrance test and a medical examination by an officer of the U. S. Public Health Service. If you pass these tests you'll be entered as an approved applicant and placed in line to await your turn for assignment to one of the seaman training schools. Such assignments are made in terms of monthly quotas for each section of the country.

Anyone interested in radio who receives assignment to Gallups Island may consider himself fortunate. Not only will he be afforded a magnificent opportunity for self-advancement and receive competent training in a high-paying field, but he'll be serving in a responsible position in a vital part of the war program. It's a job that calls for men of the highest type—and that's the kind of men they're training at Gallups Island.

JOHN BURNS has provided us with a copy of an article published in the now-defunct "Boston Evening Transcript" on Friday, July 26, 1940. The headline read: "CCC BOYS TAKE OVER GALLOP'S ISLAND; COAST GUARD TEACHES MARITIME SCHOOL." The paragraph under the picture of the original buildings on the Island reads: TRAINING CENTER. Gallop's Island 20-acre plot in Boston's lower harbor used to be a quarantine station. Today it is the headquarters of a new maritime training school operated by the U.S. Coast Guard with funds provided by the Maritime Commission. Late in June the first contingent of students arrived, 110 young men between the ages of 18 and 23 who volunteered from the top ranks of Civilian Conservation Corps throughout the country. Early this month 400 more arrived. When buildings and equipment are in shape the island will accommodate a maximum of 750 men, including the maintenance crew. These young men get no military training (*Editor's comment: Remember, this was a year and a half before Pearl Harbor.*), enroll chiefly to gain marine knowledge and learn radio operation. Their studies, which are promoted as part of the national defense program, make these men eligible for the Navy reserve and also for their able seaman's certificate. Initially these students will spend eight or nine months in training. As the school gains momentum the time may be shortened to six months. Already the island school is a hive of industry and activity. The men follow a regular daily routine, but discipline is not severe and there is plenty of time off for recreation. They get up at 6:30, eat at 7:20 and spend the morning studying, with smoking periods in between sessions. From 11:30 to 1:00 they lunch and rest. Studies are resumed until 3:30. They are allowed shore leave every other night, one group going one night and the other the next. (As most *of you remember, things changed after December 7, 1941. Note, also, they spell Gallups with an "O."*)

HISTORY NOTES —FROM ED HAYDEN'S
DEC. 1989 *NEWS QUARTERLY*

THE BOSTON SUNDAY GLOBE • MAY 30, 1993

WWII radiomen keep signal strong

Group trained on Boston Harbor island

By Kevin Galvin
STATES NEWS SERVICE

Gallops Island, a nearly forgotten islet at the mouth of Boston Harbor, holds a special place in Stan Jennings' heart. That's where he and thousands of fellow Merchant Marine radiomen trained before shipping out during World War II.

On the wall of Art Bradford's home in Duxbury hangs a hunk of shrapnel that crashed through his radio room porthole and lodged in the wall above his typewriter, a daily reminder of an exciting time when he was young and rode the high seas to war.

A half-century has passed since Bradford and Jennings learned to tap out Morse code on the island. Now the two retirees are reliving their common experience with more than 1,000 other members of the Gallups Island Radio Association. (The state spells the island's name Gallops, while the maritime service spelled it Gallups.)

A small group of friends who used to meet for an annual lobster bake in Bradford's back yard has grown since 1984 to a nationwide network of former radiomen with a newsletter published by Jennings out of his Maryland home.

Somehow, the group seems to revolve around Boston, where cadets granted weekend liberty visited friends or frolicked in old Scollay Square.

"We would frequently go to the Common," said Jennings, who was 22 when he attended the radio school in 1943. "I met some very nice Boston girls there."

Abandoned today and covered with sumac, wild roses and bayberry, the 16-acre island's history is unknown to most Bostonians.

Claimed by one John Gallop in the mid-1600s, the island was farmed for 200 years, providing fresh vegetables and dairy products to mariners and South Shore residents.

In the early 1800s, an inn opened there and wealthy Boston families visited for dinner or spent the weekend.

Eve of war

Gallops served as an encampment for Union troops during the Civil War, and later a quarantine station for immigrants to Boston opened. German sailors were briefly held prisoner there during World War I.

With the United States' entry into World War II looming, the federal Maritime Commission needed to train enough radio operators to staff communication rooms 24 hours a day aboard the ships that would supply the war effort. It opened a school on Gallops in 1940, and more than 4,000 operators graduated before it closed in 1945.

"The school was very strict. If you flunked three tests, you were out," Bradford said. "We were there for one reason only: to learn. You were the only operator on a merchant ship, and you had better know what the hell you were doing."

There were three classes in the morning and three in the evening. Students studied early radio theory and practiced Morse code until they could handle 16 words per minute, sending and receiving. Some never forgot those skills. Some may be described as radio fanatics.

"Everything has to be in the wrist," Jennings explained as he tapped out a message on a salvaged black sending key he set on his coffee table. "The wrist gives the action. The fingers don't do anything. The fingers aren't as reliable an instrument as the wrist. You just want to beat it down – a dot or a dash."

Weekend liberty

Those who passed the weekly exam on Saturday morning were granted liberty in Boston. A boat left cadets at Long Wharf, and Bradford remembered they headed to a friend's home or a Scollay Square burlesque show or the nearest gin mill.

Jennings said that as servicemen, they were often invited to dinner with Beacon Hill families. But lodging was another problem. "On Saturday night you needed to find a place to sleep," he said. "A common practice was to go to the Boston jail. They'd open up the jail. There were beds on the floor and hot food."

After an average of eight months' training, the operators shipped out. Merchant Marine ships carried fuel and food, airplanes and ammunition between the United States and troops fighting in Europe and Asia.

Bradford, who was 20 years old when he went to Gallops in 1942, saw the end of the Normandy invasion and survived a mine blast near Borneo. Jennings was lucky to have skirted danger. Many in the Merchant Marine were not. About 6,800 died in World War II.

Capt. George L. Mollison, master of the Otto Mears, a Matson-run freighter, said that even in bad weather, the inbattened holds and the kitchen, refrigerators and toilets lashed on deck would stand the gaff. The ship is so tightly loaded that it would ride the waves rather than dig into them, said the Redwood City Skipper.

Ed Frank

In Australia, Matson Line's' Bernice Bergman with Captain Mollison of the troop transport.

SPARKS OF THE OTTO MEARS

By Edward R. Frank, R-30

On December 12, 1942, a few of us, like Ray Smith, Jerry Martenbower, Hamilton Sweeney, Johnny Merrick, and me, found out we could join the Maritime Service, no matter what was wrong with us. (I was 34 years old and turned down by Navy, Army, etc.—too old.) So we found the Maritime recruiting office and signed up, and that night we were on the train from Kansas City, MO, to Boston, MA, and Gallups Island. We were in Class R-30.

I was made platoon leader and Clarence Strong was my assistant platoon leader. We got along fabulously and had a terrific group of guys. I think we started with about 125 recruits and had around 65 when we graduated in August, 1943. Three of us were assigned to *USMSTS American Pilot* at Fort Schuyler, the Bronx, NY. Besides myself there were Sol Davidowitz and Robert Mervyn Montgomery. We got genuine shipboard practice and September 15, 1943, we were released to War Shipping Administration in New York.

At the time, my wife was staying with relatives in California. I got permission to transfer to the West Coast (at my own expense) and signed up to ship from San Pedro. While waiting for assignment, I did some work in a defense industry. On March 29, 1944, I was assigned to *SS William D. Burnham.* We left San Pedro, CA on April 6, 1944 with a load of planes lashed to the deck and ammo below, for delivery to Calcutta. After zigzagging for some 60-odd days, we finally made it. By June 18, we were unloaded and then did some pickup and delivery of various supplies, at places like Visagapatam, India; Colombo, Ceylon; Aden, Arabia; Suez, Egypt; and Suez Canal, Port Said, Alexandria; passed Gibraltar around the world so to speak and arrived in Baltimore, MD, August 29. I signed off August 31, 1944. Couldn't get a plane or train out of Baltimore to get back to Los Angeles, so I hit the bus depot, sleeping on and traveling day and night, and made it home in 3 1/2 days. Buses were jammed also, so I was lucky to make connections from city to city at any time of day or night—just to keep going west and toward home.

Got some rest and visiting with friends and relatives, then was assigned on October 12, 1944, to the *SS Otto Mears* (Matson Navigation); signed articles October 17; and left Wilmington October 23, 1944. First stop was Melbourne, November 19, and then to Colombo and on to Calcutta on December 10, where we unloaded and left by December 21. Spent Christmas Day in Colombo, then to Freemantle, Australia and then on to Sidney, which became our home port. We got there January 20, 1945, and left February 8, loading and unloading at various ports through to November 24, 1945. We had covered Brisbane, Trincomalee, Ceylon, Calcutta, Colombo, Finschafen Bay, Langamak Bay. New Guinea, Jacquinot Bay, New Britain, Kokumbuna, Guadalcanal, Kukum-Lunga Point, Port Cruz, Batangas Luzon, Manila, Burnie, Tazmania (loaded potatoes!!) and always back to Sydney. Finally, we were outfitted to bring the first troops home from the Philippines. We spent so much of our time in Sidney, our ship was known as the "cupid ship".

By the way, my skippers didn't like sailing in convoy, so we did quite a bit of pickup and delivery on our own.

We stayed (my wife and I) in Los Angeles after the war. We now live in a mobile home park (since 1976) in Irvine. We celebrated our 62nd wedding anniversary in March, 1993. Rose was 83 in May, and if I can keep breathing until July, will be 85. Rose has been doing volunteer work as an accredited Medicare counselor for people with Medicare problems, and I have been available as a volunteer driver for anyone who has to visit a doctor or hospital, day or night. Due to various health problems that are catching up with us, we're not sure how much longer we can keep this up. I retired from the state of California in 1975. Worked for the Department of Motor Vehicles and some of the drive tests I administered were almost as rough as some days and nights at sea.

Converted 'Lib' Brings Vets In

(News clip from November 24, 1945, San Francisco papers.)

A rusty, gray-hulled experiment sailed into San Francisco Bay today—an experiment no longer.

She was the *Otto Mears*, first Pacific Liberty ship to be converted into a troop transport. She was remodeled at Manila by high-point overseas troops, anxious to get home by any means possible.

Her 532 cheering Army passengers left no doubt of their enthusiasm for their slow, tubby "luxury liner."

"Why, brother," said S-Sgt. David Rivelis of Brooklyn, NY, "this was like the *Queen Mary*. We had a swell time—the best chow I've had in the Army, plenty of sunshine and fresh air and free cokes and beer every day."

The *Mears* presented a strange sight as she docked in a light drizzle at Pier 45. On her deck had been constructed wooden shacks, not unlike chicken coops, to house the washing, shower and mess facilities for the troops. In two of her spacious holds two-tiered canvas bunks had been hastily nailed into wooden frames.

But despite her primitive living conditions, the *Mears* was a symbol to hundreds of thousands of overseas servicemen as a means of breaking the bottleneck of transportation in the Pacific. Though some Army official has said Liberties couldn't be used as troop transports over long distances of the Pacific, Army port officials at Manila had gone ahead and converted the Mears and 10 other Liberties into troop carriers.

The soldiers who arrived aboard the *Mears* today maintained that troop accommodations aboard most transports couldn't compare with conditions on their ship.

Officers commanding the men were enthusiastic, too, over the 28-day voyage. "We had a marvelous time," declared Major Henry D. Irwin, transport commander. Major Edward Hoffstatter, representing the Army inspector general's office, said the experiment was "completely successful," and would offer no recommendations, except that 100 additional passengers be added on future trips.

Gallups Islanders Harold Kock and Ed Frank entertain hostess of Boston canteen.

The *Otto Mears* arrived with 18 other ships carrying a total of 11,583 passengers from the far Pacific today. Twenty-five additional vessels with 10,111 aboard are due to dock tomorrow.

The *Otto Mears*, first Liberty ship converted into a troop transport by overseas veterans anxious to get home by any means available, steamed in the Golden Gate from Manila today with 534 passengers.

Docking at Pier 15, the slow former freighter—looking exactly like the hundreds of other mass-production Liberties turned out by American shipyards to overcome the war shipping shortage—was a symbol to hundreds of thousands of overseas servicemen of a possible means of breaking the peacetime shipping bottleneck which has kept them "marooned" in the Pacific despite high discharge point scores.

Maj. Gen. C.P. Gross, Army Transportation Corps commander, said that mass conversion of grubby Liberty ships into slow transports was impractical, since most of the men overseas would be home by the time the job was finished, but gave approval for Manila port authorities to try the theory on a few test ships.

The men aboard the *Mears* said that when it was too warm to sleep in the holds they were allowed to sleep on deck. It was only during the last few days of cold that even slight hardships were encountered.

"It is much better sleeping than deep down in a troopship . . . Was sure wonderful not to be crowded. . . . we ran our own dispensary, using army doctors and medics who were passengers. . . . "

Every man wished to add his favorable comment.

"Don't mind going slow, just so long as we go home," chortled one man. Most of the men had been waiting for a ship for 30 days or more.

The reconversion of this ship was on a volunteer basis. Engineers from the 821st Aviation Engineer Battalion worked on the *Mears* but they stayed behind—whether voluntarily or not is a question—to work on other Liberties. They are said to be homeward-bound now.

"I had one problem, with one of the vets who bugged me every day, insisting I make contact with Western Union to wire his family he was on his way home. Don't really know if he ever understood my explanation of why I couldn't. He just stayed upset with me no matter how friendly I tried to be."

The United States Maritime Service Radio Training course was established to provide an adequate source of trained radio operators for the vessels of our merchant fleet. Radio operators are specialists aboard ship. They are usually accorded the privileges extended to officers and are furnished quarters and messing facilities commensurate with their position.

Enrollees of the United States Maritime Service who successfully complete the radio training course at Gallups Island are qualified to sit for examination as Second Class Radio-Telegraphers. Holders of second-class operators licenses, temporary limited licenses, graduates of commercial radio schools, radio amateurs, as well as inexperienced men who desire careers in this branch of the seafaring profession are accepted for enrollment.

Gallups Island Radio Training Station has attained recognition as one of the outstanding schools of its kind. The school is equipped with the most modern equipment and facilities for instruction purposes.

To meet the new and increased demands of the shipping industry for radio operators, the U. S. Maritime Service Training Station, Hoffman Island, N. Y., has been recently converted to a radio school to supplement the facilities of Gallups Island. Hoffman Island was the first Maritime Service Training Station to be established.

The standard marine radio installation laboratory comprises at least one of each type of radio equipment now being installed aboard vessels constructed by the United

States Maritime Commission. It contains crystal-controlled intermediate and high frequency transmitters, superheterodyne sets, radio direction finding equipment, standby equipment, lifeboat transmitting and receiving equipment, small craft radiotelephone equipment and radio accessories.

The training staff stationed at the schools comprises specially selected commissioned and non-commissioned United States Maritime Service personnel thoroughly qualified to supervise and instruct the course of training which has been established at these schools.

Practical laboratory instruction supplements the theoretical instruction throughout the course. The academic part of the course presents the enrollee with a complete and comprehensive understanding of the basic principles of radio. Special emphasis is placed upon code sending and receiving.

Subjects included are:

- Mathematics.
- Fundamentals of Electricity.
- Magnetism.
- Direct Current Starting and Controlling Equipment.
- Capacitance.
- Turned Circuits-Resonance.
- Vacuum Tubes.
- Detectors.
- Transmitters.
- Radio Telephony.
- Marine Receivers.
- Auto Alarms.
- Spark and Arc Transmitters.
- Wartime Radio Procedure.
- Ohm's Law.
- Storage Batteries.
- D. C. Generators and Motors.
- A. C. Circuits.
- Power and Audio - Frequency Transformers.
- Rectifiers, Filter Systems, Voltage Dividers.
- Oscillators.
- Antennas and Coupling Systems.
- Superheterodyne Receivers.
- Direction Finders.
- Marine Transmitters.
- Practical Shop Practice.

The practical course of instruction includes the use and care of tools, drilling, tapping, wire splicing, soldering, transmitter assemblies, and the maintenance, operation, adjustment and servicing of radio equipment.

An important part of the radio operator's duties aboard ship is the responsibility for maintenance and servicing of all radiomarine equipment installed.

The phonetic method of code instruction is employed. The enrollee is taught to recognize the actual sound of the signals rather than by memorizing the dot and dash equivalents. The touch system of typing is taught. The most modern types of automatic code transmitting devices are employed for reception, insuring accurate receiving within the accepted speed limits.

During the latter part of the course, the enrollees stand actual watches. Students operate the equipment used aboard ship, keep up the radio logs and learn how to handle the radio weather signals, hydrographic reports, time signals, broadcasts and techniques of ship-to-ship and ship-to-shore operations in use.

Men who finish the prescribed course of training are awarded United States Maritime Service diplomas in recognition of their attainment.

Due to the accelerated course of instruction, graduates are offered a suitable Maritime Service Institute correspondence course in radio theory to enable them to prepare themselves for the Federal Communications Commission license examination.

Upon graduation, the enrollee is advanced to the grade of Radioman (3rd class) or Warrant Radio-Electrician.

MARITIME MEMORIES AND MORE
By Bob Clough, R-7

July 19, 1941, I signed on with the U.S. Maritime Service at the Coast Guard headquarters in Chicago as an apprentice seaman at $18 a month. Howard C. Bellman, Frank J. Ricker, Orman Myer and I all boarded the train for Boston. We had reported for service as requested on June 30, 1941. However, our arrival at Boston would have been over the 4th of July, so we were told to go home and come back the week after the 4th.

Those early Coast Guard Chief Radiomen were some of the best in the business. Many of them came up via the "Lighthouse" service and through many years of sea service. Besides Chief Summerfield and Sweeney, there were Chief Coombs and A.J. Wood who was our class advisor and several others whose names I have forgotten.

The summer of 1941 prior to Pearl Harbor was relaxed in many ways. Classes attended school for about four weeks then have a week of "duty" where the entire class was taken out of school and used to help man the Yeaton, stand anchor watch, scullery detail, and just plain work details. On one detail we removed the old urinals and fixtures from the rag shed at the east end of the island and hauled them up to help build the breakwater for the whale boats and sail boats that were at the Island then.

Fires seemed to be common in Boston. One evening in late summer of 1941 the Hood milk plant in Boston caught fire. The call came to Gallups Island for volunteers to help drag hose and fight the fire. Around midnight the "sailors" were released from fire duty and returned to Gallups. Many fellows conveniently missed the last boat and spent the night at the Seaman's Club in East Boston. Gallups had all the fresh milk needed for quite some time after that, compliments of Hood.

Thanksgiving of 1941 will probably be remembered by most of us on the Island at that time. Entering the mess hall for dinner we were amazed with the extent of the traditional New England Thanksgiving meal. Besides the usual turkey, ham and beef with candied yams, sweet potatoes, cranberries and all the trimmings, there were heaps of fruit, nuts, candy and even free cigars for everyone. How many of you remember those pictures of stuffed students smoking cigars outside the barracks? That was probably one of the most sumptuous Thanksgivings ever.

December 7 Pearl Harbor started a series of changes. The Coast Guard chiefs were replaced with Navy people starting with those in the code section. At that time some of the old time press operators came aboard the Maritime Service, the McElroy brothers, a fellow named Sullivan and others. It was always amazing to me to watch people like Steve Garcia, the McElroys and others sit down and copy 50-60 WPM code, get up for a glass of water, light a cigarette and come back without missing a word.

Six months and several work details later Class A-1, or R-007 as it was known then, graduated in late May with our F.A.A. Second Telegraph and in many cases, First Telephone licenses.

During the summer of 1942 I shipped out on an old tanker, SS *Pan Carolina*. We had some uneventful trips to Texas, Scotland and the Gulf of Mexico. On the last trip we loaded at Aruba and then went to Guantanamo Bay Cuba to form up the convoy. Coming around the east end of Cuba there was a submarine alert. During evasive maneuvers we rammed a C2 loaded with bauxite.

The bow of the old tanker was ripped open while the degaussing gear was running. It is a wonder we didn't go sky high with the load of Avgas we were carrying. After we put in to the shipyard in NYK I returned to Gallups for a visit before heading for the west coast where the water was warmer. Several of my class "buddies" had stayed on at Gallups as instructors under the Maritime Service. They convinced me to return as Chief Radioman, effective 01 Oct 1942.

Shipboard lab consisted of learning the mechanics of the various types of equipment. Code and watch standing labs consisted of learning code and watch standing procedures. However, theory was always a problem of finding suitable teaching material. College level text books were too advanced. The Radio Amateurs Handbook was one of the early basic text books. Those blue or white covered pamphlets used for study material were written by the early instructors to fit the course being given. Many times those of us instructing were only one jump ahead of the class itself.

About Thanksgiving time of 1942, Charlie Bradley, who had stayed on at Gallups as a Chief, decided to get married. The bachelor party was originally scheduled for the Coconut Grove night club in Boston. when I returned to my class on Monday morning the students were surprised to see me. There was a person with the same name as mine killed in the fire. It was a bloody mess with many service people burned to death.

June 1943 changed everything for me. I returned to Wisconsin and married my high school girl friend. We had an apartment at 50 Peterborough St. where my wife was able to finish her degree in Library Science at Simmons college while I commuted to Gallups. The commute, via bus, street car, subway, and boat was getting a bit much. That, along with changing conditions of the Island prompted me to resign from the Maritime Service, effective 20 December 1943 with the rank of Warrant Radio Electrician.

December 24, 1943, I joined the American Export Airline as a ground station Radio Operator. Part of the airline requirement was that we had to join the Naval Reserve as we would be operating a navy station on the Amazon River in Brazil. In January 1944 I became a Chief Radioman AA USNR and spent the next six months with four other R/O's at NXM in Belem, Brazil, operating navy NATS circuits for flights between the US and Africa by way of the South Atlantic.

June of 1944 the old Civil Aviation Authority sent 10 R/O's to Belem to replace us on the commercial NATS circuits. Back in NYK there were no flight vacancies so I wound up as a Radio Inspector in the ships until the summer of 1945 when Roscoe Maricle, R-0B2, bid off of flight duty to go into the radio ships. This opened the way for me to start flying as Flight Radio Officer. After seven years as Flight Radio Officer with American Overseas Airlines the R/O's were furloughed in favor of radio telephone in October 1959, when Pan American Airways purchased American Overseas.

The next 38 years we spent with various airlines in different positions. Joining the Flying Tigers Airline in 1950, I became a navigator and flew the next 20 years in that position working with a dozen different international airlines until early retirement from United Airlines in 1975. Again I was displaced by technology, this time by electronic navigation systems.

Early retirement didn't mean being put out to pasture. The next 12 years were spent knocking around the world with non-sked airlines in various management positions until taking full retirement in 1988.

Serving as GIRA Director of Region 9 for the past two years has been a most rewarding experience. GIRA is a mighty fine group of individuals come together to share common experiences through the annual reunions and on the various radio nets maintained for the ham operators of GIRA. Thanks to the early efforts of people like Art Bradford, Ed Hayden and others, GIRA has become the retirement focal point of many people. Persons such as Jim Kinkel helping to locate lost members, Tom Cruse as editor of the Spark Gap and Stan Jennings of the Islander have furthered the bonding together of GIRA into the organization we now have.

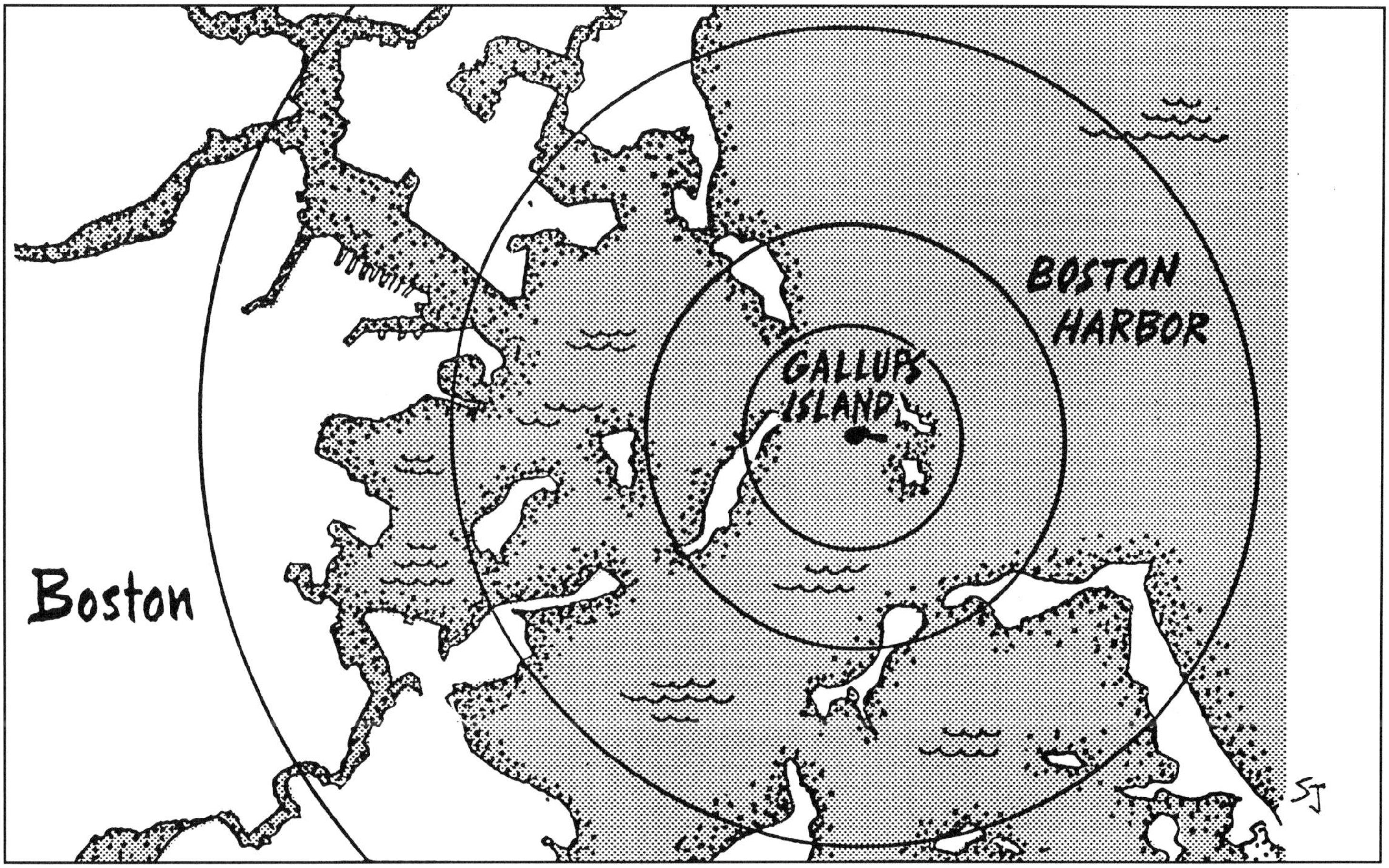

We Came From All Over, We Went Everywhere

This Was Gallups Island

by James A. Kinkel
Past President, GIRA

Patches of early morning fog hung low on the surface of the water, shrouding the islands and the approach to the land. The small brig eased her way into the bay with the smell of land strong in the air. As the sun began to burn off the fog, the islands began to appear as if by magic. They were lush with heavy growth of forests. Some islands were populated by various animals, such as deer, fox, etc. There was little human activity to be noted. . . . This was the Boston Harbor Islands in the early 1600s.

Picturing this kind of scene in your mind would immediately bring the reaction, "When I was there, those islands had very little vegetation, not much of anything that was green or lush."

The first settlement was established on Thompson Island during 1619. Here Indians exchanged furs for articles of the European Civilization. In 1626, one of the first truly permanent settlements on the harbor islands was established on Thompson Island. Gradually, as the settlements on the mainland grew in number and size, the forests on the islands were cut for firewood. Having been cleared of the forest, the islands were then put to use for farming and grazing for cattle, horses, sheep, etc.

Our island's namesake, John Gallop, was born in England in 1593. His family was part of the aristocracy. He was trained in the military science of that time in Holland. Leaving his family at home, he sailed for the New World from Plymouth, England on March 20, 1630 on the good ship Mary and John. After an arduous voyage, John Gallop and his fellow passengers arrived at Nantasket (Hull) on May 30, 1630. Later, he moved on up to the settlement of Boston Harbor, where he became well known and a well-to-do land owner. Most of the islands were owned by one or more families and were used as their residences. As a colony freeman, Gallop was granted land in Boston Harbor along the waterfront not far from Land's Wharf. He also received land on some of the harbor islands as well as ownership of our Gallups Island and Nix's Mate.'

Gallups Island was 16 acres in size and had a small meadow at the east end. The island was farmed extensively for the next two hundred years. It is hard for me to visualize the island, first as a forested piece of land, then, as a productive piece of farmland. Can you imagine the beauty that must have existed in the harbor islands during this time? It must have been awesome.

John Gallop had already gained respect as a mariner and had several vessels of his own. He was involved in the first maritime engagement against the Indians off Block Island in 1636. He was the skipper of the Colony's sloop, *Buck*, at the time. He and his small crew were successful in suppressing the Indians but he was too late to rescue a fellow seafarer. This engagement is thought of as the beginning of the Pequot War.

During the Revolutionary War, all the islands played an important part in boxing in the British shipping and troop activity. The harbor itself was blockaded by the British Navy. The Patriots smuggled many loads of illicit supplies across these islands for the small colonial forces fighting the British. I have often wondered what role our little island may have played during this war for our freedom.

Peter Newcombe and his wife bought the island in 1819 from L. Brackett who apparently had purchased the island from Gallop's heirs. They had a very prosperous farm on the island for a number of years. After her husband died in 1833, Mrs. Newcombe opened a restaurant and inn on our island. Her cooking soon had an enviable reputation around Boston. Spending time at the inn became a custom of many Bostonians. During this time the island was called Newcombe's Island. When Mrs. Newcombe died, the city of Boston Harbor purchased the island from her son. It is hard to visualize taking a boat from Boston out to the island just for a meal; she must have been one heck of a great cook. Too bad it didn't carry over to our mess hall!!

Sometime around 1855, Joe Snow purchased the inn. The island came to be known as Snow's Island after that. During the summers, once again it became a place to go for good seafood meals. Apparently, it became quite a lively place.

During the Civil War, Boston Harbor granted the Federal Government the use of the island. The Fed's built some barracks and soon there were Yankee troops camped all over the island. After the war, on July 6, 1865, a steamship, loaded with Yankee troops, the 38th Massachusetts Regiment, anchored off Deer Island. When inspection of the ship and men was complete, the troops were put ashore on our island. They were paid off later and were transported to Boston Harbor on a ferry boat. In August/September of this same year, the 54th Massachusetts Regiment was also discharged from the service at Gallups Island.

On June 1, 1866, the deserted barracks located on the island were put back into service with the relocation of the quarantine station from Deer Island to Gallups Island. During the first twenty years of service as the quarantine station, the island averaged 50 patients annually. On the hillside, facing to the east, overlooking the narrow neck of the island, is a cemetery with over 200 graves of patients that didn't make it.

The Fed's resumed control of the quarantine station on the island in 1916. It became an immigration station as more buildings were built to house incoming immigrants.

During the conflict of WWI, German sailors, who had been taken off the interned German ships in Boston Harbor, were held on Gallups Island as prisoners. After the war, the island reverted to a quarantine station and in 1937 the quarantine station was moved to Boston.

In early 1940, the USMS Radio Officer Training School was established. At the end of WWII, in 1947, the island, buildings and associated equipment, were sold at auction for some $50,000. When the salvage operation was completed, the island was basically abandoned. In the 1970s, the island was taken over by the Boston Harbor Island State Park system and converted to recreational use.

Chronologically, this is the story of our island. We can say that it has had its place in history and feel that maybe we walked, slept, and ate in the space where some of our ancestors also roamed— who knows!!

A booklet that was copyrighted in 1943 was passed out to new trainees as they arrived at Gallups Island. It was titled *USMS Radio Training School.* The opening lines of this booklet said, "We're going to help deliver the goods to your fighting men! That, in essence, is the reason for our presence at the United States Maritime Service Radio Training School at Gallups Island in Boston Harbor."

The recognized date of establishment of the training school is June

1940. However, records show that beginning in December 1939, a temporary seaman training school was set up under the supervision of the Coast Guard for the Maritime Commission. The school was short lived, as the Maritime Commission decided in February of 1940 that there was no need for it. The operation was closed down and the Coast Guard personnel were released for other assignments.

In May 1940, the Maritime Commission decided there was a need for a training school for radiomen, signalmen, cooks and bakers, and machinist mates. Gallups Island was to be reopened for this activity. Again, the training would be under the instructorship of the Coast Guard. The first trainees were to be made up of 400 CCC men, 100 other men that were under training aboard the *USMS American Seaman*, some additional trainees procured from the National Youth Association of Jacksonville, Florida, plus a small number of ham radio operators.

The initial plan was for men selected for radio and signalling training to be given twelve to fourteen months of intensive training under the Coast Guard personnel. When the training facilities were better developed and the training routine smoothed out, the time required would be reduced. This time frame was less than half of what a civilian (commercial) radio school would require. The training hours would be full of the various aspects of communications aboard a merchant vessel. It would cover code, equipment (operations and maintenance), theory, laboratory work, radio law and procedures, etc.

After completing training, the graduates would be retained as permanent personnel with ratings and pay equivalent to those of enlisted Navy and Coast Guard personnel. In the case of those training in radio, they were the first men to be trained by the Maritime Service in communications for shipboard use. To qualify for radio training, it was desirable to have had some prior electronic experience, such as amateur radio. This qualification was not necessarily maintained. When the radiomen completed their training, they were offered permanent berths as part of the Maritime Service ship's company at the school. They were also offered the opportunity to sign up for the US Naval Reserve.

In June 1940, 310 men, most being between the ages of 18 and 23, began repairing the buildings and equipment of the island in preparation for the full training groups' arrival. When they completed their task, there were accommodations for approximately 750 men.

Initially, there was to be no military type training. The trainees were to be allowed liberty on the beach every other night. A regular training routine was established but there was no "strict" discipline. There was to be time off for the trainees to do as they wished. Thus began the USMS Gallups Island Radio Officer Training school.

INITIAL PLAN . . . TWELVE TO FOURTEEN MONTHS OF INTENSIVE TRAINING . . .

As the training routine smoothed out, different parts of the curriculum were changed or modified to develop the right balance for shipboard operation. The training program stayed under the supervision of Coast Guard instructors and then under the Navy after war was declared. In the late 1942s or early 1943s the training school and the training program were taken over by the Maritime Service. The Maritime Service had their own instructors most of whom had been trained under the initial program. Again, there were some minor changes in the curriculum to satisfy the new management.

One of the major changes that came to pass after the beginning of the war was the military mantle that began to move into the program, such as formations, marching wherever you were required to go, uniform of the day, and other procedures that are associated with a military organization. Liberty was reduced to weekends if your grade average would justify a liberty card and you had not been "gigged" for some minor breach of discipline. Trainees were required to stand fire watch, anchor watch and any other watch they could think of that would keep them occupied after hours and at night. Field day was on Saturday morning before the liberty party was allowed to leave the island. Barracks and sea bag inspection was also held early Saturday morning.

With the war came more trainees and the need for more space. Additional buildings were built to provide housing and recreation. A new platoon of trainees (approximately 50 men) began to arrive nearly every week. These men had been through boot camp at one of the basic training stations such as Sheepshead Bay, Catalina, or St. Petersburg. Some had been sent to Huntington, Long Island, NY for basic radio instruction before coming to Gallups Island.

The coveted FCC 2nd Class Radio Telegraph license was somewhat illusive for some of the trainees. The initial training program was developed to provide radio officers for single operator berths; the program was amended to include "sunset classes" and special study groups for those in dire straits gradewise.

As time progressed, the FCC came out with the "TLT." This was the Temporary Limited Telegrapher license that allowed the operator to stand watch under the supervision of a 1st or 2nd Class licensed operator aboard ship. The FCC had essentially cut the grade requirements another five points for this ticket. An individual was allowed to sit for the 2nd Class ticket again in six months. Also, they had a "TLT" 3rd Class phone license. The time frame for the training was reduced from the twelve-fourteen months in 1940 to eighteen months in late 1941 and early 1942, then to six months in late 1943 and early 1944, and finally in late 1944 it was further reduced to five months.

With the reduction of training time in the late fall of 1944, several parts of the curriculum were dropped. No longer was blinker or flags taught, "hands on" activity with the equipment and laboratory work was reduced to a minimum; code was no longer taught to be copied on a "mill" (typewriter) and typing classes were dropped. The trainee learned to copy code with a stick (pencil) and that was it. Trainees who were in trouble with their grades were in deep trouble before they became aware of the situation. Generally, this caused them to miss the coveted 2nd Class ticket.

Possibly this was allowed to happen because of an over supply of radio operators and the winding down of the war. It appears the training program was again being modified to provide "watch standers" for multi-operator ships as opposed to the original concept of single operator berths for radio officers. This is not saying that the 2nd Class ticket was beyond one's reach, it just required much more effort on the individual's own part to acquire it. Many achieved this goal after they left the island and had put the six months in on the "TLT."

In August 1945, the school started closing down. Some of the trainees were returned to the stations that had originally shipped them to Gallups Island and others were sent to the other radio schools that were still open only to find that these schools were shutting down also.

At the end of 1945, USMS Radio Officer Training School, Gallups Island, Boston, MA, was history. The radio school had been recognized as the best in the country, civilian or military, bar none. Those men who were a part of this history should have a great deal of pride in what they had offered their country as civilian warriors.

Some 5,000 men received training at this station. Many were lost at sea during the war. The memorial that was dedicated on the island to these men during 1944 has been salvaged from the City of Boston, refurbished and rededicated on the beach nearby the Coast Guard station in Boston.

The island was essentially abandoned in the fall of 1945 until it was sold for salvage at auction in 1947 for some $50,000. What a terrible waste!!

Gallups Island is now covered with sumac, wild roses, bayberry, etc. The fruit trees that were down by the code rooms are still there and bear good fruit. The foundations for the different buildings can be found if you work your way through the underbrush. Some of the asphalt that comprised the drill field still shows through the grass. Your first visit back to the Island will give you a bit of emotional shock, guaranteed.

(After entering Mr. Smart's name to the membership rolls on June 4, and noting on the data sheet that he had been torpedoed, I looked up the story about this ship in "A Needless Word, A Needless Sinking." It was such a story that I wrote him, encouraging him to write the story in his own words so it could be shared by our other members. Here is his story.

Ray Anderson, Secretary-Treasurer, GIRA)

THE SINKING OF THE JAMES W. DENVER

By James W. Smart, R-C1
Calabash, North Carolina

After being on the Murmansk run for a year and experiencing all kinds of problems, I took a 30-day leave at home. After the leave was over, I reported to Joseph Rubin, dispatcher at the union office in Baltimore.

Upon learning that I had just returned from a Murmansk run and getting back in one piece, Joe said he was going to give me a brand new ship headed for a Southern run where I could get some sun and rest. My assignment was to the *James W. Denver* which was being fitted out at the Sparrow Point Shipyard at that time. I reported aboard with my personal effects and met Captain Everett W. Staley. The officers and crew were very nice and I was happy about getting a new ship which was not headed to the ice cap. We signed on March 13, 1943.

After spending two to three weeks shaking down the ship's equipment and loading cargo, we were ready to sail. Our cargo consisted of a complete U.S. Army Field Hospital; operating rooms, tents, ambulances, as well as some medical personnel as passengers. There were also some aircraft in crates and trucks on the deck.

On or about April 1, 1943, we departed from Baltimore in convoy UGS-7 headed for Casablanca. Since this was a new ship on her maiden voyage, there were a number of things that the crew was unfamiliar with. In addition, some of the equipment failed to work even if one was familiar with it. In my case, the radio room was equipped with a Mackay unit which I had been briefed on while at Gallups Island, but I had never had hands-on experience with it. At the time I remember thinking that the transmitter was a real "Rube Goldberg."

In talking with other members of the crew, I learned that they too had some problems. Eleven days out, the engine room had problems which caused us to stray from the convoy.

On April 11th, at about 5 PM local time, our ship was torpedoed. It was my practice to eat early in order to be in the radio room at sundown. I had just sat down to eat when the torpedo struck.

Best of Luck at all times
F/O W. [illegible]
4 Blossom St
York
England.

RAF SQDRON
Pt. Etiène,
Rescued Survivors
S/S James W Denver

Almost immediately there was a strong odor of Kordite, and the ship started to list to starboard. I rushed to the radio room and within a short period of time, the order came to abandon ship. I remembered that all papers and codes had to be destroyed. Captain Staley, along with myself and some of the crew, went to the boat deck to lower one of the sail lifeboats. There was a lot of confusion because boats on the starboard side were almost in the water while the ones on the port side were not able to be lowered due to the guncrew who were manning the 5" on the fantail. The sub made no appearance and dusk was fast approaching, so we abandoned ship. Captain Staley gave the order to try and stay together until morning. We attempted to tie a rope from one boat to the other, but the seas were too rough to do so.

At nine or ten o'clock there was another loud explosion. It was too dark to see what it was, but we surmised that the sub had put another torpedo into the sinking ship.

The next morning Captain Staley made a decision that passengers and some of the crew, including the Mate, take the motorized lifeboat and head northeast to the Canary Islands which was thought to be the nearest land. The boat that the Captain and I were in was going to head east to the African coast which was thought to be about 500 miles.

Twenty-five days later, on May 6, with 14 survivors, we landed in a country known a Rio Del Oro, about 90 miles north of the port of Etienne. The British maintained a base at Port Etienne and operated patrols all along the African coast.

When we landed, no one was able to walk. We managed to crawl around the shoreline and get snails off the rocks in the surf. After several days we gained enough strength to walk. I recall how cold it got at night. We could see the eyes of dogs in the dark. They did not bark or make any noise, but just stared at us.

We used tin from the lifeboat to make a griddle on which we could cook the snails, and believe me, they tasted good! I do not recall having a problem of no water.

On or about May 9th, an RAF patrol spotted us. They signaled that they would return the next day and advised us to stay where we were. We had started walking and I would guess that we were in sight of our original landing site when we were spotted.

The next morning, a boat lay off shore and two rubber dinghies came ashore to pick us up. The sea was running heavy and we had problems getting back to the boat. When we finally did, we were taken to Port Etienne.

The RAF people contacted the U.S. Army at Dakar where we were moved to an army hospital. After spending two or three days in the hospital, we were released when transportation became available to Trinidad, West Indies. From there we were flown to Miami. I arrived in Miami with only a pair of short summer pants and a cigar box with razor, toothbrush, and soap given to me by a nurse in the hospital at Dakar.

After clothes were purchased for us, we were put on a train for New York and arrived there with no papers. This caused some delay, but we were finally sent to the steamship company office at 25 Broadway. There we were paid for the time we had worked, until April 11th. From that date to June 3rd we received no pay—we were survivors and you didn't get paid for being a survivor.

Though all of this happened 50 years ago this month, to me it seems just like yesterday.

The account of the events in Captain Moore's book, "A Careless Word—A Needless Sinking", gave me information that I did not know; the sub that sunk us and its eventual fate, the boat that capsized on launching and how many died.

"Ed Hayden, Art Bradford, Dave Bulkley, Ralph Albers, Ed Pleuler, Bill Anderson tell Gallups Grads: "QTC"

This is how it all started: an ad in QST "Strays" column which read: "I would like to get in touch with WWII graduates of Gallups Island Merchant Marine School interested in a reunion in Boston or Washington, DC. Please contact Ed Hayden (R-20), K30KL, 16 Decatur Avenue, Annapolis, MD 21403."

JULY 1977. Ed Hayden (R-20) placed above "Strays" notice in *QST* magazine seeking Gallups Islanders for a reunion. After seeing the *QST* notice, Art Bradford (R-16) contacted Ed and a committee was formed. Hayden became chairman and Bradford became New England coordinator of the proposed reunion.

JULY 7, 8, & 9, 1978. First National Reunion held in Boston. Over 100 Gallups Islanders and their wives attended. A cocktail party was held on the 7th in the Parker House Hotel. A boat trip to Gallups Island on the 8th with a lobster luncheon served on the island, followed by a moonlight sail around Gallups Island later that night.

EARLY 1979. Ed Pleuler (R-3), Ed Hayden (R-20), and Art Bradford (R-16) established an informal 75-meter amateur radio net which they called The Gallups Island Radio Group.

1979-1984. Gallups Island Radio Group continued as an informal social amateur radio net.

JULY 1984. Dave Bulkley (R-37), Bill Anderson (R-72), and Ralph Albers (R-9) suggest establishment of a formal Gallups Island alumni association. Bulkley, Anderson and Albers became pro tem President, Treasurer and Secretary, respectively.

OCTOBER 1984. GIRA participated in ARRL New England Division Convention. A show booth was manned in an effort to locate Gallups Islanders.

FEBRUARY 1986. GIRA officially began. Dues established and collected. Membership drive conducted to find men who attended or instructed at Gallups Island or other USMS Radio Group.

JANUARY 1986. Ed Hayden (R-20) became Vice President of GIRA and editor of GIRA publications.

MARCH 1986. Publication of the *GALLUPS ISLANDER* and the *GALLUPS NEWS QUARTERLY* was started.

APRIL 1986. MEMBERSHIP HAD GROWN TO 67 MEMBERS.

MAY 1986. Regional Reunion held at NSS in Annapolis. Ed Hayden (R-20) was host.

JUNE 1986. Bill Anderson (R-72) Treasurer, undertook a continuing program to locate Gallups Islanders by checking Gallups Island platoon and class rosters against various Amateur Radio Callbooks.

OCTOBER 1986. GIRA participated in ARRL New England Division Convention with a show booth. Speakers from GIRA were on the convention program. Art Bradford (R-16) was the Coordinator.

NOVEMBER 1986. GIRA joined Mrs. Gladys Reed, widow of Captain Sherman Reed, former Superintendent of Gallups Island, in a program to restore and move the Gallups Island Memorial Monument.

MARCH 30, 1987. GIRA was granted Connecticut non-profit corporate status.

APRIL 1987. GIRA applied for Federal (IRS) tax-exempt status.

SPRING 1987. Jim Kinkel (R-103) undertook a continuing membership drive mailing news releases to newspapers nationwide in an effort to locate Gallups Islanders.

APRIL 1987, MEMBERSHIP HAD GROWN TO 240.

MAY 29, 1987. 1987 Regional Reunion held at Williamsburg, VA. Included was a trip to the Maritime Museum at Newport News, VA. Carl Brendel (R-7) and Ralph Albers (R-9) jointly coordinated the affair.

AUGUST 1987. News item, placed by Art Bradford (R-16) appeared in *Yankee* Magazine to obtain Gallups Island historical information and to locate Gallups Islanders. Art got 94 responses and 35 new GIRA members.

MARCH 1988. Bill Anderson (R-72) took over as Secretary pro tem.

APRIL 1988, MEMBERSHIP HAD GROWN TO 530.

APRIL 25, 1988. Gulf Coast Regional Reunion held in New Orleans. C.W. Rampy (R-5) chairman of the event.

MAY 1988. GIRA granted Federal tax-exempt status by US IRS.

MAY 20-22, 1988. First GIRA National Reunion held in San Francisco, which included a harbor cruise aboard the restored Liberty ship *SS Jeremiah O'Brien*. Al Hadad (R-13) and Alex Newbold (R-9) were co-hosts.

OCTOBER 1988. Middle Atlantic Regional Reunion held in Forked River, NJ. Bob Mitchell (R-34) put this one together and hosted.

NOVEMBER 11, 1988. Gallups Island Memorial Monument re-dedicated in North End Park, Boston, MS. (Monument has been restored to its original form as existed on Gallups Island in 1944).

DECEMBER 9, 1988. Northern Lights Regional Reunion held in Coon Rapids, MN. Ray Maurstad (R-93) sponsored.

DECEMBER 31, 1988, MEMBERSHIP AT 555.

JANUARY 1989. Art Bradford (R-16) hosted a cook-out for GIRA members on west coast of Florida.

MAY 1989. MidLantic Radio Officers luncheon held in Forked River, NJ. Hosted by Robert Mitchell (R-34) and Pete Rodgers (R-18).

MAY 22, 1989. The *SS John W. Brown* was scene of memorial services for merchant seamen lost at sea. Tom Gibson (R-17) was Radio Officer, crew member, and Treasurer for the project.

OCTOBER 1989. National reunion in St. Louis, MO. A Mississippi River cruise was the highlight of this reunion. Dave Bulkley (R-36) named recipient of the "Bolt Award."

NOVEMBER 11, 1989. William Anderson (R-72), Charles Dinsmore (R-44), and John Burns (R-7) laid wreath on Sunken Ship Memorial in Boston's North Park.

DECEMBER 31, 1989, MEMBERSHIP AT 611.

JANUARY 2, 1990. New officers elected. Jim Kinkel (R-103), President; Tom Cruse (R-50), Secretary; John Burns, Treasurer.

JANUARY 20, 1990. John Burns resigned as Treasurer. Tom Cruse added Treasurer job to Secretary and Editor of quarterly newsletter.

JANUARY 1990. John Palmer (R-13) hosted a great cookout in Florida, featuring smoked mullet.

FEBRUARY 1990. Jim Kinkel obtained the list of Merchant Marine Radio Operators that received the Coast Guard discharge.

MARCH 15, 1990. Gallups News Quarterly renamed "Spark Gap" after Gallups Island newspaper of the 1940s. Masthead from original *Spark Gap* copied for use on new newsletter. First issue wholeheartedly accepted by membership.

APRIL 11, 1990. Bob Mitchell (R-34) and Pete Rodgers (R-18) held another successful MidLantic Radio Officers luncheon in New Jersey.

MAY 1990. Alex Hadad (R-13) represented GIRA and attended memorial cruise of *SS Jeremiah O'Brien* in San Francisco Bay.

SUMMER 1990. Volume 4, No. 1 of *Gallups Islander* published. Stan Jennings (R-58) became editor.

MAY 22, 1990. Bud Rines (R-12), Ed. St. Andre (R-21), Garth Linder (R-23), Art Bradford (R-16) placed a wreath at Sunken Ship Monument, Boston, on Maritime Day. Art and Bud represented the Gallups Island Radio School at the Maritime Day Festivities at the Massachusetts Maritime Academy. Gallups Islanders donated $340 to help with the cost of operating the "Unknown Sailor Park" at the Massachusetts Maritime Academy, with the urging of Art Bradford. "Gallups Island Radio School" was added to the contribution stone.

JULY 4, 1990. Jerry McClish (A-2) and Art Bradford (R-16) returned to Gallups Island to celebrate the 50 years (to the day) since McClish first arrived at Gallups. A wreath was tossed into the water in memory of those who never made it home from the sea.

AUGUST 3, 1990. Braintree, MA, Hotel Sheraton Tara was headquarters for the 50th reunion of the Gallups Island Radio School. Over 350 attended the banquet on Friday night. 150 visited Gallups Island on Saturday morning and enjoyed a picnic lunch. Bill Anderson (R-72) was awarded the "Bolt Award" given to GIRA (non-officer) who has done an outstanding job for the Association. Bill served approximately three years as secretary.

SEPTEMBER 1990. Bill Anderson (R-72), Bob Mitchell (R-34), Bud Rines (R-12), Wesley Fowler (R-13), Ed Liggins (R-112), attended the South Portland Shipyard banquet. Bill Anderson elected to board of directors.

OCTOBER 11, 1990. Bob Mitchell hosted another luncheon in Waretown, NJ. Good food and good fellowship was enjoyed by all attendees.

OCTOBER 13, 1990. District 8 (Alaska, Idaho, Montana, Oregon and Washington) luncheon meeting at the Tyee hotel in Olympia, WA.

NOVEMBER 1990. Membership roster compiled and mailed to membership. Membership roster contained 928 names and addresses.

NOVEMBER 12, 1990. Memorial Service conducted by Armed Guard Veterans for the *SS Black Point* (last ship to be sunk in WWII) at Point Judith, RI, was attended by Art Bradford and Bud Rines.

DECEMBER 1990. Planning Committee for 1991 Reunion in Minneapolis, MN, meets and selects travel agents for August 13-14 Reunion.

DECEMBER 31, 1990, MEMBERSHIP AT 975.

JANUARY 1991. Jerry McClish held Florida cookout in Bradenton, FL. 40 GIRA members and wives attended.

JANUARY 31, 1991. 35 GIRA members received 40-year

Commemorative Medal from Russian Embassy.

MARCH 1991. Spring '91 issue of *Spark Gap* mailed to 982 members and 840 non-member Gallups Islanders with an invitation to join. Received 43 of the invitations back with check for dues.

MAY 22, 1991. Alex Newbolt represented GIRA and presented a wreath from us to the Memorial Service aboard *SS Jeremiah O'Brien*.

MAY 22, 23, 1991. Several amateur stations contacted the *SS John Brown* during the memorial voyages.

MAY 26, 1991. Ray King, Director, Region 2, placed a wreath on Sunken Ship Memorial in Boston's North Park.

SPRING 1992. Rion Dixon (R-17) elected president of GIRA, taking over from James Kinkel; Robert Lanphere (R-12) Vice President; and Raymond Anderson (R-12) Secretary-Treasurer.

AUGUST 1992. GIRA reunion at Long Beach, CA. Featured was a tour of the *SS Lane Victory*.

DECEMBER 31, 1992, MEMBERSHIP AT 1, 114.

FEBRUARY 1993. Rion Dixon, GIRA president, signs contract with Turner Publishing Company, Paducah, KY, do a Gallups Island history.

JUNE 1993, MEMBERSHIP AT 1, 215.
SILENT KEYS 377.

This chronology was begun by David Bulkley, who was GIRA president in 1988. It has been expanded by various people, according to Tom Cruse, who updated it and printed it last in the Summer, 1991, issue of SPARK GAP. The editor of THE GALLUPS ISLANDER has updated it still further with the help of others, notably Ray Anderson and Jim Kinkel.

KINKEL'S KINETIC CONNECTIONS

Jim Kinkel, past president of GIRA, has done much work seeking out Gallups Island trainees for membership in the association. He explained how he goes about it.

Working from a file of about 5,775 names, he matches first and last names with names he finds in telephone directories. If the person he locates is within 50 miles or so of the home address in his files, he sends out an exploratory letter. He says he has mailed out more than 1, 000 letters since mid-December of 1992, and received back about 300 responses. In addition to the 5,775 names, he estimates there are 300 or 400 more in platoons he has not located. Here is a breakdown of the categories of potential members or those now gone.

Gallups Island Radio School—736 located, but not included in membership. Some were once members, but dropped membership or died and didn't tell us or moved and left no forwarding address.

Silent Key—309 located and identified, who have died since the war.

Killed in Action—172. Of these located, there is no positive proof they were all from Gallups Island, but Kinkel believes they were.

Personnel from Gallups Island other than radio operators—161, of which 31 have been located and 19 are silent keys. These include people who "ran the island," including some from the Coast Guard, Navy and Maritime; some instructors; some who did ferrying. Of the 161, some have joined GIRA.

This is an ongoing search, for which Kinkel says his reward comes when he locates a radio operator and they talk, sometimes for hours. Often the man feels that just the talk with someone who has experienced similar dangers will leave him feeling a great burden has been lifted. "They feel good again, which makes me feel good, too," Kinkel says.

Always on the lookout for information, he invites tips on men or women from Gallups Island during those war years. Write: James Kinkel, PO Box 1211, Corrales, NM 87048; or radio WA5JGD; or call 505/898-8361.

Shaded lady in the center is Ellenor Yahrmarkt, guiding Shelley Jennings, Betty Jackson, Jane Jennings and Howard Jackson (R-58) to Gallups Island

SEAMAN, OHIO MAN BECOMES SEAMAN. SEES SEA. SEE?

By Roy C. Moore, R-6

I was to be a seaman from Seaman, OH. Early in 1941 there was a call for three candidates from each state for radio training while I was enrolled in the National Youth Administration (NYA) program in nearby Portsmouth, OH. The NYA had a group interested in radio headed by a graduate electrical engineer from Ohio State University. We had a ham station and serviced Forest Ranger radios.

I was given a bus ticket to Sulphur Springs, FL and arrived there 36 hours later. I didn't eat en route, so I flunked the physical exam—my blood pressure was too low—but after some food and rest I was enrolled May 9, 1941. A group of us were loaded on a train and arrived in Boston's South Station, where we marched along Atlantic Avenue to the Coast Guard pier. I remember some smells (low tide, steamed crabs, etc.) that were different from on the farm. At Gallups, I was enrolled in Class C-1, but on December 7, 1941, several of us were accelerated into Class B-2.

On February 14, 1942, I was released from active duty and given $16.15 for a train ticket to Philadelphia, PA. There a ship headed for Murmansk, Russia was waiting for me. While en route to South Station, I met a Gallups instructor on the subway. I told him I was headed for Murmansk, Russia and asked what he knew about this trip. He said, "It's OK if you don't mind taking your life in your own hands."

I sometimes wonder about my discharge from active duty. It would probably be more logical to say I was released from inactive duty and was being assigned to real heavy active duty. I arrived in Philadelphia late at night and boarded the *Hegira*. I was greeted by the Third Mate, who was quite put out with my uniform as he showed me to the Radio Room.

Later I overheard the Weyerhauser Steamship Company officials in the Captain's quarters (part of the Captain's quarters had been sectioned off with plywood and made into a Radio Room and that's how conversation could be heard through the plywood wall) trying to convince the Captain to take the ship. The Captain was arguing against it. The armor added to the bridge would make the ship top-heavy and, in heavy icing conditions, it would surely capsize. Besides that, the visibility had been cut down by the added armor. The Captain had made up his mind and left. The officials then called the Chief Mate into the Captain's quarters and asked if he would take the ship as he had Captain's papers. He said "Yes" and so the Second and Third Mates moved up and a new Third Mate was hired.

I noticed the stevedores were using copper hammers, copper nails, and copper wire to which Captain Comer replied, "Sparks, there's only one way we'll get to Murmansk, and that's if we don't get hit." His words could not have been truer, we would later learn, as the *Cape Corso* docked behind us and was being loaded with identical cargo, such as dynamite, high-test gasoline, detonators for bombs, and poison gas. We would see it disappear in a matter of seconds when it was hit with an air-launched torpedo north of Iceland. Very little was said about the ship being painted battleship navy gray.

We were armed with two 50-caliber and some 30-caliber machine guns, and a four-inch gun on the stern. Some armament to take on the German Air Force and U-Boats!

We should have done what General Patton did—wait for more supplies. Someone said that the battleship gray paint was less visible. I suppose there is merit to that—but having worked in sonar anti-submarine warfare (ASW) for over 35 years, I know that German submarine sonar depended on propeller beats and noise from the ships. Paint has no effect against sonar detection. Painting the ship navy gray and arming us did make it impossible to distinguish—a merchant ship manned by civilians—and a military auxiliary.

Philadelphia to Halifax February 1942

Once we were under way leaving Philadelphia and the main antenna was up, the receiver was jumping with SSS of ship sinkings and the shore station setting off the automatic alarm system. I was under the illusion that everyone was giving it hell, but now know that while merchant ships were being clobbered, 25 destroyers fired up and ready with 300 and 600 pound depth charges were lying idle in ports up and down the East Coast. The trip to Halifax was filled with anxiety by the constant reminder by shore stations of ships being reported torpedoed and sunk along with live SSS transmissions. When we arrived at Halifax, officials told the Captain that we were the first ship to arrive for several hours. It seems the pilot boat was sunk by a sub. The sub took over the pilot boat station. As soon as a merchant ship arrived and blinkered PSB (please send boat), the sub blinked back "pilot on way,'" sending a torpedo into the unsuspecting merchant ship. I remember going ashore with Captain Comer for a convoy briefing. The captain was given sealed orders to open if we lost the convoy. A few days out of Halifax, en route to England, we were taking green seas over the bow and a heavy truck chained on top of #1 hatch was flattened, causing it to pound like a sledge hammer on top of 88 tons of dynamite stored there. All hands were called on deck to throw the truck overboard. By the time we had dumped the truck, the convoy was gone. I was seasick and kept a bucket in the radio room to vomit in.

When the Captain opened the sealed orders, it called for me to man the direction-finder (DF) at a pre-determined time for 15 seconds on a prescribed frequency and obtain a bearing of a signal sent by the convoy escort. That bearing would be used by us to steer a course which would allow us to overtake the convoy as it zig-zagged along. This procedure was repeated for about three or four days, and we finally regained the convoy. Now, when the Captain called me to the bridge (the DF was located there), I arrived with my bucket of vomit. Needless to say, there was some scrambling for fresh air as I sat the bucket underneath the DF. The captain was so sympathetic to me that he asked the cook to give me a greasy pork chop with a long string attached so that I could retrieve it every time I threw up.

During most of the trip to Scotland, we seldom saw our British escorts. They were busy over the horizon. I was impressed with the 50-caliber machine guns as the Armed Guard practiced with their red tracers against a trial balloon. What a let-down later when those red tracers appeared to be curving away from the Messerschmitt-109's and Junkers-88's. I was often called on to do blinker. I met an electrician two years ago who was in the Navy Armed Guard, and I was surprised to learn that he was the blinker man. Well anyway, I was called out to calm the British Captain whose ship was keeping convoy station directly behind us on our stern. The excitement started when the Ensign in charge of the Armed Guard decided to practice with the four-inch gun pointed over the stern. The Ensign said he had told one of the gunners to disconnect the battery—but evidently it was not done. You can imagine what happened when they threw the four-inch shell into the gun and shouted fire. The British Captain's hair was parted as he stood on the bridge! Luckily he was only shook up and not hurt. The incident did require some quick explaining on the blinker though.

In another incident, the chief Mate asked me if I had any medical training. The only thing I knew a little about was how to diagnose appendicitis (taught at Gallups). I was asked to assist the Mate in attending to one of the gun crew who had accidently shot himself through the thigh with a 45. Luckily, the bullet missed the bone, but left a hole through his thigh. There would be no medical doctor for at least two weeks, so the Mate got out his First Aid book. The upshot of what to do was solved by the Mate, who used a long, wooden swab, with a hunk of cotton on one end soaked in iodine, and rammed it through the hole. When we finally arrived in Scotland, the doctors said that the procedure had saved his leg.

I will stop here in Scotland and continue later to Iceland and Murmansk.

For the first year, I was in British convoys. In my first American convoy to Casablanca (one Destroyer was commanded by President Roosevelt's son), there was one striking difference between British and American tactics and this was in the use of depth charges. During a sub attack, the British made the water boil with depth charges. The Americans did not do that because of survivors in the water. The difference in tactics is probably attributable to the fact that survivors were not expected to live very long in the icy water. In fact, the British advised us not to wear life jackets, but most ignored that idea.

On the way to Murmansk and while anchored in Murmansk harbor for six weeks, the ship logged 169 air attacks. I witnessed many dog fights over Murmansk and remember in particular a Russian bi-plane being shot down. According to the documentary "Unknown War," the bi-planes were flown by Russian women pilots. The German pilots made use of flying into the sun and diving out of the sun, making them difficult if not impossible to see. On the trip to Casablanca we encountered a wolf pack and the submarines launched a lot of torpedoes out of the sunset.

Scotland to Ireland

Early in 1942 we departed in convoy for Reykjavik, Iceland. Usually when a ship had to fall behind the convoy due to engine trouble or other reason, it was just a matter of hours before their "SSS torpedoed" and their position would come in over the radio. The thing I remember most about this trip was the

" . . . we were taking green seas over the bow and a heavy truck chained on top of #1 hatch was flattened, causing it to pound like a sledge hammer on top of 88 tons of dynamite stored there."

number of floating mines that were being pointed out by the escorts. That was in the daytime. At night it was just a matter of luck not to bump into one of those mines floating just below the surface. The explanation given for so many mines was that subs were up ahead of us laying them in our path.

Iceland to Murmansk

Early in 1942 we formed a convoy in Reykjavik designated PQ-15. Our first attempt was a failure for only one night out of Reykjavik the leading ships ran into floe ice with 17 ships sustaining bow damage. The convoy returned to Reykjavik. One escort had disappeared during the night. It had been run over by a merchant ship. Of course our giro failed as it always seemed to do as soon as we left port.

Our second attempt was more successful, but somewhere north of Iceland our giro failed again and the convoy suddenly had a visitor. It was a long-range German patrol plane that began flying in a circle around the convoy with a radius big enough to avoid our anti-aircraft guns. The British called it the shadow plane. It would circle us continuously hour after hour. The only way we were able to shake that airplane was when we encountered heavy fog. Heavy fog was good for losing the shadow plane but the bad side was the presence of icebergs and the need to keep each ship in the convoy spaced a few hundred feet apart, which the British Commodore insisted on doing. The Captain of one of the ships was unable to cope with this tight station-keeping. (It was said that this Captain in peacetime would start steering away from any ship showing up on the horizon.) We were running in complete blackout and this Captain started turning on lights all over the ship. I remember the blinker signal sent to his ship. "If I see another light, I am going to shoot it out." The Captain had to be restrained to stop the lights being turned on for the remainder of the trip.

We had two people on the *Hegira* who were having trouble coping. One was driving Captain Comer crazy because he saw airplanes coming, periscopes, and floating mines from one side of the ship to the other. The other was completely withdrawn and would only sit in a lifeboat with his life jacket on. The only other significant thing I had observed about the Captains was that their hair turned whiter in such a short time between Reykjavik and Murmansk.

To assist in station-keeping, each ship trailed a board about five feet long on a line about 200 feet long over the stern. The board had a funnel shape on the trailing end which caused a water spout to squirt up a few feet above the water surface and helped the ship behind to see it. I remember one time when we lost sight of the board for several hours because of zero visibility. Many times a ship would lose sight for short periods, but this had everyone on edge. The engineers were trying to count RPM to a tenth of an RPM. We were now holding position blindly. If we went too slow, we could expect a stern collision. If we went too fast, we could have a bow collision. We sweat that situation for several hours when all of a sudden the board popped out from under the bow. Our bow had been on top of it all that time.

One of the British merchant ships actually had a catapult ramp out over the bow and a fighter plane sitting on it. I don't remember what happened to that set-up, but the pilot was supposed to intercept the planes and ditch in the water.

Some of the British merchants also had depth charges. With them being so slow, they would practically lift the stern out of the water. But when we were under attack by subs, the escort would go up and down the convoy column, making the water boil with depth charges. When a man was overboard, the flag was lowered and the convoy kept on going.

Only Sub I Saw

The only sub I saw during the war occurred on this run. We often watched the escort play cat and mouse with the shadow plane. If they could only get close enough to him, maybe they could blow him out of the sky. An escort would stop his engine and go dead in the water. He had the appearance of being disabled. As the escort drifted toward the rear of the convoy, it was, of course, getting closer to the circle path of the shadow plane. When the escort thought they had pushed this about as far as the shadow plane would allow, they would let go with every gun they had. To no avail, I might add.

> ***"Americans did not do that because of survivors in the water. The difference in tactics is probably attributable to the fact that survivors were not expected to live very long in the icy waters."***

One morning, three escorts forming the corners of a triangle started drifting alongside the convoy and all of a sudden they let go with depth charges as they sped toward the center of the triangle. Lo and behold, a sub surfaced at that point and everyone cheered. Then a hatch opened on the sub and men started running on the deck. The escorts opened up with machine guns and shot them as they ran along the sub deck. We were cheering. Next, some men came out of the sub waving a white sheet or flag. Everyone was saddened when we learned that we had just sunk a friendly Polish submarine which was inadvertently under our convoy. The escort thought the men on the sub were going for the deck guns but they were running to hoist the Polish flag.

I decided to look around the receiver frequency bands and discovered a lot of CW signals being sent on various frequencies. I started a log of these signals and copied their messages, although I couldn't decipher what was being said. There were one or two special characters which I copied as a combination of the normal Morse Code. I got on the DF and discovered that the shadow plane was sending long CW signals. These signals, I found, were being used by the attacking planes out of Trondheim, Norway to fly directly to the convoy, drop their torpedoes or bombs, and return. A very efficient procedure. Also, the attacking planes were communicating by CW, and I noticed that as these signals became stronger an attack was imminent. By monitoring the strength of these signals, I was able to alert the Captain that another attack was about to start. The attacking planes usually came in over the horizon, skimming the water.

The Captain would rest by lying on top of the chart table. He was very appreciative of being allowed to rest between attacks. (When we returned to Scotland, the Captain and I were invited to an airbase by British Intelligence. The Swordfish Squadron was very interested in learning the details of my radio logs. I spent the day with the pilot of the Swordfish plane that crippled the rudder of the *Bismarck*. I have forgotten his name, but I do remember him bringing two tumblers to a small conference table. Those tumblers were at least ten-ounce capacity. He filled them up with Scotch Whiskey. He talked about flying a fighter plane at night and sleeping in the daytime during the blitz.)

Now back to the convoy. During one of the attacks, the Ensign took it on his own to drop a smoke bomb over the stern. The Captain was livid about that as he thought the stern of the ship was on fire.

When the *Cape Corso* was torpedoed, it was on our starboard side. It was gone in seconds. We could see the men on deck trying to man lifeboats which were on fire. The bow was headed down, the stern was out of the water, and the propeller was still turning as it disappeared. Pieces of ship were raining down around us for quite a while. The torpedo had been dropped into the middle of the convoy by a Junkers 88. The torpedo rudder was set so that the torpedo would go into a spiral pattern. The torpedo missed us and hit the *Cape Corso.*

When the Commodore's ship was hit, he was very anxious to get off a message. The ship was sinking fast. I was called to the bridge to copy the blinker message. (When we left the States, there were not enough steel helmets so it was decided that the radio operator didn't need one.) So, I'm out there without a helmet while shrapnel rained down like hail. (Later, after the war, I had to have an X-ray for a concussion and the doctor said to me, "Do you know that you have a piece of shrapnel in your skull?")

There was some talk of the Russian Air Force coming to aid us, so whenever we saw a new wave of planes coming our way we were hoping it was a friendly Russian. That was not to be.

We encountered some terrific storms along the way. It was sobering to see green seas ripping off steel guards and tearing away a lifeboat. At one point the Captain had to take over the wheel because the Seaman could not hold the Bow to the proper bearing.

There was always apprehension when we encountered an iceberg. Aside from the threat of hitting an iceberg, submarines were painted white and could lay alongside an iceberg without being seen.

When we finally arrived at the entrance to Murmansk and picked up a pilot, it was snowing so hard we could hardly see. The first thing the pilot said was, "It's a nice day." We would get the message when the weather cleared. Murmansk was being bombed around the clock. Shortly after we picked up the Russian pilot, a German plane, flying extremely low over us on an even keel through the snowstorm, was on fire. It was hard to believe, but standing on the wing of the plane was a pilot with what appeared to be a fire extinguisher, trying to extinguish the fire. It was said that a German pilot tried to avoid capture at all cost.

Six Weeks in Murmansk

Soon after arriving, we were sent to the ammunition dump to unload our explosives. I was on the bridge, watching the stevedores unload the sticks of dynamite. The tray had no side boards and was supported by four chains attached to each corner. A tray with a load of dynamite sticks piled on it would be hoisted about 30 feet in the air. On one load, when the tray was 30 feet in the air, a corner came loose and the tray of dynamite rained down on the pier. I was sure that the dynamite would detonate. Nothing happened. To this day, I wonder why that chain came loose. Was it attempted sabotage?

The British had an anti-aicraft ship call the *Ulster Queen*. We held a convoy conference on this ship. It was bristling with guns all over the deck and had multiple pompoms on each side of the ship. Whenever this ship went into action, it was belching gunfire from stem to stern. It was a big thorn in the side of the Germans, and I understand that they were finally able to sink her.

The Captain and I were invited to visit the British Warship *Gossimer*. I was shown around the Radio Room and introduced to some personnel. Captain Comer was told that they had not seen an egg for a year. The *Gossimer* was tied up dockside while we lay at anchor. It was easy to see the *Gossimer* from the *Hegira*. When we returned to the *Hegira* the Captain sent a case of eggs to them. The next day, during an air raid, the *Gossimer* took a direct hit on the stern among depth charges and was gone in a giant puff.

Although there was plenty of air combat during the six weeks we were there, many times the Russian fighters didn't have time to scramble because the German airbase was so close by. Often German planes would appear from nowhere before the Russians could respond to an attack.

Women ran most of the activities in Murmansk. A woman was in charge of the ships. When we were getting ready to leave, the woman in charge said to Captain Comer, "Some of your crew have not turned in their passes. Tell them that this ship doesn't move until they are turned in." Word went around fast, because everyone was anxious to get out of there. Many of the stevedores and winch operators were also women.

Murmansk to Reykjavik via Akureyri

On this trip, 12 Russian exchange students came aboard. They were going to Columbia University. They were all girls ranging in age between 18 and 22. All could speak English and other languages. We also had an extra British Captain and a British Chief Engineer.

Convoy QP-16 was formed and we had a British submarine with us. The sub was there to intercept a German task force. The maximum number of ships would occur as QP-16 going south would meet PQ-17 coming north. The raiders could then do maximum damage, picking us off by lobbing shells over the horizon. As I remember, the sub did

"The fish were piled up on deck and were about the size of four-foot lengths of 2 X 4s and just as stiff. As a result of this, I was not eating fish or prunes for a long time after the war!"

manage to intercept and cripple one of the raiders, causing them to turn back.

The convoy went along very well until, north of Iceland, it was split into one group with the bulk of ships and escorts going to England and the remainder of 17 ships and two escorts continuing to Reykjavik, Iceland. The 17 ships had formed two lines running along side by sided. The *Hegira* was on the end of one line and Lamar Sewell's ship was on the end of the other line, abreast of the *Hegira*. The next ship ahead of Lamar's was the *Hefernon* (sister ship of the *Hegira*.) The two escorts were leading the lines and could not be seen by us because visibility was poor. The *Hefernon* had women and children refugees on board and the Russian students asked Captain Comer to pull alongside the *Hefernon* so they could exchange greetings. Shortly after we had pulled alongside and returned back in line, the ships ahead began blowing up and sinking. I had a lot of confidence in the *Hegira* until I watch the *Hefernon* sink so quickly.

All of a sudden there were only two ships left, us and Lamar's. (It's funny—at one time I knew the name of Lamar's ship, but 50 years has removed it from recall.) There was a hastily-held conference between Captain Comer, the British captain, and via blinker, Lamar's captain as to what to do. We could either go by where all the ships were sunk and try to get to the escorts, or we could turn around and go in the other direction. We elected to head away from the direction of the escorts, even through our giro was out and we didn't know our position. We had been following the leader and now we were on our own.

Realizing that our position had to be established as quickly as possible, the British Captain started to work immediately. He put a chart of the area north of Iceland on the chart table and covered it with a sheet of transparent paper. He had me take DF bearings on a Reykjavik radio station every 30 minutes and had the bo's'n make soundings at the same time. He drew bearing lines on the transparency and also marked the soundings (depth and type of bottom).

After a few hours of data, he took the transparency and slid it around until the marks corresponded with the chart and steered us into Nkureyri in northern Iceland. When we saw those high black cliffs and rocky coast line on both sides, we knew it was good navigating.

When we failed to arrive in Reykjavik, we were reported missing in action. My parents received a telegram listing me missing in action.

Akureyri to Reykjavik

We were escorted by a four-stack destroyer, which the U.S. gave to the British, from Akureyri to Reykjavik. Their captain said that the ship was in bad shape and could only go six knots, so if we went faster than that, he couldn't keep up.

When we left Philadelphia we were given a three-month food supply. We were now about five months out and had to take on fish when we arrived in Reykjavik. The fish were piled up on deck and were about the size of four-foot lengths of 2x4s and just as stiff. As a result of this, I was not eating fish or prunes for a long time after the war!

Reykjavik to England to Halifax and Back to New York

There was nothing that was particularly memorable on this trip aside from meeting the RAF Swordfish pilot who put his torpedo into the stern of the *Bismarck*. We arrived in New York, according to my discharge papers, on August 3, 1942, were paid, and shared some of our pay with the Armed Guard personnel.

We lost five ships in PQ-15 and, according to the GI series, "Hell in the Arctic," five ships were lost in QP-16. I thought that only my ship and Lamar's survived, but visibility was poor and evidently the front end of the convoy got through. My hat is off to the mother of all convoys, PQ-17. I often think of Charles Hollisian and Oscar Pelligrin, who survived QP-16.

New York to Casablanca

The convoy, with an American escort, was formed out of New York during the day around the end of October, 1942, and our convoy position was compromised on the first night. An unescorted ship coming into New York at about 20 knots was on a collision course with the convoy. The escort discovered the unescorted ship coming and tried desperately to have it change course, but nothing helped. It is not easy to change the course of a convoy, but there is a procedure to do this in an emergency. Certain ships designated throughout the convoy make one blast on the ship's horn. This one blast on the horn is repeated by those designated as soon as they hear it until the whole convoy is covered. Immediately upon hearing the blast, the whole convoy makes a 90° turn to the left. As a last resort that night, the Commodore executed this command. Now, the previously impervious lone ranger, who would not respond to anything, heard all of the convoy blasting horns, and, yes, you guessed it, also made a left turn. The next thing that came over the radio was an SOS.

The loner collided with one of the ships in the convoy. The SOS gave away our position and we were in for a dance with a wolf pack. (Sounds like "Dances with Wolves.")

Nothing happened for three or four days. The convoy was moving along smoothly and the weather was good. Of course, nothing really is smooth. Often, when daylight came and we looked out at the convoy, someone would say, "What is that ship doing in our line?" and find that it had crossed through seven or eight lines to get there during the night. It was a wonder there were not more collisions!

The American escort did not use the trailing boards like the British did. I can see that the trailing boards did keep the ships in line better. But the main idea of tight station-keeping was to concentrate the fire power of the antiaircraft guns in British convoys to Russia.

Hostile aircraft was not a threat in this convoy until we arrived in Casablanca. I must say, we now had 20 mm machine guns on each side of the bridge, which replaced the 50 calibers, and the four-inch gun on the stern was replaced with a five-inch 38 gun. Around the fifth day, in the evening with the sun setting on the horizon, I was washing up for supper when I heard an explosion. I looked out the porthole and saw a destroyer fire up with a big puff of black smoke out of its stack and scoot off at top speed. At the same time, the merchant ship abreast of us was very quickly standing on its bow, the propeller sticking straight up and heading for Davy Jones' locker. I thought, "Gee, it would be nice to crank up and scoot off at 25 or 30 knots." The destroyer's Commander was President Roosevelt's son. I was waiting for the escort to start boiling the water with depth charges, but nothing happened. That was a big let-down because the British would have had the whole area boiling with depth charges before you could snap your fingers. The next evening, when the sun was setting on the horizon, another ship got it. By the third evening, when the sun was setting, a couple more ships got it. These attacks were beginning to set our nerves on end and caused quite a bit of anxiety, particularly in the Armed Guard.

"At the time, we didn't call it friendly fire, but things started getting more out of control the longer these attacks continued."

Friendly Fire

At the time, we didn't call it friendly fire, but things started getting more out of control the longer these attacks continued. Someone on the front of the convoy would shoot his four- or five-inch gun at what he thought was a periscope. The shell casing drifted back through the convoy so that ships farther back in the convoy mistook shell casing for periscopes. The original orders were, when you saw someone shoot, then all ships should shoot with both machine guns and the four- or five-inchers. As I say, this situation caused a change in orders when a lead ship was torpedoed and a lifeboat with 21 men drifted back through the convoy. Someone spotted the lifeboat and thought it was a sub and opened up, followed by all the other ships. There was only one survivor.

The next day we were given flares to light up if we had to get into a lifeboat and the orders were changed to shoot a few bursts and let someone investigate. We continued to get picked off until one evening, when we were somewhere southeast of the Azores, instead of getting picked off, a Navy PBY came out of nowhere and the sub attacks ended. The PBY was able to see the subs and started dropping flares all around the convoy. The destroyers ran over to the flares and dropped depth charges. It seemed clear that we were hounded by a wolf pack.

When we arrived in Casablanca, I asked a Navy officer why they didn't drop depth charges, and he said that the people in the water could be hurt. It was then I learned that President Roosevelt's son was commanding the destroyer. He also said that half the Merchant ships had four- and five-inch holes in their sides because when they shot at things on the water, the shell would skip and hit a ship in the side.

Shortly after docking, the booms were rigged and we were piling 500-pound bombs on the dock. Sometime after midnight when we had the dock covered with bombs, we had an air raid. The Germans hung a flare above us that seemed to light the whole area, and the bombs started raining down. An Army officer yelled, "All Army personnel to your fox hole." The officer tried to get the crane operator to climb back up on the crane and shut off the big spot light. The crane operator scooted down so fast he practically jumped, but no way was he going to climb back up there. The officer pulled out his pistol and shot the lights out. Our antiaircraft fire was a disaster. The gunner started sawing off the cargo booms, shells were ricocheting all around, and the Chief Engineer's shoulder was torn open as he stood inside the passageway before the gunnery officer could get the gunner to cease fire.

I went into Casablanca and ran into my cousin, Bob, from Ohio. He was a courier for General Patton. I lined up with Bob in the Army chow line, which was serving chipped beef on toast, but it was called sh— on a shingle by the Army. My cousin, Bob, had more passes than Carter had pills and had no trouble coming on board for a visit. He carried messages between the lines for General Patton.

I also met a French sailor and went aboard the *Jean Bart*, which was put out of commission by the battleship *Massachusetts*.

The harbor was strewn with U.S. Navy destroyer hulks lying around the banks. There was no doubt that these destroyers were expendable.

There are a lot more stories that could be told, but I have told of the experiences that are the most memorable.

After the war, I went to work at the Boston Armature Works for 60 cents an hour, the same pay as high school students. I graduated from MIT's Lowell Institute in 1948 and went to work for Raytheon. I worked in Raytheon Equipment Division until 1950, when I transferred to Raytheon's Submarine Signal Division. I worked there for 37 1/2 years and retired in 1987.

My First Class Radio Telephone and First Class Telegraph licenses have expired, along with my Professional Engineer license. I have dropped out of the Acoustic Society but still belong to the IEEE and still have an Advance Class Amateur license.

Virginia and I are approaching our 50th wedding anniversary. We have three daughters and four grandchildren.

Keep up the good work and best wishes to all GIRA members.

Whatever happened to _____? Like every sailor, this editor remembers old shipmates he'll never see again. This card from Dick Floto (R-58) cheered us many years ago. Where have you gone, Floto? Jim Kinkel will find you!

Gallups Island and the Radio School

Gallups Island Radio School, aerial.

Station Hospital

View from the water to the tower.

Officer's Mess and Hospital

Lifeboats for training, 1941.

Picture taken from the water tower, 1943.

In Class . . .

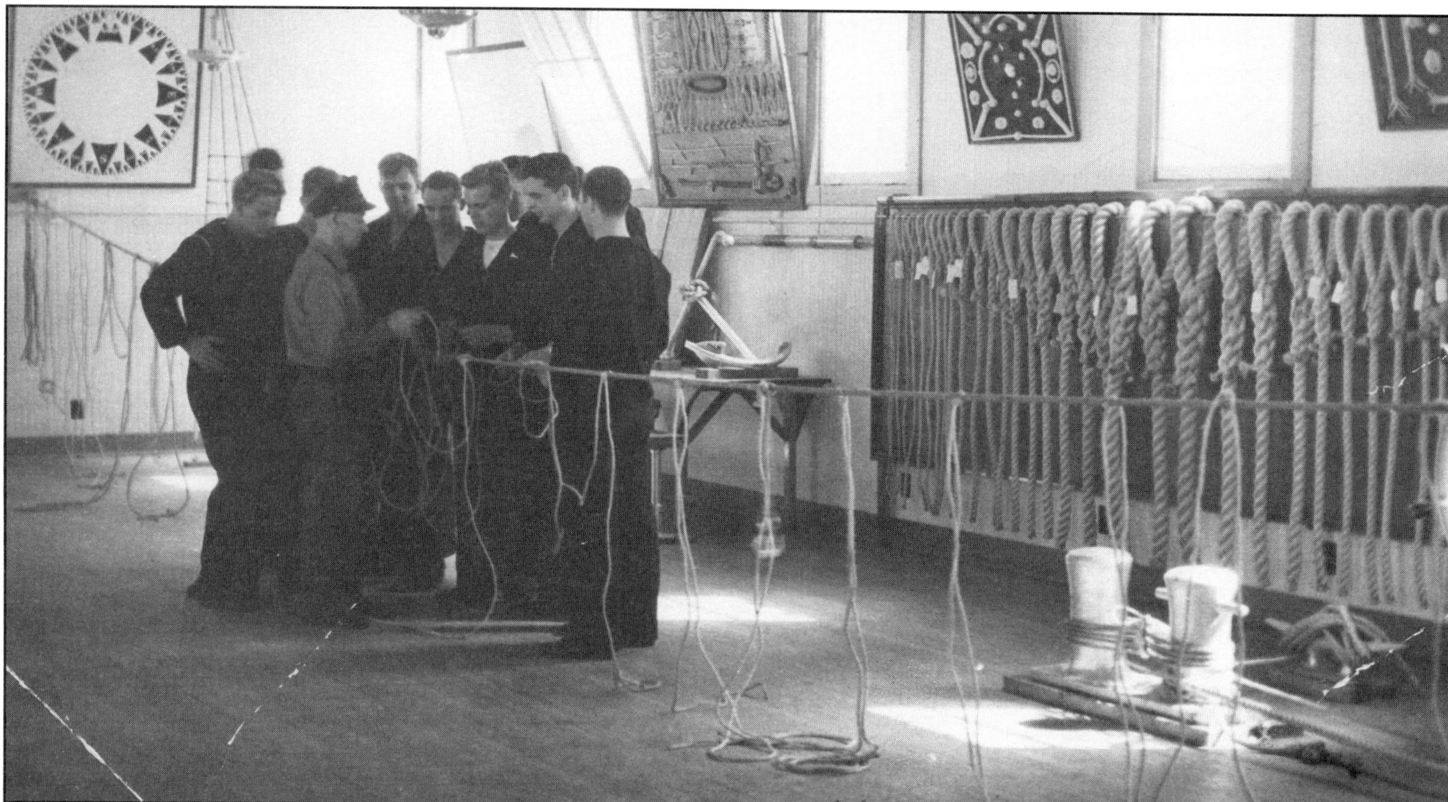

Part of Platoon R-3 in Sail Loft, 1941

Typing, a vital part of the curriculum.

Convoy procedure, Instructor Solomon Lipp, LT (jg) USN.

A class in shipboard equipment, 1944.

Ken Waller giving instruction on shipboard equipment, 1943

W/O Ed Pleuler with a lifeboat Radio RMCA ET 80.

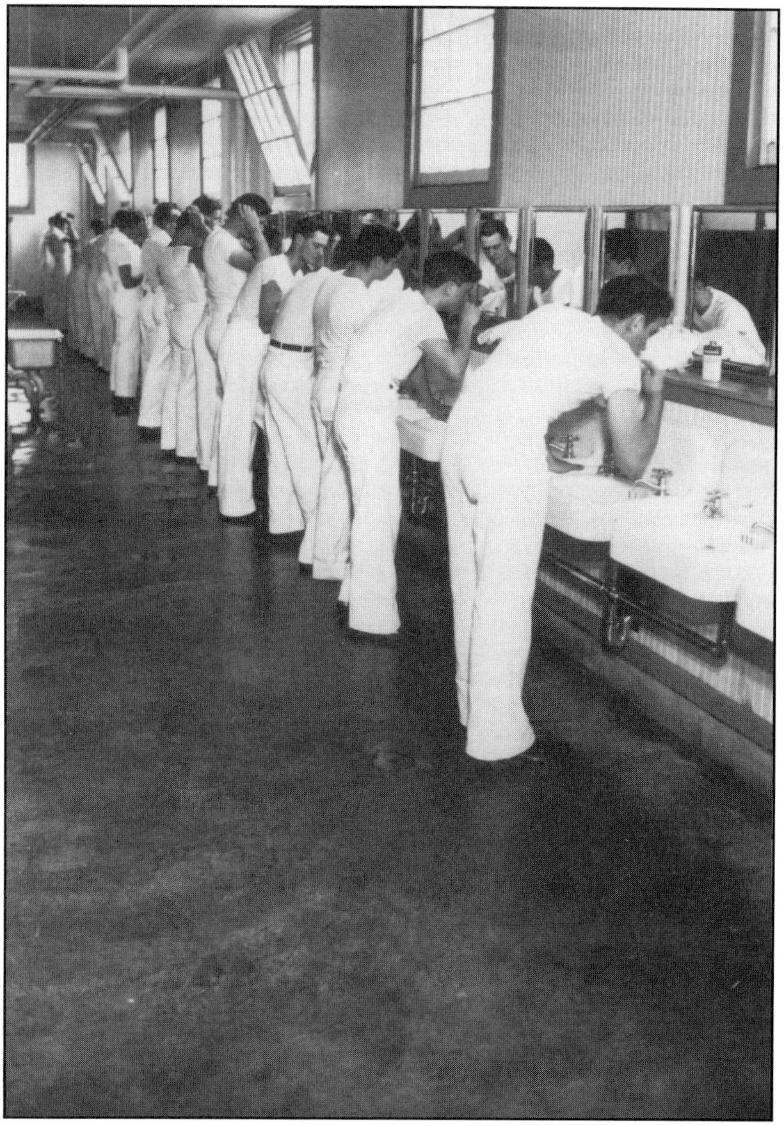

Gallups Island Graduates

Arranged alphabetically by platoon. Compiled from material sent for this book as well as the publications The Spark Gap *and* The Gallups Islander.

Racing crew. G. Gay, Coxswain. Bottom, Risen, Mizor, Vasquez, Vigil, Piper, the Ol' Kid. Top, Tencza, Pouncey, Strus.

Father's Day Service, 1944. Harold W. Arthur, Protestant Chaplain.

Platoon R-A-1

CALDERWOOD, DAVID L., R-A-1

"A Gallups Island experience and life afterwards"

I enlisted in the CCC in July 1938, soon after graduating from high school, and was assigned to Camp 3726, Clarion, IA. A CCC enlistment usually was limited to two years, so in May 1940, I was eager to apply for Maritime Service training to be a Merchant Marine radio officer when this opportunity was pointed out to me by the company commander.

During the last week of June, 1940, five CCC enrollees from other other Iowa camps, and I assembled at Fort Des Moines and went by train to Fort Dix. While there, we were discharged from the CCC and enrolled as trainees in the U.S. Maritime Service. Ony July 1, 1940, or thereabout, we traveled from Fort Dix to New York Harbor where we boarded the *SS American Seaman* for a voyage to Boston. During this voyage we were issued Maritime Service uniforms and other clothing. The *Seaman* docked in Boston on or about July 3, 1940.

The trainees aboard the *Seaman* were taken to Gallups Island on or about July 4, 1940. Another group of trainees who had been in the Maritime Service somewhat longer were already on the Island. These trainees comprised the original Platoon, A-1. The newcomers made up Platoons A-2, A-3, B-1, B-2, B-3, etc., with about 50 people in each platoon. Because my name was near the top of the alphabetical order, I was assigned to Platoon A-2. This platoon, along with A-1 started classes in the Radio School about four weeks earlier than the other platoons. Much of the remodeling of the old buildings on the Island was still in progress at the time we arrived.

The Radio School class that I found to be the most interesting was Radio theory as taught by Chief Radioman Summerfield, a former Marine. His work ethic was a good example for trainees as he burned a lot of "midnight oil" and drank a lot of coffee while preparing the mimeographed lessons which were passed out the next day to trainees in theory class.

Several months after Radio School classes had been in progress, the teaching staff determined that about half of those enrolled were not making adequate progress. This decision led to a goodly number of trainees transferring from Radio School to other kinds of Maritime Service training. Those remaining in Platoons A-1 and A-2 were combined to make up a revised Platoon A-1 (GIRA R-0A1).

Perhaps the most unique aspect of my Gallups Island experience was a somewhat early departure from the Island. This early departure was due to a shortage of Merchant Marine Radio Officers willing to sign on to ships undertaking voyages considered to be hazardous during the spring of 1941. Trainees in Platoon A-1 were scheduled to complete Radio School training around the end of June. I was among ten members of this platoon selected to take the FCC tests for a Second Class Radiotelegraph Operator License in May, 1941. I was one of the four who passed the tests and was issued a license. The others were Mac Erps, Philip Padberg and Vincent Whaley.

Philip Padberg (GIRA unlocated) was the first to ship out. He was followed by Mac Erps. My turn came next and I was issued the number 3 certificate of Graduation from the U.S. Maritime Service Radio School. Shortly after noon on June 14, 1941, I received word to prepare for a short trip to New York to become Radio Officer on a freighter that was scheduled to depart from port the next day.

I rode in the Superintendent's launch to Boston's Logan Airport where I boarded a flight to La Guardia Airport in New York. From this airport I found my way by subway and taxi to an office on Lower Broadway where I met Captain E. J. Kane of the *SS Marymar*, Calmar Steamship Corporation. After signing the ship's articles as Radio Officer, I went to Staten Island where the ship was docked. The Deck Officer on duty directed me to the radio room and the attached quarters for the Radio Officer.

The *Marymar* was of WWI vintage. Its displacement was 10,000 tons and cruising speed under favorable conditions about 9 knots. This ship was approximately the same size as a Liberty ship, but its stability in rough seas was reputed to be better. The radio room and operator's quarters were located, midship, on the boat deck.

The *SS Marymar* departed from New York on June 15, 1941, bound for Suez, Egypt, via the Cape of Good Hope. This route was necessary because much of the Mediterranean Sea and the North Coast of Africa were held by German and Italian armed forces. The voyage to Suez by this route required about 60 days. The cargo on board was equipment and supplies for the British army in Egypt. After the cargo was discharged and while the ship was heading south on the Red Sea, I received a radio message for the captain with instructions to take on cargo in India, Penang, Singapore and Java.'

After loading was completed, the *Marymar* proceeded in an easterly direction across the Pacific. The ship stopped in Hawaii long enough to take on fuel, then went through the Panama Canal and returned to New York on December 5, 1941, completing a trip around the world. On December 7th, I attended an afternoon performance of Sonja Henie's "It Happens on Ice" at Rockefeller Center Theatre. Upon emerging from this show, I became aware of the bombing of Pearl Harbor from newspaper headlines.

My second voyage on the *Marymar* also was to Suez via the Cape of Good Hope. We departed from Philadelphia on January 10, 1942, sailing alone, and we encountered no German submarine activity. Several days later after departure I heard several distress calls from ships that had been torpedoed off the East Coast. The return trip from Suez was around the Cape of Good Hope in the opposite direction. A severe gale off the South Coast of Africa badly damaged the ship. This damage made it necessary for the ship to spend a month at a shipyard in Capetown for repairs. We arrived back in New York on July 29, 1942.

Guns were installed on the *Marymar* following the second voyage to Suez and a Navy Armed Guard crew came aboard the ship. The enlisted men's quarters were built within the top level of the midship cargo hold and the Armed Guard officer was assigned the stateroom formerly occupied by the third mate. The third mate had to occupy a room on the boat deck, adjacent to the radio room and my quarters. Thus, the various third mates I sailed with were virtual roommates and frequent visitors in the radio room.

The *Marymar's* next voyage was to Paramaribo, sailing alone. All other voyages until after Germany surrendered in May, 1945, were in convoys. These voyages included trips to Casablanca, Oran, Bristol, London, Cardiff, Grimsby, Omaha Beach, Cherbourg and Liverpool. I transferred to the *SS Cape Comfort*, a new C-1 type ship, chartered by the Calmar Steamship Corporation, that was still at a Wilmington, DL shipyard for finishing touches.

My first voyage on the *Cape Comfort* was to Marseilles, France. Afterwards I made three voyages to Venezuela. My career as a Merchant Marine Radio Officer ended in December 1945. Subsequently, I applied for and received a Certificate of Service as a U. S. Merchant Marine Veteran based on 1, 148 days of service on foreign voyages during war years from December 7, 1941 to August 15, 1945.

My second career began in January 1946 when I became an engineering student at Iowa State College, Ames, Iowa. I obtained both a Bachelor of Science (B.S.) and Master of Science (M.S.) degree in Agricultural Engineering.

I was associated with the I.S.C. Agricultural Engineering Department until 1955 when I accepted employment with the U.S. Department of Agriculture (U.S.D.A.) for research in Texas on post-harvest treatment (mostly drying and storing) of grain crops. The project headquarters was located on the campus of Texas A&M University, College Station, TX from 1955 to 1962. The headquarters were then transferred to Texas A&M University Agricultural Research and Extension Center near Beaumont, TX. The crop involved in research at this location was rice. I retired from the U.S.D.A. in 1984 and since then have taken up golf as a substitute for useful employment.

Membership in GIRA renewed my interest in amateur radio. I became a member of the Beaumont Amateur Radio Club and obtained a Novice class license in December 1991. During 1992 I studied the various license manuals and upgraded, in turn, to General, Advanced and Extra class license. My call letters are AB5KM.

TONTONI, GENE A., R-1

I arrived on Gallups Island with the first group in early July 1940. From my western Maryland CCC camp, I was sent to Ft. Dix, NJ staging area. Here I was in the same tent with Leo Hower, Vern Matthews, James Flynn and John Tencza. We were all Pennsylvanians and became fast friends throughout our stay on the Island.

I had planned to go to seaman's school instead of radio school and it was Tencza who advised me "knock anyone's brains out and they can become seamen—go to radio school." That piece of advice was to determine what I did for the rest of my life. After working on ships I worked in radio/TV broadcasting and retired after 28 1/2 years at IBM in electrical engineering.

I was in NYK in February, 1942 when I read about James Flynn being lost on a tanker. Leo Hower was my best friend. He left the Island September-October 1941 and we communicated trying to get to the same port at the same time. We never made it. About September 1942, I got a letter from Leo's father telling me that Leo was lost at sea. Vern Matthews was a recent "silent key." Tencza did very well in school but could not copy above 14 wpm. He tried very hard, spending many extra hours practicing code but could not get above the 14 wpm barrier. I believe he went to seaman's school.

I do not know what the curriculum was after I graduated, but in our time, seamanship was stressed as much as radio classes. For example, we were into lifeboats every day for about two hours, rain or shine. We became very proficient in rowing and boat handling. We were able to make "eggshell landings," against a strong wind and tide. We rowed up to the dock full speed and at the last moment "toss oars port"—"hold water starboard" the boat spun with the port side against the dock. Boathookmen, fore and aft, hooked on to the pier and we landed.

Also we learned how to use breeches buoy, along with many other seamanship classes. The USCG Chiefs, mostly bos'n mates, made sure that we were seamen in addition to radio operators.

I had to go on emergency leave about mid-1941. Upon my return, much against my desire, I was put back a class, so that I graduated with the second group. I was transferred to Hoffman Island shortly after Pearl Harbor and was assigned to the *SS Santa Lucia* (Grace Lines) as Third Radio Officer. The *Lucia* was a luxury liner, swimming pool and all. Our purpose, for this trip, was to pick up German and Italian nationals in South America and take them to NYK for eventual repatriation to Germany and Italy via Swedish ships.

We left NYK on December 19, 1941 for the west coast of South America. How different the real world of code was compared to the nice static-free code of the machines in school. But I managed and tried to emulate the Chief who copied press at least four lines behind. We went through the Panama Canal on Christmas Day, 1941 and sailed south to Valparaiso stopping at many ports along the way. Northbound, as we approached the Canal, about the middle of January, we began to log many distress calls and also "SSS." The latter meant "submarine shelling us." (Why waste a torpedo against unarmed, unescorted tankers?)

Our skipper, taking all this into account, decided it was best to avoid the U.S. east coast as much as possible. So after clearing the Canal, he took an easterly course, out several hundred miles, then sailed north until we were parallel to NYK and made an overnight run into port.

It was not until after the war that I found out about the great number of ships that were sunk around the U.S. east coast, close in to shore—and, we had been in the middle of it. Bathers on the beach at Atlantic City and Miami had a front row seat to the ships being sunk.

In March, 1942, I signed on the *S/S Hybert* (Lykes Bros S/S Lines) in Baltimore. The ship was a Hog Islander type freighter built in 1920. Our cargo was "guns and butter," mostly guns, including 300 tons of explosives and ammunition. We were headed for Murmansk in PQ-16. This convoy was the first large, about 35-40 ships, summer (24 hours daylight) convoy to go to Murmansk. We had a six-day running battle with the Norway-based German Air Force before reaching port. We thought (wishful thinking) that Murmansk would be a safe haven from air attacks.

We found out; Murmansk is about 90 miles from the nearest Luftwaffe air base. After a while it seemed as though they were bombing the town every hour on the hour. We had several very close calls. The most vivid one was when we were at anchor and we were straddled by four bombs; two fell on the port side and two on the starboard. We were in port for 30 days. About the first of July, we sailed for Iceland. Just north of Iceland, on July 5, 1942, we were torpedoed and sunk. All hands survived.

About September, 1942, I was in the Boston ACA union office waiting to be assigned to another ship. Chatting with another radio operator in the office, I learned that Northeast Airlines (Boston) was hiring Flight Radio Officers. Northeast, as many commercial airlines, was a contract carrier for the U.S. Army air force; the Air Transport Command (ATC).

I had always had a desire to fly. I went to Logan Airport and was hired. Our routes were from our AAF base in Presque Isle, ME to several bases in northern and midwest Canada, ranging fro Port Churchill, Manitoba across Hudson Bay to Southhampton Island to Upper Frobisher Bay, to Goose Bay, Labrador; Gander, Newfoundland and overseas to Greenland, and Iceland. We flew in C-47's for flights within Canada and C-46's for overseas flights. The AAF bases, main bases and alternate bases, were built along the Great Circle route from the west coast of the U.S. to Scotland for the purpose of ferrying airplanes from the factories to Europe. The main route was from California or Washington to Great Falls, MT then to Goose Bay, Labrador, overseas to Greenland, Iceland and finally Scotland. All other bases were alternate bases for emergencies, bad weather, etc.

An episode occurred during my time with Northeast-ATC which became the basis for a book and a movie. An ATC-American Airlines B-24, carrying Army personnel, became lost in a snow storm over northern Labrador. Low on fuel, the pilot saw a "hole" in the overcast, dove down and made a landing on a frozen lake. They were down for a couple of weeks with only emergency rations and no gas. The AAF could not find them I was in Boston on my time off for the month when I received a call from the Boston-based radio operators' supervisor telling me that he and I were flying to Maine that evening.

We were booked on a Northeast commercial airline with priority seats. We were on a secret mission. The next day, at Presque Isle, we were told about the downed airplane and that our airline was scheduled to search for it. Surprisingly, the supervisor was assigned to a crew. He usually did not fly. After the crews were assigned, the supervisor/radio operator came to me and asked if I minded exchanging pilots because he and my scheduled pilot were good friends. I agreed.

We were flying C-47s. We found nothing in our search sector. After about four hours, the plane with my supervisor found the downed B-24. They pulled a landing on the frozen lake. The main purpose was to bring in a radio. The ground temperature was minus thirty or so. That evening my supervisor had to copy a long coded message with a pencil, and every group was sent twice. He almost froze his fingers. The message gave detailed instructions on what the AAF intended to do. Once located, both airplanes were well supplied. Then a small clearing machine was dropped to clear a runway. Finally both airplanes took off and returned to Presque Isle.

The rescue airplane had been on the ground for more than a week. I don' t remember how many times my supervisor said, "You should have been on that airplane instead of me." The lake was named for the American Airlines pilot, "Lake O'Connor."

Ernest K. Gann wrote a novel entitled Island in the Sky based on this episode. A movie of the same name starred John Wayne.

After about 18 months with Northeast I got the flying bug. I enlisted in the U.S. Navy as a Naval Aviation Cadet. I washed out of primary training flying school, I could not meet the precision flying standards set by the Navy. Since my ground school grades were very high, I was accepted at the US Navy Air Navigation School in Oklahoma. I successfully completed the course and was commissioned an Ensign, Air Navigator. After graduation I spent five months in Hawaii, real easy duty. Because I wanted to see what the real navy was like, I requested sea duty. I was assigned to an aircraft carrier, the *USS Randolph* (CV-15). I qualified as assistant navigator in about three months. My seamanship training on Gallups played an important part in qualification. I ended up in the Naval Reserve as LT (jg).

In 1947, I signed on the *SS Evangeline* (Eastern S/S Co.), a cruise ship. The chief RO was "Pappy" Moers whom I knew from Northeast Airlines days. We made a triangle run, New York, Bermuda, Nassau.

Along about September, 1948, I went with the Army Transport Service. I signed on the *USAT General Harry Taylor* (AMBX), carrying troops between Brooklyn POE and Bremerhaven. After about a year, the Navy took over all Army transports and we became the *USNS General Harry Taylor* (NXAK). AS USNS, we were attached to the International Refugee Organization (IRO) and carried displaced persons all over the world. I was aboard the *Taylor* for about 2 1/2 years.

I came ashore in 1951 and worked in radio/TV broadcasting stations until 1956 when I joined IBM. I retired from IBM in 1985.

My training on the Island served me in two ways. It started me on the path to electronics. In the time at IBM, I made the transition from vacuum tubes to transistors and finally, integrated circuits. Also, prior to being assigned to the *USS Randolph* (CV-15), as assistant navigator, I had had no navy seamanship training. All of my training was Navy air force training. Even the navigation methods were different. I don't know how many times I mentally thanked the Chiefs from Gallups for my knowledge of ships and seamanship.

Platoon R-A-2

JONES, HUGH T., R-A-2

I enrolled in the U.S. Maritime Service July 1, 1940, (along with several hundred other young men from the CCC) at Fort Dix, NJ. Some of us arrived at Gallups Island on the 5th of July and were divided into platoons. I was in Platoon B-1.

In October I had the fire watch. It was snowing and I had punched the clock with the key in the barracks, school, officers' quarter, but when I started to go down to the end of the island to the building where we stored the whale boats and sail boats, I was in snow up to my knees. I said, "to heck with this," and went back to the office. The Duty petty officer jumped me about not going down to punch the clock and said, "Come with me and I'll show you how to do it."

We got as far as the officers' quarters and the snow was up to our hips. He said, "To hell with this, it's not going to catch fire anyway, let's go back." When we got back I asked him if I could go hit the hay and he said I could. I couldn't even make it to the barracks, the snow was so deep. I just made it back to the watch tower. He asked me what I was doing back and I told him I couldn't make a path to the barracks, the snow was so deep. He told me to find a warm spot on the floor. I did just that.

In July, 1940, the FCC came to the island to give us our examinations for 2nd class radio telegraph. I passed both the code and written test and graduated July 29, 1941. I went back to Boston and while waiting for a ship assignment, I helped move Sailors Haven to the old Exchange Building at 22 Battery March St. One day, Andrew Scrudato came to the Seamans Club and asked me if I would take his ship. I said sure. We went to the ROU where we met Gene Barry, the union rep. I was to sail at 1300 that day. When everything was taken care of, Andrew headed for the South Station and I headed back to the Seamans Club.

As soon as I arrived, I was handed a telegraph. I found out I was on active duty in the Navy and had 15 minutes to get to 150 Causeway Street at the 1st Naval District Headquarters. When I arrived at Headquarters, I went to the Personnel Office right away, with telegraph in hand. The lady at the counter looked at my telegraph, went to her desk and picked up a big envelope, handed it to me saying, "Oh, yes, Mr. Jones, you stayed off a merchant ship just one day too long—this is your records jacket, DON'T LOSE IT!" I asked if I could go to the ROU and let them know they needed to get another operator to take my place. She said, "Yes, but DON'T LOSE YOUR RECORDS JACKET!"

That's how I ended up in the Navy.

I retired from the Naval Reserve February 19, 1970. I put in 19 years with Lockheed Propulsion Co., retiring from there January 9, 1976. The 17th of October, 1977, I started to work at the Jerry L. Pettis Memorial VA Hospital as electronic technician. On October 23, 1991, I retired from there with 15 years and six days service.

McCLISH, GERALD "JERRY", R-A-2

The initial classes at Gallups, did, I believe, get a longer and more leisurely and lengthy education that later groups. I would suppose that the frenzy of the war caused the curriculum. Our classes covered a very broad area. I remember that our 10-man whaleboat crews trimmed the Navy and Coast Guard units in Boston on the Charles River. I was there nearly 18 months. Finally, came Graduation Day.

The tests were administered by the FCC personnel and we had first the usual 50 or so questions that were available with answers in a book appropriately called "Questions and Answers." However, you did not know which of the questions would be on your sheet or how they would be worded. Most of us passed this. Then came the Code. Some breezed right through and some froze up with an iron fist. Finally with some good hearty encouragement, most dit-dah-ed their way through to a Second Class Telegraph License. A proud day.

I was offered a teaching job remaining on the island but the call of the sea was too great. After all, we had sat on that island real estate for many moons watching the ships ply their way down the channel—which went right by us—on their way to distant and mysterious ports. I declined to stay and shortly was enroute to New York and the United Fruit Company's Great White Fleet. My first ship was the *SS Carrillo* out of Boston enroute to the glamorous tropical port of Cortez, Honduras.

This is, to my knowledge, an insight on the early beginnings of Gallups Island. There could be much written on analogous events coupled with this saga in history. The Old Howard Theater, Scollay Square, The Seamans Refuge, The Seaman's Refuge where I slept when the liberty boat was missed, Fisherman's Wharf where we joked about the fish and the ladies who worked there having somewhat the same perfumy smell. The roller skating rink—I have forgotten its name, the swan boats in Boston Commons, Sailors sleeping on benches in the same Commons and on . . . Not many names come to my memory, Billy Waldeck, Alvis Jones are a couple. Characters I remember more vividly but I shall not go into that now.

Some of the ships I served on beginning in September 11, 1941: *Carrillo, Atlantidia, Francis R. Hart, San Bruno, Santa Maria, Juniata* and *Hibueras*. The *Hart* was a large German-built tanker. The trip after I got off, she was in trouble for stopping and giving fuel to German subs. She had a German descent Captain and a hulk of a first mate who was all Yankee. The skipper's wife was aboard and the mate played hanky panky with her, which caused quite a commotion.

The *Juniata* was a very old and very dirty coal burner. She had been converted from a sailing vessel to power. In her day (eons ago) she must have been quite stately. She had very ornate wood carvings all over.

The *Santa Maria* was once a passenger ship converted to 12 passengers. I was aboard when we went hard aground outside of Barrios Harbor (Guatemala). We were so high and dry having slid up on a gently sloping mud flat that we were nearly in dry dock. At one time, we had three ships with immensely long hausers trying to pull us off. The *Santa Maria* never budged. Finally, a salvage vessel from Merrit and Capman in New York arrived.

When our ship finally decided to move, she took on a rather severe list. I took pictures of the crew trying to launch a lifeboat from the highside. Also pictures of the purser clasping the money box ready to go. The ship could not have sunk even rolling over. Her 55 foot beam would have stuck up considerably from the surface of the water which was about 17 feet deep. During the ordeal, our captain became very despondent. Every day, he had the engineer and me, "Sparky," as pilot take him for a small boat ride.

It took three weeks to get us off with subs all around. Guess that their torpedoes would have gone aground as high and dry as we were.

PETERSON, D.H. "DANA", A-2

I was in the first original class A-1 at Gallups Island Radio School. When our class went for the FCC exam, everyone passed except me. Thus I was placed with class A-2. I passed that time.

In October/November 1941, I was hired in New York City by Pan American Airways . . . and sent to their headquarters in Accra on the Gold Coast of Africa. I was then transferred to Cairo, Egypt where we were permitted by the Egyptian government to operate a CW station. I was with Pan America about one year, then was called back to the States—the Navy showed an interest.

I was a Navy operator until April, 1944 when I became a war victim, followed by a long period in Navy and VA hospitals.

From 1948-1959 I was an engineer at WTAG AM/FM in Worcester, MA. I graduated from Worcester Junior College in 1959 and went to work for Raytheon as an engineer in microwave communications systems where I worked until 1962. I was then hired by Lockheed Electronics Co. in Plainfield, NJ, as a field engineer assigned to shipyard sites on Navy missile sub programs.

In 1966 I was an engineer with the US Navy Terrier Tartar, Talos missile systems. Was transferred in 1968 to Navy strategic systems project office in the fire control and inertial guidance section.

Retired from the federal government in 1985.

Was married in 1942. We have three children and four grandchildren. Celebrated our 50th wedding anniversary September, 1992.

Four years after I retired was stricken with Lou Gehrig disease and I am partially paralyzed; restricted to bed and wheelchair and on life support equipment and 24 hour home nursing care.

I do cross word puzzles, cryptograms and review a lot of math (i.e., algebra, geometry, trigonometry, etc.)

Would like a listing of classmates from A-1.

Hope all is well with you. I'm 74 but 73's to all.

SANDERSON, MELVIN, R-A-2

I was a hospital orderly in a CCC camp at Itaska Park, MN in the spring of 1940. One evening the Company clerk came in and handed me an application for the U.S. Maritime Service and remarked "WE are going into the Merchant Marine." I laid the application aside and the next morning he came back and asked "Have you filled out that form yet?" I replied '"No, I haven't even looked at it." He said, "Sign here." I signed. Then he took it to the C.O. office and filled it out as he had the necessary information.

In late May of 1940 on a Sunday evening, a telegram arrived telling me to leave at 8 a.m. the next day for Fort Dix, NJ. Several fellows from our camp had signed up but for some strange reason I was the only one accepted. At that time the government was trying to recruit radio operators and seamen for the Merchant Marine. They had first tried to recruit amateur radio operators with limited success, then next they tried recruiting from the CCC camps. Five hundred of us from every state arrived at Boston from New York aboard the *American Seaman.* About 100 signed up for the radio operator course and about 25 got licenses and assignments to ships.

After an enjoyable 14 months at Gallups Island, I graduated on July 29, 1941 and was assigned to the *Esso Nashville* as a cadet radio operator about a month later. On December 7, 1941 we were in the Gulf of Mexico bound for a port in Texas. When we were tying up, a man rushed aboard and said he needed the cadet radio operator for a tanker docked and loaded about 300 miles away. I left in such a hurry that I forgot my toothbrush, shaving equipment and some books.

When we arrived at the other port (probably Corpus Christi) the radio operator, an older man, was ready to leave. I peered into the radio shack and had never seen equipment like this—an old converted "Spark" transmitter and a 4 tube 01A receiver. The operator started to leave and said "I'm going back to Mississippi; you, young man, can have this war." I said "Hey, wait a minute, will you turn this thing on and show me how it works?" He turned it on and the place started to shake as the MG wound up speed.

The subs started to arrive off the east coast in January of 1942 and the largest fire I've ever seen was a tanker torpedoed near us off the New Jersey coast. The tanker was still burning the next day and a photo of it made the cover of "Life" magazine.

It took some time before we became armed, received Navy gun crews and formed convoys. The German subs were civil but were also like predators following a pack looking for the wounded or stragglers. I received a SOS from a ship drifting helplessly near Iceland because it had lost its propeller. Two or three hours later a SSS came from the same ship. Torpedoed!

In another convoy, a Norwegian tanker refueling ship for the British escort corvettes was immediately ahead of us in the middle column. About 3 p.m. three torpedoes came diagonally across our convoy, one went all the way through and never hit a ship, another hit a ship carrying about 600 Canadian troops in the column next to us. The third hit the refueling tanker ahead of us. It was smoke from end to end as seamen were jumping overboard and launching two lifeboats as we went around them. The ship didn't burn but was dead in the water and the men went back aboard. We were out on the deck watching and when she was about two miles behind our convoy there was a huge black explosion. This time the sub finished her off.

I had three 36 exposure rolls of 35mm film which Naval intelligence confiscated. He promised to have them developed and sent to my home address in Minnesota. About two years later my mother received a packet of about 20 pictures. Several of the pictures received were of survivors coming aboard our tanker in the Yukatan channel. The day before an SSS had come in loud and clear on my receiver. That was all. I told the captain a ship had been torpedoed near us. The next day we spotted a lifeboat that started to row away but our crew waved for them to come our way. Half of the crew from the sunken ship were in this lifeboat but the captain, radio officer and two of the four Navy gunners were lost. The sub had immediately surfaced but the blast had shifted the cargo of nitrate to the opposite side of the ship and she rolled over; the mast meanwhile breaking off and falling across the other lifeboat killing many of the men in it. The sub came over and wanted to talk to the captain. The chief mate, one seaman and two surviving Navy men went aboard the sub where they received a lecture in perfect English that they were sorry about the loss of life and only wanted the ship and its cargo. They bandaged one of the Navy men's legs which had a nasty wound from above the knee on down. They also gave them a couple packs of German cigarettes.

During the last year of the war I was chief radio officer on a brand new tanker, the *Esso Utica.* We loaded in the West Indies, went through the canal then pretty well followed the equator; first trip to Guadalcanal, second to the island of Biak and the third and last to Leyte Gulf in the Philippines. I remember the captain was excited one day—we were on the equator at exactly noon and the sun was 90 degrees from the horizontal direction.

I sent my one and only message during the entire war from the far east. Our ship miles from designated position and I sent a position correction. Tried to figure skip, frequency and time to send. Decided on 10 p.m. and asked the Captain to change course 45 degrees for a few minutes. Called NBA in Panama, 7,000 miles, and he came right back with "Go ahead." I was surprised.

Had a lot of respect for the captains I sailed with, especially Captain McCarthy who was 84 years old and had sailed since the age of 14. His naturalized home was Boston and he had four sons who all graduated from Harvard. On one trip we left Liverpool about the 2nd of January and hit head winds gusting to 120 mph by the Captain's log. Our convoy slowed to 2 knots for days and we lost all of the Liberty and most of the other cargo vessels as they were going backwards. After 20 odd days we made St. Johns, Newfoundland. We formed a new convoy of nine ships. The second night out of St. Johns I was awakened by the ship shuddering as we were going full astern. We were in fog and too close in toward shore. When I looked out I could see the breakers on the rocks ahead. Three ships piled onto the rocks but Captain McCarthy had saved us.

Two of the captains which I sailed with died at sea. One, a relatively young man, got tragic news in Boston that his wife had died. I saw him coming up the gangway with two AB seamen carrying cases of whiskey. Two weeks later he was dead. A suicide from alcohol poisoning. Another man, Captain Darling, died from cancer on the way back from the Philippines.

I spent the entire war on tankers of Standard Oil of New Jersey and left the Merchant Marine in February, 1946 to go back to Minnesota. From Gallups Island I had received a start in electronics which I continued in all my working life. Was in radio, electronic and pipe organs, and TV service for 20 years, then the last 15 years before retiring in 1981 taught electronics in a post secondary technical school in Minnesota.

Thanks to Gallups Island for the start.

WEBB, RUSSELL, A-2

"GALLUPS ISLAND . . . As I Remember It"

I was first introduced to Gallups Island on July 4 or 5, 1940. This came as a result of transferring from the CCC to the Maritime Service. John Lytton and myself were in CCC Company 5428, stationed at Disston, OR, about 17 miles from Cottage Grove. John was a first aid attendant and I was a supply steward. John was also the camp radio operator. He and I were close friends since Camp Beauregard in Louisiana, where we met when transferring from the 4th Corps area to the 9th Corps area.

The captain of our camp told us he had received word that two enrollees from each CCC camp could transfer to the Maritime Service. We could go either into the radio or the engineering department. John, of course, wanted to go into the radio department. I was going to go in the "Black Gang." However, at Fort Dix, NJ, we found out that the people in the radio school would go to Boston and the ones for the "Black Gang" school would go to St. Petersburg, FL. Not wanting to separate us, I decided to go to the radio school, too.

John and I went to Portland to catch a train to Fort Dix. It was there where Bobby Pouncey joined us. Bobby had been at CCC Camp Trask, Company 5477, stationed at Tillamook, OR. (Another man was a Carter from Oregon, however I am not sure what happened to him.) We left Portland and headed for Fort Dix. (Another fellow CCC member, Francis Siltz, joined us later, in Wyoming.) We were at Fort Dix about a week while being discharged from the CCC and transferred to the Maritime Service. Our CCC discharge was dated June 30, 1940. Our entry into the Maritime Service was dated July 1, 1940.

We were taken to New York Harbor, via buses, to board the *American Seaman*. What a beautiful sight! The *American Seaman* was a beautiful white ship, very clean with highly polished brass and decks. For about three days we sailed from New York Harbor to Boston, giving them time to issue our clothing. One day during that cruise I remember seeing a whale—the first I had ever seen—off the port bow.

Upon arriving at Boston we were taken to Gallups Island. The buildings were old. Latrines were a mess and everything had to be built new or renovated. There was a pretty good period of time before they could start classes. Classrooms had to be built. Equipment had to be installed. New latrines built, and other renovations made. We were put to work doing various jobs until, finally, school started. We not only studied radio in all its phases but also seamanship, some sailing, rope tying and splicing, flags and semaphores. I still try my luck at semaphores, very slowly though.

Throughout our time at Gallups, John Lytton, Bobby Pouncey, Francis Siltz and I remained good friends. I remember some of the liberties we had in Boston and the Silver Dollar Bar on Washington Street, where every time a group of us or Navy men went in, the singer, a rather heavy-set girl, and the band would break out in "God Bless America." They'd sometimes do that four or five times a night. Other places I remember include the Boston Commons, the Municipal Gardens, the band shell on the banks of the Charles River from which you could see M.I.T. Of course, we couldn't forget the Sailor's Haven with its dances and a place to sleep.

We were at Gallups from about July 4, 1940 to about September or October, 1941. My certificate of Graduation #82, was dated August 28, 1941. I shipped out with the Sun Oil Co. throughout WWII. I left there in 1946 and went with Pan American Airways for a while as flight radio operator. While flying with PanAm I studied for and got my 1st Class Radiotelephone license and went back to Louisiana and got into radio broadcasting. I did all kinds of broadcast work from engineer to engineer-announcer combo and retired as manager-salesman for a broadcast station where I live.

I married a girl from New Jersey, Barbara, and we live in Robert, LA. We like to travel and on a recent trip we went through Pennsylvania where I contacted Pete Maffett W3PBU. In the Gallups Island roster he is listed as being from A2. I thought we could stir up some memories, however his class of A2 graduated in the summer of 1942, a full year after the A2 I was in graduated.

By the way, I really enjoy the GIRA radio nets and while we were on the trip we checked in with the guys on 3919 each morning for a couple of weeks from our trailer. I sure enjoyed it. We went through New Jersey visiting with Barbara's people and on down to Solomon, MD, visiting with Bobby and Clara Pouncey. I wanted to get his thoughts on Gallups. I contacted John Lytton by phone to get his memories.

Bobby Pouncey went with Standard Oil of New Jersey, now known as Exxon. He sailed as radio operator during the war. Later he got into the engine room and worked his way up to Chief Engineer and was Chief when he retired. He lives with his wife Clara in Solomon, MD. He remembers some of the men who went with him to Standard Oil—Letourneau; Kapotsky; Reel; Risen; Trainor; Melvin Anderson; and Alvis Jones. He and I remember some of the instructors at Gallups like Summerfield, Lucas, Stewart and Curnoy.

Francis Siltz went to Standard Oil also. His ship, the *W. L. Steed*, was torpedoed and sunk February 2, 1942 off the coast of Delaware Capes. The crew was able to get in

Platoon A-2

the lifeboats but there was a raging snowstorm with high seas and before they could be rescued most of the men died from the cold and exposure, including Francis Siltz, the radio operator. Late in 1945, a Liberty ship was named the *Francis E. Siltz* in his honor. His mother came to Houston for the christening.

John Lytton went with the American Trading and Production Company but later joined the Mobile Oil Co. and remained with them for many years. He took a brief hiatus and came to Pan American Airways while I was there so we were together again after many years. He went back to Mobile after that. He is married to Olga and they live in Tampa, FL. *(Editor's Note: Russell Webb contributed his bio, written with contributions by his fellow Gallup Island classmates Bobby Pouncey and John Lytton.)*

Platoon A-3

PETRICK, ANDY, R-A-3

If my memory serves me correctly, about 96 selectees from CCC (two from each state at that time), met in Fort Dix, NJ at the end of June or the first of July, 1940. We resigned from CCC and enlisted in the Maritime Service.

We rode buses to New York and boarded the *American Seaman*, a training ship for the trip to Boston.

I remember it clearly because on July 4, 1940 I was waiting in chow line at lunch when an officer came down and selected four of us to go on deck (wooden) to hose it down as we were loading fuel. After a short period in the hot sun I became real thirsty. Being a natural landlubber, having water available from the hose, I reached down and took a good swig. I became a real "salt" and was thirsty the rest of the day.

We then were transferred to Gallups Island as A-1 and A-2 radio trainees. There were several of us left over, so we had to wait for the next shipment of trainees to form A-3 and future classes. In the meantime we worked around the new school area wiring tables and doing other odd jobs. All of this occurred during the middle of 1940.

I finished my training and was assigned to a waiting period. In November, 1941 I was sent to Marcus Hook, PA to the Sun Oil Company and was assigned to the *Texas Sun* as a radio cadet. On Sunday, December 7, 1941, we headed for Beaumont, TX. At the time, half of Texas was at war and half was not. Luckily, the captain had a Texas flag, so we were not shot at as we travelled up the canal. All of this happened before the war started.

I sailed until June, 1945 and had several graduates from Gallups Island as shipmates.

PLEULER, EDWARD F. JR., R-3

Born in Omaha, Nebraska in 1920. Graduated from South Omaha High School in 1938. Jobs were scarce so, after convincing my parents there were educational opportunities in the CCC, I joined on July 5, 1939 and was sent to Camp 761, Franklin, Nebraska. I soon became an assistant leader in the Educational Department, running training films, performing clerical duties and serving as business manager of the Camp 761 Newspaper.

In 1940 the U.S. Maritime Service sought candidates only within the ranks of the CCC for the Radio School at Gallups Island, Boston, then operated by the U.S. Coast Guard. I was chosen as one of the first students and this was the start of my radio career.

As a member of class A-3 (R-3) I was successful in obtaining an FCC Radio Telegraph License and became the 2nd Radio Officer on the *SS Alcoa Pathfinder* just 12 days before Pearl Harbor. Needless to say, following Pearl Harbor the submarine warfare went into full force by the Germans. Our ship was involved in supplying building material and equipment to various Caribbean British and American outposts from New York. On our return trip we would bring back bauxite ore from Paramaribo, Surinam, South America, to Mobile, AL.

On March 5, 1942 I became the Chief Radio Officer on the *SS Alcoa Banner*, a vessel without time to wait for armaments. With our load of Lease-Lend cargo consisting of tanks, trucks, munitions and food we sailed from Philadelphia to St. John, Newfoundland. There we joined a convoy to Reykjavik, Iceland and from there we sailed for Archangel, Russia in a British convoy PQ-16. The German airbase at North Cape, Norway attacked the convoy for six days straight every two hours. Our heaviest armament was an Australian ack-ack ship. Despite two close misses, we made it to the White Sea. There we had two days of passage through ice packed by the enormous tides of the Arctic. The Russian air cover was a welcome sight as we moved behind a Canadian and Russian Ice Breaker. We docked on June 1, 1942 with holes in our bow from the ice.

Three and a half months later, with enough cargo and military vessels to make a convoy, we departed for Liverpool, England with a cargo of lumber and chrome ore. This time a jeep aircraft carrier was part of our convoy of more military ships than cargo vessels. In order to stay as far away as possible from North Cape, we sailed north along Nova Zemlya and the northwest to the tip of Spitsbergen. This time the subs were the enemy. It finally became too much for the aircraft carrier and it left under top speed.

With a one day stop at Loch Ewe, Scotland to disembark survivors from the *SS Alcoa Cadet*, we sailed down the Irish Sea to Liverpool, England. After discharging the lumber and loading other cargo, we departed Liverpool and joined a convoy back to New York in November, 1942.

Taking some much needed leave, I married the former Mildred Clark on December 12, 1942 in Omaha, NB and we took a short honeymoon to Kansas. But back to sea as Radio Officer on the seagoing *Tug Resolute*/WKCN for 30 days, after which I was accepted back on the staff at Gallups. I became a Radio Lab/Theory Instructor. In those days the school graduated about 50 Radio Officers per week. I was the class adviser for R-81. In August, 1945 I returned to sea as Chief Radio Officer of the troopship *SS Colby Victory*/ANNR. I had two Gallups graduates for 2nd & 3rd Radio Officer.

In November, 1945 I joined Radiomarine Corporation in Boston as a Marine Service Tech. After completing the RCA Radar School I became a Resident Tech for Portland, ME in 1952. When RCA split up Radiomarine in 1958, I went back to Boston to manage the Marine Service under RCA Service Company. In 1968 I became the Chief Marine Service Representative for New York, supervising all marine field service from Boston to North Carolina. In 1981 I retired from RCA with 36 years of service.

My wife Millie and I were blessed with three sons and a daughter and now have nine grandchildren. In 1992 we celebrated our golden wedding anniversary.

In 1989 my service in the U.S. Merchant during World War II was officially recognized, and I received a USCG Honorable Discharge Certificate making me a veteran. I received the Defense Bar/Medal, Atlantic War Zone Bar/Medal, Combat Bar, Victory Bar/Medal, New York Conspicuous Service Bar/Medal, and the USSR Commemorative Medal.

I have been the Veteran Wireless Operators Association Secretary since 1983 and Editor or the VWOA Newsletter. I am also a member of the ARRL, SOWP, OOTC and a Charter Member of the GIRA and served as one of the first Directors. I am also a member of the Dennis A. Roland Chapter of the American Merchant Marine Veterans, Port Newark, New Jersey. If I find any spare time, I enjoy my hobby, amateur radio. My call is W2EIQ.

SMITH, JARED SCOTT, R-3

In the spring of '42, after a stint as a student and then as an instructor at Gallups Island, I decided I had to go to sea to be really qualified to teach. The Gallups Island staff made the arrangements for my first ship. I think they were mad at me for leaving—and being an inexperienced sailor—I signed the articles in a Philadelphian uptown office without seeing the ship.

When I did get aboard I found that she was an Italian ship, interned at the beginning of the war, and sabotaged by her crew. The receiver, for example, looked like it was built in 1920 and someone had liberally stirred its components with a crow bar, leaving a bunch of twisted coils and broken tubes. The main transmitter was a 2-tube MOFA with a final about the size of a mild bottle and the emergency transmitter was a quenched gap spark transmitter fed from an MG set. To make matters worse, there were instruction books written in Italian!

After seeing this inoperable mess I went up and knocked on the door of the captain's cabin and told him he didn't really need a radio operator because he had no radio. "Not to worry," said the captain. "RMCA is bringing a new receiver on board within the hour, which is a good hour before we sail; and, oh,by the way, remember you signed the articles." Now I knew why I signed them uptown.

Not wanting to sail sans receiver I went ashore, jumped into a dockside cab and rode up to radio row (was it Arch St.?) where I found a receiver which would cover 5 as kHz TSW. When I got back aboard with my receiver the RMCA installer was there with a new 8503 16-500 kHz TRF and was screwing it down to the operating desk.

About that time there were a couple of toots on the ship's whistle and the engine started to turn the screw slowly. The RMCA guy said, "Where is this ship going?" I said, "I'm sure that's secret, but if you look at the boxes in the deck cargo they have Russian writing on them." At that remark the RMCA guy ran down the gang plank, leaving behind a nice tool box full of tools which I was glad to use, both to finish the installation and throughout the rest of the trip.

The next thing I did was try to fire up the main transmitter to test it. As soon as I activated the big MG set it blew the circuit breakers on the ship's main AC generator. I quickly received a visit from the chief engineer who asked that I call 10 minutes before starting up the MG set so he could manually increase the flow of steam to his AC generator. This did not seem like a very fast routing for sending an SOS so I turned my attention to the spark transmitter.

While the spark transmitter instruction book was in Italian, it was fortunate I had 7 years of ham radio circuit building behind me, plus a few Tesla Coil Spark excited projects to help me at least understand the schematic. But when I fired it up it didn't work. The circuits were quite simple and I soon found that the quenched gap was internally shorted.

The quenched gap consisted of about 10 discs with heal-fins each one separated from the other by a mica washer to produce about 10 little gaps of a few mils all in series. However, the sea air had gotten in and the gaps were closed by verdigris corrosion. After careful scraping with my trusty scout knife I reassembled the gap and voila!—I tuned up on 500 kHz with 2 amps showing on the hot wire ammeter. Just as we were pulling out into the Delaware River.

By September, as a part of convoy PQ-18, we were above the north cape of Norway enroute to Archangel. We had been instructed at the convoy conference to send SSS four times on 500 kHz in case of a danger to the convoy. So when I saw 3 or 4 aerial torpedoes, through the radio room door, streaking towards our ship, I put my left arm around a stanchion post in the radio room and pulled the resistive starter over to its latch to start up the emergency spare transmitter MG set from its 24-volt battery. I sent the SSS with a message "torpedoes approaching convoy on starboard quarter." Then I rang up the bridge and gave them the same information.

Next time I looked out the door, three fish were coming straight at us—one about at our bow, on dead midships, and one headed for our stern. I clung to the stanchion and ducked down awaiting the explosions, but none came. Afterwards the deck officers told me the captain swung the ship to port causing the bow and stern torpedoes to barely miss. The midships torpedo apparently was near the end of its run and like they are wont to do, lost control, porpoised into the air, plunged under our ship, and sunk the adjacent ship in the next column.

Later on, ashore, some of the escort (Brits) W/T officers said they heard the spark message loud and clear on 500 kHz, but couldn't believe their ears that it was a spark transmitter! It was.

LEE, ROBERT H., B-2

After graduating from class B2 at Gallups Island in February, 1942, I was sent to the union in New York, then to the Bethlehem Shipyard in Baltimore, where my first ship, the *SS Paul Luckenbach*, was being modified for war duty. Our first run was to Murmansk, via Halifax, Scotland and Iceland. We saw our share of enemy action, but our convoy took fewer losses than some of those who followed. We arrived in Murmansk in early May, 1942, and were unloaded and on our way home in less that three weeks.

After returning to the U.S., I transferred to San Francisco as home port, and sailed on several ships, from "rust buckets" to Victory ships. My last ship was a troop transport, bringing military personnel home from Japan. The last voyage ended in New York in April, 1946.

I returned to my home in Colorado, and in November, 1947, I married my current wife, Stelle, whom I had met in New York. Our family eventually consisted of three sons and two daughters. Using my pre-war call sign, I was a sporadic radio ham during the '50s and '60s.

My work interests were still in radio (later electronics), and I worked on an airborne magnetometer system. This was submarine detection equipment adapted to geophysical survey. When this company disbanded, I was fortunate to find work with the High Altitude Observatory, which specializes in solar astronomy.

I started as a technician in a small instrument design group. Over the years I worked up to the Chief Engineer spot in the electronics group. I was involved in many interesting projects, among which were several expeditions to solar eclipses in "faraway places." In the early years we had an observatory high in the Colorado Rockies. At present our observatory is in Hawaii on Mauna Loa. We have observed from airplanes, balloons and more from recently from spacecraft. After 40 years with the Observatory, I retired in March, 1989, but I am still attached as a visiting engineer.

In late 1987, the GIRA locator service found me and I was able to participate in the May '88 reunion in San Francisco. The reunion and the cruise on the *SS Jeremiah O'Brien* were truly outstanding experiences. Other current interest include travel in a small motor home, and intermittent activity on the ham bands with the call N0GGG.

MARICLE, ROSCOE, C., B2/R-5

I was born December 30, 1918 in DeSoto, IA. I joined the Blooming Rose CCC camp, F-MO-9, CO 1732 in Missouri and was released in February 1941 after being accepted as a radio student by the U.S. Maritime Service. Two of us were selected from Missouri and sent by train from St. Louis to Ft. Dix, NJ. Fifty selectees were assigned to Gallups Island Radio School and shipped by train to Boston thence to Gallups Island. We had survived a week of the coldest and worst winter storms during the Ft. Dix interim period. Old WWI barracks with windows broken out, outside chow lines, circular coal heaters and continuous poker and crap games.

21 February 1941 began the most thorough radio training course for our class—the best at the time anywhere—plus seamanship. Of the original 50 students, I was one of seven who graduated in January of 1942. An additional 13 students who had been added from other classes, graduated then also.

We did not have to construct buildings. I had the privilege of a full eleven months of training. We also had the best lifeboat crew. William "Shorty" Riedle, Coxswain, Camiolo and Caldwell were crew members and we won the lifeboat race in Boston Harbor in 1941. Our prize was a mantle piece with a seaman's motif barometer, thermometer and clock. It graced the entrance room to the GIRA headquarters for some time.

Franklin B. Helsman was our platoon leader. He followed the sea throughout his lifetime with Moore, McCormick SS lines and became a silent key, W4NXW, about 1986. William Riedle followed the sea and is a retired farmer at Willow Springs, MO (amateur license KB0JHH). Camiolo died at sea very early, as did Wienstien.

I am a retired Civil Servant since March 17, 1982. I married Shirley H. Mower of Lynn, MA on January 19, 1942. I followed the sea for a year as a radio officer under ROU and Fred Howe.

My first voyage, the SS *West Kyska*, started from New York overnight to Philadelphia loaded with aviation gas, Mack truck chassis and flour. We sailed out of Chesapeake Bay. I was seasick two days off Cape Hatteras. Shorewise by the Florida Keys, we sighted a submarine at the Windward Passage. He submerged, and we ran (13 knots). We stopped at Trinidad to off-load the remains of a longshoreman from our chainlocker. He had been murdered in Philly and dumped there.

Then across the South Atlantic to Capetown to Suez, Port Said, to Alexandria, Egypt. We were there about ten days off-loading. Rommel had just been repulsed.

We just missed being in a convoy to Malta that never made it! We returned by the Suez Canal, Red Sea, Aden, to Bombay. Two human groups, of about 50 each with common head-carried baskets, partially loaded us with manganese ore.

After a week, we proceeded to Calcutta loading hemp and tea. We returned via Durban (a week of typhoon weather), thence to Capetown and Trinidad. We had lone-wolfed it all the way, joining a large convoy at Trinidad to New York the very last days of September.

In New York, just after Christmas, I sailed on the *SS Pan Maryland*, a tanker, with Captain Lund, loaded at Aruba and returned.

I transferred to American Export Airlines at LaGuardia Field, Long Island, NY as a Flight Radio Officer in February 1942. Robert (Bob) Clough, R-7, replaced me when I transferred to Maintenance two years later. I stayed with the company through a transfer of ownership to American Overseas Airlines with six months of this time as Radio Maintenance Supervisor at Chicoppee Falls, MA supporting the Berlin Airlift.

I left American Airlines in 1950 to become a Studio Engineer with ABC and was assigned to the television section at West 66th Street, Manhattan, NY. In January of 1953, I became Chief Engineer for KDRO Channel 6, a new station in Sedalia, MO. I supervised the installation, tests and then the operation, going on the air in July 1953. I remained there until July 1954 when I returned to the East Coast.

After three months with Beechcraft at Logan Airport, Boston, I changed to Teleregister Computer Systems in New York. The company had developed a passenger reservations system used by American Airlines and was developing a Railroad and Banking System out of Stamford, CT. I was Supervisory Field Service Engineer on Teleregister's Airline Reservations System at Los Angeles for one year before leaving because of hypertension and health problems.

In August of 1959, I returned to Sherril, MO. February 1960, I entered Civil Service as an Electronics Technician at Fort Leonard Wood, MO. In July of 1966, I became Production and Engineering Supervisor of the Television Training Division at the fort starting from system development to complete operations and maintenance. The system included: a 20-mile distribution system into almost 100 barracks and offices; a central distribution system starting with six channels and ampex "2-inch" video machines, which later became twelve or more with the introduction of Sony "3/4 inch" machines; a production studio producing television training programs. Later, a one-year assignment with computerized central energy control system in the cantonment area, building temperature control systems and electrical peak power control before retiring on March 17, 1982.

Shirley and I now live in Rolla, a centrally-located Missouri town with a population of 13,000, plus five to seven thousand college students at the University of Missouri-Rolla (formerly Rolla School of the Mines). UMR is recognized all over the world for its ROTC and engineering education.

I shall be forever grateful for the Gallups Island experience and the superb electronic and seamanship training. Electronics became my lifetime occupation and hobby, and I also met my ever-faithful and loving wife of fifty-one years while at Gallups Island.

Platoon R-6

BARBER, COLEMAN, R-6

Here is a thumbnail sketch of my existence since saying, "so long" on March 31, 1942 to my friends at Gallups Island, in Company C, Platoon 1.

Took a train down to New York and boarded my first ship, a small Laker-built freighter, and spent the next couple of years running the sub blockade on the East Coast, Gulf of Mexico and Caribbean down to the West Indies, Cuba and South America. Was on another Laker on the same run and then graduated to Liberty ships for more serious worldwide deliveries. My last ship loaded out of Liverpool, England for the Philippines and the Invasion of Japan but, fortunately for our side, the Japanese surrendered as we were transiting the islands to Lingayen Gulf, the invasion jumping-off point. In the immediate post-war confusion, we hung around the Philippines for about a month and hauled some GI's back to Philadelphia.

Came back to Texas in November, 1945. Worked for a while for my father-in-law in the package store business and then went into the automobile repair/body and paint business. Decided at that point that there must be an easier way, so got myself hired as a control room operator in a regionally cleared channel 550 Khz radio station in San Antonio. Spent 26 years in KTSA, the last 15 as chief engineer, then into Civil Service with the US Army Health Services Command, Media Division, Television Branch, doing medical training videotapes. Stayed there at Fort Sam Houston for 12 years and then moved into Air Force Civil Service at Kelly Air Force Base as an engineer with Headquarters Air Force Broadcasting Service, the Air Force Arm of the Armed Forces Radio and Television Service, where I am still employed and Chief, Maintenance and Engineering division.

Adelle (Vaughn) and I were married in 1943 and have three daughters who have presented us with six grandchildren and two great-grandchildren.

BREWER, W.L., R-6

My class was first known as C-1, and later was assigned R-6, something being said that the C caused some confusion with the cooking school. I had misrepresented my age when we left an NYA school in Jacksonville, FL to go to Boston. I told everyone I was 18, but I was only 17 in the spring of 1941.

I recall several things about my days at Gallups, including time spent at Sailor Haven, which was on the corner of Milk and Batterymarch in Boston. I recall the night of December 7, 1941, when we stayed there and talked practically all night about Pearl Harbor and what it might mean to us.

The only Gallups instructor I can recall by name is Summerfield. In my opinion, he was our best mentor. I remember that, just prior to one of our finals, someone had written "the shadow knows" on the blackboard. Summerfield erased it and wrote "C-1 knows." We did.

When we all graduated in the spring of 1942, I was still very green and had some difficulty adjusting to the shipboard life in the Merchant Marine. All the officers on my first ship were old and grizzled, and I felt so out of place that I left the Maritime Service and joined the Navy with a group of my contemporaries.

Naturally, I was able to advance rapidly in the Navy because of my excellent training at Gallups Island. I participated in all the major invasions in the Mediterranean, returned to the states in 1945 for a month's leave and then went to the Pacific, winding up on the South China Sea when my points where up.

My discharge came in February of 1946, almost five years from the day we first arrived at Gallups Island.

ISHEE, WILLIAM B., R-C-1

I was one of the 35 or so men graduating Company C-1 in the last days of March 1942. We began departing Gallups almost at once to take our first shipboard assignments; most shipping from the east coast ports. This was, of course, right in the middle of the "happy hunting time" for German U-boats operating in our coastal waters from the Florida Straits to the Gulf of St. Lawrence. Tankers and freighters were being sunk at the rate of five or more per day, some in sight of shore. At that point in time, we had not been able to sink one of the U-boats operating in our coastal waters.

Four of us boarded the train at Boston's South Station on Sunday morning, March 29th. On Monday morning, we reported to the USMS placement office for the port of New York, then located at 45 Broadway. The USMS proposed to put the four of us aboard two tankers and two freighters, these four vessels having already lain in port the longest of the many detained for want of a radio operator.

Since the tanker assignments were considered the more nearly suicidal, we were asked to draw for the vessel we would take. As it turned out, my first day at sea would not even be under the American flag. The freighter that I pulled flew the English flag and I was the only native born American citizen on board.

I have been doing some research on merchant shipping, convoys, U-boat and

AWS operations in the Atlantic during WWII. Recently, my particular focus has been on the use and/or misuse of the BAMS code. Of course, we were to learn after the war that the Germans read BAMS like the "Sunday Funnies" up to June, 1943. A lot of our tragedy and their success came as the result of our use of a broken code, a lamentable episode from which we can take only partial consolation from the fact that we were sometimes able to read their Enigma machine codes.

I have seen a casualty list, which I suspect may not be complete, showing the names of 156 radio operators who lost their lives in WWII. Two of the men from company C-1 are on that list. Some others are men that I knew well from other companies at Gallups. There are additional names on the list that seem to just "ring a bell."

CAMPBELL, E. SPURGEON, R-6

My first ship was the *SS Colin*. This old tub was a captured Italian motor ship which was never quite able to make it to our intended destination (usually around South Africa to the Persian Gulf). After a month of stops and starts between New York and Trinidad it was decided to return to Mobile. Since the ship was not able to meet US Maritime requirements it was placed under the Panamanian flag. The radio equipment was Italian Marconi with not a word of instruction in English. The only messages I sent were an SOS for another sinking vessel off the Florida coast and a request for a tow into Trinidad from somewhere in the Caribbean.

Between September, 1942 and March, 1945, I served on several Liberty ships and one Victory. These were: *SS Philipe de Neve*, *SS Howell Jackson*, *SS Henry Bacon*, *SS Thomas Wolfe* and *SS Cape Palmas*. We made several convoyed trips to the UK and the Mediterranean. Of course, I witnessed many ships as they were torpedoed and was in the first wave of the invasion of Sicily and the second day of the invasion of Southern France.

My most memorable experience was the trip to Murmansk, USSR during the period between December, 1944 and March, 1945. This was on the storied *Henry Bacon* which was attacked by 23 German torpedo planes on 23 February, 1945 off northern Norway on our return trip.

We had lost the convoy due to a broken rudder caused by rough weather. Our Armed Guard gunners were credited with downing five of the planes in a furious 30-minute fight, but they finally hit us with two torpedoes and the ship only stayed afloat about 15 minutes.

I sent an SOS and received acknowledgement from a shore station in UK before the captain (Alfred Carini, who went down with the ship along with some two dozen officers and crewmen) ordered me, along with some 22 Norwegian refugee passengers, and a few seamen to the lifeboat. Then I was able to rig a wire antenna and send distress messages using a lifeboat transmitter. The messages were received by the convoy escort ships. After some hours they came back and picked us up. For saving the Norwegians we were presented with the Norwegian War Medal by the King of Norway, who was living in exile in England at the time. Then in December of 1992, I, along with several hundred others who made the "Murmansk Run," were given a medal by the Russians at their Embassy in Washington.

Recently I have contacted Ken Phillips of England who was a sailor on the ship that rescued us. It was the British destroyer *HMS Oppertune*. He took some of the Norwegian babies from my arms onto his ship. He sent me recent pictures of himself along with two of the refugees who are now 50- and 52-year-old ladies. They were two and four years old in 1943 when they were passengers on our ill-fated ship.

In 1946 I left the sea and began a career in radio and television broadcasting. Then in 1955 I became involved in overseas type work. This included a project at Eniwetok and Bikini in the Marshall Islands of the Pacific. I was employed by Holmes and Narver, which was a prime contractor of the AEC during the atomic weapons testing program. Then I went with Page Communication Engineers of Washington, D.C. I was assigned as site supervisor of a tropo scatter communications project in Thule, Greenland and later as a design engineer on a Voice of America station for installation in Monrovia, Liberia.

In 1962 I was hired by Radio Free Europe for engineering service in Portugal and later at their Munich, Germany headquarters. I retired in 1982 as vice president, engineering and have lived in the small town of Cullman, AL. I have been married to the former Bea McCullar since 1945. We have two children, Steve and Linda, and one grandson, Chris, who is nine years old.

The twenty years at RFE were some of the most enjoyable and rewarding, especially as I have been able to witness the dramatic political changes in Eastern Europe and the former Soviet Union. I believe RFE played a significant role in bringing about these changes. During the last few years I have been thrilled to be able to make contact with some of my Gallups Island classmates.

NOTARO, SAM, R-C-1

I enrolled in the USMS on 6 May, 1941 at Sulphur Springs, FL and from there we travelled as a group to Gallups Island.

I'm a little confused as to my Platoon number. The roster lists me as class R-C-1 and my address label shows R-7, one of my ex-classmates tells me we were in R-6.

Nevertheless, I was at Gallups from 6 May 1941 until 1 April 1942, a wonderful experience.

After graduation I was sent to Portland, Maine and signed on the *SS Suwied* as R/O on 3 April 1942.

It was a beat up old Hog Islander with an antiquated radio shack, which also housed my living quarters. Whatever happened to the air conditioned quarters and well equipped radio shacks I had heard about?

After two days of loading general cargo we headed out to sea.

Our first port of call was Hamilton, Bermuda and then continued on to various Caribbean Islands like St. Vincent, Domenica, Barbados and Trinidad unloading our cargo.

Of course we had boiler tube breakdowns along the way forcing us to make repairs at sea and traveling on one boiler at about 5 knots. Another problem diverted us to San Juan, P.R. for repairs.

Most of my communications was done by the use of blinker and International flags, thanks to my training at Gallups.

These areas were heavily patrolled. I wasn't aware at the time of the very active U-boat action during the first six months of 1942. The Naval patrol usually challenged us with our war time code using visual signals.

When our cargo was offloaded we proceeded to Georgetown, British Guiana (now called Guyana) up the MacKenzie river to take on bauxite, (aluminum ore). The river was too shallow for a full load so we were routed to Charlotte Amalie, V.I. to top off the load.

We were now enroute, on an old Hog Islander with about three feet of freeboard on the cargo deck, to Mobile, AL. Not without boiler problems again rerouting us to Kingston, Jamaica. During the three days in port we heard about the five United Fruit Line ships sunk the week before.

Nevertheless, we left Kingston and after sailing for a full day through a tropical storm and battered unmercifully, we were torpedoed about ten hours after the storm had let up. This happened in daylight about 1700 on 7 June 1942. I was going on watch, which I usually did at sunset and sunrise.

Since my radio shack was like sitting in an oven (remember no A/C) I would keep my screen door, not my steel door, closed for ventilation until "lights out" time.

When the torpedo hit, the concussion blew me through the screen door into the passage way. When I went back in, the power was gone and the emergency batteries were under water.

I headed up to the boat deck and found my life boat was demolished. Across the boat deck our second life boat was intact. I remember stepping into this boat, screwing on the drain cap (which was my responsibility during boat drill) and we went down with the ship, releasing the davits as we hit the water.

In this lifeboat there were thirteen crew members including the Captain and the Second mate. Five other crewmen managed to release two rafts and climb aboard them. There were originally a crew of thirty-six crew on this ship.

Since it was still daylight the U-boat surfaced and circled us, taking pictures to confirm the kill.

The next evening a USN PBY flew over us and of course he didn't spot us but we could see him circling on the horizon, no doubt some debris was still floating. On his return he did see us, acknowledged, and went on his way.

The following day a USCG cutter picked us up and proceeded to pick up the other five crewmen on the rafts because we had purposely separated during the night to increase our chances of being spotted. In the meantime the PBY had landed to pick up those men but, because of heavy seas, had knocked off one of its wing floats and lost its stability. The cutter picked up our five crew members plus the crew on the PBY and sunk the plane.

We were taken to Key West and interrogated for five days and then "paid off" our ship.

My remaining WWII years were spent in the USN. With my Gallups Island radio training I was entered in the Radio/Radar Technician program. After completion of this training I was assigned to an Advanced Base Aviation Training Unit teaching PBM flying boat crewmen the radio/radar systems both in flight and ground school. The unit never left Banana River Naval Air Station in Florida (later became Cape Canaveral).

After the war I went into commercial airlines—from radio operator to technician to Chief of Aircraft Maintenance to Airport Manager—36 years of this in many different locations, retiring in 1981.

ROONES, ORVILLE I., R-6

A fellow from one of the Virginias (whose last name was Porter), Jack Vinson from Meridian, TX and I graduated in March of 1942, C-1/R-6 class, and took the train to New York City, where we checked in at RCA headquarters for job assignments. The three of us flipped coins for three openings. Porter got the coastwise coal carrier and was torpedoed and lost his life on the first trip. Jack Vinson got a tanker assignment out of Texas. I got assigned to a Moran Towing seagoing tug out of New Orleans.

We hit all the ports from Ungava Bay (Hudson Strait) to Venezuela and Panama, including all the islands in the Caribbean, in 14 months. Then I switched to cargo ships.

I saw the explosion and heard the SOS of a nearby ship when we were headed for Cuba and points south. We were headed out of Key West when we lost one of our towed barges early in the evening and had to return to port. When I went ashore the next night, I ran into Ken Morgan of Princeton, KY, who was also a graduate of C-1/R-6. He had sent the SOS I had heard, and put a book on the radio key and abandoned ship. Three survivors were brought to Key West.

In 1942 the submarines were sinking ships all around us in the Caribbean, but they left us alone on the tug. One evening, we passed right by one while running without a tow, and they did nothing.

I am still trying to get in contact with Jack Vinson from Texas. (There is another Jack from GIRA in California, I found.) I would also like to have W.L. Brewer's address and telephone number. He had a letter in *Spark Gap*, Vol.2, No.1, which I enjoyed. He was a graduate of C-1/R-6 also, and I would like to compare notes.

I don't remember if I notified your organization that John O. Bitz joined the Silent Key group a couple of years ago. Jack and I, with our wives attended the reunion at Gallups in 1978. We had signed up for lobster, but they ran out before we got any. Art Bradford felt so bad about it, the next year he delivered a large cooler full of lobsters to our house in Stamford, CT. I, in turn, took Jack and his wife out to dinner when we visited them in Reading, PA.

SIMPSON, THOMAS T., R-C-2

March 12, 1941. Gene Guarady, Camp R.O.; Dempsey Breeland, Barber; and I, Tom Simpson, left for Fort McPherson, Atlanta, GA for physical exams for a new adventure. (Following from postcards saved by my mother.)

3-15-41. Atlanta, GA. Passed exam here OK. Leave for New Jersey Sunday. Will write. Tom.

3-17-41. Arrived Ft. Dix, NJ today. Came via Washington, DC. Saw the Capitol. Sure is pretty. Have exam tomorrow, if pass go on to Boston. Tom

3-21-41. Few more days here. Our records not arrived yet. Most boys left this morning. Tom.

(At this time we got quarantined, or had already been, because of mumps)

3-31-41. No more boys taking mumps, so we will leave April 15. Tom.

Fort Dix. Long lines to mess hall, long cold barracks, cold and hungry. Before departing Ft. Dix, discharged from CCC and inducted into USMS. Left April 15 and went to Hoffman Island, NY and were issued uniforms and given plenty of good food and warm barracks.

From Hoffman we boarded the *SS New York*, an Eastern SS Co. passenger ship, on an overnight voyage to Boston. Visited the radio room; operators were very friendly to us. Next morning, bussed to wharf, boarded the *USCGC Yeaton*, for trip to Gallups Island.

Once there our group was divided into two classes, C-1 and C-2. Guarady and Breeland were in C1. Gilbert Brown from Arkansas, Melvin Nelson from Minnesota, and I were in C-2. We three became good friends. Gilbert and I visited and corresponded all through the years. I named my first son after him. We never heard from Nelson until 1991. His name was in the GIRA roster and we contacted him. Sure glad he had made it through WWII.

I enjoyed school, duties and life on Gallups Island, weekends ashore staying at Seaman's Haven, USO dances, movies and live big bands at the RKO.

Things I recall: a group of us took the speed-boat to Boston one evening to assist firemen. We rolled out hoses, etc. and stayed pretty much in the background. This was long before the ill-fated Coconut Grove fire.

One evening a seamanship class, demonstrating a breeches buoy rescue operation, almost turned fatal. After firing lyle gun and going through procedure someone went over the hill to retrieve the projectile (about 12 pounds). He came running back screaming it had hit someone on the head. And it had. He was rushed to the hospital, but was back in a few days. It had just creased his scalp! Is he with us today?

Many evenings, Gilbert and I would go up on the high side of the Island and practice semaphore, and gaze out at the British man-o-war anchored there

December 7, 1941. 7 p.m. Radio blared Japs had bombed Pearl Harbor. I was at girl friend's home in Waltham. All service men were to report back to base pronto.

January 18, 1942. We were enrolled in the U.S. Naval Reserve S1C M1, and I was discharged on August 3, 1944, New Orleans, LA, "Convenience of the Government." Did anyone get any recognition or benefit from this USNR service? I did not.

March 30th. Graduation of C-1 and C-2. Fred Hansen and I departed on the 31st by train to New York City and stayed at Seamens Institute. Contacted War Shipping Administration. No job as of April 1st. April 2nd p.m. two jobs: a tanker and a freighter. A coin was flipped and I chose the freighter—Fred got the tanker. We said goodbye and good luck and I never heard from Fred again. Recently Bill Ishee told me Fred's ship was torpedoed, all hands lost. That was a shock for me.

Boarded the *SS City of Philadelphia*, a Laker. Had to identify ship by blinker to USCG outbound. Since we had been standing watch on 500 KC at Gallups, copying SOS was not new. Auto alarm went off before we got out to sea. Radio room was on the outside corner of midship housing, starboard art. First run was not easy!

SOS every night at sea. Arrived Galveston, discharged general cargo and loaded same for Cuba. April to December: Gulf to Caribbean, Cuba, Haiti, Dominican Republic and Puerto Rico. General cargo down and raw sugar back to NOLA. Proceed daylight and anchor close in to coast for night. Leave Key West in the morning, make Havana by dark and vice versa Only messages sent were via WLO once we got inside Mobile Bay. One night, small naval craft escorting us to Guantanamo Bay got his commands mixed up .and rammed us port side head-on. Filipino steward department ran to lifeboats with suitcases! Little damage to us. Two to three days at Guantanamo Bay. One night, around 11 p.m., while anchored off the Gulf coast, my room lit up bright. A tanker had been torpedoed—probably a gasoline cargo. We heard no SOS and the captain chose not to break radio silence. Why didn't he shoot us? Glad he didn't.

September, 1942. We went into Tampa Shipyard for repairs and armament. Was this ship old? It had a spark-gap transmitter. What a souvenir! No way to get home. Mounted a WWI 3-inch on bow and a 50 caliber machine gun on each wing of the bridge. Loaded for bear and that's about all! Back on same run until middle of December. Arrived in Philadelphia and was paid off. Saw and received my first $100 bills and went home for Christmas, 1942.

Joined ROU NOLA and was assigned to a new Liberty ship, *Simon Willard*, December 27th. Loaded at NOLA for Cape Town, South Africa. Convoyed to Port of Spain, Trinidad. Departed southeastward. One anxious night, sighted oncoming vessel. There were German raiders in area, but this was a friend. Arrived Cape Town. Not the Africa I was expecting!

Crossed the Atlantic to Bahia, Brazil, and loaded stinking cow hides for East Coast, USA. 1943, Voyage 2: New York City. Loaded full load bombs plus 200 tons mustard gas cylinders on deck. US Army officer and five men in charge other than gun crew. Convoy to Halifax, thence to Europe, with us in the coffin corner. No shower!

Arrived Glasgow. Were refused entry. Arrived Belfast. Were refused entry. Arrived Cardiff, Wales and discharged cargo. Port area covered by balloons attached to wire to ward off low flying planes. I visited London. No picnic there, either. Voyage 3: New York City. Attended convoy conference with Captain. Afterwards, I came aboard. He stopped at a bar. 0200. He woke me crying in his beer—had lost briefcase with convoy papers. 0400. Federal Security Officers boarded with briefcase and a new skipper. Sailed.

Arrived Liverpool. Stayed ashore at Adelphia Hotel, city blacked out. Strange shoes outside door to be shined . . . we were tempted to take them, but didn't. Asked a waiter for hot chocolate-cocoa. He said, "You Yanks! Cocoa!? The bloody queen don't have cocoa!" Back to New York City.

Fall 1943, Voyage 4: New York City to Savannah, GA. Full load of bales of cotton for Liverpool. Discharged and chartered to British. Full load of army supplies for Mediterranean Sea. Arrived Gibraltar prior to Christmas 1943. German planes and subs met us. Read blinker message from destroyer. . . . Torpedoed. . . . Sinking. Saw U.S airmen in water. Proceeded to Agusta, Sicily, huge anchoring depot. Thence to Naples, discharged by Italian POWs.

Shuttled from North Africa to Naples via Sicily; once through Messina Straits, saw Mt. Etna glowing in the night. Air raids in Naples, hospital ship off-loading wounded from Anzio. Rain and cold. Visited Pompeii.

Awoke one morning to sound of "Ave Maria." Thought I was in heaven. Looked overboard and saw a boat load of children singing, cuts your heart out. Glad U.S. women and children don't know war.

One trip to Bari, Italy on the east coast. Port almost destroyed in December air raid, direct hit on ship loaded with mustard gas bombs. Burn victims did not get proper treatment for many days. (Paperback book, Disaster at Bari, very good description.) We knew about the Siege of Malta and passed nearby but never went in. Our final anchorage at Agusta. While departing for home, bombers arrived, but we escaped. May 1944, arrived Norfolk. Paid off and went home.

June 19, 1944. Liberty *SS Wm. F. MacLennan*, Galveston to Calcutta—a ten-month voyage via Suez Canal. Stopped at Aden, thence to sea. Gun crew sighted sub and fired the guns. I sent SSS and seaplane came out and circled for a couple of hours, then left. With one other radio operator (Navy) and instructions to stand watch two hours on and two hours off. That didn't work. We put cot in radio room and each stood watch until request relief. That worked.

Stopped Colombo, Ceylon, thence up to Calcutta. First sight of real poverty and over population. Another stop Colombo, thence to Lourenco Marques, Mozambique. Africa is as I thought. Boat trip up river, saw many wild animals. Big barbecue for ship's crew. Loaded ore. Thence around Cape of Good Hope and to Monteviedo, Uruguay to discharge ore. Saw the German battleship *Graf Spee* [?] scuttled. Buenos Aires next, met heavy weight boxer at night club. Loaded corn. Departure and another encounter with a sub. More firing, but no help from shore. Sent SSS. April 1945. Arrived Portland, ME. Paid off, went home.

May 10, 1945. New vessel: *MV Wall Knot*, Beaumont, TX. Diesel with one reefer hold. Loaded Gulfport, MS, supplies for Pacific islands. First trip through Panama Canal to Pearl Harbor. Gilbert's brother-in-law took me around the harbor in a speed boat. Thence to Marshall and Caroline Islands group.

One island we met King John. His greeting was, "I am a Christian." His picture and and article was in Pacific edition of Stars & Stripes. We were arriving Tarawa Island the night war ended. We had our victory party and suffered the next day.

Up to Wake Island with Navy vessels and we towed a small craft. We came face-to-face with Japanese soldiers. Hate exuded from them. I go a Jap rifle and large hand-key in under ground bunker radio station. Our last duty, load of aircraft wing-tanks to Bimini Island. Marker buoys for vessels to anchor for the atomic bomb. Thence to Guam, load steel plates for USA. Articles expired, Captain chomping at bit.

Arrived San Francisco. Paid off, June 11, 1946. All hands left except Captain and Chief Engineer.

I continued sailing, retired January 1, 1984 with 35 years of service.

Platoon R-7

BUCKLES, NELSON, R-7

I learned about Gallups Island Radio school from an article in "QST." I arrived at Gallups the early part of July, 1941 in Class A-1, now called R-7. Boston was quite an experience for a 19-year-old boy from Kansas. I registered for the draft while I was there. Turned 20 the day the Japs bombed Pearl Harbor. Remember working blinker from the island water tower to a signalman on the *HMS Rodney* which was anchored in Boston Harbor. Was quite a thrill at the time.

After graduation (June 4, 1942), took the train to New Orleans. Signed on the *S/S Oneida* KDJO/ANAK as combination radio operator/purser. It was an old Ford Motor Company laker. The skipper's only question for me was if I knew the code.

We loaded rice in Lake Charles, LA and took 17 days going to San Juan, Puerto Rico, anchoring nearly every night around the coast or in a port. Spent the 4th of July, 1942, in San Juan. We left there to go to Preston, Cuba to load nickel ore. We were to take it to Pensacola, FL but never made it; our ship was torpedoed and sunk about noon on July 13, 1942. Those of us who survived floated ashore on a raft a little east of Guantanamo, Cuba. We lost seven of the crew.

After going back to Kansas to wait for duplicate papers, I returned to New Orleans and signed on the *S/S Crawford Ellis* LCKO. The ship was under charter to United Fruit Company. Spent a year and nine months (46 voyages) as the only American on the Norwegian ship. We were on a banana run from Tampa to various ports in the Gulf of Honduras. Got chased by a sub several times and saw lots of stuff floating from ships that had been sunk.

The remaining ships I sailed on were either owned by United Fruit Company or operated by them for the War Shipping Administration. Left New Orleans on June 6, 1944 on a new ship, the *SS Samuel A. Fabens* KUUZ. We were dead in the water in the straits of Florida a day or so later. The steam valve on top of the boilers stuck and they had to cool to repair the valve.

Made six voyages on a sugar run from Cuba to Boston. I remember the distilled water jars freezing in the winter in a metal bin on the radio room roof. Nothing could be down until we went south again and the mess thawed out.

Made six trips as chief operator on the *S/S Benjamin Rush* KETT, a Liberty ship, to Europe in convoy with cargo for the war, including locomotives on deck. My last voyage on the *Rush* was to Belem and Fortaleza, Brazil in 1945.

The only time I ever heard from the draft board was in Antwerp, Belgium. I received a letter from them to report on a day that had already passed. I wrote them saying if they could arrange transportation, I'd be happy to come to their office. Have never heard from them since.

Brought troops back from Europe and the Mediterranean on the *SS Central Victory* ANMV in 1945 and 1946 as chief operator (seven voyages). Remember relaying traffic for the *Queen Mary* and *Isle de France* to and from WBF. Spent a short time on the *SS Pacific Victory* AOER in Baltimore.

Then on the *SS Fiador Knot* ANXY (two voyages) from New Orleans to ports in the Caribbean with general cargo. Went through the eye of a hurricane in the Gulf of Mexico and chewed a hole in the stem of a pipe I was smoking.

Married a hometown girl in November, 1946 and moved to New Orleans. Took the *La Perla* KDCK from Mobile to New

Orleans. Then on the *SS Esparta* AOGM (six voyages), a new United Fruit Co. refrigerated ship, on a run through the Panama Canal from New Orleans to Golfito, Costa Rica. Returned with a load of bananas.

Ended my time at sea in 1947 on the *SS Antigua* KDCJ, a United Fruit Co. Passenger ship (3 voyages). Sailed from New Orleans to Havana, Cuba to Puerto Barrios, Guatemala and returned. We also carried 45,000 stems of bananas each trip.

Came back to Kansas and spent 30 years in the meat processing business. Sold it in 1976. Spent 12 years as executive director of the Kansas Meat Processors Association and finally got someone else to take over the duties in 1991.

My wife Zelda and I have traveled extensively through the years, visiting meat plants in many different countries and many places of interest throughout the world.

CLOUGH, ROBERT F., R-7

QST carried an article in the spring of 1941 extolling the virtues of a radio school on Gallups Island in Boston Harbor run by the Coast Guard. Having expended my meager saving on two years of college, this sounded like an attractive way to go.

Five of us "hams" from the Chicago area arrived at Gallups the second week of July, 1941 to join one of the first "QST" classes where most of the fellows were already licensed amateurs. The second A-1 started with 53 people and graduated 38 fellows a year later as class R-7 in June, 1942.

After a couple of trips on an old tanker, the *SS Pan Carolina*, I returned to the island as radio theory instructor until January, 1943. At this time I joined the Naval Reserve and was assigned back to American Export Airlines under a NATS contract to help operate the commercial circuits of a Navy radio station at Belem, Brazil on the Amazon River.

The summer of 1945 I started flying with American Export Airlines as a radio operator. In 1950 I went with Flying Tiger Line, but as radio operators were "short timers," I got my flight navigator's license and spent the next 25 years as a navigator with various worldwide airlines. Retired from United Airlines in 1975 only to go back through the ranks of several non-sked airlines in management positions until finally retiring for good a couple of years ago.

HALL, DAYLE, R-7

As a member of the second class of A-1 which graduated as class R-7, I was probably the youngest of the group at age 18. This class was recruited from hams with an article in QST, and some 38 or 39 were selected. For the most part the thrust was to have one ham from each state, but several states had more than one representative.

Received notice on July 1st to report for a physical, and after a short delay reported to the cutter *Yeaton* on a Saturday evening. I shared a hot dog and bean supper with the crew, watched their poker game and arrived at Gallups Island about midnight.

One particular memory stays with me of those months on Gallups Island. This was December 7, Pearl Harbor day. I had not gone ashore for the week-end and was spending a leisurely Sunday afternoon listening to my radio; had ham bands as well as broadcast. I had a lower bunk, and the radio was suspended by cords from the springs above. Enjoying the music of Sunday Serenade with Sammy Kaye's orchestra when the announcement of the attack on Pearl Harbor broke in. Likely I had the only radio at that time because it seemed that 40 or 50 warm bodies were massed in the aisles and atop adjacent bunks within minutes, and that continued for most of the afternoon.

Upon receiving FCC license and graduating, the father of one of the class members who lived in the Boston area drove four of us including his son, to New York to pick up our ships. I was taken to the ROU office and accepted a ship that had been waiting for days in New Orleans for a Sparks.

Immediately a taxi was obtained, and an Eastern DC3 was held for my arrival. It was rather emotional as the wife of some US Senator had been deplaned so I would have passage on a higher war priority. She was one hot and loud lady! Arriving about dusk, I took a cab to the docks and was soon on my assigned ship. I had a big surprise, for it was the *MS Norfolk,* owned by Ford Motor Company, a twin diesel and the largest craft that could navigate the Erie Canal.

The Norfolk was about 300 feet long, 49 feet in the beam, and when loaded drew five feet forward and eight feet aft. The wheel house and mast was designed to telescope into the deck, and the aft engine stacks to fold over so the vessel was able to duck under the fixed bridges that crossed the Erie Canal. The radio room was on the port side forward. It was triangular with the point just wide enough for a doorway, and the transmitter, the bed was across the aft bulkhead with the auto alarm above it. There was room for a swivel chair and a sink mounted under the one porthole. A small desk on the forward wall held the LF receiver. Then energy batteries were on a shelf on the other side of this bulkhead sharing space in the chain locker.

The captain signed me on early May 9th. The ship sailed at noon down the Mississippi. During that night, a tanker was torpedoed not too far off our southerly course set for Aruba. The following day at dusk we underwent a torpedo attack. The tin fish passed under our shallow draft vessel. Probably, evasive steering tactics coupled with our small size caused the sub to break off action. The two member Armed Guard crew got off one shot from the 4-inch gun that resulted in the gun jamming in full recoil and froze in place. I recall checking it—the base was 1892, the barrel 1898, and the recoil mechanism 1904. As a result, none of it fit very well.

Supposedly, ship movements were secret, but the *Norfolk* had been in Havana on an earlier voyage, and as we passed the Morro Castle inbound, a bum boat came alongside seeking the ship's cook. It seems that he had formed a liaison with a daughter of the Harbor Master and she was waiting for him. No sooner had we anchored than he was down the ladder and the blinds on the bum boat were dropped.

After two days we were cleared to sail, but we routed via Dry Tortuga to Panama City, FL where we overnighted to avoid submarines. As we were approaching Mobile, a sub surfaced behind us, and we raced to get past the nets. A coast Guard auxiliary boat tried to bar our entry as it was against orders. The captain advised them that if they would engage the sub (about two miles behind us), he would proceed on our original course. Their answer was full speed north into Mobile harbor, the *Norfolk* followed.

After other ships and other adventures I returned home, and included in my mail was a letter from American Airlines offering jobs as Flight Radio Operators. A few days later in New York City, I interview at their headquarters at La Guardia Field and as hired as a FRO trainee. I was to proceed to Memphis, TN where they were assembling crews for Air Transport Command service world-wide. Having to drop by my Ohio home to exchange my sea going clothing for more suitable attire, I reported on the 26th of October in Memphis. I spent all of 1943 and two months of 1944 working point-to-point radio, ground-to-air radio, and aircraft radio repair.

In addition, for several months, I moonlighted by working radio for Warner & Tamble Barge Lines in Memphis which operated a number of tow boats on the Tennessee, Mississippi, Ohio and Missouri Rivers as well as intracoastal waterways to Galveston, TX. This work was enjoyable, but working a10 p.m. to 6 a.m. shift six days a week for American became too burdensome so dropped the tow boats for rotating shifts.

Returning to blue water on March 23, 1944, I signed on the Liberty ship SS Tristram Dalton operated by Bull Lines of Baltimore, MD. Sailing in convoy from Norfolk, April 3rd, we arrived in Oran, Algeria the 19th. We departed for Naples, Italy overnighting in Augusta, Sicily harbor and arrived in Naples on the 3rd. Cargo was discharged and we sailed for Palermo, Sicily on the 10th. This voyage was notable for viewing three volcanoes erupting the same day. Mt. Etna had a minor eruption as we passed through the Straits of Messina; Stromboli was tossing up fiery lava as we passed about three miles to the east; and Vesuvius was still belching some fire and dust following an earlier major eruption in March.

My final voyage was on the *SS William K. Kamaka* (named for a heroic Merchant Mariner), operated by the South Atlantic Steamship Lines. I signed on at Baltimore, July 18, 1945. We loaded at Norfolk with cargo for Pearl Harbor. Passed through the Panama Canal August 3rd, arriving September 11th. War in Pacific over. We loaded supplies for Marine detachments in China. Departed Honolulu about ten days later, arriving in Buckner Bay Okinawa just after the October 1945 typhoon devastated that island.

Our ship lay hove to in the vicinity of the Iwo Jima Island group. Wave heights were logged at 75 feet and the winds at more than 133 mph The waves did not break as the wind blew their tops off and the few times one could see across the sea, it looked as if the waves had been planed off. Many Navy ships were upside down on the reefs around Buckner Bay, and it was days before we could get our orders.

About October 20th, we set sail for Tsingtao, dodging many mines in the Yellow Sea. The Armed Guard set off 10 or 12 with rifle fire. After only a couple of days in Tsingtao, we were driven out by small arms fire from Communist troops, and went to Taku anchorage which was the port for Peking, the capital of China. We off-loaded supplies into lighters for an American Marine detachment in that part of China.

Once the cargo was discharged we proceeded back to Okinawa and after some delay set sail from Norfolk via the great circle route. We did not touch land until we reached the Panama Canal. Enroute, we rendezvoused with a Navy oiler to replenish or fuel and take on a few supplies. We arrived in Norfolk January 2, 1946. I paid off ship January 4th.

So ended my career as a shipboard operator. Two voyages as a single operator, one as Second Radio Operator and two as chief Radio Operator.

Returned to the airline, radio and operations work until 1962 when I transferred from Cleveland, OH to American's Maintenance Headquarters in Tulsa, OK, retiring 1983. Hobbies include: Music—was charter member, President and other offices, Tulsa Organ Club (now in its 28th year); Ham Radio—a DXer, lacked only three Zones on 80 mtrs and 3 on 40 mtrs to achieve 5 Band WAZ. In 1959 became involved in raising and exhibiting pedigreed cats, publishing a monthly newsletter devoted to cats and clubs for 14 years. Upon retirement I took up genealogy, and have traced most of my ancestors to arrival in America.

Not in love with the triple digit temperatures of Oklahoman summer, nor much caring for the ever present tornado threat, my wife and I moved to southern Indiana in late 1985. Both raised in Ohio, we had family members there, and because her parents were in their 80s, requiring supervision closer than two days drive, we opted to not go where too many snowflakes fly, but still be within three hours driving distance. Now the elders are gone, and my wife and I are the oldest surviving family members. Truly Senior Citizens.

HICKS, ROSS, R-7

In July, 1941, I hitchhiked from Oklahoma to St. Louis, where I went to the Coast Guard recruiting office to become a "QST" enrollee. It was then that I joined a great bunch of truly dedicated men.

During my days at Oklahoma University, I made a trip on a Liberty which sailed from Galveston, TX to Marseilles, France. Although I made the trip simply because I needed the money to stay in school, I will always remember how great that two and a half months trip was.

In 1992, I let my USCG and 1st telegraph lapse, but the memories sail on. It was a great life and I'm glad to have been a part of it.

TEEGARDEN, HARRY L., R-7

I was born in St. Joseph, MO in 1921. In February, 1939, I passed the amateur radio license exam and became W9FSI. In 1946, after the 10th call area was created, I became W0FSI.

In the spring of 1941, I read an article in "QST" magazine concerning the USMS Radio School at Gallups Island. I applied for enrollment and was accepted. I arrived at the island in July, 1941 as a member of class R-7. Upon graduation in May, 1942, I shipped out on my first voyage. I sailed until January, 1946, on six different vessels, covering most theaters of operation.

Upon returning home in 1946, I entered the radio/tv broadcast field and pursued this line of work until 1957, when I joined the federal work force. I was employed by the CAA, which later became the FAA. I worked in St. Joseph, MO, Las Vegas, NV, and Grand Junction, CO, where I retired in 1981.

I was married in 1950; my wife, Ormah Jean, passed away in 1990. I have two daughters and three grandchildren.

Platoon R-8

CANTOR, HAROLD, R-8

"Shanghaied to Africa"

I graduated July 9, 1942 and was sent to the War Shipping Administration office for assignment July 15, 1942. The ship was the *SS Cathlament*, of the American West African Lines. It was a 1918 Hog Island type, jokingly called a Liberty Ship in WWI. The radio shack stood alone on the boat deck, about 10'x12', surrounded by a thick wall of cement blocks to protect against machine gun bullets. Inside was my bunk, small closet, the radio equipment, emergency batteries and charger. The tools were a hammer, screwdriver, pliers and a three-range voltmeter. There were no instruction books on any of the equipment.

By the time I found the radio shack, the ship had pulled away from the dock so I could not get off. It seemed that there were many operators sent to the ship from War Shipping and the radio officers union, but one look and they all scooted off quickly.

Later on, I found out that the Captain had given orders to the Second Mate that as soon as another Radio Operator showed up, to cast off quickly. Unluckily, it was me. So there I was "shanghaied to West Africa."

When I got to the radio shack, I found that the transmitter had been taken apart and was literally in pieces, strewn all over the place. No instruction book. No spare parts. The transmitter appeared to be one of the first tube transmitters that evolved for the original spark sets. When I considered my situation I sort of went into shock. Here I was, a green operator on my first trip, a graduate of the finest radio school of its type in the world, with the best instructors, the latest and most modern equipment, but I never expected anything like this.

To top it off, Captain Huntley then made his appearance. He made it known to me that he had nothing against me personally, BUT, he hated radio operators. According to him, radio operators were slackers and were not needed. He described himself as a sailor of the old school and he had only grabbed me as I came on board because the government would not let him sail without an operator. As to why the radio equipment was in such unusable shape, his answer was he did not know and did not care. The responsibility was mine and I was not to bother him about it. It took me a long time, but I put it together the best that I could, hoping I would never have to use it.

Our crew consisted of about 34 officers and seamen plus one U.S. Navy lieutenant and eight U.S. Navy seamen to man four machine guns on the boat deck and a 3-inch gun on the stern. Most of the cargo consisted of supplies for trading posts in the jungles of Africa. The ship started the run in Freetown, Sierra Leone, then Monrovia-Ghana and ended in Doula, Cameroons. We picked up and carried natives of the KRU tribe who rowed the cargo ashore in oversized rowboats which we carried on deck. We seldom docked anywhere but anchored off shore and had the natives row the cargo ashore. On this ship, we had four mates who were used to handling this type of cargo. Luckily we did because the captain was always drunk and mostly in his cabin.

When we arrived at our first stop after leaving the U.S.—it was Trinidad—what we called "bum boats" came alongside selling bad liquor to the sailors. That night, one of the sailors went berserk, running wild on the ship wanting to cut someone's wrist and make him a blood brother. Needless to say, we all locked ourselves in till things calmed down in the morning.

These were the days, 1942, when merchant ships were sent out without any protection; no such things as convoys. Survival was a matter of pure luck. I remember going through the Windward Passage, between Cuba and Haiti, and hearing the explosions of ships and seeing fires on the horizon; wondering if we were going to be next.

It took us about two weeks to get to Africa as the best speed we could ever do was about 8 knots. When we anchored in a small harbor in Liberia, I came down with a bad case of dysentery. I was carried ashore on a stretcher by natives into the interior to reach a Firestone rubber plantation that had a small hospital. I was the only white patient with 200 natives down with malaria. I was told a white doctor came to see me but I do not remember him as I was delirious at the time. He prescribed intravenous feeding for me and left some bottles but the native girls did not know how to apply this. As a result, my veins were punctured so many times that to this day I have difficulty raising my arms above my head or making blood donations for hospital tests.

After a week, I was released in a very weak condition and sent by truck to a Pan American Air Base in Robertsport, Liberia. In the meantime, my ship was given a small tugboat escort that had radio equipment and proceeded without me down the coast.

After I had gone through much questioning at Robertsport, it was decided that I was to be put aboard a plane that was carrying about 40 U.S. Army radio operators who were on their way to India. I was asked if I thought I could recognize my ship from the air. I told the pilot I could recognize that old rust bucket anywhere.

The pilot had instructions to proceed down the coast until I spotted my ship. When I did, he landed in the next clear spot he could find, which was about five miles south of the *Cathlament*. I kept walking north on the beach hoping that the ship would not leave without me. I finally spotted it off shore and came to the trading post we were then supplying just before it was getting ready to pull out.

The sailors on the ship were glad to see me back and to show off their new possessions. To my surprise, they had been buying chimpanzees. They had heard that they could sell them back in the states to Frank Buck's zoo. Some of the chimps were tame as they had been raised as house pets by the locals in the hope someone like ourselves would come along and they could sell them. Of the six the sailors bought, only two survived the trips as we got back to New York in January and they could not take the cold weather. The two were sold to a Chicago zoo.

Our next port was a place called Takoradi/Sekondi in Sierra Leone. The English were in charge here and a dockmaster came aboard and told us flat out he was in charge and, I quote him when he said, "There is no war in this part of Africa and I'm in no hurry to get you to unload." As it was, we were stuck there for two weeks. Food aboard the ship was horrible and I took the opportunity to take a room in a small, inexpensive hotel on the beach and enjoyed the food which was mostly chicken.

A native black boy got to following me around and I took a liking to him. He was about 10-years old, an orphan. He lived on the streets, depending on handouts whenever he could get them. The food I was getting at the hotel was plentiful and I talked the hotel owner, a likable Greek fellow, into letting the boy stay in my room with me and share my food.

The boy, John Amuah, had little education but had been discharged from school because he could not pay his tuition. The local grade school was run by a Methodist Mission and it meant buying a school uniform, school books, supplies and paying tuition. I decided to give him a chance in life and paid his tuition for six months. The hotel owner helped by promising he would give him a job cleaning up and would see that the boy was fed and had a place to sleep.

The English used the native constables for their police who were hated by the villagers. The constables would only walk by threes through the village for their safety. The native villagers soon heard about what I was doing for the boy and they were very friendly to me, inviting me into their huts to talk.

It was then I decided to give a gift of a white sport shirt to the boy's teacher at the school and ran into a lot of trouble. Those of us who went to sea remember there was always a "slop chest" aboard ship mainly to supply basic needs like shirts, pants and shaving supplies to the sailors who sometimes came aboard with very little when they signed on. In our case, our Captain decided that the slop chest belonged to him and was his source of income. He had no intention of selling anything to the merchant sailors and only sold things to the natives for exorbitant prices.

When I went ashore carrying my gift for the boy's teacher under my arm, I was arrested by a native British Customs guard who accused me of smuggling and put me in jail. I was brought before a white British official who asked me what I was doing with the shirt. I explained about giving it as a gift to a black teacher. He almost exploded, saying that sort of thing is not done in this part of Africa and if I still wanted the shirt he would impose a heavy fine of ten times the worth of the shirt which I had paid $3 for. He almost died of a heart attack when I paid him $30 and proceeded to give the shirt to the boy's teacher who was very appreciative.

After leaving Takoradi, we proceeded down the coast picking up a cargo of ore for the trip home. The last port was Doula, a French colony in the Camaroons. From there we went to Freetown to prepare for our return to the states.

I went with the captain to the U.S. Navy Conference to get radio instructions for the next port which would be Trinidad. They gave us code books but the captain took possession of them and refused to give them to me. While we were in the office, merchant seamen covered with oil from a tanker that had just been sunk, were brought in. The naval officer told me that there was a submarine laying off shore that sank every ship leaving Freetown. There was no chance of us getting through but his instructions were to keep sending ships out.

He warned us that when our ship was torpedoed to jump off the opposite side of the ship because they lost many seamen who were sucked back into the hole. He asked us what kind of life preservers we had. We explained that we had the old type with blocks of cork. We left there after he warned us that when we jumped off the ship to jump with our arms linked above the cork, otherwise when we hit the water the cork would hit our chin and knock us out. He wished us "good luck" and promised to have a small tug standing by to pick us up.

The pilot took us out of the harbor and our next port of call was Trinidad. Any minute we expected to be hit by a torpedo, but nothing happened. After seven days of this we were still only half way across. I kept listening to the BAMS traffic list schedule for our area, put out by the radio station in Rugby, England. All I knew was our secret call letters were KLAH and had been told by the Navy that there were very few merchant ships in our area at that time. If our secret call letters were heard, we were to answer at a specified time and frequency, and if Rugby heard us they would break their traffic list and acknowledge.

Suddenly, for the next three traffic lists I kept hearing KLHA repeated on the list but nothing else. This was almost our call but the last two letters were reversed. I spoke to the captain, but he just about threw me out yelling I was a green kid, fresh out of school, did not know the code, and refused to give me the code books. I talked it over with our Navy gunnery officer whose additional duties included being communication officer and liaison between the Navy and our captain.

We decided we had to get those books, at least to find out whether a mistake was made and the message was meant for us. The captain was, as usual, drunk, locked in his cabin and refused to open up. At the point when we almost had the door broken down, he finally opened up and gave us the code books. The message when deciphered said just four words, "GO BACK TO AFRICA." It did not name the ship or say that there had been a mistake. We then decided to code up a message asking them to name our ship, and if a mistake was made. The dreaded time came when I finally had to fire up the transmitter, not knowing if it would work, to send the message out and see what would happen.

I did this and the next traffic list from Rugby had my correct call letters KLAH, did not say the name of the ship and the message when deciphered said "GO BACK TO AFRICA." When I reported this to the captain, he went into a rage, calling me a spy, and threatened to have me shot before a firing squad. The Navy officer sided with me and told the captain he had to turn back. He finally turned the ship around, heading back to Freetown. The next week was the most miserable of my life, I kept doubting myself and wishing I never had started this. After all, how could the British make a mistake like that. It must have been me. So I told myself.

When we reached Freetown harbor, the pilot came on board, the same man who had taken us out two weeks before. The first thing he said was, (and I quote), "Captain, what are you doing here, you were supposed to be back in the States." As anyone can imagine, I just about died at that point. The captain left the ship and did not come back for hours. When I saw a tug coming back with him aboard, I actually expected an armed guard to be with him, but there was just himself. When he came aboard and saw me, he never said a word. But, at dinner that night, he turned to me and just said, "You were right, Sparks," and nothing else.

It was not until a few days later, when we were at the Navy conference ashore to get instruction that I found out what happened. The Germans had called all the subs in, to form a huge wolf pack around Trinidad, so no ship could possibly get in or out. By doubting this mixed up message, I had actually saved the lives of all aboard.

The trip back to New York was uneventful and the captain treated me a lot differently on the way home. There was one more incident that happened when we were getting paid off which I will never forget. When the captain was paying me, he turned to me and said, "Sparks, how would you like to sail with me again?" knowing him, this was a very difficult thing for him to say, but it was the equivalent of getting the United States Legion of Honor. I still remember my answer to that: "Oh, no captain—get yourself another boy!" As to our ship, the *Cathlament*, it never sailed again. It was condemned by the Navy as unseaworthy and junked.

I sailed as chief radio officer on many ships after that, and had many experiences including a Liberty which was turned over to the British Navy that carried British troops into the Normandy invasion; running back and forth across the English channel from the very first day of the invasion.

While I was at Gallups, we normally got about three liberty weekends a month. I always liked music and dancing, so I usually went to the USO at North Station in Boston. I was attracted to a beautiful girl who played the piano for sing-a-longs. When I found out that we had exactly the same birthday, I knew it must be fate that we met. When I proposed, I never could figure out why she accepted me because I had nothing to offer, just a $12 engagement ring. I must have done something because we are still in love and married over 50 years. Our 50th Anniversary celebration was on a cruise of course.

After the war, I got a job with RCA Global Communications as a Transmitting Technician. It was at Rocky Point, NY. It was the world's largest radio station with 100 transmitters. I worked for RCA for 34 years and am now enjoying retirement in Delray Beach, FL.

JEFFERS, CLARENCE G., KZ1R, R-8

One late July 1986 morning, a Boston Harbor Island State Park water taxi was making good headway through the swells and chops, ferrying a quintet of "Gallups Islanders" and two XYL's on a nostalgic visit to the Island. Graduates of the USMS Radio School at Gallups Island, Boston Harbor (1940-1945), Art Bradford, W1EQK; Ralph Albers, W4ER; Ken Palmer, K2FJ; Ray Sutcliffe, W1RND; and Clarence Jeffers, KZ1R, had abandoned small talk and harbor reminiscence to peer into the haze, awaiting the looming land mass that had figured so prominently in their lives.

In the early summer of 1941, a somewhat depressed 20-year old (the author, then W1LZR), operating 2-1/2 meter mobile, had a fate QSO and resulting ham shack visit with W1KOY. Bob was exuberant over a June QST article entitled "QST Visits Gallups Island," pointing out that this was a great opportunity for a youth with an amateur license. His possession of an application startled the 20-year old, since Bob was well past the qualifying ages as well as a successful insurance executive with a beautiful family, (Incidentally, a few years later the author was gratified by a QST picture of a happy W1KOY, an officer in the U.S. Army Signal Corps, having an "affair" with a Jeep.)

In August, the 20-year old, enjoying a complete turn-around in his life, was enrolled in class R-8 at Gallups Island, the beginning of WWII involvement with shipboard and aircraft radio and government service (FCC). A letter home contained a smug "I told you so," referring to his firm conviction against parental doubts, that radio held the key to his future.

Any gathering of "Gallups Islanders" is replete with similar accounts of rather despairing circumstances having been reversed as a result of the Gallups Island Radio School. Although the Great Depression of the thirties was past, the early forties still saw a depressed job market and education beyond high school for most was not yet a tenable goal. Many youths had turned to the government sponsored Civilian Conservation Corps (CCC) which became the source for the formation of the first 1940 radio school classes.

Arriving at Gallups, the visitors stepped onto a pier with a large gazebo and quietly surveyed the changes of the past forty years. The island was completely overgrown with shrubbery and trees without a single remaining structure. Footpaths maintained by Boston Harbor Islands State Park management were the only visible influence of man. Within the gazebo were documents relating to Island history. The radio school representation was sketchy with some chronological error. The radio school graduates receiving a hearty welcome from the Island managers (two summer working college students) were invited to comment. All were eager to fill in, correct or further explain the historical details.

PULEO, ROSARIO, S., R-8

After graduating high school, at the height of the depression in 1937, I held a job as an electrician in the Garment District of New York City. Later on I got interested in radio theory and Amateur Radio and enrolled in an evening course in radio theory at a Brooklyn High School; I taught myself Morse Code by converting everything I read into dits and dahs.

One evening while buying radio parts at a Lafayette Radio store, as fate would have it, I missed the elevator going down. While waiting for the next elevator I wandered over to the book section of the store and spotted the June 1941 issue of QST magazine. I bought it because the cover indicated that it had articles about Amateur Radio. I had never heard of the magazine and would not have bought it had I not missed the elevator.

As all Gallups Islanders know, this particular issue of "QST" contained the article about Gallups that induced many of us to apply for admission to the school. I showed the article to my evening class instructor asking him if he thought I would be qualified. He wrote a glowing report about me to the United States Maritime Service and as a result I was accepted.

I was never away from home before. My first morning at Gallups, being awakened by the shrill bugle call, and surrounded by strangers, was a traumatic experience I will never forget. I was brought up in an Italian ethnic community, but now being thrown in with boys from all parts of the United States, who for the most part were better educated, was very intimidating to me.

Most of the students were Radio Amateurs with a good knowledge of radio theory and code, and some had electrical engineering college credits. It was July of 1941 and the United States had not entered the war yet, therefore there was no immediate need for radio officers. We were told that the course would run for a full year, with exams every Saturday morning. It was a competitive situation, and those not making the grade would flunk out. In the beginning I studied constantly, but after about four months I gained enough knowledge and confidence to compete comfortably with those of us who survived. Because of the intensive training, the FCC exam was a snap, I even got a First Class Radiotelephone and a First Class RadioTelegraph license without even studying for it.

After graduating in July of 1942, my first assignment was to join a freighter in Philadelphia, the *SS Pennsylvanian*. When I came aboard, the Radio Officer was already packed; he shook my hand, wished me luck and left immediately.

I have never been aboard such a large ship. Again, because of the extensive training at Gallups, I was able to figure out quickly how all the equipment worked.

We set sail within a few hours after my arrival on the ship. I was fully aware of all the ships being sunk off the Atlantic coast, and I expected the worst. I immediately put on my life jacket, and prepared to send an SOS. Going to sleep that night was out of the question; I sat and waited for the torpedo to hit. After awhile I fell asleep in my chair. To my surprise I awoke the next morning to find we were still afloat. As I was never out of sight of land before, it was a strange feeling being out in the middle of the ocean.

The next day we entered New York Harbor where they made repairs and loaded cargo. A mate who was leaving the ship took me aside and advised me to be reassigned to another ship as my ship was loading cargo for the deadly Russian Murmansk Run.

I decided to stay with the ship because no trip was safe. Again, as fate would have it, the Navy stopped all shipping to Murmansk because of extremely heavy losses. Instead we were sent to the Persian Gulf via South Africa where we encountered submarine wolf packs operating out of Capetown. Our cargo was unloaded in Persia (now Iran) and transported overland to Russia.

Subsequent trips on Liberty Ships took me to many countries around the world in which we participated in the invasions of North Africa, Sicily, France and the Pacific Islands. I also made the Russian Murmansk Run in which our small convoy was used to lure out the German battleship Scharnhorst which was sunk by the British Navy.

I had my share of submarine attacks and dive bombings but, as fate would have it, I never got my feet wet.

After the war I married, had two children, worked for Pan American Airlines and ITT. I attended Hofstra University evenings and graduated with a degree in electrical engineering. I then formed Puleo Electronic manufacturing company in which I am still active.

SEIGLER, WILLIAM C., R-8

I was a ship's officer in the Merchant Marine during World War II. When Japan agreed to an unconditional surrender on August 14, 1945, my (Liberty) ship, the *SS William L. Gable,* was discharging military cargo at the port city of Tacloban on the Philippine Island of Leyte. Shortly after discharging all cargo, the ship was modified and outfitted to accommodate Japanese prisoners of war for repatriation to Japan. All U.S. Navy Armed Guard personnel got off the ship and U.S. Army personnel came aboard to guard the prisoners. The Army personnel consisted of approximately 25 troops including a major, a captain (medical doctor) and two first lieutenants.

From August, 1945 to May, 1946 my ship was used solely for the purpose of repatriating thousands of Japanese prisoners of war from the islands of Leyte, Mindanao and Luzon in the Philippines, and Okinawa to Yokohama and Tokyo. All the prisoners, including officers, were quartered in the holds of the ship.

The Tokyo Bay area was still heavily mined and we were allowed to enter port during daylight hours only. This operation was strictly under the control and direction of the U.S. Army. We merchant seamen aboard were subject to the discipline of the military during this period which extended further than the August 15, 1945 cut-off date.

So, you see, I know from personal experience that there were U.S. Merchant seamen who sailed under military control beyond the August 15, 1945 cut-off date. I was one of them.

(Editor's Note: This excerpt is from a letter dated June 19, 1991, from Seigler to syndicated columnist James Kilpatrick. Seigler was responding to Kilpatrick's column which dealt with WWII merchant seamen receiving veteran's status, specifically changing their arbitrary cut-off date from December 31, 1946 to August 15, 1945. In his letter, Seigler quoted "The opponents principal objection is that after August 15, 1945, merchant seamen were no longer subject to military control. The answer to that is yes and no." In reply, Seigler wrote in support of Kilpatrick's opinion; the excerpt is Seigler's reasons why.)

TROY, JUNN, WING, R-8

The first month at Gallups was like living in a country club. Code speeds were not threatening as yet, but typing on a typewriter without visible letters drove many of us to smoking stronger cigarettes. Then came the "day of infamy" which exploded the "heaven" myth. It took quite some time after President Roosevelt's declaration to settle down, ponder, and adjust to the new reality.

I recall several of the upper class students had resigned to join the armed forces, hoping to be in a position to throw a stronger retaliatory punch. By Christmas recess, plans were already in place to shorten the school term. We all worded a bit harder and adopted a more serious, or possibly a more mature attitude. Our instructors worked with us, often burning late evening oil. No questions were considered trivial and none were left unanswered. There were not very many answers for our type of questions in Terman's "Radio Engineering." Among those who tolerated our group were Sweeney, Summerfield, Hilliard, Garcia and Sussman. In later years I even learned to appreciate being knocked off my high horse by Ashkins, the boatswain's mate aboard the *Yeaton*. After all, he had warned me not to spill any #!@*$! disinfectant on his #!@*$! deck.

Almost everything I was taught while on the Island was put to good use during my very first voyage at sea. During a six-month trip, I coxswained lifeboats during drill, taught nautical phraseology to seasoned salts, and once even relayed traffic while in convoy by semaphore. Proficiency in blinker came about ever so naturally.

A week or so after graduation on July 10, 1942 from platoon (A2)R-8, I posted my late model 2nd Class license in the radio shack of the *SS Zarembo* (KICS). The ship was a Hog-Islander built in 1919. Top speed was 9.3 knots, and could peak slightly over 10.0 if the safety valves were intentionally disabled. My bunk and living quarters were inside the radio shack along with my wash basin, complete with running water but no drain plumbing.

We were usually the slowest ship in convoy and we were often straggled. During mid-August enroute to Trinidad, we were attacked by an Italian sub and sustained minor damage from depth charges thrown by the Y-guns of our Free-French escorts. From Trinidad we inched across the Atlantic to Lagos, Nigeria where we discharged our cargo of war-machines to be assembled in support of the North African campaign. Bouncing back and forth along the African coast, we finally loaded sufficient cargo for our trip home. In addition to cacao and palm oil, we carried animal hide (phew!), live monkeys and snakes. We sailed to Freetown, Sierra Leone to rendezvous with two other vessels, formed a "spearhead" formation and sneaked out to sea accompanied by a British destroyer escort.

On the morning of the third day, our escort flashed a "Cheerio, God's Speed," and made a u-turn. Our little group of three ships separated per instructions and were soon out of sight of one another, heading toward our common destination, Trinidad.

Later on that leg of the trip, during my afternoon watch, we sailed into tropical heat lightning. All at once a strange noise covered the entire radio spectrum to at least 30 Mcs. It started with random, short duration spikes, which within the half hour evolved into a high intensity blanket of noise. The BBC was masked on all their scheduled frequencies, and NSS as well as GBR no longer seemed to exist. This phenomena continued and by midnight of the following day, I had missed five consecutive BAMS schedules.

I had never missed more than one per day. I dared not sleep, but Morpheus triumphed the second night. Suddenly, some sort of yet to be explained stimulus caused me to spring to my feet; clearing my eyes while staring at the clock, I read 0100 GMT. I perceived NSS through the noise which had only diminished somewhat and recognized my ship's secret call. Waiting for my turn at bat, I copied bits and pieces of traffic intended for the other ships before copying our own.

None of the individual 'bits' made sense but in combination and with some intuitive interpolation, the Captain and I decoded the following: "TO MASTER S/S ZAREMBO - CHANGE COURSE IMMEDIATELY NORTH TO 15 DEGS NORTH LATITUDE THENCE WEST TO POINT Y - END." Point Y was further identified in our top secret, wax sealed, envelope, as the US Virgin Islands.

By the time of the next schedule at 0711 GMT the noise had subsided. I checked the traffic list. There were to be no repetitions. The 'bits' I had copied were indeed the last of several blind transmissions. We were later to learn that the ships were diverted to avoid heavy enemy submarine activity off both entrances to Trinidad harbor.

Neither of the other vessels received their revised destination. I can only surmise and hope that they both fared well in the face of danger. (I would appreciate information regarding these two vessels. Would like to know how and where data or records can be obtained, copied or simply viewed. I can supply additional details which may help. Thank you.) Almost immediately upon dropping anchor in Charlotte Amalie, the Captain and I were summoned for consultation with congratulations given by both the Admiralty and the US Navy. We completed the voyage, safely returned to New York City on December 8, 1942.

Although not exactly a "lid," I have received my share of QLFs from more experienced operators suggesting that I should proceed sending with my left foot. I did manage a lucky 28 wpm once in a while at Gallups. If any skills were brought to bear in the experiences I have related, I was taught those skills while at Gallups Island, and credit is due the instructors. Moreover, I, as well as the civilian crew and gun crew of that voyage, owe a debt of gratitude. Specifically, to those original master telegraphers who had the foresight to formulate and structure the course of instruction at Gallups to included reading traffic under interference and/or high noise environment.

The Captain, Chief Engineer, "Chips" who rose to the rank of Chief Mate, and eventually to captain, the Captain's messman, and I, formed a bond during my eventful first trip. We sailed together continuously to the end of WWII and beyond to 1946. The *Zarembo* was one of the original vessels of the American West African Lines. The members of our team were reassigned to a new C2 type ship, the *SS Sirocco* (WQU#).

After two years, the vessel was sold from under us and we were placed on the *SS Cathlament* (KOLP). (The five "angry" men were: Captain Samuel M. Thorne. How did he know we nicknamed him "Whiskers?" The brain of the group. Chief Engineer Frank Betron. He could diagnose a problem with an up-and-down 3 piston engine just by listening. Chief Mate Robert F. d'Arcy. Probably a true genius and a wonderful friend. Best wishes to you and Julia. John Pilat, Messman. Hope to see you again, buddy. Junn Wing "Wayne" Troy. Flattered to be counted as one of the group.)

I then took a year-long vacation, later returning to the sea for a peacetime tour of the Mediterranean. After a short stint with the United Nations as a point-to-point operator in Israel and Lebanon, I retired my chrome-plated Vibroplex and reverted back to a low-pay landlubber. Spent years sweating an engineering degree and went directly into microwave research with the US Navy. For 34 years I chased the ever elusive electron, I caught a few, but was never able to make them dance in tune, so I joined my "bug" in retirement.

Platoon R-9

ALBERS, RALPH E., R-9

Between November, 1990 and January, 1991, Ralph served as radio officer on the tanker, *MV Chesapeake Trader*, plying between Anacortes, Washington and Valdez, AK. Next, he accepted an assignment from the Radio Officers Union on the C4 freighter *SS Cape Archway*. The ship is chartered to the U.S. Navy to transport ordnance. She served in the Gulf War right before Ralph came aboard.

During this assignment, ending August 28, 1990, the ship traveled to Rota, Spain; Sounda Bay, Crete: Augusta, Sicily and Cartegena, Spain, where Ralph completed his assignment and flew to Frankfurt, Germany. His wife, Birdie, met him in Frankfurt and they spent a month traveling around Germany. They visited the eastern German cities of Berlin, Magdeburg and several others. High points of the German sojourn were: travelling freely about Eastern Berlin, walking through the Brandenburg Gate and a really exciting trip from Munich to Frankfurt on the famous "Schnell Zug"—the high speed train travelling about 125 mph.

Ralph expresses special appreciation to John (N2RU) and Polly Orr for their gracious hospitality during the many weeks his ship was loading and awaiting orders in Earle, NJ.

Operating on the modern ship is nothing like it used to be. Ralph did not sent a single CW message in four months—everything is sent through satellite by teletype, fax or telephone. He did talk once to Don Gagne (KERL/W2LID) from the ship on 454 Mhz. One still stands an eight hour daily watch on 500 kHz. News also comes in daily via the satellite.

One really important new skill of great use, if not a real necessity today, is computer literacy: Ralph's SITOR equipment was controlled entirely by a Toshiba lap-top computer. The captain found great use for Ralph's knowledge of computer software, such as Lotus 1-2-3, to help in some of the ship's paperwork.

The captain must have been satisfied. He asked Ralph to return to the ship for another voyage!

ABEYTA, GREG, R-9

I joined the CCC in early 1940 and was assigned to the Soil Conservation Camp near Magdalena, NM. While I was in line to get my tools I kept hearing the guy giving the tools out ask, "Do you know how to type?" Everyone kept telling him no, because some old timers had told us not to volunteer for anything. When it came time for me to get my tools, he asked me and before I knew it, I had said yes. He told me to follow him and I got really scared that I had made a big mistake. But no, he told me they had to update all the inventory cards and that I could do that while the other guys cleaned around the camp. Fortunately, typing was my forte and I was able to do the work he assigned me.

In the next two weeks I had met the Company Commander, the camp instructors, the cooks, and the camp doctors. When the old guys completed their hitch and left the doctor requested I work with him as his assistant and I accepted. I remained at headquarters where I got to see how the Camp was run. I was promoted to Assistant Leader, which meant I received $6 a month more and I didn't have to pull KP duty.

The next six months were rather routine. In fact, I was thinking of getting out. One morning after mail call, the Commander called me to his office and asked me if I would be interested in attending radio school. I told him I knew nothing about radio. He went on to say that it was a beginner's course in radio. communications. There was a need to have radio communications at remote camp sites in New Mexico and the surrounding states. I agreed and was sent to the radio school in El Paso, TX. There were about thirty other boys from camps around New Mexico and Texas. We completed the training in about six months.

I was sent to Santa Fe, NM as a replacement for the radio operator there. I stayed at that camp for about four months and then I was transferred to Melrose, NM, to set up a radio transmitting station at that site. It was there I received a message addressed to the Camp Commander asking for recruits to join the US Maritime Service as radio students. The Company Commander said he would recommend me and I accepted.

I reported for a physical in Dallas, TX in early September 1941. There were about 15 other boys from Texas and Oklahoma. I was the only one from New Mexico and the CCC. We were sent to Boston, MA, arriving at Gallups Island around the 15th of September 1941.

Imagine my surprise when we reported to the barracks to find out that a few of the fellows were from the CCC. In fact the Barracks leader was a CCC man. Small world!!

So you can see that I landed at Gallups Island Radio School through no effort on my part. It was mostly a set of circumstances that led me there.

ALEKSUM, JOE, R-9/10

On July 25, 1941, I arrived at Hoffman Island for training as a deck hand. It was a USMS training school run by the USCG. I was in GR18. While there, a notice was posted (for high school graduates with 2 years of math and 1 year of physics (minimum)) about Gallups Island Radio School in Boston. As I was from Boston, I applied and was accepted with about 16 other trainees.

I arrived at Gallups on October 31, 1941. I was either in R-9 or R-10 at that time.

As time went on, many trainees either dropped out or flunked out so they combined platoons R-9 and R-10 and it became R-9/10. We have a class picture of 37 graduates at "USMS Radio School, Gallups Island, Class R-9/10, August 1942, Boston, MA." We also had class rings and mine has the inscription, "G.I. RADIO SCHOOL USMS C2 '42."

I guess you can see by now it is even more confusing as on my mailing label from you people, I am listed as R-0A3. I suppose if I had a preference it would be R-9/10, but it really doesn't matter.

MORRIS, RONALD C., R-9

I started on Gallups Island on November, 28th, 1941 and I was in the original class A-3; which later became R-9. I failed to graduate—I didn't fail to graduate—I joined the Navy a few weeks before graduation and stayed there from July or early August, 1942 until December 10, 1945.

I was aboard both the (North African) shore stations.

I was also on minesweepers, subchasers and patrol craft. We made the Invasion of Anzio; I was on the sub-chaser at that time. We were lead ship for the assault boats coming in. Then went up into Southern France in June, 1944 for the invasion. We left in late November or early December, for stateside rotation.

After the war I went to sea in the Merchant Marine. I sailed on two Army transports, one was the *USAT Acadia*, sister ship to the *Evangeline*, from New York harbor to Japan via Korea. We carried quite a few hundred women passengers. It was a very interesting trip to say the least.

I made numerous trips on the Acadia and then transferred to the *USAT General C. G. Morton* in 1947. I stayed with the Morton until June when I went on the beach and married my present XYL and went back to the Long Island from the San Francisco area.

I went to work for Ma Bell in '47 and stayed there a little over a year before joining the Air Force. We stayed in the Air Force until June, 1965 when I retired. From there I worked a couple of years and went into the civil service, the FAA, in 1968 and retired in 1979 for a total of 32 years of federal service, all together.

NEWBOLD, ALEXANDER, R-9

After leaving Gallups Island in August, 1942 I was assigned to my first ship out of Mobile, AL, the *SS Domino* KIZB, a medium-sized freighter. Operated in the Caribbean. Two trips. Brought back aluminum ore (bauxite) from Georgetown, British West Indies and Paramaribo, Dutch Guiana. Signed off the ship.

Joined the Navy in March, 1943. Sent out to San Francisco and assigned to *USS Flint* CL-97, an anti-aircraft cruiser as a radio op. Five major assignments in the Pacific.

Discharged, Boston, 1946. Drove from New Hampshire to Stockton, CA in May, 1946 in a 1937 Chevy coupe. Married Esther. We now have four sons and seven grandchildren.

TWA, ground radio operator, 1946-1958; Aeronautical Radio, Inc. (Arinc), radio operator, 1958-1984. Retired in 1984.

I am a member of the Masonry (Blue Lodge and Scottish Shrine) and Amateur Radio organizations.

NEWTON, HUBERT C., R-C-1

I enrolled with the USMS at Sulphur Springs, FL on May 6, 1941, recruited from an NYA Camp in Charlotte, N.C. I was sent from there, along with a group of others, to radio school at Gallups Island. I spent the next 11 months studying radio with the finest group of young men that I ever was associated with. Our platoon was C-1, later changed to R-6. I eagerly watch for news of any of them in the Spark Gap.

Arthur Sheddan and I left Gallups Island together on April 2, 1942 and reported to the Radio Officers Union in New York to be assigned to ships. At that time two ships were available, one a tanker for a coastwise voyage to Jacksonville, FL and the other one was a freighter scheduled for the Red Sea. I wanted the trip to the Red Sea and Sheddan wanted the Jacksonville voyage because his home was there. The flip of the coin gave me the Jacksonville ship and Sheddan the Red Sea ship. Fred M. Howe, the ROU official who made the ship assignments, sensed that we were both dissatisfied and arranged a swap for us which pleased us both. Sheddan's ship left port first and I later learned that it had been torpedoed on the way down to Jacksonville. I was pleased to learn that Sheddan had survived, especially since I was the one that the flip of the coin said should have been there. (Thanks, Sheddan, for taking a hit for me.) I boarded the *SS Cavalcade*, a Danish freighter sailing under the Panamanian flag, and headed for the Red Sea. My ship stopped at the port of Trinidad on the way and while I was there I ran into my good friend and classmate at Gallups Island, Spurgeon Campbell. We filled each other in on our experiences up to that point and my ship headed on its voyage to the Red Sea while his ship, I believe, had to stay there awhile for engine repairs. That was to be my last contact with anyone I knew from Gallups Island for the next 49 years when I was surprised by a telephone call from Bill Brewer. Since then I have learned the whereabouts and activities of several of my classmates and instructors through a platoon roster that Bill sent to me and through articles in the Spark Gap and Gallups Islander.

My wartime experience consisted of service on the following five ships: *SS Cavalcade, MV Winkler, SS Charles H. Windham, SS Samuel de Champlain* and *SS Sea Nymph*. The *Winkler* was sunk by enemy action in the North Atlantic on February 23, 1943. She was the victim of a "wolf pack" of German submarines. We were in a convoy of about 60 ships that had formed in the British Isles and was sailing for the east coast of the United States and were somewhere near the half-way point when the submarine attack began. They would only attack at night. On the first two nights of the attack, six ships were sunk. On the third night our number came up and we were hit by a torpedo in the bow of the ship which slowed us down to where we couldn't keep up with the convoy. This happened at about 3 a.m. After we fell behind the convoy about 3 hours later, we were hit with two more torpedoes which caused the ship to sink like a rock. Being an empty tanker it seemed impossible that it could sink so fast. There was no time to launch lifeboats. The only means for the crew to get off was to float off. In very short order that was the manner that the crew used to abandon ship and all who survived were left floating in the cold water of the North Atlantic. We depended on our life jackets and hung on to life rings to keep afloat. A nearby escort ship, a British corvette, chased the submarine and dropped depth charges and then returned to pick us up. They took us to the Argentia Naval Base in St. Johns, Newfoundland and from there we returned by train to New York. About a third of our crew lost their lives. I took a little time off at this point and then returned to ROU in New York for another assignment. This time I was sent to the west coast and spent the next several months sailing the Pacific, seeing very little enemy action, only an air raid in New Guinea that didn't even come close to our ship. I returned to the east coast in early 1945 and continued to sail the Atlantic until March 15, 1946 when I signed off the *SS Sea Nymph*.

Awards received: Merchant Marine Emblem, Atlantic War Zone Bar, Pacific War Zone Bar, Mediterranean-Middle East War Zone Bar and Combat Bar (with star).

After the war, I returned home and married the girl of my dreams, Grace McClure whom I had known since my high school days. We have now been married 47 years

We have four children and eight grandchildren.

My training at Gallups Island had a direct bearing on what I was to do the rest of my life. I was intensely interested in the theory end of radio while I was there. When the war was over it was my hope that I could get a job utilizing what I had learned about radio theory. I wasn't quite that fortunate but I came close. A job in radio was unheard of in Marble, NC but I was able to migrate into the Knoxville, TN area and get a beginners job in the Electrical Maintenance Department with Alcoa. I worked there for a while, was laid off due to lack of work, entered the electrical construction field for awhile and then hired in as a maintenance electrician for Kaiser Aluminum and Chemical Co. at Ravenswood, WV. I was promoted to Electrical Power Supervisor after about seven years service. I worked for this company for 28 years and was Electrical Power Supervisor in charge of the distribution of electrical power to all parts of a plant that used as much as 350 megawatts of electricity at the time of my retirement in January of 1987.

WESTLEY, RICHARD N., R-9

I was in class R-9. I always thought it was A-3 but somewhere they changed it to R-9. I came from Grand Island, NB to Gallups Island. I had the ham call W9YEW. It was a Class C license. I got it deliberately so that I would be able to qualify for Gallups Island. I traveled from Grand Island to Omaha where the Maritime Services had offices. They gave me my ticket and sent me on to Boston in November, 1941, and I graduated with R-9 in August, 1942.

My first and only vessel was the *S.S. Totico* for the Mexico Shipping and Trading Company. She was built in 1916 and had sunk two subs in WWI by ramming them. I spent 13 months at sea and never went back to sea again. We had a gentlemen's agreement to serve one year, and as soon as my year was finished and the boat was in the harbor, that was the end of it.

When I quit going to sea, I took a job with the PanAmerican Grace in South America as a flight radio operator and I flew for them for 18 months up and down the West Coast of South America. I quit PanAmerican Grace and started flying for American Export Airlines out of Flushing Harbor, Long Island. It was later to become American Overseas Airlines and I was on the North Atlantic run as flight radio operator for five years—flew in the Berlin Airlift and all of that mess. When PanAmerican merged with American Overseas Airlines, all the radio operators got a severance notice and that was the end of that.

So then I flew for Flying Tiger Line and I flew around the world with them several times as the radio operator and as flight navigator. Then one day along came the seniority list and I was out of a job. That ended my flying career and also my radio career.

I went into business for myself out in California, a little gift shop in Apple Valley, and I forgot all about my radio licenses—let them all lapse. I didn't think another thing about radio until I retired from IBM in 1983, when I was spending some time in a little resort park in Florida. I went to a class and got my novice ticket and upgraded about every three months until I got the extra class ticket, now AA5.

Platoon R-11

HAGEWOOD, JOHN, R-11

Like many others, I thoroughly enjoyed Gallups Island. I heard about the radio school in a pamphlet called "QST Visits Gallups Island," reprinted by the U.S. Maritime Service in June 1941.

I left Nashville the 2nd or 3rd of January, 1942 with three others from Tennessee. When we arrived in Boston, on the train of course, we spent the first night on the Coast Guard cutter *Yeaton*. What a thrill!

Next morning we sailed to Gallups Island and got checked in. Our first pay was based on $34.00 per month. Originally the classes were to be one year in length, but, I believe they started cutting back to nine months with our class (R-11).

I graduated in September, 1942 and sailed on the tanker *SS Bohemian Club*, Liberty *SS Daniel Webster*, *SS John Hopkins* and *SS Iowanda Victory* until February, 1946.

I've lived all my life in "Music City USA" (Nashville) and worked thirty years in the electronics industry. Retired now!

I have the full original class picture of Class R-11 made September 4, 1942. The names and everyone's signatures are on back. Several have the ham call signs. I also have most of their addresses, at that time, in old address books I kept.

(Editor's Note: Hagewood kept other mementos of his days on Gallups Island in addition to the class picture, including the above mentioned pamphlet, a page from the "Quincy Patriot Ledger," February, 14, 1942, with pictures of and a story about Gallups Island. He also has photos he took on the Island of class members.)

RELLER, MAURICE, R-11

I came to Gallups Island during the first week in January, 1942. By the middle of February, our class, R-11, had come up for a week of "general detail." My assignment was to be part of an auxiliary crew and flunky for the Coast Guard ship *Yeaton*, which was being used as a ferry for transporting passengers and supplies to the Island. I believe it made two round trips a day and always tied up at Pier 1 Commercial Wharf at night. This meant shore leave for the crew members every night with the exception of one crew member who had to stay on board for firewatch and security.

By this time I already knew that I was in trouble with my math and radio theory classes and I saw a great opportunity to stay on board every night to study.

In exchange for favors and other things, I volunteered to take everyone's watch, every night. The duties consisted of making regular inspection rounds, checking for fires and also taking up slack or letting out on the mooring lines as the tide changed. I believe Boston has one of the highest tides on the east coast, but that did not especially lay heavily on me as a responsibility.

As the night wore on, I was comfortable in the wheelhouse, while totally engrossed in my studies. At some point, I became aware of a kind of pleasant rolling or rocking as the cutter gently rubbed against the pier with the incoming swells. Then, there was a rhythmic "squeak-squeak" as the mooring lines were tightening. I did not come to the realization that I had not tended the lines until I began to have trouble staying in my seat. It hit me like a fireball that I had not been conscious of the tide going out and that I had this ship hanging on Pier 1.

I ran from mooring bit to mooring bit, trying to find a line where I could get some slack. It seemed to be no use, as even the spring lines were tight. But, finally, with panic and adrenalin, I managed to get some slack, and back and forth I raced, slowly lowering the boat back into the water. I don't remember ever telling anyone on Gallups Island about this. But now, I have confessed it to the world.

My exposure to enemy action was quite simple. From October, 1942 to May 1946, I travelled on ships through the Gulf of Mexico, the Caribbean, the Mediterranean, the Persian Gulf, the English Channel, and the North Sea.

Circumnavigated once, hitting every continent (except Antarctica), the three big oceans and the coast of Normandy. The enemy action always occurred before I had arrived or after I had left. Watching buzz bombs fly wildly over Southampton, England is all I can claim for enemy action. At the time, I felt cheated, but now I know I was lucky.

Platoon R-12

ANDERSON, RAYMOND "RAY" C., R-12

Born and schooled in Staples, MN. Heard about Gallups Island Radio School through an article in the local newspaper. Sent in an application in November 1941 prior to Pearl Harbor. Was accepted and left in early February 1942, entering Class R-12.

Upon graduation in October was sent to New York and from there to Baltimore. Received first assignment on a Hog Islander, *SS Felix Taussig* for a trip to Brazil. Saw first enemy action in Caribbean.

Following was a Liberty, *SS Joseph Hooker*, on which I made four trips to Europe and Mediterranean. Other ships were *SS Robert Luckenbach*, *SS Charles A. Dana*, *SS Sweepstakes* and the *Fort Lane*, a tanker.

Left the sea in May 1946 and returned to Minneapolis where I have resided ever since. Joined Midland National Bank and worked in the Operations and Systems Departments. Also served as Auditor and Security Officer.

Retired in May 1985 as Assistant Vice President of the bank and Vice President of Midland Realty Company. Joined GIRA as soon as I heard about the group from Alex Newbold on an SOWP net.

HOXIE, WILBAR, M., R-12

From time to time I have occasion to recall my own experiences in the Coast Artillery Corps, USA, at stations not greatly different from the Island of your Alma Mater. Now, a third of a century later, many of these posts are all but forgotten, but their vestiges do exert an impeding curiosity which results in calls for information. One of these, in fact, was for your own Gallups Island, now stricken with blight and abandonment.

I went there, searching for something that might hold a clue to its former uses, and found little. The printed record does mention a training camp during the Civil War, of strictly temporary construction. Some of the foundation remain of the Radio School, but too little to discover alone what they represent now. As I mentioned, now is the time to accumulate all that we can of knowledge concerning this School, and I ask your associates to help me sketch, in as bold outlines as can now be done, what the Island looked like when it was a living, vital force for our nation.

Please ask all of those you can reach to call or send me any information, particularly sketches, drawings, photographs of the structures. If none still exist, perhaps the collective memories will tell us what these buildings were, will afford rosters of the several classes, the names of the instructors, dates of beginning and ending of classes, through the years. It is not unthinkable that together we might produce a book, large or small as the information permits, telling all who will listen of the importance of this Station. Perhaps we may get it listed as a Site on the National Register of Historic Places.

Platoon R-11

Platoon R-12

Platoon R-13

LANPHERE, ROBERT C., R-12

My sea career began on my 20th birthday—November 20, 1942 in Port Arthur, TX and ended on April 16, 1953 in New York City.

I was employed on the following ships: *SS Torres, Joseph N. Dinarnd, Smith Victory, Robin Sherwood, Robert Treat, John Greer Hibbern, MV Anacapa, SS John Barton Payne, Bara Meeker, John Isaacson, Fairport* and *SS Jeff Davis*.

I was working on the *Jeff Davis* in 1953 when I met my wife who was a passenger. We have two daughters, Nina and Milette. Nina has a dog training business right here in Las Cruces. Milette lives in Monroe, WA and is into Wild Life Rehabilitation. No grandchildren.

My civilian employment history began in 1953 when I went to work for Rich Industries, a toy manufacturer, in Tupelo, MS as a machine operator. I was also looking for work in the electronics field at this time. From 1954 to 1960, I worked as a broadcast engineer with WELO Radio in Tupelo. During my employment with WELO I co-supervised the building of a new 1,000 watt transmitting plant for the station.

In 1961, began working as a senior technician in the Electromagnetics Section of the New Mexico State University Physical Science Laboratory. My work there included every phase of designing, fabricating and performance testing of various antenna types. I also instructed graduate and undergraduate students until my retirement in 1985.

I really enjoy the GIRA amateur net. I wasn't active on the ham bands for 20 years. I belong to the Mesilla Valley Radio Club here in Las Cruces where four or five of us monitor 28435 Mhz daily. My rig is a Swan Astro 100 BXA into a AV3 Cushcraft Trap Vertical Tri-bander 10-15-20 (meters). I hope to attend the reunion in Long Beach as I have never discussed my seagoing days with anyone. Louis J. McCourt, who now lives in Arizona, was my best buddy at Gallups Island.

McCOURT, LOUIS J. (JIM), R-12

I attended Gallups from March to October, 1942. I started out in R-13, but was moved to R-12 a few weeks after we started—I guess they were hurting. I never was a real radio operator like the many amateur and ham operators that went to Gallups. I kind of fell in to the place, not really knowing anything about radio. I became a good operator, but in school they lost me when they started on vacuum tubes.

I got my license by memorizing the Q&A's. Had the chance to do some coding and transmitting and used the DF, but fortunately the equipment didn't fail or I'd have been in real trouble.

Had three close friends, at one time, but haven't seen or heard from them since 1945: Frank Raines, maybe Raynes, lived in West Roxbury, Bob Lanphere, Tupelo, MS, and Carl Duarte, lived in Cambridge last I knew.

My wife and I are going to Boston at the end of May and I'm going to try to take the trip to Gallups which I understand starts running the 1st of June.

As the next step in Gallups Island Day, I'm going to fill out the GIRA application and submit it.

WYNNICK, NICHOLAS B., R-12

I was informed by mail to report to 45 Broadway, New York City on February 9, 1942 for enrollment in the U.S. Maritime and being transported to Boston, MA to attend the Resident Radio Schools at Gallups Island.

We departed late and arrived in Boston after dark. There was nobody to greet us and Travelers Aid had no idea of who we were or where we were supposed to go. After numerous phone calls they learned we were to go to Commercial Wharf. I don't recall how we got there but when we did, we found a cold drafty warehouse. The Coast Guard Cutter *Yeaton* was tied up along the wharf, for many months that was to be our liberty boat on weekends. I recall there was a pot-bellied stove and somebody had set up cots for us to sleep on. We stripped down to our underwear but I am sure that others besides myself put everything back on because of the bitter cold. There was no way on earth that pot-bellied stove could beat the drafts. The *Yeaton* did not take us to the island that night because the submarine net, beyond which G.I. lay, was closed.

The next morning the *Yeaton* took us out on its first trip of the day. We had our breakfast on her. The pancakes were all right but the syrup must have been molasses and the coffee was in a world I had never been in. But it was warm aboard the *Yeaton*. Boy, was it cold at Gallups! We were issued our sea bags and uniforms and settled down in a barracks. We were R-12 and there were 94 of us. My uniforms, dress and undress, were used previously, so it was apparent that there was a high dropout rate. Our platoon was issued the sailor suits. One fellow from Louisiana was issued a uniform that someone must have washed in bleach water because it was a mid shade of blue and stood out. He protested but to no avail. At our first inspection before weekend liberty the C.O. told his aide to get that man a proper uniform.

To me, the most remarkable thing about Gallups in a negative fashion was the cold. We were each issued two army blankets from 1918. Apparently, after taps they would shut off or turn down the heat. I just couldn't sleep the first three months even with full uniform, sweater and peacoat on. I don't know how I was able to absorb all the theory they laid on us in warm classrooms. A lot of time was devoted to lifeboat rowing and learning the various commands until we were quite good at it. But again, that was cold even with the kapok life jackets on.

In March or April I came down with a severe respiratory problem that put me in the island hospital for over a week. Another fellow from R-12 was in the same room as I with the same problem. All I recall of him was that his name was Dick from California. He and another fellow from R-12 dropped out and joined the Navy after about five months at Gallups. They were given Radioman 2/C ratings and assigned to the battleship *USS Massachusetts* which was being fitted out in Boston. The treatment in the hospital was simple. Stay in bed and drink cold water. Golly, how my head, throat and breathing always hurt! They gave us a spit kit consisting of a small rectangular metal box with a hinged lid. Inside was a waxed paper container into which we spit after coughing up globs of yellow stuff.

After we returned to the barracks, two new practices were instituted. No more bunks head to head and after breakfast outdoors calisthenics.

The good part of Gallups was devoted and dedicated instructors. For those who were hams prior to Gallups, it may have been comparatively easy. But to those, as I, who had to learn to read and send code, grasp theory and learn touch typing, it was a struggle. Those fine instructors gave everybody all the help wanted. They also warned us that the movies at night were for those who were attending the cooks and bakers schools at the island. We were supposed to go down the hill and practice code and study theory. Boy, were they right when they said classes of 94 were unwieldy and that many would drop out. My recall is that 24 of the original 94 made it.

It was some time in the summer of 1942 that the Yea*ton* was "gone to war" and the fishing boat *Manuel Roderick* became our main vessel to and from the island. Somebody wrote in one of our newsletters that the *Roderick* had come to an untimely end, but many years after the date cited I saw and photographed her in some fishing port.

Now comes the fun part!!!

After doing well on the final exam at Gallups, I was sent in the early group to take the FCC exam at the Custom House in Boston. Having passed that also I was all set to ship out with the R-12 Platoon. Up popped a problem. Up to that time I had been using a baptismal certificate for I.D. Now the F.C.C. required a birth certificate. I wrote to Harrisburg, PA for one and was told no such certificate existed. Getting leave, I hitched hiked to Sunbury, PA where I was born and learned all vital statistics had been sent to Harrisburg. Arriving at Harrisburg, I was told there was no certificate for me. I had learned from the doctor's widow of his full name so I asked the clerk if I could search through the records. She would have none of that but did take all the information as to date, city of birth and the doctor's name. I went back to the island. Shortly, a letter arrived from Harrisburg stating that I should get an affidavit from school authorities at the elementary and high school I had attended that I was a male. They had found a certificate for the only child delivered by that doctor, on that date and in that city for a girl, Mary. Which was my mother's name. The certificate had my mother's marriage name wrong. I learned how this came about. My mother told me the flu epidemic was still raging during her pregnancy. Her doctor would often stop in to check on her but was exhausted. Sometimes he would sit on a rocking chair and my mother would put a blanket on him and he'd sleep before going on his rounds. He did not make out the certificate the day he delivered me at home and made errors. So for over 22 years I was legally a girl—undoubtedly the only girl to graduate from wartime Gallups.

Our R-12 photo shows the date as October 8, 1942. That may be the date R-12 left the island but actually it was taken earlier in the year. Secondly, I was the last R-12 member to leave the island and the date was October 22, 1942. Because I was the only member left, I answered no muster or roll call. I did study for Radiotelephone Second class ticket, Element Three. This was on my own.

Students were required to take the early boat back to the island after weekend liberty. One Monday morning I overslept and took the later boat. So what? But the master-at-arms, he was a handsome fellow but overly strict. Probably a flunkee. I explained to him but he put me down for Captain's Mast. At the Mast I explained why I was still there, with no roll calls, and that I was studying for Radiotelephone Ticket on my own which would reflect favorably upon the school and staff. Penalty was keep studying. F——'s face fell. I passed the exam; my tickets came and I shipped out.

Platoon R-13

CONWELL, FRANK, R-13

One interesting thing that happened to me: Arthur Fillebrown and I were down at the docks one day looking at the bulletin board, and there was a notice that said all men of Jewish faith would have a liberty pass for Jewish holidays.

I looked at Fillebrown and said, '"That ain't right. We didn't get our holidays off." And he said, "No, we sure didn't." Then I said, "I feel pretty Jewish," and he said, "Me, too."

So I signed Frank Conwell and he signed Arthur R. Fillebrown. We did it on the spur of the moment and thought we'd get our wrists slapped or something like that.

A few days later a list came out that said: "The following men are granted liberty for the Jewish holiday." And there was "Goldberg," "Finklestein," "Conwell," "Fillebrown," and so on. It was during the week when we lined up to get our liberty passes. They called out our names, we got our liberty cards, got on a boat and went ashore.

On the way to shore, Goldberg, who was the first Jewish fellow I ever met, said to me: "Father Conwell, you're going to the temple tonight?" I said, "What else?'" Everything was fine.

So anyway, I graduated in early '43. Fillebrown was in R-12, but we had the same barracks. I went to sea on three different ships. One was the *Artemis Ward*, another the *Efem Brevard* and I don't remember the third.

I got hurt over in the Mediterranean. I heard that the Navy was training people for IFF radar, so I went out to Little Neck, Long Island, and they trained me for IFF radar. My first assignment was in Brooklyn Navy Yard, then they sent me to Orange, TX.

After the war was over, I worked for Houston Radio Supply which sold link radio. I sold it and installed it to police and sheriff departments. I worked at that for about a year before going back into broadcast. I built a radio station in Houma, LA and was there six years before moving over to Mobile, AL, to work and manage Channel 5. I worked there for six years.

FOWLER, WESLEY L., R-13

Born February 1, 1920. Graduated Brockton High School in Massachusetts, 1938. Attended Northeast University in Boston and obtained bachelor of business administration in engineering and management in 1962. Licenses: radio telephone, radio telegraph and general amateur.

During World War II, served on cargo ships, sailing the Atlantic and Mediterranean oceans. Also served during two hair-raising trips to the Murmansk.

Employment: 1951-1958, engineer, WBET (5kw) radio station, electrical equipment, Aldan Products, Brockton; 1969-1988, electrical designer on nuclear and fossil power plants, Webster Engineering Company.

While working in Brockton went back to college at Northeastern University in Boston and got a bachelor of Business Administration in Engineering and Management in 1962. Now retired and a member, IEEE and GIRA.

Have two daughters and one son.

HADAD, AL, R-13

Was born and raised in Toronto, Canada. At age 17, I left high school and went to sea aboard a Panamanian tanker as an ordinary seaman. A highlight of my year aboard this vessel occurred in May, 1940 while unloading avgas in the port of Southampton, England.

I was able to view the evacuation of Dunkerque across the English Channel in its entirety. We expected the Germans to follow almost immediately thereafter. The situation was so critical, that we were cleared to leave port the first week of June without enough time to clean tanks nor have an armed escort to protect us from submarine attacks. We headed for the Panama Canal and were quite fortunate to make it without being detected.

I moved to Chicago after signing off the *Beaconlight* in February, 1941. I applied for advanced seamanship training at USMSTS Hoffman Island in December, 1941 but was not called for duty until February, 1942. At Hoffman, I was bored stiff. I was appointed a platoon leader and, instead of receiving training for my AB ticket, I was giving marching orders for enlistees who had never been to sea. . . . march to the barracks, march to the gunnery school, march back to the barracks, then march somewhere else. I was excited and pleased when I was accepted as a radio trainee at Gallups Island, arriving at India Wharf on April 1, 1942. I was surprised to see that the Liberty *USCGC Yeaton* was an exact duplicate of the *Kimball* that plied the waves between Hoffman Island and the Battery in New York. Graduation occurred November 11, 1942, Class R-13. My friend, Carl Thenemann and I rode the overnight to New York and didn't sleep a wink. We never saw one another again.

My first ship was a Laker. A small freighter, she was destined to become a hospital ship serving in the South Pacific. I spent four days aboard her, just long enough to travel from New York to Baltimore. I was told that six Army operators were taking over. I was flat broke. I had no winter civvies. But to the rescue came "Shanghai" Joe Rubin, the ACA agent in Baltimore. He lent me enough cash to carry me over until I could get an advance once I was assigned to another ship.

Until December, 1945, I sailed on a variety of vessels to North Africa and Iran. I made three trips to Iran, carrying supplies to the Soviets. This was the southern door to Russia, and although not as dangerous as the Murmansk Run, it was a lot longer. It took 50 days from New York to Bandar Shapur via Liberty, but less than 30 days via C-2. I can't forget how eager the Russians were to board the ship after docking to buy everything we had in the slop chest with greenbacks. And some of the Soviets were good-looking gals and even they were buying skivvies!

The ships I sailed on out of the East Coast Ports (New York, Baltimore and Charleston) were: *SS City of Dallas* (laker), *SS Pan Carolina* (tanker, WWI converted Hog Islander), *SS Caleb Strong* (Liberty), *SS John M. Harlan* (Liberty), *SS Mormacswan* (C-2) and *SS John Murray Forbes* (Liberty).

I signed off the *John Murray Forbes* on November 26, 1945 after transporting 600 troops from Naples to Newport News, VA. That was to be my last seagoing job until I returned for relief work aboard the *SS Texaco Massachusetts* in July, 1989. How great it was to have non-blacked out portholes and a lighted ship at night.

In January, 1946 I went to work for TWA in Toledo, OH as a senior ground radio operator. For the next 13 years, the company transferred me eight times while working as ground radio operator, chief ground radio operator, flight radio officer and ground radio technician. But the future looked dim by 1958, so I decided to seek advancement elsewhere.

In March, 1959 I was accepted by Philco Corporation, as field technical representative (TechRep). I was sent to Guam and assigned to the USAF AACS 1958th Command Squadron. In 1960 I was appointed as the TechRep team leader and assigned to Air Force and Navy contracts on Guam. I transferred to Fuchiu Air Station, Japan in 1962 and returned to the U.S. in August, 1963.

I could not force myself east of

California, so that is where we settled. Until my retirement, I worked in systems engineering for aerospace ground equipment that supported the Military space program. In 1969, I settled permanently in San Jose and worked on numerous military programs concentrating on the design and implementation of telephone switching systems. Retired for Ford Aerospace Corp., in February, 1985.

In July, 1989, I sailed aboard the *SS Texaco Massachusetts* as a temporary REO, having been coaxed to do so by fellow GIRA member Ray Maurstad. I had an exciting time operating the old CW system and the newer satellite modes. There was a lot to do since watch standing is not all that the REO was required to do. Now he is a combination Purser, Cashier, Secretary, and Communicator.

I am a member of ARRL, IEEE, AFCEA, VWOA, SOWP, and ARES and am serving my second term as Region 9 director, GIRA.

In conclusion, I must say that my term at Gallups set the course I sailed for my entire working life. I have no regrets at all.

JONES, K.C., R-13

Silent Key, May 8, 1991

Born June 2, 1921, K. C. Jones. Graduated from Jefferson City High School in 1939 and from the University of Missouri-Columbia, with a degree in business administration. In 1958, founded and became president of Cole County Oil Company; also a co-founder of Fill 'N' Wash in Jefferson City, MO.

Among other community service work, Jones had served as lieutenant governor, Missouri Optimists; charter member and past president, Christopher Casey Jefferson City Chapter of the Sons of the American Revolution; and member of Sons of the Confederate Veterans.

Jones graduated from the Gallups Island Radio School, serving as a radio officer in the Merchant Marines during World War II.

Jones married the former Pat Ward on June 11, 1949. They had three daughters and two grandchildren. He died on May 8, 1991, in Jefferson City.

(Editor's Note: Jones' widow, Pat, submitted his obituary to GIRA, from which this information was taken. She wrote, "He was so proud of your group and did so enjoy the reunion in Boston last summer (1990). He had planned to go to Minneapolis this year. Best wishes for another great time there. He and I shall be there in spirit."

PALMER, JOHN, R-13

I was enrolled in the University of Florida at Gainesville—I lived in Gainesville—and as a freshman I was having a (hard) time with my grades and with having enough money to handle it. So I went down to the Florida State Employment Agency to get a part-time job. They didn't have any for me, but on the counter they had a pamphlet about Gallups Island up there in Massachusetts, and it had a picture of a ship on it. I asked them for that pamphlet because of the picture of a ship on it. They said: "Well, do you want to go to the sea? That's a sea job." I said, "No, I don't want to do that."

But they gave it to me and I read it. It appealed to me because they would furnish everything I would need—books, clothes, medical care, my food, my lodging—and they would provide me with training, so I took advantage of that.

At that time, you did not have to be a ham to apply.

PARCE, JAMES W., R-13

I've been interested in radio since I was knee-high to a grasshopper. I figured that was my thing. I have always liked radio. I was an old SWL guy, you might say; I used to listen to all the broadcast stations and where they were from. The radio bug bit me when I was about 13 years old and just stayed with me.

Somehow, I got word about the Maritime Service, so I signed up in December of '41 and they sent me to Hoffman Island for basic training. I got to Hoffman in January of '42 with Al Hadad—and it was colder than the inside of an icebox.

Before we completed our basics at Hoffman Island, I said, "Hey, fellas, when I signed up I asked about getting into radio school. Now how do I get in?" So they put the necessary papers through, and I wound up on Gallups the first week in April of '42.

I didn't know anything about Morse code when war broke out, so I came in green as a grape. The first couple of weeks I was very slow. But then things started clicking. My speed picked up and my knowledge increased. Like I was telling my friends up in Vermont, when I graduated I was doing 25 words per minute—no sweat. I have no problem at all with the code.

The first ship I got was in November of '42. The first cruise was to Casablanca and back. Then it was all over the bloody place. In spring of '42 we went back to Casablanca, took a side trip down the coast 100 miles to Southea and then home.

In the summer of '42 I wound up in England. In the fall we wound up in the English Channel at Hull, England. In '44 I spent all the time in the Mediterranean, Iran, Naples, Southern France—all over the bloody place. In '45 it was back to England, then to Murmansk, Russia after Germany quit.

On the final cruise, I went to Naples to bring the boys home. We brought back 450 or so on the Liberty and took them into Boston Harbor.

RUNMARK, DON, R-13

I left home on Easter Sunday, April 4, 1942 . . . left my mother's skirts for the first time. I suppose I went through Chicago on a train and on to Boston. Had to stay overnight to get the ship, and got the Coast Guard boat to the Island the next morning, about April 7.

I took my seamanship training at Gallups Island; some of the other fellows took their seamanship elsewhere, I understand.

About the second week in November I went down to New York and signed up for a ship. A couple of weeks later I was assigned an old WWI freighter. On our first trip we went to Casablanca and arrived there December 30. From there we went to Gibraltar, took a load of sulfate to London, went around Scotland, down to London, and back to the States over the stormy North Atlantic.

Somewhere around March, 1943, lost the convoy right out of Liverpool and went all the way across alone. I have never experienced a worse week than I did that particular week. It was a painful one. From there I went back to San Francisco and shipped out of there the rest of the war.

I spent most of my time shuttling back and forth between Australia, New Guinea, and the Philippines and spent a whole year at anchor at Lady Bay, in the Philippines. Took a load of ammunition back through the Panama Canal, back to New York, New Jersey at the end of the war.

Went back to Minneapolis, my hometown, and ended up working with the Bell Telephone System in Minnesota, 'til I retired a few years ago. My wife, Dolores, and I have two children.

Platoon R-14

HALVORSON, ROBERT L., R-14

"Born Again"

For our particular bunch of seventy year olds, everything started with a thing called Gallups Island. You were, after this great event in your life, like, born again. The only important item before Gallups is where you came from and when you arrived to be born again. Once there you were born again to the code, Morse Keys, Hartley and Colpitts, chow, bunks, radio tubes, direction finders, antennas, and eventually ships and radio rooms, mates and engineers, and THE SKIPPER, YOUR BOSS AND MASTER OF THE SHIP! If that isn't being born again, I don't know what is.

You were born again to war and the risk of death by bombers, subs, and the caprice of weather and sea. That is a lot of reborning to get used to in a few months—like when you were first born. Then, you looked around and said to yourself, "Boy, this is an interesting world; I hope I survive!"

Born in Duluth, MN in 1922. Reborn on Gallups Island from April, 1942 to December, 1942 in class R-14. First ship was the *SS Derblay*, an Alaska Steamship Co. vessel, carrying dry cargo to Seattle in January, 1943. Made two trips to Alaska on the *Derblay*; she was built in 1918, and had a wooden boat deck. The radio shack was a stand alone 10' x 20' deck appendage that included my bunk, wash stand, and water bottle. I peed in the sink at times during rough weather.

The transmitter had a single big tube about a foot in diameter, and the marble slab front was held on the iron frame by brass nuts. When I pushed the key down to test the transmitter, all the lights on the ship dimmed to half brilliance. Third Assistant Engineer didn't like that a bit!

Ship survived the war, sold to Chinese 1946, re-named *YuChang*, later *Eastern Mariner*, scrapped in 1959. Oh yes; the receiver was the National SW-3 with plug in coils and a B battery, remember?

Other ships I sailed on: Liberties, *Felix Hathaway, Simon Bolivar, Edmund G. Ross, Samuel K. Barlow*, and (then a good name, now not) the *Yugoslavia Victory. Victory* took aboard 7,000 tons of ammo at Long Beach, CA for invasion of Palau Islands, August, 1944. Water was too deep to anchor; steamed in tight circle night and day for three weeks with ducks taking 75's and 105's to dedicated guns on the beach. First Marine Division suffered 6,000 casualties.

Back ashore September, 1946: University of Minnesota, Bachelor in Electronic Engineering June, 1950; Westinghouse Electronics Corp. 1950-1955; Collins Radio 1955-1960; other manufacturers to 1968. Formed Halvorson Assoc. 1968, which survives to date. Registered Professional Engineer specializing in audio tape recording.

Nixon Impeachment Inquiry: April, 1974 to August, 1974; Technical Consultant to U.S. House of Representatives Judiciary Committee assisting with the White House tapes.

Watergate Cover up Trial: August, 1974 to December, 1974; Technical Consultant to Special Prosecutor Leon Jaworski on White House tapes, and played evidence tapes in Judge Sirica's courtroom at second Watergate trial November, 1974 and December, 1974.

Post Watergate: Things have been kind of dull since then, but I continue in civil and criminal litigation as Technical Consultant in tape recordings and expert witness in court trials.

Hobbies: N3CZF (ex W3DOF & K0DZX), old Volvo's, Lyman Islander power boat, guns, camera, 2500 volume library, tape recording classic great books—Aristotle to Leo Tolstoy, 100 titles/800 hours.

Born Again? You bet; every day!

MILLER, WALTER T., R-14

After graduating from USMS Gallups Island Radio School (Class R-14: April to December 1942), I made one foreign trip on the *SS Robin Gray* from New York City to Casablanca, Morocco to support the resupplying of the American and British troops with ammunitions and tanks in Africa. I also made two coastal and four foreign trips on the *MV Blenheim* from New York City to London and Liverpool to support the buildup of supplies for the Invasion of Europe. The *Blenheim*, a ship built in Hamburg, Germany, was captured by the U.S. Navy on a run from South America to Germany early in WWII. The ship was modified for use by the U.S. Maritime Service and assigned to the Waterman Steamship company. After leaving the *Blenheim*, the vessel was damaged by a V-2 rocket bomb while unloading cargo in Antwerp, Belgium. I knew when to get off this ship . . . my apologies to the replacement radio officer. While five crew members were injured, none was killed by the incoming bomb. The *Blenheim*, which was badly damaged, was repaired and put back in service.

I left the Maritime Service to join the U.S. Navy after taking the Captain Eddy Test. After completing Boot Camp at Great Lakes, I attended three radio schools (Hugh Manley Primary Radio School in Chicago, IL; Del Monte Secondary Radio School in Del Monte, CA; and the Aviation Radio/Radar/Navigation School in Corpus Christi, TX) and a crash course in instructor teaching techniques. After the end of the war, I was an instructor at the U.S. Navy Aviation Radio/Radar School and at the U.S. Navy Aviation Fundamental School in Jacksonville, FL, until my discharge six months later.

My first serious job after the war was as a radio/electric mechanic for United Air Lines at LaGuardia Air Field, Long Island servicing DC-3, DC-4, and DC-6 aircraft; and shop overhaul of aircraft navigation and communication equipment. I met my wife, Vivian Provost, who was working as a Ground Radio/Teletype Operator for U.A.L.

After one and one-half years in New York City, we went to Cincinnati, OH where I became the first television repairman and antenna installer for a large department store. Before leaving the store (approximately five years) I became the shop leader of approximately 25 repairmen and TV installers and a customer relations representative.

I left the store and became a Nuclear Instrumentation and Field Liaison Specialist at the General Electric Company in Cincinnati and Idaho Falls, ID, working on the instrumentation and controls for a Nuclear Aircraft engine. Prior to the cancellation of the program by Congress to shift more money to the Missile field, GE demonstrated the utilization of nuclear energy to operate two GE J-47 engines for a period of 100 hours without shutdown at the Idaho Test Site. During the nine years at this department, I took many computer courses relating to nuclear instrumentation and electronics and evening courses (at least 2 nights a week) at the University of Cincinnati studying Electrical Engineering. I even found time to teach radio and television classes at the Ohio Mechanics Institute, a vocational/trade school, one to two classes a week depending upon my other studies/work loads.

When this operation shut down, I went to work as an Instrumentation/Electronic Specialist working on the first Nimbus Weather Satellite, Ground Station and Launch Equipment Vans at GE in King of Prussia, PA. After the first launch and extensive modifications to the Launch Equipment/Tracking Vans for the second Nimbus Spacecraft, I was transferred to the GE reentry Systems Department in Philadelphia. After fifteen years of evening classes at U.C. and Villanova University in Villanova, PA, I obtained my degree in Electrical Engineering.

Platoon R-14

Through meetings and computer network, we worked closely with other co-contractors and government agencies involved in the development and testing of Missiles and Supporting Ground Test Equipment for the numerous Minuteman and Peacekeeper Reentry System's Programs carried out at the Air Force Eastern Test Range in Florida and the Air Force Western Test Range in California.

After an early retirement from GE in 1983, I worked as a job-shop Electrical Engineer establishing the qualification of electrical parts, equipment and controls for use at Nuclear Power plants throughout the country.

My wife, Viv, worked at a local hospital as a receptionist to help put our two children through college. Our daughter (Kathy Miller Locker) is presently a grade school counselor in the Portland, ME area, while our son (Dr. Walter T. Miller III) is with the University of New Hampshire as an EE Professor and Robot/Artificial Intelligence Researcher. We have two outstanding grandchildren (Brendan Locker, 18, and Erin Locker, 10).

Some exciting moments during WWII that I remember include the loss of steering on the *SS Robin Gray* when a large bolt fell into the steering gears causing us to leave the convoy; air raids while docked in English towns; a near miss by bombs from a German aircraft while traveling up the North Sea coast; the maneuvering of many ships in a convoy approaching Newfoundland in a dense fog; riding out a hurricane in Halifax while anchored with many other ships; a large bomb bouncing around in the hold of our ship on a return journey from Newfoundland; submarine alerts/activities off of Bermuda, Halifax and Newfoundland; a breakdown (blown diesel-engine gasket causing an evacuation of the engine room) of the ship near a New York harbor mine field; loss of electrical power due to a heavy cable separation (cable was determined to be partially cut by German Merchant Seamen before they surrendered the ship to the U.S. Navy); and a minor collision with another ship in the Boston harbor causing a leak which plagued us during the trip abroad and return to the States but which may have saved our ship from enemy action out of Halifax. We had to remain in Halifax for hull inspection by a diver while the convoy sailed into a wolf pack engagement.

POLLITT, BOB, R-14

Signed on the collier *Suffolk* in December, 1942 after graduating from R-14. She was built in 1911, did a good eight knots with all laundry aloft, truly a real bucket. The radio room was added on in the late '20s and was about 10'x8', everything included: equipment, bunk, washbasin and clothes locker, etc. Water on the deck when we had heavy seas from the port side, which made for minute shocks from the ear phones.

First night out in Buzzards Bay on the way to New York, the alarm went off and we had a fire in the after section; coal dust around the stack had ignited. Warmed up the transmitter, however sent nothing as the crew soon got things under control. In New York harbor we had lifeboat drill and inspection. The inspector suggested we avoid using the starboard boat because of our perpetual list to port, making the launching of the starboard job about impossible. This condition (the list) was always blamed on the radio shack being built on the port side of the boat deck.

From New York went to Norfolk for a load of coal then to Cuba and on to Trinidad in an eight-knot convoy. Every four hours we would straggle from the convoy as the fireman took out the clinkers from two of our eight fire boxes. We were kicked out of convoy once for making too much smoke.

At Port of Spain a deck hand got a bit eager and chipped into the 2nd mates quarters, which indicates what shape the ship was in. The bulkheads below decks were completely shot full of holes. After discharging our coal we proceeded to British Guiana and shuttled from there to Trinidad with bauxite for about five months. Off the Orinoco River we would often run aground in the mud and just wobble about to get free. Arrived in the USA, Mobile, AL about June 1, '43.

The *Suffolk* went down with all hands on December 11, 1943 (my birthday) floundering in a northeast gale approximately 20 miles south of Montauk, Long Island. A picture of her is in Captain Arthur Moore's book "A Careless Word—a Needless Sinking." Only two bodies were recovered, from a life raft, one Navy and one Merchant.

My remaining time was spent on two Liberties, the *Hannibal Hamlin*, and the *James K. Paulding*. Sailing the North Atlantic to Britain, eight trips to Normandy with troops, and then to Cherbourg, France when the Jerry's had vacated. After the Germans quit we went on to the Pacific with the load of bombs, which nobody seemed to want. Went to the Admiraltys, Solomons, New Guinea, and about ten ports in the Philippines. Broke down in mid-ocean on the way home and limped into Hawaii at two knots. Made Frisco in May 1946.

Since then have spent my time in the lace business as a draftsman, and went into banking at age 47 retiring at age 60, mainly because of my wife's ill health. Except for a little house wiring I did not use my Gallups training.

I live in my grandfather's house and it overlooks Block Island Sound; until 1983 it had been used as a summer cottage. It's surprising how many Liberties went sailing by decades after the war—not bad for wartime emergency ships!!

For hobbies: I garden, mostly vegetables; fishing whenever I have the chance; travel—especially to England, where I have a girlfriend. My wife died in 1988. Spend a lot of time fixing up the old (100 years plus) house, a great labour of love. Until two years ago I was on the volunteer rescue squad and held an EMT-A license from the state.

SCHIFF, RICHARD, R-14

At the end of 1941 when the war broke out I decided I wanted to get involved with electronics, and as long as I was going into the service anyway, I volunteered to go to Gallups Island, Boston, and enlist in the radio school. The only stipulation I told them was that because I have a bad heart, I can't stand cold weather, and I could not be shipped any place where there was cold weather.

When I got out of Gallups, they signed me to a ship out of New York and written right on the ship was "U.S.S.R." I thought for sure that ship was never going to north Russia, because I had told them that I couldn't stand cold weather. Well, we got out to sea and opened our convoy orders, and sure enough it was Murmansk, Russia. It was a horrible experience going there, because we never made it.

I'm hoping someday I'll be able to get together with some of my former classmates. Unfortunately, every time I pick up a Gallups Island new report, I read something like "Silent Key," which means another one of us has died. Let's hope we can get together before all this is all over.

SLATER, LLOYD, R-14

My major accomplishment as platoon leader of R-14 (March, 1942-December, 1942) was to officiate at a very raunchy graduation party held on Long Pier Wharf. Dick Schiff took some memorable pictures which have been revealed to drinking buddies but safely hidden from the moral majority.

My debutante voyage UK-wards on the Liberty *Eleazor Wheelock* saw me lose the convoy on a confused course change signal from the Commodore while under sub attack. So we sailed alone for a few days like a skittish hen and thus dumbly avoided action. Thereafter I was chief operator on seven other ships, with my final, a post-conflict voyage on elegant C-4, carrying the first commercial cargo to China in early '46.

Highlights: Attended an Oxford University short course after mid-ocean collision required a new bow for *SS Eugene Field*. Liverpudlians did a record job so I was pulled out of class by a gorgeous Admiralty wren and rushed back to catch my ship under way in mid-stream. Captain Sullivan thereafter chided me as "Slather, Balliol '44." Our convoy made up at Loch Ewe and sailed for Molotovsk in USSR, serving as lure to pull a pocket battleship out of a Norwegian fjord and observing a great British naval victory astern. Fell in love with a petite Russian lady but unable to bring her home. U.S. return convoy (in March '44) lost a dozen ships to newly developed sound-track torpedoes. We shed a steel protection net with one and lived for a week in rubber suits.

Most memorable event in England: Bought the only available new sink in Liverpool (via a search on bicycle) to replace one destroyed by our cloddishly drunk ship's purser at a party I arranged at lovely Welsh girl's home (parents away!). Parents returned to find outrageous monster sink in small loo. They adopted me and named their new artistic plumbing "Sparky."

My C-2 ship, *Santa Elisa*, went from Atlantic convoys to sail alone at 20 knots in Pacific war in late '44. Followed invasions towards Japan while off loading cargo amid Kamikaze attacks. Met my lady Marine wife-to-be when she came aboard in San Francisco to welcome returning Marine Officers. Married in '45, we launched son, Peter, and I went East to join *SS Benjamin Lundy* headed for France to load materials for conquering Japan. In August, along with countless similarly laden ships facing the Panama Canal, the bomb dropped and war ended. Despite this we all squeezed through to then wander the Pacific to return months later to unload.

An exciting moment: In April '46 the *Marine Leopard* ran at 20 knots onto sand bar in North China Sea near Tientsin. The stevedores on a barge sent out to unload us were a raffish lot and I was reluctantly issued a gun to stop pilfering and reclaim stolen cargo. I was sent down into the dark, dank and opium smelling depths of the barge crew quarters and must confess I never brandished my weapon, nor spoke in anger to the surrounding villains. Instead, I made my tribute to the new China from a tumbler of stolen bourbon from our highly irrelevant "commercial" cargo. A few weeks later, on the first major U.S. commercial cargo to Shanghai, I saw each morning small children dead of starvation in doorways while Kuomintang functionaries eased our bric-a-brac cargo into the black market.

What have I done since? Given my studies background (biochemistry and food science at Cornell '40) and wartime radio, Honeywell took me on as an engineer to develop instrumentation for the food industry. A few years later I switched to journalism to become Midwest Editor for McGraw-Hill's "Food Engineering," later Managing Editor of its new "Control Engineering."

In 1957 I launched a foundation in New York to help develop the academic field for the fast emerging technology of automation and computing. The foundation accomplished its mission in seven years and I went on to be Associate Director, Research for Case Institute of Technology. I joined the ex-President of Case as his assistant when he became President of Associated Universities, the Washington based consortium which manages National Scientific Labs (e.g. Brookhaven, National Radio Astronomy).

An assignment to help establish a National Tropical Marine Lab took me to Puerto Rico where I was asked to direct a new Institute in Boulder, CO. When the five year program was completed I ran an International Academy, based at University of Colorado, turning wise elders onto world problems.

Retired in 1985 and down to seaside Florida. I must say that my war experience (radio gave time for much reading and visits in 23 nations) deeply shaped my personal outlook and future work. One empathized with people under wartime stress and glimpsed inequities from political mismanagement and greed. I'm certain that many of my fellow Spark-Gappers gained a similar humanistic outlook which shaped their later lives.

Platoon R-15

BEAULIEU, WILFRED J., R-15

Going to Gallups Island after signing up in the Maritime Service was one of the best decisions I ever made. When you came from a small town in Minnesota there were not a lot of opportunities for someone with a high school education.

After graduation from Gallups Island, February 8, 1943, in platoon R-15, and with a brand new 2nd class telegraph license from the FCC, I went back to Minnesota for a two week leave. I traveled to New York and checked in at the ACA Union hall. There was only one problem—my luggage with all my clothes did not arrive on the train. They never did find it.

I was finally reimbursed for the loss, but by that time no ships were available, so I went back to Boston and checked in at the ACA union hall. I signed on the only ship available at that time, as a relief operator on a coal collier for a month. I had to get something fast as I was running out of money. I thought it would be an easy run. What I didn't realize was that there were a lot of subs along the East coast of the U.S.

After the coal collier, I signed on the *SS Freeman* for our first of many trips to British Guiana carrying bauxite ore to ports in the U.S.

I had a relief operator take a couple trips for me. While on shore I took the test for the Army Air Corps, as they called it in those days. I passed everything except part of the eye test. I went back to the ACA and signed on a tanker, the SS *Richard Johnson*. We carried oil from Caracas and Aruba to ports on the East coast of the U.S.

On March 1, 1945 I signed on the *Freeman* again. When I was on the ship in 1943, I became friends with the 3rd mate. In 1945 he became Captain of the ship. He got in touch with me and asked if I wanted to come back as R.O. I stayed on the *Freeman* until June 1946.

I got a job ashore as a supervisor of a test department in a small electronics firm in Somerville, MA. I got married in 1946 to a girl I met while I was at Gallups Island. I started going to Northeastern University nights in 1948. I then got a job as a foreman of a test section at Raytheon Co. After six years of night school I got a degree in electronics.

In 1954 I took a job as a junior engineer. We had a daughter in 1957, and I was promoted to a QC engineer. I worked on various missile projects including the Patriot missile. I did a lot of traveling for Raytheon on different projects. I retired from Raytheon in 1986. I now spend my time skiing in the winter, building and flying radio control model airplanes, and traveling for pleasure the rest of the year.

BOOREN, LYNN C., R-15

My first ship was the Liberty *Benjamin Harrison*, which I boarded in Norfolk, VA and sailed in convoy February, 16, 1943, supposedly headed to Murmansk, Russia. I received my salt water baptism about 1,500 miles out of Norfolk when the *Harrison* was torpedoed and our life boat capsized while abandoning ship. We lost the emergency radio transmitter and the bilge pump, so we hung on to the half-submerged life boat for about seven hours. Around midnight we were picked up by a Norwegian freighter and taken to Casablanca, Algeria where we caught a troopship back to New York.

I boarded my second Liberty ship, the *Crawford W. Long*, in Savannah, GA and sailed to Jackson, FL to take on cargo. We sailed north, stopping at Charleston, SC, to join a convoy heading east. We docked in Liverpool, U.K. From there we went through the Straits of Gibraltar to discharge cargo in Oran, Algeria. After returning to Philadelphia in August, 1943, we again loaded up and caught another convoy to the U.K. We discharged our cargo at Cardiff, Wales and returned to Norfolk on November 9, 1943.

My next ship was the Liberty *Richard Yates*, which I signed on in December 1943 at San Francisco. We sailed alone to Pearl Harbor, loaded up, and sailed on to the Marshall and Gilbert Islands, returning to San Francisco on March 27, 1944.

I stayed with the *Yates* and we left April 5, 1944 for Pearl Harbor, where we took on cargo for the South Pacific. We sailed to the New Hebrides, New Caledonia, and on to Guadalcanal. From there we went to New Zealand, visiting Auckland and Wellington on the North island and Christchurch and Little on the South island. After spending six weeks around New Zealand, we loaded up with potatoes which we delivered to Guadalcanal, Bougainville and other ports "up the slot." Best cargo I ever rode on as most of the time we carried bombs during my trips.

Our ship went on to Biak and Hollandia, New Guinea, where we joined the first convoy after the Philippine invasion to sail to the Leyte Gulf area without air cover. On the trip up we were attacked by Jap torpedo planes. The crew's "artist" painted the planes on our stack and we felt we were sailing on a warship of the fleet.

We laid at anchor in Leyte Gulf for 30 days and then moved to Samar where we became a floating bomb warehouse for the Liberator bombers that were bombing Formosa. During this time our food went overboard—traded for "jungle juice"—with the consent of our alcoholic captain. Next thing we knew we had a mutiny on our hands.

While I called ashore with the blinker light, our mates kept the crew off the ladders to the top deck with their guns. The Coast Guard took over the ship, removed the instigators in handcuffs and leg-irons and sailed us back to San Francisco, where I was discharged April 20, 1945.

After a welcomed leave, I signed on the Liberty *Morgan Robertson*, and we sailed to Pearl Harbor, Tinnian and to Guam (where we took on more explosives) about the time the Japanese surrendered. However, SNAFU sent us up to Okinawa through a tremendous typhoon. The only consolation in that trip was that I met up with two of my brothers on the island. We returned to Guam where we unloaded part of our Okinawa cargo and then sailed to Guadalcanal to transport excess war material back to the States. We docked in Long Beach, CA where I was discharged on February 8, 1946.

I returned home to Minnesota, and after I cleared up my new 1A draft classification, I went on to Optometry College in Memphis, TN. After graduation in March 1950, my wife and three children traveled with me to Craig, CO where we've spent the past 43 years. I retired from my optometry practice in September, 1987 and now we spend summers in our homes in Craig and Fort Collins, CO and most the winters in our condo in Fountain Hill, AZ.

BROWN, MARWIN, R-15

I sailed out of San Francisco after leaving Gallups Island. I served on all Liberty ships. The first three trips to the Pacific were on the *Robert G. Harper* KEVO, left her on December 2, 1943. My second ship (also to the Pacific) was the *Uriah M. Rose* KVTK, on shuttle runs around New Guinea and islands from January 20, 1944 until January 29, 1945. Third ship was the *Walter Wyman* KWXO. Left her and the sea on December 5, 1946. All my trips were to the South Pacific, and I saw all ports and islands from Brisbane, Australia to Osaka, Japan, except Guam and the Philippines.

I remember many incidents. Once we came within 30 feet of ramming a surfaced sub in the fog off the California coast. Don't know to this day if it was ours or Japan's. Another time, we had a Navy gun crew member jump overboard in calm seas off Oahu. We were able to rescue him.

Townsville, Australia was an unlucky port for me as our ship, loaded with munitions, caught fire at the dock and so with a skeleton crew we had to leave port to fight the fire in case it blew. We got the fire out in time. On another trip to Townsville, a storm tore loose the mooring lines to the dock. Because "Sparks" had his hand in his pockets watching the proceeding (and thus had the driest hands and matches), I got him to light the fuse on the sperry gun to shoot a line to shore to secure a tow line and pull us back to dock.

Had the usual submarine scares and bombings, but our ship was never directly hit. However, I was in two memorable storms while at sea. The first was a storm in Okinawa. We had one earphone on 500 KC and the other on the VHF voice circuit. That storm had to have set a record for SOS and Mayday signals in a 24 hour period from ships in distress. I will never forget all the calls, especially on the voice circuit, as no one could help anyone else during the storm. The second was off Alaska, in which our ship lost steerage way and got broadside to the wind and rolled deeper than any ship is designed to take.

I returned to Wessington, SD after the war and bought and operated a Red Owl Agency grocery store for 11 years. Married a local girl, Adela Zvonek. We have two children and four grandchildren. Sold the store about 1959 and went into banking with First Bank System, Inc. of Minneapolis. Worked in Gettysburg, Huron and Platte, SD; managed banks in Corsica, SD; Sauk Centre, MN; Windom, MN. Then on to bank headquarters in Minneapolis where I supervised a group of small banks. Back to South Dakota, and Sioux Falls, where I retired from the system at age 65 on January 1, 1987.

We presently spend six months in Scottsdale, AZ where we have a double wide mobile home, and six months in Sioux Falls each year.

73's to all.

COATS, DAVID W., R-15

Early in 1942, I applied to Maritime Service for Gallups Island. I could not get into any of the Armed Forces due to an atrophied left leg from polio when I was three years old.

I was living in Tucson, AZ and they said they would notify me when to come to San Diego for my physical. I didn't wait to be notified but went directly to San Diego for the physical. They gave me the physical but said I should have waited as there were 13 ahead of me.

I went home to Tucson and three days later they called me; said they had an opening and the number one applicant had disappeared. I would have to go immediately. I grabbed the opportunity and went back to San Diego where they put me on the train to Boston and changed my life forever.

After leaving Gallups Island, I returned to the West coast to ship out. I shipped out from San Pedro. My first ship was the Liberty, *Edward W. Scripps*, operated by Alcoa. Our destination was Calcutta, India with a cargo of Jeeps and thousand pound bombs. Forty days to Tasmania as we steered far south to avoid Japanese submarines.

On board was an AB seaman who complained about the food. (It seemed very good to me.) Finally he announced that he was going down to beat the hell out of the Steward.

Since I was in charge of the Medicine Chest I was soon called and found the AB bloody and unconscious lying on the deck of the Steward's cabin. The Captain arrived and said, "He was looking for trouble and he found it, wash his wounds with Boric Acid and throw him in his bunk."

The Steward told me that when the AB threatened him he tried to talk him out of

Platoon R-15

it to no avail. The Steward then grabbed the water carafe and pounded the AB on the head, all the way to the deck. We treated him with Boric Acid and put him in his bunk. We did so and he recovered but his head was swollen grotesquely for some time. He did much mumbling that he would beat up on the Steward, but it was very noticeable that he stayed away from the Steward.

We went from Tasmania to Calcutta, Madras, Ceylon, and Perth, Australia to Panama and New York.

I then shipped out of New York on the *Robin Locksley*, a C-2 (Steam Turbine) operated by the Seas Shipping Company. The ship's Captain, McKay, was hostile toward Radio Operators because the last one had fouled up a message and brought the *Locksley* into the wrong port.

Later we were making up a convoy in Ireland. To attend the Convoy Conference, the Captain and I had to walk along a country road. Suddenly the Captain had a seizure and fell in the ditch. I helped him up and after a rest we continued and he seemed OK. He was too proud to ask me not to tell anyone but I knew he wouldn't want it to be known, so I never told anyone. When he realized that, he became very friendly toward me. When I thought of getting off for one trip, he asked me to stay and I did. A few months later, Captain McKay died from a heart attack.

During the Battle of the Bulge the Army flag flew in many dog teams from Alaska. They were used as rescue teams. We brought them back. The dogs were tied all over the deck. At night they would howl causing the other ships in the convoy to wonder what was going on . The depth charges would really set off the howling.

I spent a great deal of time on North Atlantic Convoy trips—although some times we were cut loose because of our fast speed. The U-boats were active and we lost a few ships and frequently shook from depth charges.

Once while unloading in Antwerp, I hitch-hiked up into Holland and located some relatives (by marriage). Had a nice visit and to this day we are close.

After the war we went back to the *Robin Locksley's* peace time runs; South Africa, Capetown, Port Elizabeth, East London, Durban and Laurence Marquees.

Later under pressure from my wife, I quit the sea and went to South America as a Flight Radio Operator for Panagra Airlines. After a few years that job was eliminated and I returned to the US and worked in broadcast at WQXR New York, United Nations New York and KCBS, San Francisco.

Next I flew again as Flight Radio Operator for Seaboard World Airlines. Seaboard was non-scheduled cargo airline flying all over the world and the Korean War Airlift. After a few years that job was also eliminated.

I needed a job that wouldn't be eliminated and I went to work for the Burroughs Corporation in computers. I stayed with Burroughs for 25 years in various management positions. I retired at age 59, but went back to work as Regional Field Engineering Mgr. for Computer Consoles, Inc. After three years I moved over to Sun Electric Corporation as Regional Field Engineering Mgr. Three years later, at age 65, I retired again. This time for good.

Jack Murray and I have kept close all these years. He also was employed in Panagra as a Flight Radio Operator in South America and various management positions in Burroughs Corp.

I am presently single. I raised two sons. One is a structural Engineering Ph.D. at the Livermore Radiation Laboratory and the other owns and operated an advertising agency in Atlanta, GA. I have four grandchildren.

Since retiring I have done some leisurely traveling. Spent some time in Thailand and Europe and plan to do more. I am really enjoying these reunions and the opportunity to see the old gang from R-15 once again. I hope we have many more. I was in touch with Eddie Skinta for many years until he passed away a few years ago. (Remember Eddie, who used to play the piano in the bars.)

I became aware of GIRA while I was on board the *Jeremiah O'Brien* when GIRA members were there. I joined right away and haven't missed a reunion yet.

COOL, CLAYTON, JR. , R-15

Silent Key

My brother, Clayton Cool, Jr., W8WYP, died 30 years ago on June 18, 1961. I wish that Clayton, or Ike as we called him, could be alive to tell his own story. He had such a brilliant, inquisitive mind. Who knows where it would have led him. As his younger brother (by six years), I will relay a brief history of what I know about his life after he graduated from Gallups Island.

For the duration of the War years he served as a wireless operator on a number of ships. He came home for a short furlough in 1943 to marry his childhood sweetheart. Unfortunately, they were divorced less than two years later, while he was still at sea.

Returning home for several years, he operated his own Le Roy Radio shop. He then returned to the ships for a year or two. He once told me that he had sailed every sea and ocean and visited every continent.

Later, on the east coast in New York, New Jersey and Pennsylvania, he worked for the government in top secret atomic submarine systems While in this job, he contracted a swift cancerous melanoma and died shortly after his 38th birthday. I have often wondered if he could have been just one more victim of our country's inadequate precautions in their fledgling atomic industry.

Ike had no children and no surviving widow.

He was an avid stamp and coin collector, which he stocked up on at every port he visited. I still have some of his coins, but his fabulous stamp collection, sheet after sheet of new stamps purchased during his travels, was ferreted away from his widow by a so-called friend.

I know that, if living, Clayton would be attending your convention. I would be interested in receiving an account of your gathering to put in his bibliography. (*Editors note: This was taken from a letter from Cool's brother, Mike Cool. His address is 5557 N. 210th Avenue, Leroy, Michigan 49655)*

CUSHMAN, GEORGE W., R-15

Forty-nine years ago, April 1942, a group of future radio operators left on a train from New York to Boston. In this group were Joe Considine, Edgar Burns, Al Landon, John De Meo, me and some others whom I have forgotten. On arrival in Boston we reported to the Seamans Club and that evening we visited the famous Scollay Square. Next morning, out to Gallups Island, from Commercial wharf on Atlantic Avenue aboard the Coast Guard cutter *Yeaton*.

My time at the school was a turning point in my life. I had never been away from home, and after a short period of nervous anxiety as to my capability to learn code, I really enjoyed my days on the Island. I have fond memories of touch football with Leland Schultz, Hoyt Engen, Vern Hegge, Gerald Berger, Johnnie Davis and others; and of walking back in the dark from the canteen at the end of the island. Other memories: detail week, dock watch, fire watch, cold winter nights in the barracks, and the early morning bugle calls. Boat journeys through the submarine nets in the *Roderick* and on bad foggy days when the nets were closed, the walk over Long Island. Let's not forget the seamans classes in the long boats, code classes and study groups to help pass the exams.

After graduation in January, I married a local girl from Malden, MA whom I had met in October on a stroll through the Boston Gardens with classmates Hoyt Engen and Vern Hegge. For me, the city boy from New York, falling in love with a beautiful New England country girl was the crowning point of my time at Gallups Island. Iris and I just celebrated our 48th anniversary last January 31st.

In March, 1943, in New York, R-15 classmate Frank Perry and I were assigned to the tanker *SS Northern Sun*. This was a three-operator ship. The *Northern Sun* left New York for Galveston, TX to load up with oil and then was to go on to Corpus Christi, TX to complete loading. We ran alone along the coast from New York to Jacksonville, FL where we formed a six-ship convoy to make the trip to Galveston. I heard more submarine sightings and submarine attacks during this time than at anytime during the War. We left Texas for New York to join a convoy for a voyage to England. The ship was old, and we had engine troubles and would have to stop to make repairs. We straggled once on the way over but fortunately, we were able to get back to the convoy before dark.

Dean Cook, an R-15 classmate, was lost on the *MS Sunoil*, another old Sun Oil tanker, in April, 1943, when his ship had engine problems and straggled from the convoy and was unable to rejoin it. The ship was never seen or heard from again and went down with all hands.

We arrived in Morcombe, England. After discharging, we went to Gurock in Scotland and then returned to New York in June, 1943.

My next trip was on an old Sprague Co. coal boat, the *SS Coastwise*, which ran along the coast from Searsport, ME to Newport News, VA. Iris was now pregnant with our first child and was due in March, 1944. She was having a very hard time when her doctor asked me if I could get shore duty for time. We were living in Malden, MA and I thought of Gallups Island. It just so happened they were looking for an instructor in code and ship's procedure. So from October, 1943 to June, 1944 I worked with a great group of instructors and students—also a very good instructor and classmate from R-15, Joe Brown. I left Gallups in July, 1944 and waited in Houston for assignment.

My next ship, and my first Liberty, was the *SS Howard T. Ricketts*, an ammunition ship. We sailed from Boston to Halifax, Nova Scotia to form a large convoy coming from New York to Baltimore. My junior operator, a young American-born Japanese, was on his first trip and was seasick most of the voyage to Halifax. Our destination was Hull, England, where ships were unloaded by priority of need for the invasion forces in France. We were completely discharged in three days. We sailed from Hull to Loch Ewe, Scotland. At our anchorage, a heavy storm cause our ship to lose its holding, and consequently before we could get up steam, we collided with a Yugoslavian ship. She stripped off our starboard lifeboat, gun turret and raft. We went into the Clyde shipyard for repairs and then made the trip home in September, 1944.

My next ship was the Liberty *SS James McHenry*, a two-operator ship. I was fortunate to have as my partner Hoyt Engen, an R-15 classmate. Our cargo was general military equipment, including Sherman tanks. On the 6th of October, 1944 we left New York for Loch Oban, Scotland. We were delayed a week, as British soldiers had to clear German troops out of Walchern Island in Holland. We would be the first convoy into Antwerp, Belgium. The ship in front of us was struck by a mine and grounded.

All the time we were unloading (about a week), we were under attack by V1 Buzz Bombs and V2 rocket bombs. We were very lucky not to receive any damage during that time. When we left Antwerp and entered the English Channel the ship behind us was attacked by German E boats and caught fire. Sailing the North Atlantic in December on the *McHenry* with slack ballast, made for the stormiest sea voyage I have ever experienced. The trip home took 21 days rather then the normal 14 days.

From February to August, 1945 I was the single radio operator on a small bulk carrier, a former Great Lakes ship called the *SS Samuel Mitchell*, running along the Atlantic coast. On one trip of interest we carried gypsum from Virginia up to the basin the the Bay of Fundy in Nova Scotia. We carefully docked our ship at high tide. Four hours later, there was not a drop of water under the ship. There is a 28-foot tide change here. The tide comes in at a three-foot-high wall. Good fishing.

My last trip, a most memorable one, was once again with Hoyt Engen. In August, 1945 we left New York in the *SS Woodstock Victory* bound for Karachi, India for our first port of call and then on to Bombay, India for final discharge of cargo. We next loaded up with 55 gallon drums of aviation gasoline. We left Karachi for Shanghai, China. From China back to Calcutta, India to load a cargo of guns and ammunition for our return voyage to Charleston, SC in January 19, 1946. Hoyt and I purchased 30 gifts; we had made sea bags to bring it all home. We arrived back at the states on January 19, 1946.

With the war over, a family to support and no college money available for the Merchant Marine veterans, I had to find work. I was unable to find employment in radio so I started out as a credit interviewer for a department store. Two weeks later I was working in the printing industry.

I was an apprentice in the electrotype trade. This was letterpress platemaking. I worked for New England Electrotype Co. for 24 years as a finisher. A new printing method called "offset" was replacing letterpress, so the company asked for volunteers to learn this new method. I was a volunteer for the new trade at 50-years of age. I went to work for Arlington Offset Printer as a cameraman, stripper and platemaker. Two years later they went out of business. I then worked for Middlesex Printers for two years, and then for VanBuren Printers as production manager. After 11 years there I retired in January, 1987. I am enjoying retired life very much. I am looking forward to getting my ham radio license in the near future. Iris and I have four children, three girls and one boy, eight grandchildren and four great-grandchildren.

DAVIS, JOHN, R-15

I heard about Gallups Island from a gentleman at an employment office. He told me they had a radio school going and that it might be good if I got into it because of its advantages. Prior to that, I had been to what they call CMTC, Citizen Military Training Camp, in Alabama. On the certificate I received from there it said if I ever wanted another government job to use the certificate. I did. I was accepted and went to Boston.

From Boston I shipped out to Australia a couple of times, up around the Solomon Islands and New Caledonia. Made one trip to Italy, visited the Leaning Tower of Pisa and came back.

After the war I went to my little hometown of Tuscumbia, married a girl from there and settled down.

I worked as an electrician and retired from Rentals-Alloys Metal Company, located in Sheffield, AL.

HEGGE, VERNE D., R-15

Left Gallups Island with a group of about twenty R-15 graduates with orders and train tickets for San Francisco. Upon arrival we began reporting daily to a downtown Maritime Service officer. Assignments were slow and like most of the others, I became anxious to get a ship. Was finally assigned to a tanker whose name I do not remember, only to come down with the mumps several days later. Was pulled off the ship, spent several days in bed and several more taking it easy and then began the whole process over again.

After some delay was assigned to the *SS John S. Sargent*, a new Liberty ship. We sailed alone for Brisbane, Australia. We did a lot of zig-zagging and made several course changes to avoid areas of sub sighting. Arrived in Australia after an uneventful but long crossing. After unloading, took on cargo for Milne Bay in New Guinea. That proved to be the first of several trips to New Guinea and to a number of ports as far up as Biak. We ran aground in Finchhafen, thanks to an Australian pilot whose confidence exceeded his knowledge, and came to the attention of Tokyo Rose. Visited several Australian ports, Brisbane and Townsville several times and Sydney for dry dock and repairs. I spent a week or so in an Army hospital in Townsville where I was visited by my one of my brothers who was with the 41st Infantry and back in Australia for a rest. Also got to see Sam Hucke in Milne Bay. We got together by sending call signs with signal lamps illuminating low flying clouds.

Returned to the States and after a short leave was assigned to another new ship, the *Claremont Victory*, which we boarded at outfitting docks in Portland. That trip took us to Hawaii and on to Guam for the invasion.

Returned to the States and was assigned to another new ship, the *Golden West*, a C-2 and a great improvement over a Liberty or a Victory. I don't remember all the ports and islands we visited, but the highlight of my time on that ship was a visit to Noumea, New Caledonia, where I spent several days with my oldest brother who was with the 25th Infantry Division. They had been at Guadalcanal and were staging for the Philippines.

Later, we put into the small port of San Fernando on the island of Luzon, and learned that the 25th Infantry was engaged near Bagio, less than 100 miles away. I managed to contact my brother by phone and made plans to visit the 25th to see him, but the Captain heard about the plan and vetoed it because of enemy activity in the area.

We returned to the States and after a brief leave, I was reassigned to the *Claremont Victory*. We were loaded with small bombs and ammunition with orders for Okinawa, when, to our great relief, the War ended and our orders changed. We took on different cargo and returned to the Philippines, were we spent months moving from port to port and waiting for no one knew what. We eventually returned to Manila and off-loaded. We received orders for China but they were canceled a day or two before we were to sail. We returned to Long Beach where my days in the Merchant Marine ended.

Although some experiences, events and places remain vivid, many have faded in my memory. I see in my discharge, which was issued by the Coast Guard, that I am entitled to the Merchant Marine Emblem, Pacific War Zone Bar, Combat Bar (with stars) and an Honorable Service Button.

None of that impressed my Montana draft board during the build up for what became known as the Korean Conflict. After receiving a notice to report for induction I enlisted again to gain some control over where and how I would spend time in the Army. I attended a school in switching and long-line telephone equipment and was then shipped back to Guam. Shortly after the invasion at Inchon I got orders for Korea but they were later canceled and I spent the majority of my Army time on Guam.

After discharge from the Army and some additional schooling, I went to work as an Electronic Engineer for the Army's Research and Development, Test and Evaluation Command at the Electronic Proving Ground at Fort Huachuca, Arizona, "where the sun spends the winter."

My work for the Army involved considerable travel and a number of flights in and out of Boston. I would always look for Gallups Island and have seen it several times from the air, but never got back there for a visit.

My wife, who is from Columbia, SC, and I met here in Arizona. We were married in Phoenix in 1957, and have lived here ever since. We have one son, a daughter-in-law and granddaughter who live in Phoenix and another son who infrequently calls from some place or other.

I have spent some time working in real estate since retiring in 1979; and have done considerable traveling, much of it in conjunction with Masonic Lodge activities. I served as Grand Master of Masons in Arizona in 1986.

The 50th reunion of my high school class conflicted with the reunion in Minneapolis. The 1992 reunion in Long Beach marked 47 years since I signed off my last ship there in 1945.

HUCKE, SAMUEL T., R-15

"After Gallups Island itinerary & brief life review"

Original home - Shepherd, MT—pre-WWII

Current home - Fayetteville, AR—since WWII

Period of Gallups Island—May 1942 to January 1943 (R-15)

Spent three months, first part of 1943, in San Francisco on WSA, recruiting and manning rolls awaiting assignment. Finally gave up and joined ROU and was shortly assigned through union hall. The ships I sailed on in chronological order are: *SS Baldhill*, May '43-June '43; *SS Thomas C. Power*, August '43-September '44; *SS Samuel F. Dewing*, October-December '44; *SS Thomas Kearns*, December '44-January '46; and *SS Marine Fox*, February-May '46.

I attended USMS Radar School, Sheepshead Bay, NY from May to July '46. After graduation I did not take another ship assignment, but took a flight radio operator job with American Overseas Airlines. I was assigned to an Air Transport Command contract, flying from Westover Field, MA to Paris, France, with some flights on to Berlin or Rome.

But all good things have to come to an end, and in December, 1946 I quit to go to college as a pre-med student. One of the Kings Point Cadets on the *SS Thomas C. Power* was a pre-med and influenced me to make that decision. I went to the University of Arkansas at Fayetteville for four years, then the U. of A. School of Medicine in Little Rock, for four more years.. I then took a year of internship in Tulsa, OK. At that time I held a commission in the USAF reserve as a medical officer. While I was in my internship (1954-55) the Air Force sent me orders to active duty, following completion of the internship. The Korean War period of draft was over for people my age, but I was considered eligible for call-up because of the doctor's draft that still existed at that time.

After 26 months of active duty, I returned to civilian life and took a residency in general surgery at the VA Teaching Hospital, Memphis, TN. After completion of the surgery training I transferred to the VA hospital in Fayetteville, and remained there until my retirement in '86. I also remained in the Air Force Reserve, and retired from the Air Force at age 60 with 29 years of service. So now I am a triple-dipper—drawing a little Social Security retirement, some military retirement and civil service retirement.

My wife died four years ago and I live alone here in Fayetteville, AR. Two of my three children are doctors, the third is a social anthropologist and writer and is married to an MD. So, if you don't like medicine don't visit me.

I never became a ham, but did work as a radio broadcast engineer while I was in college, and did go back to sea during summers on two occasions ('47 and '50) while I was in college.

Currently I have just completed chemo-therapy treatment for abdominal lymphoma (60% chance of cure) and two years ago I had radical prostatectomy for cancer of prostate. When it rains it pours, but there's a fairly good chance both are cured. If any of you reading this have not had a Prostatic Specific Antigen test for cancer of prostrate, you should do this. The test is fairly new and some doctors haven't been ordering it on their patients.

MORGAN, ROBERT, R-15

Marwin Brown called Robert to see if he would be going to the Minneapolis Reunion and to get a synopsis of his life after Gallups Island.

Robert was not able to attend the reunion because of ill health. He has a nerve degenerative problem that has him bound to a wheel chair and he cannot write. Although he can type with two fingers he asked me to write up our visit together for the rest of our R-15 classmates.

Robert's first trip out of New York on the *SS Mark Twain*, a very appropriate ship as he is from Missouri. He sailed on Liberty ships, Victory ships, troop ships and tankers, spending about seven and a half years at sea between his 20th and 29th birthdays. He took time out from sailing after the War to work as a lighting engineer in New York where he worked with the New York I.E.S. president. His last electrical work was designing the top lighting on the Transamerica Building in San Francisco.

He sailed to North Africa, Iran, England, Europe, the Caribbean and Gulf ports, South America, Australia and New Zealand. He believes his best living conditions were on the old Standard Oil, now Exxon White Fleet, going primarily between South America and U.S. Gulf ports. He also hauled copra with the old Lykes Lines.

One of his closer calls was when a torpedo was sighted coming at his ship after leaving Gibraltar in convoy. With quick reactions on the bridge and a hard left helm, the torpedo just missed their fantail and hit a French munitions ship, blowing it up.

He had a 2nd Lieutenant rating with the Maritime Commission when he left the sea.

Hard work and good investments have allowed him to live comfortably in his own home. He extends his best wishes and 73's to all his old R-15 classmates.

MURRAY, JACK M., R-15

As a ranch-reared guy from the Colorado high country, the sights and sounds of Boston's North Station and the launch ride to Gallups Island were the uneasy but exciting beginning of a four-year USMS experience that helped shape the rest of my life, as I'm sure it did for us all in one way or another

After R-15's ten months of dit-dahs and tubes, I shipped out from Los Angeles on the *SS Edward Livingston*, a Liberty ship whose designers and builders I will always hold in highest respect. After a stop in Sydney we took a route through the "roaring forties" south of Australia enroute to India with ordnance, aircraft, etc. for the China-Burma-India operation, then in full swing. This turned out to be more of an initiation to life on the high seas than I had bargained for. Trapped by hurricane force winds with seas breaking over the upper decks, that Liberty withstood 10 days of pounding that surely defied the normal laws of physics. Fortunately so for all aboard. We finally limped into Albany, Australia for temporary ship and crew damage repairs and later into Freemantle for more extensive work. Having lost virtually all of the deck-lashed aircraft, we were patched up and sent on to Karachi, Bombay, Colombo and Calcutta with the pieces and other cargo.

The return trip to Baltimore was around the Cape of Good Hope via Lourenco Marques, Durban, Capetown, Rio De Janeiro, and Trinidad. Fortunately, there were no further tests of bolts, welds and crew.

Having thoroughly established my good luck and seaworthiness on this first voyage, the following ones were comparatively routine (or as routine as wartime at sea can be). These included several in the CBI, Pacific and Mediterranean theaters on the *William Wolfskill, John S. Bassett, Rensselaer Victory* and *Tucson Victory*.

It was aboard the latter that I copied the BAMS message canceling blackout regulations, radio silence and manning of guns as of 0001 GCT, September 6, 1945—it was over!

It wasn't long until the lure of the skies overcame what little lure of the sea remained. Dave Coats (R-15), whom I first met on the train enroute to Gallups Island in early 1942 and kept in periodic touch with during the War years, had heard of openings for Flight Radio Operators with Pan American-Grace Airways (Panagra). Consequently, we spent the next several years over and on the South American continent.

As airway communications and navigation technologies advanced during this period these jobs gradually became obsolete. By then, the serious business of raising a family had become a priority anyway and that was not the part of the world in which to do it.

Back in the U.S., Gallups Island basics again paid off with a stint as a civilian instructor for the Army Signal Corps and then in the rapidly evolving radio and television broadcasting field for some very interesting years in high power directional arrays, microwave links and early color television station installation. Also, the then embryonic computer field was beginning to emerge from the labs and becoming a valuable technology for commercial and military uses. Spurred on by the technical fallout form the Mercury, Gemini, Apollo and other space programs it was obviously an exciting field to get into.

It was thus that Dave and I again made contact and began a life long career in an industry which, in a short span of 30 odd years, has grown exponentially from its technically primitive roots to its all pervasive presence today—a heady time and experience we are not likely to see repeated anytime soon.

Although now retired, I still maintain many contacts in the industry and enjoy occasionally spirited discussions with colleagues in these and other technical organizations. No less enjoyable was the opportunity to attend the GIRA reunion at Braintree and visit with many comrades from those earlier days. It has indeed been a long trip from the Island to here, but an interesting one, to say the least!

So there it is—a "this is my life" with a real sea story thrown in.

73s

OSIKOWICZ, JOHN, R-15

Here goes a smattering of what I can recollect; it's been a long time.

The ships I sailed on were the *Sabine Sun, James R. Ramsall, Clinton Kelly, Samuel Parker, Montery Cooper* and at least one other that I didn't get a record of from the Coast Guard.

It was quite hairy after leaving Gallups Island. Although no ship I was on was torpedoed, I had many close calls. I was in the Atlantic, Pacific, Mediterranean, South Atlantic and North Sea. In Europe, North Africa, South America, India, South Africa and the Philippines.

The high point was the Normandy hostilities. I made many runs between Southampton and the beachheads until we ran out of supplies due to being so generous with the other services. We were ordered back to the States in October, 1944.

After leaving the service in May of 1946, I attended the University of Pittsburgh for one semester. Lack of funds drove me out. I took employment with the Pennsylvania RR and then transferred to a subsidiary, the Long Island RR, where I remained for 33 years. I retired in 1982. I also took R-7 courses, which consisted of TV and broadcasting, for the RCA Institute. However, I did not pursue employment in that field.

I have lived in the New York area since 1948. Since my wife is from this area we enjoy living in the Long Island. I have a youngster, age 29, living at home.

I will always cherish the fond memories of Gallups, Boston and all the fellows that have gone on and those that remain.

PEREZ, CHET, R-15

"Where I've Been"

After graduating (R-15) I shipped out, sailing the Atlantic and Mediterranean mostly on tankers.

After the war ended I worked at Republican Aviation, Long Island, New York; as a textile salesman in New York City; and spent two years with United Geophysical Co., in Venezuela's Amazon (Delta Amacuro). I was an observer on a water and portable (with donkeys) crew. I monitored and recorded on film and tape dynamite blasts looking for faults in the subterranean shale and rock formations. I worked at Maracaibo and near the Colombian border (Cucuta). In 1958 there was a revolution and that was the end of that job.

When I returned to the U.S. I went back to Republic Aviation, then to C.B.S. Labs in Stamford, CT.

I became a VTR operator in West 57th Street, New York City when they had a strike. Being in the engineering department, I taped the morning news, soaps and a great lot of commercials and "promos."

From there I went to Machlett Labs (Raytheon) testing X-ray tubes and giant Voice of America output tubes. Working with over 250 KV and about 100 MA was scary. I left and went to Ft. Lauderdale to work with Motorola in Plantation. After 10 years I retired in 1986. I plan to take another 2nd class test.

My dear friends that I remember at Gallups: Wil Beaulieu, Dave Coats, and John W. Davis, to name a few. If you remember me, please drop me a post card. I currently reside in Fort Lauderdale, FL.

PERRY, FRANCIS, R-15

The following is a rather brief account of almost 50 years since leaving Gallups Island. Time has dimmed a good part of what happened after graduation.

After October 1946, I married and stayed ashore until one trip to Italy with a load of coal during the Korean War.

After Gallups, I served on the following vessels: *MS Northern Sun*, March-June 1943, with George Cushman, R-15 (we exchange Christmas cards); gasoline to Scotland and England on the *SS Jonancy*, August-December, 1943; a coal boat from Virginia to New England; many trips on the *SS Walter A. Luckenbach,* January-October, 1944; gasoline in 55-gallon drums to New Guinea aboard the *SS Kenmar*, November 1944-July, 1945; explosives to Wales and then gasoline in Jerry cans back and forth across the channel, *LL Benjamin Goodwine,* September, 1945 to February, 1946; and *SS Marcus H. Tracy*, March-October, 1946.

I have spent my time in the same city I was born in. Many jobs, but nothing in radio. Now retired. I wish you all well.

SCHULTZ, LELAND D., R-15

I am from a little town called Plum City, WI and I enrolled in the Gallups Island program through a fellow that was working at the vocational school at Eau Claire, WI. He said that you would eat well in the Merchant Marine, so I signed up.

That happened before Pearl Harbor and I went to the Island May 11, 1942.

After taking the FCC test early in January of 1943, I reported to the ACA union office in New York. I signed on a Liberty, the *SS John Vining*, which had a Murmansk destination. This first trip was by far the most hazardous trip I had. We hit extremely rough seas near Iceland. The deck cargo, which appeared to have been poorly secured, came loose and we dropped out of convoy and became stragglers. It was at this time that I prayed a lot. Four other Liberties, that were also carrying ammunition, blew up when torpedoed. We were able to make it to Reykjavik, Iceland where we received much needed repairs. Fortunately, our orders to Murmansk were canceled.

Later in that nine-month trip, I met three members of R-15: Fred Hallach, John Early and Hoyt Engen. It was like seeing my immediate family. We had a great time together and it helped each of us immensely. Fred Hallach and I continued on to the Persian Gulf. I have not seen or heard of him since that trip.

After sailing through the war, I left the Maritime Service and went to school at the University of Wisconsin. After my first semester I married Maxine, whom I had met before the war. (We now have four sons and nine grandchildren). We both graduated from the University. I worked for a city/county health department for about three years after graduation. I then worked for the USDA for eight years, went to Borden Food Company for five years and then joined the University of Minnesota where I stayed for 22 years.

I retired in 1985 and enjoy retirement and have now reactivated my license and am enjoying that hobby immensely.

STOUGH, RAYMOND, R-15

I came down from Boston in March, 1943 to my first ship, the Liberty *William S. Graham*, I think. It was in Brooklyn's Eric Basin—a wasteland then, worse now—waiting to take me on a nice trip to Algiers. The *Graham* took me to North Africa, Casablanca, in convoy, of course. I was never out of convoys, except for my last two trips and one out of Portland to the Philippines, where I learned to body surf. All my trips were on Liberties and most to England.

I knew Liverpool, Birkenhoven, Chester, North Wales and South Hampton well. I ran into Lynn Booren in Liverpool when both of us were heading for a highly spirited mid-summer's night dance. I don't remember whether we danced together, but I know we danced with everybody else. I made two trips, I think it was two, to England and eventually to Europe for the invasion, with classmate Leland Schultz. I think we spent the winter in every little coal port in Scotland from Oban to Leven (which is not recognizable now—I went back to look) waiting to get loaded for the Invasion. We brought freedom, or at least some of its goods, to Cherbourg. Then we learned about Calvados and LeHavre. The Bulge undid shore liberty in LeHavre and sent our ship home for more stuff. I did some troop carrying somewhere in there too.

After the last trip we came back empty with half enough ballast. I was sure I was going to be a casualty of a capsized Liberty, what with the rolling, starboard to port and back almost dipping the rail each time. Either that, or (hopefully) survive on a tip of our Liberty from its cracking in two—what with the snapping of the entire hull, as the screw came out of the water and the bow crashed down on the crest of the rollers ahead. I think that was the Ingram. I humbly thank the welders, of whatever sex, who put it together in whatever shipyard. The ship got back in one piece and so did I.

I had one trauma left—no Murmansk, but comparable. I had to break radio silence and talk to the shore station to get our ship a port, a berth, a pilot and a time. This happened half-way back from Marseilles. Got a sense of hopelessness and embarrassment. Three years of radio silence, convoy life, taking BAMS religiously and filing them away didn't prepare me for this. I want to bless the shore station operators, too. I rank them right along with the welders. I still don't know if it was Eric Basin, but we found a berth somewhere in New York.

After that trip I had enough time to say ashore, since the war was over. I did consider an RCA brush up school to learn to be an RO and do a little peace time sailing, now that the seas were safe. Instead, I chickened out. It was the Christmas season, and I was alone in New York, so I took the train (remember them?) from New York to Dakota to plan my future. I taught school for six months in my North Dakota hometown and then moved to New York to act and teach. Although I did quite a lot of each, and enjoyed both, I never got rich or famous, as you know.

I did get married and then divorced—that's all American—no kids. After 20 years of teaching public school in New York, I retired. I still live in New York and like it, believe it or not. With my teacher's pension, I am now free to chase whatever acting work there is for us wise, alert elders.

The last three summers, four including this one, I have done work sponsored by the North Dakota State Historical Society. I wrote the show and have been acting five different historical characters (half-hour monologues) associated with six different historical sites in North Dakota. I have my eye on a move to Florida when I find the right place. It must have a hot tub of course, but not the shuffle board, I think.

By the time this is published, I should have two new knees.

WENTZEL, MANARD "SKIP", R-15

I graduated in January of 1943 and was sent to the WSA pool in New York. We stayed the night at the YMCA and my money eventually ran out. It looked like it would be some time before a ship would be available. At one of the sign in's I heard a rumor of Liberties coming out of South Portland, ME; so with my remaining bucks I got a train back to Boston. I registered with the WSA in Boston and met a couple of other Gallups Islanders who advised me to join the ROU. This office was located in a walk up dingy building in the market district. The man in charge worked a night shift at the McElroy Mfg. Co. He was a good friend of Ted McElroy and when I told him I was a student under Howard McElroy at Gallups Island, I was welcomed and asked to make myself useful around the office, which I did.

I got a call a couple of weeks later and was told to be at Sprague Steam Ship Offices at 9 a.m. the next day to join a crew of officers enroute to Portland, ME. We were assigned to assist in the fitting out of a brand new Liberty, the *William Pierce Frye*. We spent a very enjoyable two weeks outfitting the ship and then sailed to Boston for loading. The captain, chief engineer, gunnery officer and I became very close and I was really looking forward to shipping out with them. The *Frye* loaded munitions and heavy equipment and was to sail to Halifax to join a Murmansk convoy. I remember there were several wooden launches on the foredeck for deck cargo.

The night before we were to sign Foreign Articles I got a call from the union rep. He was very apologetic, but I had been bumped by a senior member of the union. He told me about this radioman, a Carl Dalbey, who had spent some 37 days in a lifeboat in the South Atlantic, and after being rescued was acclaimed a hero by the other crew members. It seems his leadership had a great deal to do with their survival. The union agent didn't have much of the details, but it seems the day before, he was invited by Eleanor Roosevelt for tea at the White House and a commitment was made that his wish to be assigned to a new Liberty be honored. I was to lose my ship. The agent asked for me to assist Carl in every way and promised the next new Liberty out of Portland.

Carl arrived the next morning in a taxi loaded down with a rubber life saving suit, sextant, compass, canned foods and all kinds of charts. We had lunch together and I took an immediate liking to him. He told me about his days in the lifeboat, how they were sunk without warning and had no time to put food or gear in the boat. He steered by the stars and they were eventually rescued off the African coast. He mentioned that he was considered for the Merchant Marine Distinguished Service Medal.

I got my assignment to the *John Trumbull*, another new Liberty out of Portland the next week. We were ordered to New York and loaded Army equipment for the African invasion. I made three round trips to the Mediterranean on the *Trumbull* with all the usual torpedo attacks, air raids, and even a mining episode in Gibraltar. The *Trumbull* was a lucky ship and we survived. Incidentally the gunnery officer was an Annapolis graduate and brother to Admiral Sheridan. He was a little gung-ho but had the gunnery crew in good shape.

I returned home after three trips to the Mediterranean and volunteered for the Navy. After a couple of years I was discharged as a RM 2/C. I spent most of this time in the South Pacific on a PC.

After the War I contacted my friendly ROU agent and got a Liberty, the *Theodore Parker*, and sailed for most of 1947. I returned to enter Northeastern University. My friend, the union agent, told me that the *William Pierce Frye* was lost with all hands. It wasn't until I told this story to my fellow officers on the *Parker* that one of the engineers said he had sailed with a survivor of the *Frye*.

I finally tracked this man down and had a phone conversation with him. He was the second mate and remembered me. The *Frye* left Halifax in convoy and several days later had an engine breakdown and was torpedoed. The second mate had a friend in the gun crew from his home town and knowing the conditions of the sea, didn't have too much confidence in the lifeboats. He told his friend to cut the lashings on one of the wooden launches, put some food and water aboard and make for it if they were sunk. I assume this is what they did since they were the only survivors. You can understand he was a bit reluctant to talk about this part of it. When I asked about Carl

Dalbey he praised him and said he was a fine shipmate. The last he saw of Carl was him swimming toward a lifeboat hanging on to the emergency radio.

I often think about Carl Dalbey and wonder if he ever received credit for his heroism.

As for my later adventures, I went into electronic distribution, working for Lafayette Radio, Harvey Electronics, Arrow Electronics and up to my retirement a few years ago, assisted a friend in a new start up.

73's.

Platoon R-16

ANDREN, RICHARD, R-16

I consider the time I spent in the Merchant Marine as some of the best years of my life. During that time I visited 40 different countries and saw another 41 through a porthole while being paid at the same time. What a vacation!

I did not want to be a foot soldier in the army, so when my draft number came up, I tried to enlist in the Navy and then the Coast Guard, but both rejected me because I was color blind.

The Merchant Marine indicated they would need a waiver from Washington for me to be accepted. My waiver came through stating I could be accepted for cooks and baker's school only. To me that would still be better than being a foot soldier in the army.

The Minneapolis office sent me to Gallups Island, but the cooks and baker's school had moved elsewhere three weeks before. I asked for a transfer to radio at the suggestion of another student from Minneapolis. The school undoubtedly did not check my records or perhaps were just in bad need of radio operators.

On completion of my training I asked for, and obtained, a ship out of Wilmington (Long Beach, CA). It was a new Liberty ship and I was the only radio operator.

Here I was, on my first ship and the first trip at sea—the auto alarm bells sounded! I dashed into the radio shack but could not find any distress signals. I checked the timer on the auto alarm and it was going at the proper speed. I went back to my bunk only to have the same thing happen two hours later. Again, no distress signals and nothing found wrong with the equipment. The third time it happened and nothing was found, I am sure the captain wished I had gone to the school for cooks and bakers.

On the second night it happened twice again. On the third night, I found the motor to the auto alarm was overheating, which in turn slowed the motor down to make the alarm bells sound.

The McKay equipment had the auto alarm motor inside a small door and with a new motor it overheated. I tied the door open and that solved my problem.

My first two trips took me west to India and then continued west until we reached New York. I had enough of India so I transferred to New York in hopes of going some place else. My third trip took me back to India again.

I often wondered if I had flunked out of Gallups Island, would the Navy have taken me in with color blindness?

BOND ROBERT, R-16

Upon completion of Resident Radio School at Gallups Island in Boston Harbor, I was sent to San Francisco and assigned to a new ship *(SS Dashing Wave)* as Chief Radio Operator. The maiden voyage was with cargo to Pearl Harbor and returned to San Francisco.

In San Francisco, the *Dashing Wave*, a C-2 cargo ship was modified to carry troops. First part of voyage was to Port Hueneme, CA to load a battalion of Navy SeaBees and their equipment. With other cargo aboard, the trip continued to Guadalcanal. Jap planes attacked en route without success on their part.

Bombs were dropped at our bow as well as at some other ships. Fortunately the bombs were duds and did not explode, only splashed water on the gun crew. In Guadalcanal Bay, cargo was unloaded on barges during daylight hours. Frequent bomber attacks were made at night. Returned to San Francisco.

This voyage was a return to New Caledonia with SeaBees and cargo for Espiritu Santo Islands. No enemy attack on the return phase. The ship carried wounded and other troops stateside. En route, traveling with darkened vessel, another ship collided with us at about midnight—striking #4 hold and caving the side in about six inches. Voyage was continued as no leak occurred. Next port was San Diego and thence on to San Francisco.

From San Francisco (after repairs), to Port Hueneme for SeaBees and cargo. Thence to Pearl Harbor where considerable time was spent at anchor near Ford Island. Finally we joined a convoy bound out to the Atoll of Tarawa. Arriving a few days after the main battle. Small arms fire was continuing and there were nightly enemy aerial attacks. Back in Honolulu harbor the news was out that we had been sunk with all hands lost! After regrouping, the ship was routed back to San Francisco to pick up cargo. Just outside the Farallons Isles we came upon a Navy float plane drifting out to sea. The two crewmen were the sickest I had seen. The aircraft engine mounts had broken sending the unit into the sea. This was on Christmas Night. Undoubtably the BEST gift the two crewmen received.

With cargo, a return trip was made to Hawaii and took on a Regiment of army personnel and sailed to Eniwetok Atoll for the second time. While cargo was off loaded, the army set up camp on shore. When cargo was cleared out, the army unit came back on board and returned to Hawaii. Our ship then returned to San Francisco for more cargo.

With cargo on board, it was back to Pearl Harbor and more waiting. Shuttled army troops between Islands. In June, we joined a convoy which was bound for the Invasion of Saipan. We arrived there shortly after the battle had begun. We endured nightly enemy aerial attacks—presumed to be from Guam. After a few days of cargo unloading into landing craft all Merchant Vessels were ordered to depart in haste—even without closing cargo hatches. A convoy of about a hundred ships formed and sailed to the east many miles. In a very large circle we all sailed for about a week, then returned to Saipan to complete unloading cargo. Nightly bomber attacks continued. With cargo off loaded, the Dashing Wave returned to San Francisco. The crew was relieved as the navy took control of the ship.

From history I have learned that the reason for the odd departure was the time of the "Marianas Turkey Shoot" carried out by our navy carrier task force.

Being relieved from the *SS Dashing Wave* after 18 months, I took leave to my home in Texas. After leave I went to New York and on to Boston to join the *SS Hubert Bancroft* as Chief Radio Officer.

The *Bancroft*, a Liberty ship, was an ammunition ship. From Boston it sailed to Searsport, Maine Ammo depot. After loading 7,000 tons of HE shells and deck cargo we returned to Boston. Shortly, we sailed in convoy to Halifax, Nova Scotia and on to Belfast, Northern Ireland in convoy again, on to Le Havre, France to unload. Return to New York and Philadelphia for a second load.

From Philadelphia Ammo Depot, we sailed in convoy, first to Wales then on to Cherbourg, France to unload and return to New York.

New York to Houston/Galveston, TX expecting to load ammo, but plans changed and we took on 5,000 tons of Sulfur and food stores plus deck cargo. Sailed back to Norfolk, VA where a convoy was forming for the Mediterranean. Stop over at Marseilles, France, where deck cargo was removed overnight. Next day we sailed onto Toulon, France where all cargo was off loaded. Cargo of sulfur carried to Sete, France. Sulfur was considered dangerous cargo, similar to ammo. WWII ended in Europe while in Toulon. We returned to New York.

Took leave home to Texas. Journeyed to New Orleans and joined crew of *SS Calvin Austin* as Chief Radio Operator.

SS Austin, a Liberty ship, took cargo, mostly beer, and construction equipment on deck to the Philippine Islands via Panama Canal. While en route the A Bomb was dropped and Japan ended War II in Pacific. Returned stateside with some homebound troops through Canal on to Baltimore, MD where I left the *Austin*.

Traveled cross country to San Francisco to Portland, OR, joining the crew of the *SS Sea Flasher* (a C-3 troop/cargo ship). From Portland we sailed direct to Manila, where we embarked approximately 2,000 home bound troops. Delivery was to San Francisco, CA. In February, 1946 I also ended service and returned to my home in Texas.

BRADFORD, ART, R-16

Art started his career at sea two months before Pearl Harbor. He shipped first as an officers' messboy, then as an ordinary seaman on the *Esso F. Q. Borstow*. After sailing around South America for seven months, he heard about the USMS Resident Radio School on Gallups Island. He landed on Gallups in May, 1942. and graduated with R-16 in February, 1943.

From messboy to ordinary seaman to radio officer!

His first ship after graduating was the *SS Covalt* which got stuck on a rock outside Montreal. The next, the *SS Peregrine White*, got rammed in the North Sea—had to send an SOS for that—and spent six weeks in Edinburgh, Scotland for repairs.

Went to Normandy after the invasion on the *SS Grace Albot*, with 7,000 tons of ammo. After VE Day, Art sailed on the *S.S. Joseph Carrington* out of New York. She hit a mine in Borneo which destroyed the engine room, and the ship spent five months there. After being towed to Manila, Art had to hitch a ride back to Duxbury.

His first ship after the war was the coal collier, *SS Berkshire*, and he didn't find out until 39 years later that he had relieved none other than Bill Anderson, N1BWP.

Art qualified for the Merchant Marine Defense Bar, the Combat Bar, and the Mediterranean, Pacific and Atlantic War Zone Bar.

CLAMAN, ARNOLD Y., R-16

"Where It All Began"

In early 1942, while working in the Boston area, I first heard about Gallups Island and the possibility of going to sea as a radio operator. As a boy, my parents had taken us to France on the old *Leviathan* in 1928, and I was hooked on ships ever since. This was my chance to combine a new career and fulfill my military obligation, since we all became members of the Coast Guard Reserve when we were sworn into the USMS in May 1942. I joined R-16 and my life afterward always had a maritime influence.

Having attended the University of Southern California just before the war, I signed up for the West Coast and shipped out of San Francisco on the *Amos G. Throop*, a U.S. Lines Liberty named for the founder of Cal Tech. I was to remain on her for a year and a half including a voyage to the Southwest Pacific and Chile, three North Atlantic convoy runs and the Normandy invasion. Our skipper was an old time British naval type who had commanded a gunboat in, of all places, the Tigris south of Baghdad in WWI.

The radio room had the usual Radiomarine unit and to break the monotony of copying BAMS schedules with their warnings to navigation and long stretches of dead air on 500 KCs, I would type up the ship's newspaper after listening to news reports and closing down the watch each night. It proved to be very popular, especially a Walter Winchell column on the past and present activities of the crew.

Our first destination was the New Hebrides where we arrived some 23 days later after zig-zagging across the Pacific and being presumed sunk when our naval escort failed to meet us south of Fiji. It was the practice in the Pacific to send ships out unescorted on different routes to reduce the risk of enemy action. At a Havana Harbor, the big naval anchorage below Espiritu Santo, we unloaded into lighters with the help of work details off the *USS Maryland*, who did a good job of breaking into our cargo of Lucky Lager beer.

(Today the New Hebrides is the independent nation of Vanuatu, and the then dusty waterfront of Port Vila, the capital, has been transformed into an international vacation resort and high-rise off-shore bank building.)

After several inter-island runs to Espiritu Santo, the staging area for the Solomon campaign, we headed for Antofagasta, Chile, to load a Grace Line cargo of copper for the East Coast. The *Throop* had been diverted back to San Francisco on her first voyage, so I was doubly alert for any BAMS messages as we crossed the South Pacific below Tahiti. This did not prevent some anxious moments on arrival 27 days later, when the pilot coming on board said we were not expected and the captain called me to the flying bridge to ask, "Sparks, are you sure you did not get a message?"

It developed we had set a new record for the crossing due to good weather and our naval attache in Santiago only sent up the names of impending arrivals 24 hours ahead of time to prevent possible leaks. Many years later, while on a business trip to Santiago, I related this story to a banker and he introduced me to the attache who had retired to Chile.

On discharge in New York, most of the crew from the West Coast signed off, but I remained with the *Throop*, making three convoy runs to the UK between October 1943 and March 1944 before we were detached for the Normandy invasion. The first trip took us to Manchester, up the old ship canal from Liverpool, where we had to lower the masts and the crow's nest to get under the railway bridges. After unloading we were ordered back to Liverpool to load for Murmansk, when it was discovered the *Throop*, having been built in California, did not have a reinforced bow for ice.

Back in New York for Christmas, we took on a cargo of bombs and ammunition, at Caven Point, just beyond the Statue of Liberty, for the Eighth Air Force and joined our convoy out of Halifax. While it was a typical January crossing in heavy weather, we had some uneasy moments New Year's Day 1944 when our engine broke down and the corvette covering us announced she was leaving to return to the convoy 13 miles ahead.

We spent the rest of the day getting back into formation. A day or so later, we had a near miss right in the middle of the convoy when a tanker loaded with high octane passed within a few yards of our bow out of control. North Atlantic convoys were made up of formations ten to twenty or more ships across and five or more ships deep. In foggy weather, intermittent blasts from the ship's horn would warn other ships of your position.

We discharged at Hull, after picking up a convoy at Loch Ewe in northeastern Scotland, and on returning to Loch Ewe we had the frustrating experience of watching the homeward bound convoy sail without us when we were unable to raise our anchor which had become fouled with somebody else's. Loch Ewe was the jumping off point for convoys to Russia, the East Coast of Britain and the North Atlantic. It did not take long to fill up and we were out of there in a week.

This time we returned to Boston, my first time since leaving Gallups, and the sign of coming events appeared when we were fitted out with bunks for troops in the #4 hold. We sailed for the Bristol Channel with a cargo of Sherman tanks, railway locomotives and heavy equipment. We unloaded at Barry Dock, near Cardiff in late March.

We spent the next two months at anchorages in the Clyde and South Wales until June 5 when we loaded 800 troops and their equipment. We were at sea, rounding Land's End, when the news of the D-day landings came over the radio and we arrived off Utah Beach the morning of D-day plus 2. We dropped anchor near a U.S. destroyer shelling a small town on the water's edge and watched it knock out a church tower. The village was Grandcamp and I made a point of exploring it on a visit to the Normandy beaches in 1980. Only close examination disclosed where the tower had been restored.

Late that afternoon, a group of Luftwaffe fighters flew over strafing the beaches chased by our Ninth Air Force unbeknownst to us. The gun crews on a thousand ships, who had been waiting a long time for this chance, opened up in a hail of fire and when it was over, nine planes had been shot down, five of theirs and four of ours. The next day, the order came out—no more shooting by merchant ships.

We returned to Southampton with wounded from the 101st Airborne landings and began a series of runs to Utah and Omaha [Beaches]. We took General Middleton and his 4th Army Corp and later brought fifty Red Cross girls with their vehicles.

On the morning of July 25, as we were dropping anchor again off Utah, the entire U.S. Eighth Air Force was flying overhead to begin their bombing runs for the breakout at St. Lo, which brought the Normandy campaign to a close. It was an awesome sight.

The cross-channel runs came to an end in August after our propeller hit a log off the Isle of Wight and we had to be taken out of service. We were ordered to Tilbury Docks near London for repairs and on a brilliant day passed through the Straits of Dover in a small coastal convoy. The sky was active with V-1 flying bombs roaring overhead and

Spitfires, which had the speed to shoot them down, giving chase. Despite the planes and barrage balloons, a number got through and headed for London. When we got there, we found half the city had been evacuated and it was only after Montgomery had overrun the launching sites in northern France a short while later did the threat disappear. The advent of the V-2 rockets shortly thereafter gave Londoners another tough period since they came in without warning.

After signing off the *Throop* on our return to New York, I shipped out on another U.S. Lines Liberty, the *R. F. Peckham*. We made two voyages to the Mediterranean, interrupted by an emergency run from Norfolk to Boston through Long Island Sound and the Cape Cod Ship Canal to keep the Boston Edison Company in coal. In November 1944, we sailed from Baltimore to Palermo and Leghorn, where we maintained a watch for enemy frogmen who had been reported active in the area.

In March 1945, we made our final transatlantic crossing to Marseilles returning via Oran. The war in Europe ended as the biggest convoy we had ever been in approached Hampton Bays and huge flags were broken out to celebrate.

By the end of the war, every ship on the North Atlantic carried three radio operators to maintain a round-the-clock watch. As I recall, only one was a Gallups Islander. The rest were assigned by the Navy and one memorable old-timer was a former Chief, who had been at Pearl Harbor. He had worked on the project to break the Japanese code, and could take 35 wpm while holding a conversation.

The *Peckham* wound up its wartime career carrying a Lykes Brothers cargo, including 5,000 tons of sulphur, from Tampa to Melbourne. On a 44 day voyage through the Canal and across the Pacific, we saw no ships and only sighted the top of New Zealand in a raging storm. From Melbourne we headed up the coast to Sydney when the bombs were dropped and the war ended.

The celebration for a country that had been at war since 1939 and saw most of its men overseas was tumultuous. I recall all the restaurants were closed, but you could get all you wanted to drink. The local longshoremen, true to Australian tradition, immediately went on strike and we finally headed for Brisbane, where we finished loading a cargo of wool and kangaroo skins for the American Pioneer Lines.

The highlight of the voyage home to Boston through the Panama Canal was the end of radio silence and sailing with lights on and portholes open. Our most important function was bringing back twelve National Guard officers, who had been stationed in Australia and New Guinea since the start of the war. There were big family reunions on arrival.

After the war, I returned to the University of Southern California, which I had left six years earlier, paying my own way as we had no G.I. bill. After graduating in 1948, I went on to Columbia Law School in New York and became a member of the Bar in 1951.

By this time, I was married to my wife, Barbara, who by coincidence had been a radioman in the Waves. The Korean War was under way and experienced seamen were in demand. I renewed my license and signed on the *Western Trader*, a U.S. flag Liberty operated by PHS Van Omeren, a Dutch group. We carried coal from Newport News to Rouen, but as we were proceeding up the Seine estuary above Le Havre, we were rammed by a French freighter coming downstream. She opened a large hole in our forward hold and we settled in the middle of the river. Although we were in no immediate danger, we began listing sharply as the tide went out. I managed to get out a distress signal after clearing the antenna from the booms.

We held up all shipping in the Seine for two days before we could be lightened and moved out of the channel. We were later towed ignominiously backwards upriver to Rouen, where we spent six weeks under repair. Today, the entire crew would have been flown home.

Barbara met me in Philadelphia and our seagoing trips since then have been limited to cruises. On my return, I joined an admiralty law firm in New York, where I practiced for the next seventeen years. Among our cases was the *Andrea Doria—Stockholm* collision. Our clients included Lloyd's of London and I transferred to the financial insurance side of the office.

One of the banks we insured was Citibank and in 1968 I became head of the Bank's Potential Loss Department, retiring in 1987. In keeping with my maritime history, many of our cases covered familiar territory—ships, bills of lading, letters of credit and foreign exchange. More than one vessel arrived at some far-flung scrapyard with insufficient documentation for payment.

In retrospect, while I did not follow-up radio engineering as a career or pursue amateur radio as a hobby, Gallups Island was the catalyst for all the good things that followed. Barbara and I just celebrated our 43rd anniversary and our three daughters, their husbands and our six grandchildren are the ultimate reward.

HAMMES, EUGENE F., R-16

I grew up on a farm near Sidney, MT. It was 160 acres with a variety of crops; sugar beets being the main cash crop. I heard of one of the neighbor kids going from the CCC camps to Gallups Island and it sounded like something I would like to do. After Pearl Harbor I signed up with the USMS in Minneapolis and waited at home for awhile to be called, then went to Tacoma, WA where I worked at a shipyard helping build destroyers when I got the call. I quit and went to Minneapolis and joined the groups going to Gallups Island. This was May 1942 and I was in R-16; graduated January 30, 1943.

WSA sent a group of us to San Francisco where I waited for a ship, then they needed us in Seattle. I was finally assigned to the *SS John Burke*, a new Liberty just back from its maiden voyage. I spent the rest of '43 sailing to Alaska because they were making the push to take back Kiska and Attu. We made five trips to Alaska and the Aleutians where we hauled supplies for the war effort. I was even "drafted" to help unload ammo and 50 gallon drums of avgas. I was not forced but they made it clear my help would be appreciated. The *Burke* had great cooks who were able to put out meals like a first class restaurant due to our quick turn-around and the availability of fresh supplies.

At the end of 1943 I signed off and stopped in Montana to see my family on my way to Boston. I had to get back to see my girl friend, Doris Cunningam, in Dorchester. Maybe I should thank Doris for getting me off the *Burke*, because a year later she was off Mindanao in the Philippines and was hit by a Kamikaze and blew up with all hands and the load of ammo.

On the train to Portland, ME , I met my neighbor from Montana, Mac Erps, going to Presque Isle, ME. He was a flight RO ferrying planes for the military. He had been in one of the earliest Gallups Island classes and was the same person that prompted me to go the same route—Gallups Island Radio School. Thanks, Mac.

In Portland, I became chief radio operator because I had a Navy "striker" for 2nd operator. On the Alaska run I was the only operator. We didn't get very far out to sea and we had to get boiler repairs. The Boston Navy yard for de-gaussing, then to Halifax for a shipload of wheat for England. We were in a 100 ship convoy and being 1944 the U-boat menace was not such a menace although the escorts did drop depth charges at times. So far as I know our convoy survived intact going to London and back.

Then we loaded our cargo in Camden, NJ and formed our convoy outside Little Creek, VA because I recall we sailed on my birthday May 22, 1944. We were headed for the Mediterranean and Iran and on June 6, 1944 (D Day) were three days sailing time west of Gibraltar and I got all the news and posted some news bulletins about the war, etc. So when we passed the Rock of Gibraltar, I knew we were not going to be a part of the European Invasion.

We went on to the Suez Canal and the Captain had me send a radio message while we were in the canal—I made my first dental appointment by radio—had a tooth pulled in Suez.

To make a long story short, after this trip I went back to Montana for a visit and got drafted into the Navy. Not too exciting, managed to get scarlet fever and went to a couple of schools. Was offered destroyer duty and had a choice of that or school and the war was winding down.

Had a lot of free time and managed to find a wife, Beth Simonson; she was a clerk steno for the state of California in Sacramento. Love at first sight. So, got out of the Navy and worked for Pacific Telephone for 38 years. Retired in 1984 and joined the Olympic

Torch Caravan for a week as a radio operator. Now enjoying our periodic mini-reunions with other GIRA people and the nets.

73s and may thanks to all you good people who are working so hard keeping GIRA afloat.

HOLLAND, RAYMOND J. "RAY", R-16

I read an article in the end of 1941 that said: "Learn how to be a radio officer in the U.S. Merchant Marine." I sent away for the application and four or five months later I was accepted. I took the train up to Boston, went on the boat to Gallups Island, and from there it was all history.

HUTCHINS, G. B., R-16

Reading various accounts of the adventures and mis-adventures of the Gallups Islanders it can truly be said that this interlude in our lives has certainly influenced our various destinies. I went to Gallups quite accidentally, with a college education and having just washed out of the Navy aircorp. At the time it seemed to be an almost insurmountable tragedy; but in retrospect, it appears that someone upstairs was keeping his eye on me and had bigger things in store.

After the war, I took the international route and, except for a couple of years when I went back to school, I spent all my years overseas. I retired and lived eight years (as a widower) in a little fishing village in south Portugal. I finally returned to the States and married the girl I was going with while I was at Gallups.

I was employed as a vice president of ITT, and ran a company of theirs in Brazil. I got caught in a revolution, was imprisoned and brutalized, and consequently made a cash settlement with the company and retired early.

As therapy, I went back to sea, starting out with a drunken slob of a captain on a victory ship. This bastard deserved the worst, and that's exactly what he got when he got me. I learned, however, and when I settled in Europe I served the ARA as a stand-by when the RO didn't show up for departure, and picked up a ship to the next port while they found a replacement.

I found it all very enjoyable and, frankly wish I could still do it. But I have diabetes now and a host of other ailments that are turning me into a crawling wreck.

KUZYK, HARRY, R-16

It is difficult to remember specific routines on the Island, but I recall our radio classes, studying Morse code and radio procedure, having life boat drills and watching movies at night at the canteen. My class number was R-016 which had approximately 53 students. I wish we had taken a class picture at graduation.

After graduating in February, 1943, with my FCC license in hand, I got my first assignment, in New York City. What a disappointment! My vision of sailing on one of those new Liberty ships was shattered by this rusty old freighter, going under the name of Mathew Luckenbach, and looked like it came out of the junk yard. It really didn't matter. We sailed out of New York on the March 9 and ten days later it had two huge holes in it on the port side. Final resting place was the North Atlantic.

I followed the procedure taught at the Island—such as, in an emergency make sure you get permission from the bridge prior to sending a distress message. This meant you could break the seal on the transmitter, install the final tubes, fire-up and get your message out pronto. Just prepare a distress message in advance and have it available when needed.

I was glad when they did away with this business of sealing the transmitter. After I followed this procedure, I carried the life boat transmitter out on deck and into a life boat. Then I noticed that the other crew members had their rubber suits on and I didn't. After all, it was March and the North Atlantic! When I got back on deck, after donning my rubber suit, all life boats were full and away from the sinking ship. The Chief mate and I seemed to be the last of the crew on board. We found an oval life raft and used it for our floatation device. We were taken aboard the Coast Guard Cutter *USS Inghram* with no casualties.

During my time at sea the ships I was on visited various ports of call including Marseilles, France, Oran, North Africa, Panama Canal Zone, Pacific Islands. We were also one of the first ships to enter Bremerhaven, Germany—after the fall of Germany. One trip I recall was on the Liberty ship *SS John Alden*. After loading at the Boston Army Base and another stop at Charlestown, SC, we headed for Guantanamo Bay, Cuba to make up a small convoy which was heading for the Panama Canal. We set a course for what was called the "southern route," and were traveling alone at the time—no convoy. This was puzzling until some Army officers appeared on deck dressed in what I would call the forerunner of the space suit. The only difference was that their suits were not as bulky. They were going down into hatch #3. It had something to do with those huge storage tanks that had been loaded aboard back in the States. Their sole responsibility was to inspect those tanks on a regular schedule for leaks. It was anyone's guess what those tanks contained and the Army men never volunteered any information.

One day while we were tied up at a dock in New Guinea, some Army brass came aboard and had a long meeting with our Captain. When the meeting was over, the Captain called us all together and informed us that we were selected to carry the 1st and 2nd field hospital units on our next assignment. This included approximately 28 female nurses. The cry went up! "Women on our ship?!!" Where would we put them? What about showers . . . bathroom facilities . . . . meals . . . ??? The questions were endless. The Captain assured us that every detail had been planned out. I don't know when I was so proud of my fellow shipmates as I was at that time when they all offered to share whatever meager accommodations they had.

The boat deck on both sides of the ship was converted to sleeping quarters for the nurses; using 2"x4"s to build the structure and canvas material to cover it. All male Army personnel, both officers and non-coms were assigned space in the holds. The boat deck and above became "off limits" to all unauthorized personnel with MPs assuming that responsibility. A workable schedule materialized for all other facilities.

It wasn't long after the field hospital personnel started to arrive that the word was out that we were going to participate in the invasion of the Philippines. Several days went by for everyone to get organized, and then we set sail. Enroute I was both flabbergasted and pleased to see so many Navy vessels going in our direction! A taped message by General Douglas MacArthur informing the Philippine people that we were returning, was the signal to start the invasion.

We arrived in Leyte Harbor D-Day plus 2. That night a typhoon struck the islands. Everyone was scared. We had the two forward anchors out and they couldn't hold the ship from dragging. The Army people said that they were in a life threatening situation and they couldn't even dig a fox hole! The following morning our ship was given dock space and everyone went ashore.

All during this time we were continuously subjected to air attacks. During one of these attacks, a Japanese plane came over toward our ship as if on a strafing run. Our Navy gun crew did their job and that plane crashed several hundred yards beyond. I was curious that there wasn't that familiar sound of "hailstones on a metal roof" from the action of the enemy plane. Intelligence reported later that the plane was just loaded with camera equipment. We were lucky.

The last word we got about our nurses was that they were established in a school house just outside of town and everyone was down with dysentery! They said that they missed our ship and wished they were back. One interesting side line to this story was that the head nurse in charge of the group was one of the nurses that had to be evacuated from Corregidor. One pleasant memory that has stayed with me is the game of cribbage. All of the nurses and doctors seemed to play this game all the time. You could wander into the dining room at 2:00 a.m. and find several little groups playing and one was encouraged and invited to play. Now when I play cribbage against my computer, my mind wanders back to what seemed like a care-free existence when I first learned the game.

After we unloaded, we returned to New Guinea where instruction awaited us to go to San Francisco. It was good news indeed!

My wife and I took early retirement to enjoy life a little. We settled in Maine after 26 plus years with a Massachusetts electronic company as an environmental tech/quality control engineer. I'm in the slow lane now, doing gardening during the summer with wood carving and stamp collecting as back-up hobbies.

I have no regrets with the time spent on Gallops Island or sailing to the many ports around the world—only pleasant memories.

MOFFATT, ISAAC RAYFORD "RAY", R-16

Heard about Gallups Island—said to be the best radio school in the country. Each naval district could appoint so many candidates to each entering class at Gallups. Four of us applied—myself, Jay Turner, Ralph Malloy and a fourth whose name has faded. We took our physicals at a marine hospital in Mobile. Three of us were assigned to R-16 and the fourth, Ralph Malloy, to R-17.

We walked onto the pier at Gallups on May 23, 1942. The events to follow in the next four years created a period that I consider the best years of my life.

Academic training was intense. Morse code and radio theory five days a week, eight hours a day. Examinations each week. I don't remember what grades were required but I do know that not all who started calls finished. Since I had attained Morse code speed of 20 wpm, I was excused from most code practice sessions. That was probably the only thing that saved my neck—it gave me time to concentrate on theory. The course content and quality of the instruction was excellent. After the war the school provided me with a course description and my grades for each element of the course. This was evaluated by the Engineering Department at the University of Kansas and I received 18 hours of technical elective credit—a full semester's worth of credit toward my degree in electrical engineering.

Other memorable experiences—those 12 a.m.-4 a.m. fire watches—that long lonesome walk down to the rec hall on the tip of the island. Seemed like it was always foggy and the sound of those mournful fog horns had to be the most eerie I had ever heard. As you continued on along the seawall to punch in on the seaward side of the island, I would imagine German subs sending small boats ashore to capture Gallups Island—no matter what they would do with Gallups if they had captured it—a lonely, scared 17-year-old from Mississippi wasn't thinking that far ahead. KP duty was a welcome diversion—almost like a vacation.

Some attempt was made to give us a least a smattering of ignorance about seamanship. I never did find a use for those knots we learned to tie, so that seemed like a useless exercise. Downright stupid was the effort made to get us proficient in launching, rowing and docking lifeboats. I still break out in a cold sweat when I think of us 17-year-old farm boys who had never seen a lifeboat floundering around in a high wind in the outer harbor off Boston. The regular navy boatswains who were instructing us were putting their lives on the line and should have received medals.

There must have been a shortage of radio operators in early 1943 because they picked several of us to take the FCC license exam several weeks earlier than the rest of Platoon R-16. Took and passed the exam on January 28, 1943—three months later I received my 2nd Class Radiotelegraph license certificate. I chose to ship out of New York and was given a train ticket and $21 to get to New York. Checked in to the Sloan House YMCA on 34th St.—fifty cents a night—later upgraded to a room with a private shower for seventy-five cents.

The War Shipping Administration paid us $7 a day for room and board—you could live like a king on this amount. These payments ended when you were assigned a ship. Reported in to the American Communications Association (ACA) hiring hall on Beekman St. The ACA was a CIO union which was rumored to be communist dominated. I don't know whether this was true or not, but I do know the union hall was decorated with banners demanding the U.S. open a second front in Europe.

In a short time I was assigned to the *SS West Nilius*, and old WWI ship built in the Hog Island Ship Yards. She was tied up to the Bush terminal in Brooklyn. After loading 10,000 tons of ammunition the captain and I attended a convoy conference at 17 Battery Place. Taken by tug back to the ship and sailed early the next morning. Interestingly, I found I had a Navy radioman as an assistant. I copied my first BAMS schedule that morning containing our sailing orders. Was pleasantly surprised that everything worked just like we practiced back at Gallups.

WAR TIME VOYAGES:

1. *SS West Nilius*, Hog Islander, February-April 1943.

This was to be by far my most dangerous experience. Sub attacks started off Ambrose Ship Light almost in sight of New York. I awoke the first morning by the sound of depth charges. These attacks continued almost daily. We were headed toward the Mediterranean in a convoy of 45 ships and 6 US Navy escorts. We were loaded with 10,000 tons of ammo—I remembered the forecastle was packed with drums labeled "Delousing Agent." Later I was told this was actually poison gas that we wanted on hand—"just in case."

We lost 15 ships in several nights near the Azores. The sub packs attacked about dusk each day. They came out of the sun and on the surface. At that time our US Navy had not acquired the skills and techniques learned by the British. Also, we had no air cover. I didn't see all the ships that were sunk but we compared notes with other crews later and reconstructed a pretty accurate account. In any event, it was an exciting experience. I marveled years later when I realized I never suffered from seasickness—I know now that I was so scared for so long on that first trip I never had time to get sick.

Years later I was asked the question, "When did you first consider yourself a man?"—easy—the first trip to sea, 18 years old and the entire crew depending on me for communication with the rest of the world.

2-4. *SS Robert Dale Owen*, Liberty—May, 1943-April, 1944.

Norfolk, Oran, Algiers, Bizerte, Tunis, Licate, Cagliari, Naples, Bari, Taranto. Got well acquainted with the Mediterranean on those trips.

5-6. *SS Lot Whitcomb*, Liberty—May-October, 1944.

New York, Oban, Loch Long, Cardiff, Swansea, Southampton, Omaha, Cherbourg, New York. Sub activity greatly diminished. Also had new destroyer class escort vessel—very effective.

7-8. *SS Walter Hines Page*, Liberty—November 8, 1944-March 16, 1945.

New York, Casablanca, Liverpool, Southampton, Cherbourg, New York. War winding down in Europe—time to head for the Pacific.

Went back to Kansas City on leave and married Ruth Ann—only smart thing I did in WWII.

9. *SS Calvin Victory*—April-August, 1945.

New ship from San Pedro Ship Yards—18 knots, wow!! First time I had an assistant USMS radio operator. Loaded at Port Hueneme thence to San Francisco, Eniwetok, Ulithi and convoy to Okinawa. War ended en route back to the States.

10. *SS Bowling Green Victory*—October, 1945-April, 1946.

War was over and had enough "points" for discharge but they said nothing doing till all the boys are brought back. We loaded war materials and proceeded once again to Okinawa. Spent Christmas hanging on the hook off Naha. Then ordered to Philippines-Batangas, San Fernando. Still hadn't touched the cargo. Finally orders to Hawaii thence through the Panama Canal and back to Boston where I started out four years earlier. Would you believe the idiots then ordered us to Europe with a cargo on the deck that was all rust?! The crew put on a little demonstration, got help from the press and we were finally paid off in Boston. As soon as we were paid I headed for Logan Airport and never looked back.

Went to work for AT&T-long lines in Kansas City in August, 1946. I started night school working for a degree in electrical engineering. Later went on night shift and drove to Lawrence, KS six days a week until I obtained my MS in electrical engineering in August, 1952. Served in various management positions around the country. Went to Iran for two years with AT&T as director of installation and testing. May have still been there had it not been for that nutty Ayatollah. Retired in 1980 as division engineer in Kansas City. Then went on a two year consulting assignment with the Ralph Parsons Company on a project in Yanbu Saudi Arabia.

OWENS, LAWRENCE T. "LARRY", R-16

I signed up in the Maritime Service at Louisville, KY, and they sent me to Hoffman Island, NY. I went to the school on Hoffman until we were ready to take our seaman training on the school ship.

There were nine of us in our last class at Hoffman when we went from New York up to Boston. I remember the first time I went over to the Island. There was a little boat by the name of *O'Roderick* that everybody used for shore leave and for whatever reason to get away from the island. That was in '42, because I graduated in February of '43. Anyway, it was about three weeks before our class R-16 started, so we had liberty just about every night and led the life of Raleigh, you might say.

I nearly flunked out after the second week of RFF amplifiers. Mr. King called me to his office and encouraged me to leave, but I talked him into letting me stay and graduated OK. There were about eight or nine of us that ended up getting our 2nd class license, and the rest of the guys got their PLTs.

I met my wife in Cambridge, MA, while going to school, got married and lived together for 43 years and had three children. She passed away in '87. I'm now happily remarried, and enjoying life again. I'm back on the air, retired from the Merchant Marine in '74, and went to work at WLO in '76.

STRUZE, WALTER H., R-16

I had a friend in one of the early classes—I think he wrote me a letter and said that this would be a good way to go to sea if you wanted to be a radio operator. I applied to go to radio school, went downtown in Cleveland and took the physical exam and subsequently was accepted.

I came ashore after the war, that would be in '45, went to college and finished my degree from Western Reserve University in Cleveland. Then I worked for Armour and Company, a pharmaceutical company. From there I worked on a research project up at Union College, in Schenectady. Eventually the Korean War came along and I went back to sea and stayed there for the rest of my career. I put in 31 years and I figured that was enough sea time, so I quit in '79.

YORK, JOHN J., R-16

I became interested in radio when my father got a one tube breadboard in the 1920s. Later it was a three tube Sparton TRF, when Amos and Andy was popular. In the early 30s I built a one-tuber using an 01-A tube and rebuilt hot-shot batteries for power. By using the ignition spark coils from Model T Fords as spark gap transmitters we boys were able to communicate.

With that background, I jumped at the chance to answer a USMA ad looking for people to be trained in radio operations and maintenance. The other services would only offer one or the other, but not both, and then could not guarantee either one. Enlistment and tests were at the USCG Station in Buffalo so . . . May 1942 found me at Gallups Island.

I started in R-15 but after several months an attack of appendicitis sent me to Brighton Marine Hospital. I was later thankful for being ill while ashore, because once, in a convoy we had to transfer a crewman with such an attack via breeches buoy, to a Navy ship. Since I lost several weeks, I was placed with R-16 for the duration of training, Art Bradford and Bob Richelson became my best friends. Art's mother had Bob, me and others at her home for a New England Thanksgiving dinner. We lost track of each other for forty-seven years, until GIRA came into existence.

Two light memories stand out about Gallups Island. As new recruits came in ,we upper-classmen told them that they would have to stand sea gull watch up on the water tower with a broom to scare the gulls away. The newcomers were gullible but so were we upper-classmates. In January, 1943, when it was about 10 below zero, with ice on the salt water, one of our classmates said he was going swimming. We were gullible enough to put up money on the dare, not knowing he was a member of the Boston Polar Bear Club. He was stiff as an icecicle when he came out of the water. We wrapped him in blankets, gave him a cold, then warm shower. Thankfully, the administration didn't learn of the Saturday night prank! I only lost 20 cents. He made $2.00, so . . . everyone was happy. I have been grateful for Gallups Island because it opened doors to technical education and a technical career.

My first ship was the *SS Virginia*, a tanker, the largest afloat at the time. We were in a small fast convoy, unescorted most of the time. We shuttled from New York City to Aruba to England with AV gas and deck loads of planes. I sailed alone, which required 24 hour watches manning the DF during sub warnings.

My next ship as a Liberty, *SS George M. Campbell* and ran the shuttle from Oran to Naples from October, 1943 until June, 1944. We carried over troops and took a general and his staff to Italy. Also carried some Muslim soldiers who only had a woven cloak for protection from the winter. They were always stealing white bedding and clothing from the crew. They were called "gooks." Their duty in Italy was to infiltrate behind enemy lines at night, alone, with only a knife. They were paid a bounty for the right ear taken from each enemy killed. They slept on the open deck with no protection; they were really tough. Only one and a half berths were open in Naples, because the Germans had scuttled and sabotaged the ships beyond repair.

The next ship was a Hog-islander, *K.I. Luckenbach,* fully loaded with bombs. In the largest convoys at the time, we would drop out of convoy during rough weather; it was safer than being rammed by cumbersome LSTs. The radio shack was tiny, my feet stuck out of the end of the bunk into the sink.

I then signed onto the Liberty ship *Benjamin Schliesinger* in September, 1944. When they added double radiators to the cabins, issued overboard suits and felt boots, we learned we were headed to Murmansk. With almost daily sub and plane alerts, we arrived in Archangel, where no newspapers or magazines were allowed ashore and the radio equipment sealed. The private citizens were very friendly but they were deathly afraid of the KGB. I observed that the Russian military were brutal to the slave laborers, mostly women, who were the stevedores.

On my second Murmansk Run, we went to Murmansk. When forming single file to enter Kola Inlet, the ship who was to precede us, developed engine trouble and dropped astern. We moved forward and that ship was torpedoed, as well as a Canadian Corvette. On my last run to Murmansk, the captain came out of retirement and was 77 years old at the time.

After VE Day, I went to the Pacific on the *William Tyler*, returning Christmas of 1945. I then started college, married a young lady whom I met through the Purser on the *SS George Campbell.* I then taught Radio at Berea College, Berea, KY; a school run for the underprivileged students from Appalachia. 1953 found us at IBM in Pughkeepsie, NY where I set up and ran an Electronic Technician Apprentice School for testing large scale computers, maintaining the test equipment and maintaining the computers on the National Defense Dew Line in the Arctic. Have enjoyed retirement since 1981 with many hobbies and volunteer activities.

My highest regards to Gallups Island, the best radio school in the country during WWII. In fact, I still have all of my school class notes on theory.

Platoon R-17

BESCH, GORDON, R-17

Attended Ohio State University, PhD, Science Education, Physics, Mathematics Education, 1969. Harvard University, EdM, Science education, Mathematics Education, 1954. United States Merchant Marine Academy, BS Nautical Science, 1947. University of Wisconsin, BS Physical Sciences, Mathematics, 1951. Massachusetts Institute of Technology, Physics, 1952. Case Institute of technology, Physics, 1954. University of Chicago, Solid state Physics, Atomic/NuclearPhysics, 1960s. United States maritime radio School, Gallups Island, 1943, Diploma, Second Class Radio-telegraph License.

Employed 1969-present University of Wisconsin-Superior, Professor of Physics. 1972-1976, UWS, University Consultant. 1958-1968, Concordia Teachers College. Associate Professor of Physics. 1961-1968, North Park College, Lecturer Physics. 1955-1957, Luther High School, Physics and chemistry teacher, 1953-1954, Harvard University, Teaching fellow Science Education. 1947, Third mate, *S.S. Genevieve Lykes*, South Africa, Germany. 1945, Cadet-Midshipman, USNR, *SS Musa* (New Zealand, Tinian, Saipan). 1943-1944 Radio Officer, *SS Sebastian Vizcaino* and *SS Beauregard* (New Caledoni, Chile. Ecuador, Panama. Wales, England, Scotland. Newfoundland, Omaha Beachhead.) (Liberty ship cracked at #3 hold near Cape Race, NF.)

Publications, The Evolution of Some Major Concepts in Science Education, 1969, OSU. Abstract, Automated Protein Activation Analusis, 4th Annual Conference of Application of Small Accellerators, G Besch and M. Jones, 8/6/75.

Research—As USW consultant set up unique neutron activation analysis operation for crude protein.

DIXON, W. RION, R-17

I was born in South Carolina on October 12, 1923, and graduated from Greenville Senior High School in 1940. I worked my way through the last two years of school in a newspaper company and started to Furman University in Greenville, SC.

On December 7th I was studying and listening to Sammy Kaye's orchestra when the announcement came through, "Pearl Harbor has been bombed by the Japanese." On December 8th I left for New York to join the Royal Canadian Airforce. I had received my pilot's license at age 14. The RCAF accepted me, but said there would be a 30-day waiting period. I hitchhiked back south and decided to await my call in Washington. I was working as a soda fountain employee at Peoples Drug Store at 14th and Massachusetts, the only 24-hour drug store in Washington, D.C. Two weeks later the U.S. stopped all citizens from going to other countries to serve. I signed up for the U.S. Marine Infantry and was turned down due to poor eyesight. I went over to Baltimore and signed up for the U.S. Maritime Service. They sent me the same afternoon to New York to Hoffman's Island Coast guard boot camp. I was there until June 1, 1942 when I graduated as an oiler with papers and assigned to a ship. On the last day they asked for anyone who had been in college prior to the war. Art Downing of R-18 and I responded. They asked if we would like to go to Radio Officer's School and we responded in the positive. The next day we were sent to Gallups Island. I was appointed Platoon Leader of R-17 because of my prior training at Hoffman's Island.

No radio experience made the course very hard for me. The duties of platoon leader made it even harder. My platoon mates helped me in many ways: Donald Boyle, Roland C. Croats of Minnesota, Henry Priebe, and others with ham licenses helped me or I probably would not have passed.

In February, 1943, R-17 graduated. Donald A. Boyle and I went to Baltimore to ship out. The next day I was sent to Savannah, Georgia, to pick up the infamous *Roger Williams* of Guadalcanal news stories: "Sick Marines unload ship due to strike of crew." This was read at Gallups. When I arrived on board, they were taking handcuffed seamen ashore. Sparks had been long gone. I wish I knew who he was.

The *Roger Williams* sailed four hours later with the tide to New York to join a convoy to Russia and to England. When I went to the Radio room there was a big McKay transmitter just like the one we had at Gallups Island. This put me at ease. The armed guard officer was a lawyer from my home state of South Carolina and one of the gunners was from my high school class in Greenville, SC. We left Savannah alone and on the way up the coast we were diverted into Wilmington, NC, and the Chesapeake, due to subs in the vicinity. We loaded ammunition and armaments in New York and left to form in convoy and join others from Boston and must have totalled nearly 100 ships of all types and speeds. Off Halifax we hit dense fog for three days. When the fog cleared we were in the North Atlantic all alone. I heard many SOS calls from other ships. We diverted into Halifax until we could join another convoy a few days later. Again we hit fog and for two days we were alone again. We went into Newfoundland this time for a week.

We joined the next convoy and made it to Liverpool uneventfully except for sub harassment and losing a few outside ships. The Liverpool docks were bombed almost every night the week we were there. We formed convoy going back to Belfast. The second day out we were alone again. Someone turned seawater in the feed water for our boilers. We turned back into Belfast and had the salt water pumped out and fresh pumped back in. The British authorities arrested an oiler of German origin who had signed on in New Zealand. We all knew him because he didn't associate with the others and only had a pet rat as a friend. We went with the next convoy back to New York and loaded out for South Hampton. Half way over, a German submarine surfaced in midst of our convoy after we had depth charged it for sometime with other ships around us participating. I took some pictures, but they were confiscated. We arrived back in New York without problems.

I signed off the *Roger Williams* and went home for 10 days to visit my parents. On the train down I met Art Sheddan of R-6 who had just returned from Murmansk with many dangerous experiences. He spent a few days at my home. I would like to get in contact with him. My two brothers and father had joined the military: my youngest brother to the Naval Academy, my second brother to Naval Airforce, and my father into the 17th Airborne Division.

Back to New York and the *Maria Mitchell* to Hull and through many small German speed boats to London. We were bombed every night in the East Ham docks. *Maria Mitchell* was made Commodore's ship

Platoon R-17

on the trip back. We went through a severe hurricane with no cargo two days out of Norfolk. This was my most frightening experience. My typewriter went through the air to the chart room. We loaded in Norfolk for North Africa and Sicily. We loaded in Bane, Algeria, with snake infested cork and a few American soldiers who had been held as prisoners by the Germans. We reloaded in Norfolk for the Persian gulf. A day after passing the Canary Islands we were bombed through our smoke screens several times. Through the Suez Canal, Red Sea, and into Aden for supplies and water. We off-loaded in Khoramshah, Iran, and reloaded 55 gallon drums of aviation fuel for the Air Force base on the Island of Masira, Oman. On our way from Masira to Aden on February 22, 1944, we were in the largest submarine attack I experienced in WWII. Many tankers and freighters were sunk. I would like to know if the subs were German or Japanese. We were diverted to Port Sudan for repairs and then back to Norfolk, Virginia. We loaded out for Italy and Anzio. On this ship I had two U.S. Navy operators as assistants. We stopped in August, Sicily, for military reasons for a few days. Swimming in the harbor, I ran into Henry Priebe from Kent, Ohio, of our R-17 platoon. I haven't seen or heard of him since Sicily.

We returned from Italy to New York where we loaded for England and "D-Day." We were in Belfast waiting. On June 4th, we left in the darkness of early morning down the channel. We didn't get too far before the Canadian ship *Whiteshell Park*, rammed us and put us in dry dock in Belfast for ten days of great fun.

On November 8th, 1944, I signed on the new *Granville S. Hall* in Panama City, Florida as chief operator with two merchant marine operators to help. We loaded out in New Orleans for New Guinea. We crossed the Panama Canal and during this passage our purser fell down the ladder and broke his back. When we arrived in Hollandia we formed up for the Invasion of Leyte. Again we were the convoy commodore ship. We landed at Orange Beach, I believe the town was Dulog. We were bombed for days. We watched the Japanese bomb the strip less than 500 yards away. We returned to Finchhaven in New Guinea and converted to a troop ship for next Philippine invasion. We took MacArthur's rear echelon group and several hundred soldiers to Manila. The sunken Japanese ships still were manned and fired at us for several days. We left Manila for San Francisco.

I returned to the East Coast to be married. On the second day of my honeymoon a phone call asked me to report to Charleston, SC to sign on the *Alfred I. Dupont* for Brazil and Argentina. On the way down early one a.m. we saw a ship on the horizon blow up. We raced to rescue the survivors some 15 miles away at the mouth of the Amazon. We carried what was left of the crew to Rio. The ship was the *Norte Llyode* of Brazil loaded with Brahman cattle and machinery for Minaus. The ship was burning so the armed guard officer ordered the bulls shot. I was very sad. I feel the ship hit a mine because we didn't see or hear from a sub. We went on to Buenos Aires and loaded flax seed for the U.S. Four days north of Trinidad a fire started in the forward holes. We pumped water without success. We were routed into Philadelphia and their fire department put out the fire. We went up the coast alone to New York in total fog by direction finder. This kept me up, between the fire and the fog, for too many hours. I was near breakdown. The *Alfred I. Dupont* went back to Santos, Brazil, and back to the US with lights on for the first time.

On August 15th, 1945, I signed off and returned to college at V.P.I. for a degree in electrical engineering. My wife was dying of cancer and her family asked that we move back to a college in her hometown. We moved to Furman University where I graduated. I traveled for the world's largest shoe corporation and became sales manager, then vice president of retail corporation. I became a Christian about the time I could retire, so I left the industry.

There was an article in TIME about me as a business man. I studied until I had my Master's in Divinity and Doctorate of Medicine. I had remarried and now had five children. We packed off toCentral America, South America, Mexico, and Haiti where we built and operated several hospitals. My wife, Ann, is a registered nurse. In 1975 we returned to the U.S. to Darlington County, South Carolina, to make a home and provide an education for our children. I was Chief of Staff of three hospitals, Chief of Surgery for two hospitals, County Medical Society President for five years, and Chief of Staff of both big hospitals for many years. My favorite sports are sailing and flying. I am still practicing, but would like to slow down and work more with the fine people of GIRA and friends.

DOERHOFF, RAY, R-17

"WWII As Seen Through the Eyes of an Ozark Schoolteacher"

The unique thing about Gallups Islanders is that each of us found our way to the island through a wide variety of routes and methods. I was a neophyte rural school teacher in the Missouri Ozarks and looking for a "summer" job during the summer of 1941. The state employment office informed me that the only thing available was a U.S. Maritime Service Radio Training School in Boston that paid you while you were in radio training.

I filled out an application because I felt getting paid $60 a month for going to school might be better than getting $60 a month for teaching. Especially when you had to build your own fire in the stove, sweep your own floors, pump your own water and use an "outhouse" 200 feet from school—fully equipped with winter-time ventilation, a splintery seat and last year's Sears Roebuck catalog. My application was "still pending" at the end of the summer so I resumed my career as a "rural educator." Then came December 7, 1941, a day that none of us will ever forget.

In May of 1942 I received my induction notice from the county draft board, but on the same day I received a letter from the U.S. Maritime Service saying my application had been approved and they were in urgent need of radio operators to man merchant ships. I took both notices to the draft board and asked them what I should do. They told me I could take my choice since I would be serving my country both ways. Since radio school was an unknown quantity, but the draft notice was more of an unknown quantity, I chose the troop to Gallups Island and have never regretted my choice.

My memories of Gallups Island include the weekend trips to Boston in the hold of the *Roderick*, a converted fishing trawler, that always retained the aroma of its previous cargo. Since I had previously played in a family dance band, and one of my enjoyments was big band music, the RKO theater was my favorite spot each weekend to hear Goodman, Miller, Kyser, Dorsey, etc. They scheduled a different band each weekend, much to my pleasure. I was also in Boston the weekend of the "Coconut Grove" fire, in which so many servicemen lost their lives. Luckily, I was at the RKO that night.

R-17 graduated in February and after passing all FCC exams we were given our choice of ports to ship from. My closest friend, George Hord, better known as "Little Missouri," chose New Orleans, while I ("Big Missouri") chose San Francisco. George had married his hometown sweetheart during Christmas break the past December, but on his first trip the ship was torpedoed and all hands lost. Five members of R-17 chose San Francisco and were assigned a suite at the Bellevue Hotel to await assignment to a ship. To my surprise, on a recent trip to the area, I found the Bellevue still in existence.

I was assigned to a new Liberty, the *SS Clarence King*, and along with a skeleton crew helped bring her from the shipyards. We zig-zagged our way to Brisbane, Australia with a load of bombs and vehicles. From there we shuttled back and forth between Australia and the fighting front in New Guinea to such places at Port Moresby, Milne Bay and Lae, loaded with either bombs or high octane gasoline. Port Moresby had the only dock facilities in New Guinea so most of our unloading was done into army "ducks" who kept shuttling from ship to beach 24 hours a day.

In Boston I was lucky in not choosing to go to the Coconut Grove the night of the deadly fire. When we graduated I was again lucky in not choosing to ship out of New Orleans and getting assigned to my close friend George Hord's ill-fated ship. My luck again held in New Guinea when another Gallups Islander and I went swimming on a remote beach and he had to rescue me from an undercurrent that I could not overcome.

Many of us picked up Japanese ammunition from the caves on New Guinea to take home as souvenirs. Aboard ship, each of us in his own "ingenious way," was trying to remove detonators, primers, etc., to make these souvenirs a little "less deadly." The Navy Ensign in charge of the Armed Guard brought a large Jap projectile, along with two large wrenches, to the radio shack and asked me to help him turn the detonator out of the nose of the projectile. We tried but couldn't get it to budge. He left to find the gunner's mate, who was bigger and stronger than I, and a few minutes later I knew what happened when an explosion came from the fantail. Both were hurt seriously enough that the skipper had me break the seal on the transmitter and call for a doctor and seaplane from Townsend, Australia to evacuate the two. For the fourth time I was very fortunate that I was not any stronger and avoided another casualty.

After nine months of shuttling to New Guinea we returned to San Francisco. The union office said there was a shortage of operators on the East Coast so I boarded a train to New York, signed on another Liberty, the *SS Cardinal Gibbons*, formed a convoy and headed for India via the Mediterranean. We hugged the North African coast but a group of German torpedo bombers found the convoy and hit it with deadly accuracy. Again our ship was fortunate to escape a hit. After this, I concluded that there was no comparison between the accuracy of the German pilots and that of the Japanese pilots around New Guinea, and if I had a choice I was headed back to the Pacific.

I returned to Baltimore. Since my hometown girl friend and I had previously decided on marriage after this trip if at all possible, I headed for Missouri. We were married on August 10, 1944 and we now have just two years to go before our 50th anniversary. We spent our honeymoon in an upper berth on a 3-day train trip to San Francisco.

I signed on another Liberty, the SS *Edward N. Westcott*, while my wife found a job with the Federal Reserve Bank in San Francisco. The *Westcott* went to New Guinea and formed a convoy heading for the first invasion of the Philippines at Leyte Gulf. The gulf was literally crowded with merchant ships waiting to unload while Major "Bing Bang" Bong and his squadron of P-38s kept the sky clear of Jap planes. While the P-38s were in the air, all ships were ordered not to fire. This gave us a ringside seat for the dogfights overhead. We finally had our cargo unloaded and picked up another load in Australia. We formed a convoy and headed for the Lingayen Gulf invasion in the Philippines.

This time we were unlucky enough to be assigned one of the two "coffin corner" spots at the rear corner of the convoy. Just outside of Manila we had a quick introduction Kamikaze tactics. Ten planes approached from the rear and nine immediately went into a steep dive. Our gun crew hit two of these as they headed for the ships in front of us and aside of us. Both planes were on fire when they missed their targets and dove straight into the ocean.

I always rigged a long extension on my earphones, so as usual I was standing watch on the catwalk behind my cabin and the Captain's cabin and had a ringside seat. The tenth plane turned away but our rear gunners burst a 5" shell directly under him and he started trailing smoke. He immediately turned back toward our corner of the convoy and I suddenly realized that those flashes coming from both wings of the plane were from his gun muzzles so I headed back to the shack which was protected by reinforced concrete. He strafed the decks all the way in but our gun crews stayed on him causing him to explode just before hitting the ship. After the hit I looked out on deck and on the same catwalk, not five feet from where I had been standing, was the engine from the Jap Zero. The pilot lay crumpled on the deck below with pieces of plane and deck cargo scattered everywhere. We had injuries, mainly due to the strafing, but no deaths.

After unloading in Lingayen Gulf, we returned to San Francisco with three Jap flags painted on the ship's funnel, of which our gun crew was rightfully proud. We had a crew that "stuck to their guns" when we needed them.

I signed on a C ship, the *Cape San Blas*, and headed for the Philippines to await further orders. Two weeks out to sea I was copying press when the news about the A-bomb attack on Hiroshima came in. This news traveled through the crew faster than "Ex Lax" and I had to keep my "ears on the wire" constantly until the second bomb hit Nagasaki. Two days after that we received a message to pull into the nearest harbor and await further orders. We anchored in some isolated lagoon for four days, then received orders to return to San Francisco still loaded with our cargo.

Shortly thereafter, my wife, Rosetta, and I returned to our hometown of St. Elizabeth, MO where she taught Biology and I was Superintendent of schools for 41 years, along with teaching Physics. We have five sons who found five wonderful wives, resulting in fourteen wonderful grandchildren. They all help to make our retirement very enjoyable.

GIBSON, THOMAS R. R-17.

Arrived at Gallups Island in mid January 1943, having had about nine months experience as a radio amateur prior to the start of WWII, three months experience in a broadcast station and on quarter at the University of Minnesota. Picked up a Second Class Radiotelegraphic license at St. Paul, MN the week before leaving for Gallups. On arrival I notified Cdr. Clark that I had a license so after a few days with class R-32, I was moved to class R-17 for shipboard lab and practice watch standing classes. Graduated with class R-17 on February 20, 1943 and assigned to San Francisco.

San Francisco RMO office had a surplus of radio operators at that time, so I got a part time job with Don Lee Mutual Broadcasting System handling remote pick-ups in the evenings.

On June 16, 1943 was assigned to a new Liberty ship, *SS Irving M. Scott*, for a voyage to Australia and New Guinea plus a few other islands. Back to San Francisco January 10, 1944. Again shipped out of San Francisco on February 12, 1944 on *SS Joseph Lane* and back to Australia, New Guinea and up to Eniwetok Atoll for loading for the invasion of Guam. Navy ran an LST into the side of the ship and we were declared unseaworthy. After the invasion fleet left a destroyer tender crew welded a plate over the hole and we were sent back to San Francisco empty arriving November 13, 1944.

Back to San Francisco and the *SS Cape Saunders*, a C1 turbine ship, to Western Australia and India this time. Returned to San Francisco on May 14, 1945.

Decided to try New Orleans next time to see a different part of the world. Assigned to a T2 tanker, not quite completed in shipyard, so I was assigned a 3-day trip on the *Edwin Thacker*, a (made out of) concrete ship moving from New Orleans to Mobile. Ten days later, having missed the T2 because the Coast Guard kept us tied up six days at Pilot Town, east coast of the U.S., and Europe, finishing at New York, May 1946.

Back to University of Minnesota and an Electrical Engineering Degree. Went to work with Northern Pacific Railroad installing and maintaining VHF radio, then the Wabash Railroad as Communications Engineer followed by the New York Central as District Supt. of Communications at New York.

NELSON, MICK, R-17.

I left Gallups Island some time in January 1942, for New York. Immediately, I was sent over to Brooklyn to sign on a tanker. "Shanghaied" would be a better term. We left immediately and I had nothing with me but the clothes I wore. Soon we were in the Gulf Stream and I saw my first flying fish. One day out of Aruba (Dutch West Indies) a sub surfaced behind us and there was lots of excitement as shells were exchanged between us. We shuttled between Aruba and Freetown, West Africa, until August.

Next ship was a Liberty at Philadelphia heading for the Persian Gulf. We were torpedoed near Bizerte, Tunisia. A British Corvette picked us up and took us to Bizerte where we were guests of the 7th Army, soon to become the 5th.

Back to Baltimore, where I signed on another Liberty and back to the Mediterranean, shuttling between North Africa and Naples. Managed to be up at the Vesuvious Crater when it erupted around January 1944. Back to Baltimore again and then to Philadelphia where we loaded bombs for England. Managed a few days in London (invasion of Normandy started) and in time to watch the V-1s (buzz bombs) coming in. Back to Baltimore and on an ancient tanker—pre WWI. all mahogany and the messmen were all in white.

My last ship, another Liberty, and we were on our way to the southwest Pacific in time for the invasion of the Philippines.

Sat in Subic Bay for several months. Nothing to do but write letters. One letter went to a girl I knew back in 1940 and when I returned to San Francisco about June 1945, her letter answering me was there.

We were married the following year. Back to school, Marquette University, 1945-46. Married moved to Pittsburgh and transferred to Duquesne University. Back to Marquette, graduating in 1949. Moved to Astoria, OR as a YMCA Secretary.

The Navy came after me when the Korean War broke out and I was back at sea once more. I got involved with some of the events over there. Then back to the Mediterranean, carrying NATO supplies to north Africa, Greece, Turkey and Yugoslavia.

Back home and quit the sea for good. I found work as a miner in Butte, MT, sold insurance, magazines and used cars. Moved back to Wisconsin to work for Wisconsin Power and Light, retiring after 32 years.

ZAHRN, FRED J. R-17

Gallups Island Radio School was one of the greatest experiences of my life. I was far from home (Chicago) and on my own but never alone. Platoon R-17 was filled with a bunch of super guys who took care of each other no matter where you were from or how smart or dumb you were or how big or small. There are so many names I would like to mention, but at my age (72) I would forget a few so I will only mention one: Rion (Dixie) Dixon. He was a superior platoon leader who was younger than many of us, but had the respect of all of us. I have been delighted to see him serve as president of GIRA.

After graduation (February '43?) three of us were sent to New Orleans for assignment: George Hord, Bill Naney and me. We had several extra days before reporting to Nola so Bill invited me to his home in Flora, IL for a couple of days where I met the finest, most beautiful girl in the world. (We were married in 1945 and will celebrate our 49th this year.) IN New Orleans, George was assigned to a filthy, scrubby bauxite carrier, Bill to a Waterman ship out of Mobile and I was assigned to a new Liberty (*Newton D. Baker*) just built at Panama City, FL. George married his home town sweetheart after graduation from Gallups. His ship sank and George went down with his ship. Both Bill and I were at sea and heard his distress calls loud, clear and with a firm hand. I am sure he was trying to get help for his shipmates who were in the water. The brutality of war is a horrible thing.

I have always been a very lucky person. None of my ships were torpedoed or bombed although there were a few close calls. I crossed the Atlantic several times, into the Mediterranean twice and to the British Isles twice. Some of the ports in the Med. were Algiers, Oran, Beni Saf, Arzew, Palermo, Naples and Ajaccio, Corsica, and of course, Gibralter.

I was involved in the invasion of Sicily and Corsica, but the latter was a bungle of allied intelligence because the Germans had completely evacuated before we arrived and the waterfront bar was open and waiting for us. My ports in the British Isles were Glasgow (with 10,000 tons of sugar) and Hull (with 12,000 tons of bombs). Mr first ship went to Glasgow after loading the sugar in Havana. When leaving Glasgow we got lost in the fog coming out of the Firth of Clyde and eventually ran aground near Bangor, Northern Ireland where we were stuck in the mud for three days. I took six or seven direction finder readings which showed we were heading for Northern Ireland, logged the reading and gave them all in writing to the captain and first mate. The captain decided to ignore my warnings, much to his sorrow and mine. At the hearing in New York the captain testified that I had not taken any d.f. readings. Fortunately, I had logged the readings and the first mate confirmed my report. The captain lost his license and was sent back to Mobile.

I went through the Panama Canal twice into the Pacific to Eniwetok, Manus Island, Philippine ports, Darwin (which had been completely evacuated except for service personnel and aborigines) and Balikpapan, Borneo. On one of the trips we carried ten tons of dynamite caps in our cargo so they would not let us enter any ports until we arrived at our final destination in the Philippines and the caps were off-loaded before we went into port. What if we had blown up going through the canal? On one of the ships we had a piano which the Red Cross had put aboard but no one could play it. At Manus, we traded it tot he Navy for thirty cases of beer. It was a long negotiation and I still think we could have got fifty cases, but our fellows were awfully thirsty.

Every once in a while, from Arzew, Algeria to Eniwetok, I would run into an old friend or acquaintance but none of my R-17 cronies. A lot of new friends were made on board and on shore. I left the sea in November, 1945. Bill Naney and Ralph Milloy were the only R-17 men I saw after the war until Rion Dixon came to see me in 1991.

I had two years at the University of Chicago (on scholarship because my family was poor) before going to Gallups so in December of 1945 I took my wife to Boulder, CO where I received my B.S. in Accounting from the University of Colorado in 1947. Then I went to work for Armstrong Cork (flat broke, in debt, one child) in Lancaster, PA. After five years I left Armstrong and went to work for Dan River Mills in Danville, VA, then New York, then back to Danville, then to Greenville, SC where I ended up as chief financial officer and member of the board.

I retire in January 1983 and moved to Seabrook Island, SC and retired from the board in 1985. Since retirement I have played golf, fished, travelled, etc. and, like all of us, wondered how I ever found time to work. My wife has had a severe balance problem since 1990 so our activities are now somewhat limited but we are happy and have fun. We are proud of our two sons, their wives and our three grandchildren. I hope all of you have been as lucky as I have.

I had two years at the University of Chicago (on scholarship because my family was poor) before going to Gallups so in December of 1945 I took my wife to Boulder, CO where I received my B.S. in Accounting from the University of Colorado in 1947. Then I went to work for Armstrong Cork (flat broke, in debt, one child) in Lancaster, PA. After five years I left Armstrong and went to work for Dan River Mills in Danville, VA, then New York, then back to Danville, then to Greenville, SC where I ended up as chief financial officer and member of the board.

I retired in January 1983 and moved to Seabrook Island, SC and retired from the board in 1985. Since retirement I have played golf, fished, travelled, etc. and, like all of us, wondered how I ever found time to work. My wife has had a severe balance problem since 1990 so our activities are now somewhat limited but we are happy and have fun. We are proud of our two sons, their wives and our three grandchildren. I hope all of you have been as lucky as I have.

Platoon R-18

ELLISON, FRED F., R-18

"What a way to end a war!!

I guess, like airplane landings, any war you can walk away from is a successful one, at least from the individual point of view. Between March, 1943 and May, 1945, I made fourteen trans-Atlantic convoys in which a lot of things happened, and I saw a lot of things—not all of them pleasant, as any man who sailed in those times will know. We all have our tales of ships and men—of empty lifeboats and worse, of boats not empty, but with no life aboard. Of the thunder of depth charges and torpedo detonations reverberating through the ship, and the clamor of alarm bells, so vivid that for years afterward a telephone bell in the night would have one on his feet before he even woke up. After nearly fifty years these memories are still with me, but I think the voyage I remember most was the last one I made during hostilities with Germany.

This tale begins on February 14, 1945, when the *SS Lucius Fairchild*, a Liberty ship, sailed from New York with a general military cargo destined for Marseilles, France. This was my second voyage as Chief Radio Operator on this ship, having "shipped over" along with most of the crew, after a rather long stay in the Azores, from September, 1944, through February 9, 1945. She was a good ship, and we liked her captain, John Larch.

The trip across was noteworthy in that several nights before we entered the Mediterranean, our lookouts reported a dim "bluish" light some distance astern of the convoy. Eventually, we all saw it. Since we were the last ship in the column, this was reason for concern, and we reported it to the

Commodore of the convoy. We later learned that the convoy was probably being shadowed by a sub using the snorkel device, about which we had heard only rumors. Evidently he couldn't summon up enough friends for a major attack, for none developed, although the escorts did drop a lot of depth charges.

Marseilles was a mess! Large sections of the city near the waterfront had been leveled, reportedly by German artillery, firing "point blank," in an effort to root out French Resistance forces. Although the city was free of organized German forces, there was still fighting in Southern France, and there were still Germans occasionally in the city, as well as French collaborators, both of which the Free French were seeking just as zealously as they had been sought by the Germans! We therefore spent little time ashore after dark, because they exhibited a marked tendency to shoot first, and then look for identification.

Those French were wild, and the effect on the German POWs who were working on the docks was little short of comical. Ordinarily the POWs seemed quite happy with their lot. They were cheerful and worked hard, under the supervision of their own NCOs. At the end of the day they happily marched off to their prison pens, singing at the top of their lungs. I remember one, however, who somehow managed to get left behind at the end of the day, when his group marched off. That was one freaked-out German! He literally flew down that dock, about a block behind the squad, hob nailed boots clattering, yelling like crazy for them to wait for him. German troops on the march wait for no one, but the prospect of being found alone in a German uniform with all those "Grease Gun" equipped Free French around prevailed: He caught 'em!

The Free French notwithstanding, during our stay in Marseilles I did one of the biggest damfool things I've ever done in my life. I still can't believe I did it! Our Gunnery Officer, A. Walton Nall, of Atlanta, and I had become good friends, and one day he told me he had learned that two schoolmates and friends of his were doctors with the Emory University Medical Unit, then based at Aixe-en-Provence, about thirty-five miles away. Would I like to go with him to visit them? Like an idiot, I thought it was a fine idea—just the thing to break the monotony, so we hitch hiked across a section of Southern France where no one knew who the hell ruled the roost, except for the French, who were certain they did.

We got the OK, and the Boland brothers, Joe and Kels, were glad to see us, but shocked to realize that we had come without any protection what-so-ever. Well, Walton did have his .45, but I had nothing. When we realized that the hospital was operating using both U.S. and German doctors, nurses, and medics, and that casualties were still coming in from both sides, it dawned on us that we had not "erred on the side of caution!" It's interesting to note that the U.S. forces took over a hospital being operated by the Germans, and when the German forces were overrun, the German medical personnel elected capture to leaving their wounded and worked with the injured from both sides. So they weren't all bad. We enjoyed a nice visit, but the Bolands made sure we were in safe company on our return.

As we returned to the States, the *Fairchild* was detached from the convoy as we approached Cuba, and directed to enter Puerto Padre, Cuba for a load of sugar. Puerto Padre is situated on the northern coast of Matanzas. It lies several miles up a river in a landlocked bay, and the dock is on an island in the bay, connected to the mainland by a causeway about a mile long. There was a little dockside village and, to our delight, a club house used by the sugar company officials, built out over the water with a swimming area protected from barracuda and other such nuisances. Best of all, it was put at our disposal, and there was a little Cuban who would come down and open up whenever we wanted to use it. They had a piano, and we had a signalman who had played in a night club in happier times. Old "Flags" looked at that piano the way most of us would have looked at Marilyn Monroe! They also had about four years back issues of Colliers and Saturday Evening Post magazines, and—best of all—cold beer and cokes! We thought we'd died and gone to heaven.

Our entry into Cuba, however, was saddened by the news of the death of President Roosevelt. We were surprised, by the depth of the Cubans' sorrow. Cuba, in those days, was far different from what we know today. The Cubans were enjoying relative prosperity which they attributed to Roosevelt's personal efforts to give them a major portion of the U.S. sugar trade. Hence their sorrow. We found them to be very hospitable. The best rum was twelve dollars a case, and the finest cigars were five bucks a box.

We had a good time in Cuba. There were few watches to stand for any of us. I had none, so every day we'd go down to the club, with our little Cuban right behind us, swim, read the Colliers and Posts, drink beer or cokes, and listen to "Flags" play the piano.

One day, being an avid fisherman, I hired a little Cuban kid and his sailboat to take me fishing in the bay. Well, the fish weren't biting, and those that did were small, so it wasn't much fun. We headed back for him to drop me off at the club, which should have ended the expedition. Not so! Unfortunately, the Chief Mate and several others had been at the club just long enough to get thoroughly tanked up on beer. They met the boat and hi-jacked it with me aboard, but minus the kid who knew how to sail it, and merrily we sailed away, with them assuring me (who knew nothing about sailing), that they were, indeed, expert sailors. And so they were, as long as the boat was sailing down wind. Up wind, it developed, was another matter, and we soon found ourselves down wind from the dock, on the mainland side of the bay, in a mangrove swamp, and none of those idiots knew how to sail. There was only one way to get out of that mess—the water was only about four feet deep, and someone had to get out and tow the boat.

The bottom was shell, mud, and mangrove roots, and guess who was the only one wearing shoes. My best pair of Florsheims, no less, and it was about a mile to the causeway leading back to Puerto Padre. So there I was, pulling the boat, while those jokers just laid back in that boat having a ball, with me swearing every step of the way. They didn't know anything about sailing a small boat, but they sure could give advice!

Eventually we reached the causeway, where we were met by the kid, who had been anxiously watching all the time. We paid him off, and I went back to the club to wash the mud off and try to cool down. The rest went back to the ship. When I returned some time later, I realized that the rest of those characters had not only been shoeless, but shirtless, and that old Cuban sun had really done a number on them! I mean world class sunburns—second and third degree burns that put them in their bunks so bad I could hardly even get a feeling of satisfaction, until I remembered how they'd bragged about being such good sailors.

Then we left Cuba. By then things were quiet on the Eastern Seaboard. There had been no reported sub activity for some time, and it was evident that things were winding down in Europe. We were, therefore, ordered to proceed independently to New York. We got almost as far as Cape Hatteras when I received a radio message ordering us to turn back to Charleston immediately, which we did. When we got there we were told that a sub was operating off the Outer Banks of North Carolina, very close to where we had turned back. I now know that this sub had sunk SS Belgian Airman, the last ship sunk in our waters during WWII, on April 13, about 80 miles east of Nags Head, NC. If we had not turned back, we would have been his next victim.

They held us in Charleston long enough to assemble a convoy and *Fairchild* was honored by being designated as the Commodore ship. I felt it to be a somewhat dubious distinction, since I had to give the Commodore my quarters, but there he was, a Retreaded Navy Commander, with his staff, including a rather "snotty," I thought, lieutenant who was his communications officer.

So off we went again, this time with an escort. We passed Cape Hatteras, about a hundred miles abeam, during the night of April 29, 1945. About 1:30 a.m. on the 30th, east of Cape Henry, Virginia, one of our escorts, *USS Natchez* (PF-2) made sonar contact and sighted a U-Boat snorkel. She tried to ram, but the U-Boat managed to dive, and *Natchez* called up *USS Coffman* (DE 191), *Bostwick* (DE 103), and *Thomas* (DE 102). *Coffman* made three attacks; *Thomas* fired a creeping Hedgehog barrage; and *Bostwick* fired a pattern of depth charges. About eight minutes later members of the escort heard deep sea explosions. At 2:07 *Natchez* dropped a pattern of depth charges and a huge oil slick resulted. Forty five minutes later *Thomas'* sound crew heard a huge underwater explosion, and there was no further contact. The sub was down—about a hundred miles from his last victim, the *Belgian Airman.*

The sub was subsequently identified by the Germans as *U-548*, under Captain Krempl. She was a type IX-C40 boat, an improved version of the long range boats, which represented a large portion of the U-Boat fleet. She displaced 1120 tons, surfaced, had a surface speed of 18.3 knots, and a submerged speed of 7.3 knots. She had six torpedo tubes.

After the ball was over, a rather unusual event took place which involved me personally: to his acute embarrassment, the Naval Communications Officer had to ask me if I had contact with a shore station. I've never understood this, and he wasn't about to give me, a mere "civilian," any explanations. It's a fact, though. Here we were, almost in sight of one of our major naval installations, Norfolk, and they couldn't pass a message! In retrospect, I have to wonder if their communications were tied up (ashore) due to some kind of a major flap going on over the fact that one Adolph Hitler shot himself on that same day!

Could I pass a message, within a few hundred miles of my main shore based contact, Station WSL on Long Island? You bet! So we encoded a message to the Commander of the Eastern Sea Frontier, reporting the sinking and asking permission to detach a "killer" group from the escort, just in case he was playing "possum." He wasn't. Incidentally, I personally transmitted that message.

The actions of the escort vessels which I have described are based on information about the individual ships found in the Dictionary of American Fighting Ships, James L. Mooney, Ed., Naval Historical Center, 1981. The information pertaining to U-548 is found in the book U-Boats Under the Swastika. There's just one thing that puzzles me: I don't know what the other ships in the convoy were doing, since I was on watch, and radio shacks didn't have portholes, but at some point in this fracas the *Fairchild's* after deck gun was involved and fired 58 shells. The Commodore must have ordered it, but it's rather unusual for a Merchant ship to fire when the escort is engaged. It seems someone either saw, or thought he saw, a periscope. That wasn't unusual at the time.

So that's how the last U-Boat on the Outer Banks was put down. Just a week later, the last enemy submarine in American waters was destroyed off Rhode Island. Maybe it's a little poetic that the sinking of *U-548* was reported by one of her intended victims.

And that was the end of my war! Oh, there were still months of sailing for me, through December, 1945, bringing troops home, and taking food to Europe, but for me, the war was over. Unfortunately time had erased names, but not the faces, of most of the men who were with me on that last voyage. If any of them ever read this, I hope they'll contact me. I'd like to buy them a drink—even the mate who conned me into towing that damned sailboat!

NORBERG, LAUREN RAY, R-18

I first heard of Gallups Island and the need of radio operators through an article in the amateur radio publication "QST." I responded and in due time, was accepted to go to Gallups Island Radio School in Boston.

I was assigned to the class of R-18. During our "work week" when each class took their turn at working, I was assigned to wiring the code tables in the new code room. I probably was selected because I had done electrical wiring when the REA came to Nebraska. Many of my classmates did kitchen and maintenance work.

During shore liberty I would visit many of the early American historical sites. This brought new meaning to those dry and stuffy history classes in high school. I spent a lot of time at the "Buddies Club" on the Boston Common. I went to the historic Park Street Church and of course, Scollay Square.

I began to realize that Gallups Island was a good radio school, but not aware of its academic excellence until a few years later, when the Dean of Admissions, at Illinois Institute of Technology, evaluated my academic work at Gallups Island. It was a good school.

Following graduation on April 6, 1943, I was sent to New York City for assignment. Later, we were transferred to New Orleans, then on to Mobile. At Mobile I joined the crew of *SS William Windom* for the next eight voyages, leaving her February 13, 1946, in Los Angeles. I have often wondered what happened to her before she was scrapped in New Orleans during August 1964. During those eight voyages I had become attached to her.

After loading cargo on the *Windom*, we sailed in convoy for Liverpool. This was my first experience as Radio Officer. I was the only one aboard. This was common in those days. I doubt if the Captain had much confidence in me, being a new "sparks" on his first trip at sea. I was apprehensive. The convoy was new to me. It was remarkable to see the collection of ships with precious cargo all sailing in orderly rows and columns while being buffeted by the North Atlantic. We arrived in Liverpool August 6, 1943. This was my first time in a foreign land. It was interesting to note how differently they do things—especially when speaking English. We were back in New York by the end of August.

Voyage three was more eventful. We left New York during September, sailing in convoy for England. The crossing was uneventful. We arrived safely and discharged our cargo. Then things began to

Platoon R-18

happen. We were selected to go to Murmansk. Although not told, the scuttlebutt and preparations so indicated. Degaussing coils were installed. In addition to its normal use, the coil could be used to melt ice.

By mid-October we were loaded and outfitted for cold weather. We left the Clyde River sailing north to Loch Ewe. We waited until a convoy could be assembled. Our captain had survived a previous trip to Murmansk, so we had an experienced leader. This provided some confidence in a scary situation.

Our convoy, JW54A, sailed from Loch Ewe on November 15, 1943. Enemy planes came out daily to look and plot our progress, but kept a respectful distance. As we sailed northward the temperature became cooler. The nights became longer and longer until our day was just a little twilight at noon. The trip was safe. Due to the weather, we were able to sail into the White Sea discharging our cargo new Archangel.

The fathometer had developed problems. Russian technicians came aboard to repair our unit. I did not know the Russian language but was able to recognize some of the technical vocabulary. We could almost communicate using the schematic diagrams. They stayed on the ship longer than expected. We offered them food, but they declined at first. Later, they sent for permission to eat aboard. When permission was granted they devoured our food as only two hungry men could.

The "Culture Center" was certainly not like the Boston "Buddies Club." They tried to make our stay interesting, but everything was so difficult under war time conditions. I got a fresh new insight into the ravages of war. I also heard about the virtues of Communism. The people were kept informed by loudspeakers at most street corners.

The convoy procedure was for the British Admiralty to give protection to an outgoing, fully laden convoy. The escort was usually in two parts. There was close escort, usually sailing right with the convoy. There was distant escort cover, usually much heavier and sailing at a distance, but within range of help. When the outgoing convoy had passed the area of danger, they would meet a homebound convoy and escort them home.

Our convoy, RA55A, sailed in mid-December and was well under way by Christmas Day 1943. We didn't realize it at the time, but we were the pawns or objects in an important Naval engagement of WWII. The German battleship *Scharnhorst* was sunk during the fighting in the afternoon of December 26, 1943. She went down at 72° 16'N, 28° 16'E just a few miles off our port side. We did not participate in the action, but were just there. I am thankful that the *Scharnhorst* was prevented from doing her assignment as a convoy raider. The sinking was a real Christmas present.

I served my country for a little over three and a half years. First Gallups Island, then at sea aboard the *SS William Windom*. I was Chief Radio Officer for eight voyages. I began in the Gulf, crossed the Atlantic twelve times—once to Murmansk, twice to Cherbourg, through the Panama Canal, and crossed the Pacific four times. I had grown attached to the ship. I liked her. I had a lump in my throat as I left. When leaving, I turned for one long last look at the *Windom*. I never saw her again. Records indicate she was scrapped in New Orleans August 1964.

This was the end of a chapter of my life. I returned to Nebraska, never to go to sea again.

ROGERS, P. M., R-18

"Patton's Prayer and the Liberty Ship *John Cropper*"

"Almighty and most merciful Father, we humbly beseech Thee, of Thy great goodness to restrain these immoderate rains with which we have to contend. Grant us fair weather for Battle. Graciously harken to us soldiers who call upon Thee that armed with Thy power, we may advance from victory to victory, and crush the opposition and wickedness of our enemies, and establish Thy justice among men and nations. Amen"

The story of Patton's Prayer, as I think it applies to the Liberty ship, came about because I signed on with U.S. Maritime Service Radio School at Gallups Island Radio School, Boston, MA as of July, 1942.

Because of the great training I received I was able to make a contribution to the U.S. of America.

This story is written to give a belated thanks to General George C. Patton. I feel to this day that his famous prayer during the "Battle of the Bulge" had something to do with keeping the *John Cropper* from being blown to hell on three occasions.

It was around Christmas of 1944. World War II seemed to be as undecided as ever. The German Army had initiated a counter-offensive that was soon to be known as the "Battle of the Bulge." Patton, beleaguered not only by the German forces, was also suffering from the effects of bad weather, and a gasoline shortage. With the weather in their favor the Nazis were threatening to burst through allied lines, and a possible serious defeat of Patton's forces. General Patton prayed for clear weather so the U.S. Air Force could counter the massive blows of the enemy. He also prayed for gas for his tanks.

The Liberty ship *John Cropper* set sail from New York Harbor in late November carrying a cargo which consisted only of 5 gallon cans of gasoline from stem to stern; her destination Antwerp, Belgium.

We entered the mouth of the Scheldt Estuary about the first week in December, '44. Needless to say the *Cropper* was one of the ships kept further from other ships in the convoy because of the nature of her cargo.

As the *Cropper* steamed into the estuary she struck an uncharted sandbar and came to a dead stop. As we lay dead in the water, a ship behind us was ordered to go around us and proceed to Antwerp. An E-Boat hit her with a torpedo. The ship was carrying no ammo or explosive; her cargo consisted, I was told, of mainly food-stuffs.

That E-Boat would have triggered one of the biggest explosions of the war had she hit us. Not only that, it would have caused a huge dent in Patton's gas supply. This was one incident.

We managed to free ourselves of the sandbar thanks to our own engine and help from the tide, and proceeded to Antwerp.

The second incident occurred when the *John Cropper* docked in Antwerp. We were at the dock about two hours when a V-2 hit about 500 feet from us. The force of the explosion blew some of our port-holes out and busted doors on the ship. It also scared the hell out of every man aboard.

Incident number three occurred the next day. I was on the flying bridge at the time. I saw a Buzz-bomb come roaring right at us. The world turned unreal and everything seemed to run in slow motion. I froze. I prayed. The Buzz-bomb's propulsion engine stopped. It was as close to kissing the mast of the jumbo boom as it could get without touching it. The silence of the Buzz-bomb meant she was about to drop on us. For some unknown reason the engine sputtered to life for about five seconds (those seconds seemed like an eternity). It was enough time to keep it skyborne long enough to land on a clear area of the docking complex. There was a great explosion, but it wasn't us. I did not move for a long time.

So, there you are. The *John Cropper* was nearly blown out of the water three times. To this day I think it was Patton's prayer at work.

Platoon R-19

FELLS, W.E., R-19

Early in 1942 while employed as a test technician with Westinghouse Electric in Baltimore, MD I began to think of becoming a shipboard operator. Later, when a co-worker of mine, Bob Whittaker expressed the same interest we went to apply for training. We were told that we must first join the U.S. Naval Reserve. This was apparently to keep the Army from drafting us while we were at sea! Unfortunately we both failed the eye examination. The Navy doctor said that we were probably tired since we had been working the so-called "graveyard" shift. He left the room for about 15 minutes. When he returned, we both passed the test. Miraculously our eyes had become rested, no doubt from staring at the eye chart in plain view.

In any event, on a cold and snowing morning in October we stood on the dock in Boston awaiting our first sea voyage (to Gallups). Except for standing anchor watch at 10 below zero and a few forays into Scollay Square we survived the winter. Since I already had a ham license and was reasonably proficient I was able to skip the code classes and much of the theory. I was given permission to concentrate on "Shipboard Lab". In July of 1943 I graduated, by myself, so I don't know what class or classes I was officially registered in. Incidentally in another article by a Gallups grad someone mentioned knowing a "tall skinny kid". I was 6 ft 3 in, 160 lbs at the time so I met the description.

I was assigned to San Francisco and spent several months "on the beach" along with another Gallups graduate, Hank Brown, from Broken Arrow, Oklahoma. Unfortunately Hank was killed when a kamikaze plane hit his ship somewhere in the Pacific. I couldn't get a ship so I joined the ROU and was immediately assigned to the *SS Vachel Lindsay*, a liberty being built in the Oakland shipyards. Shortly thereafter, loaded with bombs and other goodies we sailed for Noumea, the Russels, and Bougainville. I particularly remember the thrill of setting foot on foreign soil for the first time, in New Caledonia. I guess this set my itchy feet for a life of wandering the globe. I made another trip to the South Pacific aboard the *SS John Banvard* before I returned to the East Coast.

It was in the dead of winter and I was assigned to the *Gov. John Lind*, a coal carrier which had been converted from carrying molasses from Cuba! The RO's quarters were in the radio shack which was on the deck aft of the flying bridge. Instead of portholes the shack had two 3-foot square windows with the glass missing! On the trip from Boston to Norfolk (we were carrying coal) I stuffed pillows into the windows. When I complained to the port captain he said that "fresh air was good for the sparks." For some reason I decided to part company with the Bull Line and was assigned to another, the *SS Arthur St.Clair*, bound for the UK. Of course I didn't know the destination but it was my first experience travelling in convoy, which didn't reassure me too much. Except for watching "V2s" sail over us in Southhampton the trip was uneventful.

My final ship was the *David S. Terry* where, after several Mediterranean stops, we went into the Black Sea to Poti, Batum and Novorossisk. The latter city was totally devastated with hardly a building and the population living in cellars. Although we were very popular, people were afraid to be seen with foreigners. While there, we celebrated VJ day in the harbour dodging flack from hundreds of big guns firing in joy.

My final trip was to Mozambique and thence to Calcutta and Colombo. This time we celebrated VE day in the Seaman's Club in Calcutta.

Since the war was over and by this time I was married, I had no desire to stay at sea. The National Bureau of Standards was looking for men to man their radio propagation research facilities in the Pacific. We wound up on Palmyra Island, 1500 miles south of Hawaii, where we spent three wonderful years and our son was born. During that time we went through a serious tidal wave and our house burned to the ground. I received a citation from the U.S. Navy for maintaining emergency communications with my amateur radio station KF6AB during the tidal wave.

In 1949 I transferred to the FAA and we lived in Honolulu, Hilo and Midway Island until 1951. By this time I had my fill of civil service and we returned to Baltimore where I went to work for Bendix Radio as radar field service engineer. My first assignment was at the 775th AC & W Squadron in Cambria Pines, CA. After a few months I was transferred to McClellan Air Force base where I ran the radar repair facility for the western US. In 1955 we went with Bendix to Wiesbaden, Germany supervising radar field service activities in Europe and North Africa. In 1957 we returned to Baltimore where I became a communications equipment sales engineer.

In 1959 I went to work for Collins Radio in Cedar Rapids, Iowa as manager of their Ground Communication Product Line. In 1960 I was made General Manager of the Collins subsidiary in Frankfurt, Germany and in 1963 transferred to Beirut as Vice President for the Middle East and Africa. Unfortunately as a result of the famous "six days war" we were forced to evacuate Lebanon. In 1967 I was moved to Bangkok in charge of operations in the Far East, most of my time being spent in Viet Nam and Korea.

I will never forget my experience as a sparks and am very grateful that I did not have to make the ultimate contribution that so many of our fellow shipmates did. It would be nice to think they know that their country finally officially recognized their contribution.

JORGENSON, RAY, R-19

I grew up on a ranch in western Montana and graduated from the University of Montana in 1942. With the war under way it was time to pick a service. I saw an ad for the Gallups Island Radio School and I thought t'is the seafaring life for me.

After the necessary paper work, I was soon on my way . . . On the train enroute to Boston I met Bob Wilson who was also headed to Gallups Island. We entered the class R-19.

R-19 graduated about the end of March 1943. (My FCC Radio License is dated April 9, 1943.) I then took a train to Seattle and there I signed on as radio operator aboard the SS *Samoa*. This was an old, old freighter. It looked like a hammer could put a hole through the hull most any place. My bunk and the radio equipment was in one small cabin, about 8'x10'. I stayed with the *Samoa* until August of 1944.

Ports of call included various Alaskan ports such as Ketchikan, Juneau, Dutch Harbor, and Bethel. Our South Pacific Ports of call included Funafuti, Tarawa, Kwajalein, and others. The aftermath of the Tarawa battle was a gruesome sight.

On September 5, 1944 I signed on the Liberty ship *SS John W. Cullen*. The cabin and radio room were a great improvement over the old *Samoa*. The *Cullen* hauled various military cargo to such ports as Hollandia, New Guinea, Lingayen Gulf in the Philippines, Ulitaha, and Naha Harbor in Okinawa. While at Okinawa a typhoon hit the area. We

Platoon R-19

were ordered to sea to ride out the storm. As we were about half loaded we rode out the storm fairly well. We did see some other cargo ships that were nearly empty that were having a rough ride.

Japanese planes were occasionally overhead while we were anchored in the harbor. One night we were ordered to unload cargo—we had our lights on and a Jap plane dropped a torpedo that missed us but hit a nearby cargo ship that had their lights out. On September 21st, 1945, about one month after the Japanese surrender, I signed off the *Cullen*.

In October of 1945 at Portland, I signed on a brand new tanker the *SS Victory Loan*. This was a fine ship. It had a large cabin with a large adjoining radio room. We picked up a cargo of oil in California and headed to Nagasaki, Japan. The second atomic bomb was dropped on an industrial area several miles out of town. We were not allowed out to the bomb site, but we did see numerous Japanese with burn marks on their arms or face.

From Japan it was a fast trip to Manila—then to Bahrain in the Persian Gulf for a new load of oil—then back to Manila and on to Honolulu. We were on our way home to San Francisco when we received a radio message to proceed to Panama. We went through the canal to Aruba for a load of oil—then on to Bristol, England, and Bremen, Germany. We returned to Galveston, TX for discharge on May 30, 1946. This ended my seafaring life.

After returning home I bought the family ranch—married a local gal, Ruth, and we raised three boys and one girl. I also spent 23 years working for the U.S. Department of Agriculture. I retired from the USDA in 1981 at the age of 60. In September of 1987 I saw a news item in the Daily Missoulian about the Gallups Island Radio Association. From this I contacted James Kinkel and joined the GIRA. Ruth and I attended the reunion in 1988 at San Francisco and also the St. Louis reunion in 1989. We enjoyed these very much.

LEMMA, ARMAND, R-19

When I was a youngster- about 12 years of age- I received a radio . . . I was quite into it.

Then of course . . . (there was) quite a big to-do about the Merchant Marine—operators were needed; radio operators. And at that time, they claimed they wanted you to have some type of experience in order to send you to school. So I went for my physical, but I failed.

They were going to send me home and they said 'Go home and put on some weight and come back in three weeks.' . . . I (finally) got to Gallups Island.

MOORE, MURDOCK C., R-19

"Long time no sea"

GIRA bios being in vogue, herewith my nautical notes. Bred, born, schooled and reared in Boston, I at least had somewhere to go on weekend leave. May 30, 1942, was my 18th birthday; six weeks later I was on Gallups Island as a member of R-19. I had given up the security of my 44-hour $12-a-week job in the Parker House Hotel print shop for the life on the ocean wave. In April of 1943 I was safely ensconced in Baltimore as Chief (and only) radio officer of the fair to middling Liberty ship, *SS Lewis Morris*. Then it was off to the bounding main.

Thus my 19th birthday was spent in passage through the Straits of Gibraltar. The next few months were spent roaming North African ports and taking part in the invasion of Sicily and Salerno with a side trip to Malta for an "individualized dive bombing attack." Back to the USA and on through the Panama Canal for a winter's sojourn visiting assorted Pacific islands. To the US by way of Chile (for nitrates) and the Canal and then over to merrie England for the invasion of Europe. (I spent my 20th birthday in Cardiff the week before.) Back and forth across the Channel to the beachheads and Cherbourg for the summer, and in the fall once again to the far Pacific to visit more islands for the winter.

Signed off in San Francisco and took the first train home to Boston. I immediately signed aboard the *SS Matthew J. O'Brien* and sailed across the Atlantic to Antwerp to be greeted by assorted V1 and V2 missiles. Thence to US to leave Norfolk on VE day (no celebration for us). Spent the summer again touring the Mediterranean (Italy, Yugoslavia—21st birthday in Greece—Sicily and France). Loaded "Invasion of Japan" cargo in Marseille and carried same through Canal (V-J Day in Panama—WOW!) over to Okinawa and back to Stockton, CA. Signed off last voyage December 19, 1945. Returned to high paying job as printer (now $18 for 44 hours). Married prettiest girl in Medford, MA. Moved west to Ohio, and in time became longtime production manager of daily newspaper.

My active radio operating career (other that standing night watches so I could listen to the Armed Forces Network) consisted of two ETAs to the Panama Canal Zone (a grand total of about five minutes), and decoding two BAMS messages for my ship. Saw my share of death and destruction, but only thing I really feared was a raging storm. Never got seasick or sick of the sea.

My deepest respect and admiration goes to those merchant mariners who survived and sailed on through the carnage of 1942. Beyond that point chances of survival for merchant seamen rose dramatically, notwithstanding we still sailed in harms way! Harms way took me to 66 ports of call and 25 countries!

SMITH, JAMES S., R-19

Left Gallups on March 30, 1943 for San Francisco. Assigned to the Liberty *SS Louis Hennepin* at the Richmond, CA. Kaiser Yard on April 8, 1943. Left San Francisco on "World Cruise" in mid April. Scheduled and unscheduled ports-of-call were in Wellington, Hobart, Melbourne, Perth, Port Tewfik, Suez, Benghazi, Tripoli, Malta, and Gibraltar. Arrived in Norfolk in early October. Left last week of October via Gibraltar, Malta, Taranto to Bari. Survived "Disaster at Bari" in December, and with temporary hull patching, arrived back in Philadelphia in February, 1944.

Back to San Francisco and assigned to *SS Britain Victory* at the Kaiser Yard in Portland, Oregon on March 2, 1944. Was designated as a fleet support unit, usually ammunition, and was involved in Guam, Leyte, Iow and Okinawa operations. Aboard until February, 1946 when she came back from western Pacific to Baltimore. Back to San Francisco and assigned to *Haiti Victory* first week in March, and made two trips from Mobile to San Juan, Ponce, New Orleans and Mobile.

Became involved in special Signal Corps unit in New Orleans and spent two years with this unit throughout the world and ended up being based at Fort Monmouth, NJ, assigned to Carlisle Barracks, PA (At that time it was the home of the Army CID).

Left the Army in July, 1948, was married and went to the University of Washington to complete education. Went to Glendale, CA, in 1950 and had an offer from old KFI (Earle C. Anthony Radio) because I had my first radiotelephone ticket. Didn't work out because "police action" started and I was recalled to active duty. I was fortunate, because one year and nine days later I was invalided out. Went to work for IBM in 1987 as national account service executive. Again, I was fortunate in my experience in IBM. Had the opportunity to witness and to be a part of the evolution of computing from vacuum tubes to 4 meg integrated memory chips.

Have raised and educated five children (so far eight degrees among them and working on two more).

Have to believe tour at Gallups contributed positively to my opportunities in later life.

WALLACE, KEITH E., R-19

I was at Gallups Island when it was closed and was transferred to Hoffman Island for a couple of weeks where I received my license and shipped out on the *S.S. William Blount*.

I graduated from South Dakota State University in 1949 and taught Vo-Ag at Barnard and Groton, South Dakota. I then attended graduate school at South Dakota Southern University, married Marilyn Page from Haverhill, Massachusetts in 1953, and in 1954 received an MS degree in agronomy from South Dakota Southern University. I accepted a position as extension weed specialist for Washington State University at Spokane, Washington. In 1976 I was promoted to full professor at WSU. In 1977 I took sabbatical leave in New Zealand, Australia and the orient. In 1982 I accepted a position as agronomist with WSU at an agricultural research station in Lesotho, southern Africa.

I retired from WSU in August 1984. I still live in Spokane and have been contracting to do research on weed control since retirement. I have three children: Dee Ann Fagan, Bruce and Loretta.

WARD, JOHN "JJ", R-19

From the hills of West Virginia I arrived at Gallups Island on the old schooner *Rodney* in late June of 1942, just in time to make R-19. It was my first encounter with an ocean, albeit a bay or shoreline. Still green as the seas, I shipped out of Galveston, TX on the *SS Alcibiades*, an ancient Panamanian registered tanker built in Scotland in 1908.

We were supposed to deliver it to the U.S. Navy in Australia but spent almost a year refueling the Seventh in New Guinea and neighboring islands. The navy finally accepted the cockroach-ridden thing in Brisbane and the crew was returned to SFO on a Dutch transport.

Two or three weeks later I shipped out of New York on a T-2 tanker, the *SS Red Canyon* that returned to the Pacific Ocean where we spent the remainder of WWII bringing avgas from West coast ports. I was in Manila on VJ Day.

Back in Los Angeles I took a Liberty ship on its last voyage to a NJ boneyard, then spent the winter on the *SS Wahoo Swamp* taking South American oil to England.

In the autumn of 1947 I went to South America (Lima, Peru) as FRO with Panagra until they furloughed me in late 1948. Sensing the twilight of the trade, I enrolled in college. During the 1950 summer break to replenish dwindling resources I shipped on the *USNS Sappa Creek*, a T-2 resupplying Arctic airbases (Greenland, Newfoundland, and Labrador).

The *Sappa Creek* was headed back to New Orleans (Algiers Navy Yard) and reserve status and I back to college classrooms. Near the mouth of the Mississippi we got news of North Korea's invasion of the south . Naively believing the resulting war would last for only months I decided to stay aboard. We loaded in Baton Rouge and headed for Pusan, then spent boring months shuttling between the Far East and the Persian Gulf with an occasional break to Europe and the U.S.

After a brief vacation, I joined the *USNS Tamalpais* for more of the same—we did make two complete trips around the world (one in each direction). I would never have believed one could spend so much time at sea and so little in port. At most of the stops there was either no shore leave or no point in it.

After Korea there was the cold war (literally) when we hit a huge iceberg out of Thule, Greenland, and almost sunk. About then I got married and bid the seas adieu.

I worked at a soul-shriveling and low paying ground radio job with Braniff before signing on with TWA on their Europe and North Africa flights. Long after we were really needed, TWA furloughed the entire FRO force. I shipped out in Hawaii on a coastwise tanker for three years while awaiting a call from the FAA. Then to Utah, Nevada, and finally, to Arizona where I spent more than twenty-two years in air traffic at Phoenix's Sky Harbor Airport.

In the early 60s I used accumulated annual leave to sail on a Viet Nam bound freighter. Due to the primitive conditions and enemy fire, unloading required more than a month and I was weeks late getting back to my airport job, to the FAA's great annoyance. My final FAA years were relatively uneventful. Despite rotating shifts I managed to finish work for a degree and enjoyed an avocation writing and teaching.

We now live in the foothills some twenty miles north of Phoenix on two acres where we raise Corgis, pheasants, and all sorts of fruit and vegetables.

I work part-time with the the University of Arizona Extension Service dispensing horticulture advice via articles and telephone. Never would I have guessed that worlds would be opened up for an impoverished hillbilly by answering a Maritime Service ad in a pulp magazine. My first choice was a deck cadet at the Kings Point Academy, but I had no money to buy the required uniforms. What a fortuitous break being broke turned out to be when I settled for the Island instead.

My wife, whom I met in college, is a test engineer with Honeywell Avionics working on navigation systems for the French-British Air Bus and McDonnell-Douglas (MD-11). She is also a Captain (weapons controller) with the Air Guard.

Best wishes and 73s.

Platoon R-20

HAYDEN, ED, R-20

I was born at home in Danville, VA, May 26, 1923. I grew up in Delaware, along the coast, just below Cape Henlopen, in Rehoboth Beach. I was exposed to the maritime world by the Delaware River pilots who lived in the area. And when the war started, I got first hand reports of the tankers being sunk off shore from the river pilots and from my next door neighbor who was a nurse at the Beebe Hospital in Lewes, DE. In fact, I knew about the *SS China Arrow* and how Ken Maynard, the radio operator, rigged up a temporary antenna that helped save the crew, long before it ever became public. Also, I later met Maynard at Gallups where he was then a code instructor.

I graduated from high school on June 4, 1942 (the day the Battle of Midway was fought), and shortly afterwards joined the Maritime Service. I was assigned to the training ship *Allegheny*, which was later named the *American Seafarer*. From there I applied to go to Gallups Island.

The reason I did it that way was that a neighbor heard of my interest in Gallups Island and introduced me to Mr. MacDonald who ran the ROU office in Baltimore. He advised me to join the Maritime Service first, then apply to Gallups, and I would get there that way. He was right.

I arrived on Gallups the first of August 1942 and graduated in April 1943. It started out with warm weather and great fun. But the winter got cold, and the work was demanding. We had our share of laughs and good times in Boston, along with the tears—I was pressed into service to help the firemen at the Coconut Grove fire in November, 1942.

My first ship was the *MV Surprise*, a C-2. We made a quick trip to the UK (Liverpool) with a load of canned hams for the military, and brought back 40,000 cases of Scotch whiskey. We traveled alone on a zig-zag course, reasonably uneventful, except for coming close to an iceberg and firing on a suspected surface raider.

My next ship was the *SS Herman Melville*, a Liberty. I signed on in August as Chief Radio Operator with a Navy operator as an assistant. We started loading ammunition in Red Bank, NJ. I also noticed a special Army crew of men in strange looking outfits. I soon learned we were, in addition to the ammo, taking a cargo of mustard gas. All were assigned gas masks. I was even given a special decontamination suit (served well in bad weather).

It turned out to really be a long voyage, because we were gone for 17 months—August, 1943 through January, 1945. We were a permanent ammunition ship for that period. Also, during that time, we were assigned to the British for a year—December, 1943 through January, 1945. We plied between Canada and the UK, as well as numerous convoys around the UK, including the stretch on the East Coast from the North Sea down to the English Channel, better known as "submarine alley."

Prior to the invasion, we had civilian aircraft spotters from Scotland training on board. They stayed with us as we supported the British beachhead at Arromanches. We supported Omaha Beach as well, and I went on board one of the sunken Liberty ships at the "Gooseberry Breakwater." The *Melville* was the first cargo ship to go into Cherbourg after it was captured in July, 1944. We were in Rouen when the Battle of the Bulge started in December, 1944.

One story I have to mention: The captain's cousin was one of the crew, and he was allowed to bring his Harley Davidson with him. When he got ashore in France, he

one problem, however—Paris hadn't been liberated yet. He was cheered by the French who thought he was part of the advanced guard. He got back safely, believe it or not.

I won't go into the air raids, the Buzz Bombs, including the one that malfunctioned and missed us by inches. Most of the fellows had similar experiences. Some had much more.

My next ship was the *Oakley Wood*, another Liberty and an ammo ship. We loaded again at Red Bank and started for Antwerp. We arrived in the English Channel the night of May 4, 1945 when we got word the Germans were going to surrender. We were a 20-ship convoy. There was a British convoy of about the same size heading down the channel. All we can figure is they got word, too, celebrating possibly, and not being alert. The two convoys collided. We missed a huge British tanker by a hair and it struck the ship astern of us. Most of the ships limped into the waters off the Isle of Wight for repairs.

Since we were not damaged, we were re-routed to the Philippines by way of New Orleans where we picked up telephone poles for deck cargo and a third radio operator. It was a slow trip across the Pacific because the engine developed trouble and, at most, we could only make six knots. Sometimes the engine stopped and we drifted for hours. It took two months to get from Panama to Manila. The negotiation for the surrender of Japan was under way, and we watched the Japanese VIP plane fly in with the negotiators. We returned to Seattle, arriving there the last of October, 1945.

My next ship was the Liberty *Jared Ingersol*. We loaded grain in Baltimore and went to Edinburgh. It was on that trip that I got married in Shirehampton, England (near Bristol). I met Pat in 1944, and it took two years to get back. I went back again as Chief on the Victory ship, *Bardstown Victory*, taking German prisoners from the States to the UK so they could help clean up some of the mess their countrymen had made. They seemed like good Americans to me. Most loved the U.S. and vowed they were coming back as soon as they could.

I went to radar school at Sheepshead Bay, and sailed on two tankers: the *SS Fredericksburg* and the *SS Mobile Fuel*. I closed out my career at sea in December, 1947.

After the war and my sailing days, I worked for TWA and United Airlines in the communications department, leaving United as Supervisor of Communications in 1961. After that I worked as a field engineer and regional sales engineer for a few years. In 1970, I set up my own small manufacturers' representative company selling high-tech electronic equipment to the R&D labs, the government and the defense industry.

I retired in 1983. I have four kids, five grandkids, and a great wife.

LaROCQUE, EDWARD V., R-20

I had been a Radio Operator in the Marine Corps before going to Gallups Island in September, 1942. . . . having received a medical discharge from the Corps after spending a year in the Navy Hospital in San Diego with a badly broken leg . . . so I found the code and theory very easy. After two or three months on the Island Howard McElroy wanted to see how fast I could copy. He spent about an hour sending to me with the "Bug" and said I was copying well over 40 words per minute. He said he could get me the rank of Chief in the Coast Guard for the duration if I wanted it. I declined, saying I wanted to finish school and ship out. I never did regret that decision.

Gallups Island certainly must have been the best school for sea going radio operators in the U.S. The classes and the method of teaching were very different from the Marine Corps Radio School I had attended in San Diego. But everything jelled and classes were very easy for me. On the weekly theory examinations I scored 100% on about thirty tests in a row.

The first ship they gave me was a non-union United Fruit ship—no overtime, etc.—and the looney Chief Engineer rationed the water the whole trip.

After discharging cargo at a small dock in Arzew, Algeria, we had to make accommodations for 800 German prisoners—crack Panzer soldiers. Wooden framework was extended out over the sides of the ship to be used as their toilets. When we finally got under way for the trip back to the States, toilet paper kept flying all over and we had to duck when on the deck.

These prisoners had been told that New York and Washington D.C. had been bombed as part of the propaganda they were fed. One day these prisoners were on deck and sang so beautifully we all applauded. We later found out they had been singing the "Today, Germany—Tomorrow, the World" type of songs. Sure was nice singing though.

From then on every ship I took was out of the ARA Union in New York.

At times I had Navy Radio Operators who could barely copy the code and other times I had an operator from the ARA who could. Sometimes the Armed Guard thought he could order me around—but it didn't take long to straighten him out.

Had to do all the electrical repairs on the ship. On one ship I had to repair the electrical system on the 3" bow gun and the 2" on the stern. From the shape the wiring was in, they must have been sabotaged.

Was gazing out the port hole of the shack one day at about noon when the ship in the next column went up in one huge explosion. It must have been loaded up to the water line with high explosives. The tanker aft of that ship started burning and black smoke rolled a half a mile in the air. Since it was broad daylight everyone figured it wasn't a sub that did it—probably a floating mine or something aboard the ship just went off accidently. The concussion gave me quite a jolt.

I think the most dangerous cargo a ship I was on ever carried was a load full load of DUDS. We brought them back to be examined to try and find out why they had not exploded. We had to anchor a couple of miles out of New York while they were unloaded onto lighters. The longshoremen got triple pay but our bonus money was stopped!

I am glad I shipped out when I did. I'm also glad I quit shipping when I did. If I hadn't I'm sure I would have contracted some incurable disease, or had been crippled in an accident while drinking, or killed in some forgotten country.

Memories, memories—and all because I went to the Gallups Island Radio School!

While I was in the hospital last year some nurses overheard me talking about making 20 crossings of the submarine infested North Atlantic during the War. One of the nurses asked if the subs were Japanese, the other thought they were Russian. Can't really blame them, I guess. It was a long, long, LONG time ago.

By the way, I have the Conspicuous Service medal from his honor Mario. Why did they wait almost 50 years to give us this medal? And for what?

I am now 75 years old and when I look back it seems like another life . . . but, like Voltaire's Candide, I must say, "It WAS the best of all possible worlds . . . "

STAMMEYER, LEE, R-20

Some recollections from a 72-year-old mind concerning events that happened to a young mid-American small town boy 50 years ago.

I knew I was going to be drafted but I also figured I might have some control over my destiny as to what branch of service would suit me best. I had friends already drafted who regaled me about ditch digging and marching. I made up my mind this was not for me. I was sure the Navy did not dig ditches, so I formed a plan.

I informed my father that I should get some formal education in radio technology as I had been helping him as an electrician. I figured that a diploma of sorts would help me in the service.

With this intent, I enrolled at Coyne Electric in Chicago in the fall of '41. At Coyne, I (and hundreds of Army personnel studying there too) learned radio theory and repair. With my electrical background the course was easy, but I sure wonder how the Army ever got things repaired judging from the GI's in my class.

The day I graduated from Coyne (incidentally, I had to borrow $300 from my grandfather for tuition and work in a girls' rooming house for board and keep) some gold-braided USMS officers addressed the class concerning an opportunity to continue schooling and become radio operators aboard ship. This sounded good to me so I signed up.

They paid for my rail ticket to Boston and I got to Gallups Island in August of '42. This was all high adventure for a kid from Iowa who had only been 300 miles from home before.

The first nights in the barracks I

was exposed to a relapse of history. After lights out the words "Yankee" and "Rebel" became very outspoken and I thought that I was in a time warp back to the Civil War. However, reality prevailed once classes started.

My time at Gallups was quite pleasant. I was used to discipline, and code was the only thing that needed more diligent study. Shore leaves on weekends were great. I realize now that I should have spent more time visiting historical sites instead of the Silver Dollar bar but I blame this on my friend Ed Larocque. Ed had a medical discharge from the Marines and was more into bars than I. Through GIRA, I got in contact with him again and learned that he led a very interesting life. I hope that he puts it in print for all of us to read.

I graduated from Gallups in April of '43. At the same time I was threatened by my draft board to report. I explained this to the commandant and they took some of us to the Naval Station in Boston and swore us into a special class Naval Reserve. At first they told us it would be the Coast Guard, but in about a ten minute conference among the brass present, it was the Navy. At the end of the hostilities the Navy sent me a discharge called: "At the convenience of the government." I also have the new Coast Guard discharge. They ought to give us an Army discharge too, for all the Army personnel and material we hauled for them. And how about one from the Air Force for all the bombs we delivered!

In early May of '43 several of us graduates were sent to San Francisco to ship out. I was there several weeks and signed on to a new Liberty ship, the *George Washington Carver*. The USMS had intended to staff it with an all black crew but could find only a few to work below decks. The same for the Armed Guard. They had to settle for two apprentices.

The first time I walked up the gangway of that ship I was overwhelmed by the magnitude of the fact that I would be the only radioman aboard. The ship sure looked huge to me. The skipper had come out of retirement, as well as the chief engineer, so we sort of formed a mutual admiration society and got along well.

It wasn't long before we were loaded with Army material and even two locomotives secured to welded pads topside. The skipper and I picked up the code books and he also presented me with the first aid books and the keys to sickbay. He said, "You got more time to study this than anyone else, so you're the doc." I made friends with the Gunnery Officer real quick and told him that he had to help, too, as long as he had a crew aboard. I never did get to operate on anyone, but I sure feared that the day would come. However, I did put two Armed Guard men in sickbay for a week after they stood guard in shorts and got a severe sunburn on their legs.

We left alone from San Francisco about May 25th in the late evening. About four hours out the bearings on the high pressure piston rod got hot and the engine had to be shut down. The swells off Frisco are huge and we wallowed around in them broadside. Luckily, the old chief broke out the wrenches and they got the inserts loosened up and we were under way again by morning.

The skipper then opened our orders and it was for a zig-zag course clear across the Pacific to Hobart, Tasmania. The same morning my radio went dead. Luckily, I had a Simpson along and in a little while I found a poor soldering job on my antenna receiving coil. This built up my confidence and things did not seem so overwhelming.

It took us 32 days to get to Hobart and we never saw another ship the entire time. I did, however, hear a lot of BS stories at night on the top deck from the second mate. He could really lay it on to the kids in the Armed Guard that stood watch topside.

We finally arrived at the Red Sea and were piloted through the Suez Canal to unload at Port Said. We spent a good deal of time here as we were loaded again for the Invasion of Italy. We made a rail trip to Cairo and had our picture taken at the pyramids to prove a point. We had a Greek derrick barge tied alongside to unload the locomotives and they invited us over for a cup of coffee (demitasse). It had the consistency of syrup and the kick of 90 proof.

We loaded English equipment and a company of Ghurka troops from India with their English officer. The Ghurkas ate their own cooking, mostly lentil beans, which they cooked on little stoves on deck. They were small in stature, but with a rifle and a curved knife strapped to their waist, they were formidable allies.

We were in a small convoy headed for Taranto, Italy and just as we approached the coast the German dive bombers dropped over the mountains and in a few minutes it was over. We were not hit, but an English ship to our starboard was hit in the #2 hold in which English troops were quartered. I never found out how many were killed.

Our voyage ended on November 7, 1943. We had been gone about six months, and our folks had never gotten a letter from us in that time so there was a lot of relief for parents and girlfriends when we got back.

My second voyage was on the Liberty *Bartholomew Gosnold*. We loaded bombs in Baltimore and proceeded in convoy to Barry, England. Everything went according to plan and we were back at Baltimore on February 8, 1944.

About this time we were seeing baby flat-top carriers with our convoy escorts. They were dangerous to land on; several times I saw planes go over the end and the pilots consequently killed.

My last trip was on the *Albert Burleson*, another Liberty. We sailed from New York to Avonmouth, England where we loaded a complete hospital to take to the Pacific front. We proceeded to Batangas, Philippines where on liberty, we traded something for a big bunch of ripe bananas. They tasted like ripe strawberries to me, and I got the worst case of trots I ever had in my life.

From there we went to Eniwetok in the Marshall Islands and then to the Ulithi anchorage to wait. Well, we didn't have to have long before it was over with Japan and we were ordered back to San Francisco.

I went back to Iowa and my Naval Discharge was waiting for me, so I figured it was time to settle down.

I held many jobs in the post-war years, mostly with electrical maintenance, and I finally went to work for a public electric utility. I retired from this company as an electrical structural and maintenance engineer in 1983.

I have remained active, doing carpenter work as a hobby. I also enjoy fishing in Canada.

My home in West, Union, IO is always open to any GIRA member that might be coming through here. I might take one of you up on going south for a while in the winter one of these years.

God bless you all.

STEFFENS, BEN, R-20

As I was rearranging some books yesterday, I pulled a folded scrap of paper from a long-unused Bible, unfolded it, and suddenly found myself, sweating and scared, back aboard the *SS Richard J. Cleveland* off Unalaska Island in January, 1944.

Torn from a yellow composition tablet, the slip held a brief message: SOS 54 17 North 166 25 West Vessel breaking.

My hand shook. I tried to hurry. When I tuned, the transmitter showed only 20 percent of normal output, and I went into shock for a second. *What was I doing wrong?*

I tuned again with no better results. "The hell with it," I thought, and started sending. With the first set of Ditditditdadadaditditdits, my fist smoothed out. After repeating the message twice, I turned around to yell, "Who's got a watch with a sweep second hand?"

Someone stuck his wrist in front of me. Only after I sent the four second dash, one second space series that would set off alarms in every ship within hearing did I notice what was going on behind me. The youngest Navy operator was vomiting into the wastebasket. The second navy boy stood ready to abandon ship, his most precious possession, a dress uniform of "tailor-mades," stuffed under his lifejacket. The third mate, who had offered his watch, stood in the doorway. Behind him in the passageway anxious faces were watching.

"Jeez, Sparks, how long does it take for someone to answer?" someone whispered. That was what I was thinking, too. So I sent the distress again. And then we waited.

Sooner or later, every seaman draws a sour voyage or a sour ship. This was my first. The *Cleveland* was a Liberty tanker, the U.S. answer to the slaughter ships during the first half of 1942. On my first trip in '43, I was the only operator aboard on easy shuttles from San Pedro to Pearl.

Then things changed. Two young

Navy operators came aboard to provide 24-hour watches. And one day as we were loading in San Pedro, and old AB said, "Ready for a change of climate, Sparks?"

That was secret stuff. As far as I knew, even the Old Man didn't know our destination yet. But guys with 30 years sea time have ways of knowing things. I watched his wrinkled grin.

"You know that big sealed carton they just took up to the bridge?" he asked. "Well, it's full of fur-lined jackets and face masks."

Tradition has it that the three roughest seas on earth are met at Cape Horn, the North Atlantic in winter, and the Gulf of Alaska. This trip the Gulf tried to prove it. The voyage to Dutch Harbor in the Aleutians was one long storm with no sun sights for the navigators. One day a rogue wave smashed the port side, carrying away a doorway on the boatdeck, framing and all, and doused the galley stove. Frozen spray covered the bow gun tubs and spread aft over number one and two holds. The ship's interior got colder and colder . The plastic radio room clock stopped. The *Cleveland* rolled as only Liberties can. What little sleep we got, we got fully dressed in clothes and woolen long johns.

The Old Man—actually, he was under 30—was really good. He felt our way through Akutan Pass. Then the ocean really went crazy. For the last 20 hours I had spent off-watch time taking dubious radio bearing, but felt good because we would close Dutch Harbor the next day.

They came together, the concussion and the jolt. Gone aground? But the ship continued to roll and pitch. On watch in the radio room, I wondered whether some fool had somehow fired the stern gun. There was confused yelling. Then steps pounding up steel ladders to the bridge.

"Sparks, Skipper wants you in the chart room." The first mate's eyes were huge. I ran.

As I entered, the young captain tossed down the dividers and pushed a yellow pad at me. "Write this." He gave me our position twice, to be sure. "Send SOS. This bucket's splitting in half. Wait! Say, 'Vessel breaking!'"

Before my transmitter was warmed up, the ship's whistle began the "Seven shorts"—the signal for abandon ship stations. In cases like this, that doesn't mean the radio officer. As my tough first Skipper told me, "Sparks, you stay until you get a position out, no matter what anybody else does. Remember, it's your neck you'll be saving."

Dutch Harbor Navy answered first. I never heard a sweeter signal. They took charge. They had us on radar. A rescue vessel was on its way, steaming at seven knots. Only 35 miles away. Even closer was the neutral Russian *Sovietskya Latvia*.

"Go to the *Cleveland*, spread oil and stand by to pick up survivors," the Navy directed her.

Listeners in the passageway passed the good word to the bridge and to gutsy men in the engine room far below the waterline. I felt the propeller turns slowing, the ship now easing into pounding seas with bare steerage way.

At first, time flew. The Russian's English was good. I took bearing on him, then sent long dashes so he could take bearings on us. We couldn't miss. . . .

Crewmen were just voices behind me, offering news. The crack was amidships right through the bosun's quarters . It was getting bigger. It wasn't getting bigger. The gun crew had been moved forward out of their aft quarters. The Skipper ordered the columns of ice chipped off the lifeboat davits. Somebody nearly slipped overboard. The Skipper delayed chipping because no lifeboat could launch or live in those seas. Blackout was off. Forward cargo lights on the ice made us look like a huge frosted cake. The Russkies couldn't miss us. . . .

So where were they?

Time began to drag. No ships appeared. Talk stopped as morale soured. As long as rescue was coming quickly, they could handle the feeling that the crack was slowing working itself the rest of the way around the hull. It had to be, as long as the ship labored up the steep seas, trembled, and then plunged down again. And again.

A messman passed the word that our usually foul-mouthed chief engineer was huddled in a corner, praying.

"He thinks we've had it."

When the sixth and seventh tense hours had passed, the crew spoke only of the crack. Arguments flared. The Russkies are lost. So what? The hull is holding . . . The Navy lied about how fast they were coming. There's no way out of this . . . But doesn't it feel like the weather is easing a little? . . . Metal fatigue is what it is. You know every minute the hull is getting closer to snapping again . . .

I turned in after 30 hours. Not to sleep, but to lie in my overboard suit, listening wearily to the ship. What got me on the deck was the yelling. Happy voices. Cheers. Someone saying, "Okay, dam you, now sink!"

Lying close aboard in the freezing dawn was a Coast Guard ship. I love that service to this day. They had taken a terrible pounding to get to us. Six crewmen had been injured, three with broken limbs. The gale eased as we limped beside her into Dutch Harbor.

Ashore they were kind to us. Aboard they welded chains and plates across the crack. They would let us go home by the Inside Passage to a Seattle drydock to install huge longitudinal stiffeners.

And they found answers. Because freezing oil can't be pumped, we began to warm the cargo tanks with steam lines two days before ETA. As temperature differences grew between the outside hull and the interior, stress in the steel multiplied until it snapped in the heaving, twisting seas. As for my transmitter problem, small leaks were everywhere, including the tall steel box protecting the antenna on deck. I had been sending through six inches of salt water.

On my next trip, I sailed as chief aboard the *New Zealand Victory*. I didn't go near a Liberty tanker again until the *Alan Seager* in 1947.

Had a really close call from a Kamikaze while aboard an ammo ship at Okinawa. And as I look over this hasty document, I didn't explain visibility was too bad for the Russians to find us. That's something a landsman might not understand.

73's

VALDEZ, ALIPIO "AL" , R-20

I had an awful lot of fun. When you're young, you have fun anywhere—but Gallups was particularly nice. Now, I'm an old geezer in Tampa and own a marine electronics depot. And I'm having a great life.

What was good about Gallups Island? I guess what I remember—almost everything . . . (especially) freezing my left ear in 14 degrees below zero.

Platoon R-21

BERINGER, BERT, R-21

I made my first three trips to the East on the *Horace L. Luckenbach*, which was a twin screw. It was the largest ship in the Merchant Marine fleet and it was built right after WWI. It was a fantastic ship.

The Liberty ships would crack in half in the high seas. I remember once when the Skipper told the Navy gun crew never to fire the 5-inch guns unless necessary. But they went ahead and practiced with them anyway. Sinks came down off the bulk heads and pipes broke in the kitchen and showers. Other than that, we sailed right along.

The biggest job I had was (with the help of the third mate) repairing the antenna when it snapped in the high seas.

CASH, RAYMOND, R-21

I was in R-21 at Gallups Island in '42-'43. I went up to enlist in the Navy at 42nd and Broadway in New York City in '42 and the Navy up there told us we'd have to wait several months; but they said they did have opening for radio operators training at Gallups Island and any of us with a high school education could go into it. So I went to Gallups from there.

After graduation I sailed until December, 1945. I was on four ships during the war. . . . The first ship was the *Glenn White*—that was the coal burning ship with the Greek engine room gang and a Yugoslavian deck gang. I was the only one who spoke any English on it. After that I sailed with the *F&M Stanton* for three or four voyages to the North Atlantic, Mediterranean, Suez Canal, Gulf Bay near Barrier Island, the South Pacific, Admiralty Islands, New Guinea, invasion at Lydia.

I sailed on the *Andrew* in 1945 in the North Atlantic, Mediterranean, to Argentina and Belgium. In the last months of the war I sailed on a ship carrying troops home.

In 1945 I terminated my time, returning to work for my father.

CLAING, MARCEL-HENRI J., R-21

Graduated from Hartford Public High School, Connecticut, June, 1942.

Hitchhiking trip to New York City. On lower Broadway, saw US Maritime Service Recruiting office. Stopped in and asked questions. End result—signed up and accepted for Gallups Island Radio School.

Arrived in Boston August, 1942, assigned to R-21. Various events . . . learned . . . passed exam . . . graduated June, 1942. Home for the holidays. To New York and joined ACA. Assigned to *SS Pan Georgia*, a converted Hog Islander built in 1917.

Radio room/quarters had curtain separator between cot and equipment. Too young to know any better. Various exciting moments at sea, on approaches and at anchor. . . . Collisions . . . breakdowns . . . towing operations . . . etc.

Discharged to USPHS hospital in Galveston, July, 1944 with malaria.

From hospital to union. Assigned to *SS White Bird Canyon* at Swan Island Ship Yard, Mobile, AL. New T2-SAE1. Sailed from Texas to UK, South America, Caracas, Aruba, NWI, New York, UK from July, 1944 through December, 1944.

Married in January, 1945. To Mobile again, new T2, *SS Crow Wing*. To Texas for a load of avgas, to New York convoy to UK, South America load Colombia to UK. Submarine attack on Good Friday, March 30th. Mass confusion during convoy turn. Collision in stern tank and engineer area caused major damage, deaths. Straggler with escorts to Swansea, Wales. Offload. To Avonmouth for temporary timber repairs. To U.S., Baltimore, Sparrows Point yard.

WWII end near. To Texas, load transit canal for Pacific. Ulithi, Mariannas, Spain, Tinnian. Sail from Saipan, WWII ends four days east bound. Death and illness in family. Pay off Canal Zone. Hop NTS air to states, USAAC B-25 to north.

September, 1945 New York hall. Assigned to *SS Mill Spring*. Postwar Aruba, Puerta la Cruz, Venezuela, Colombia, to New York, Delaware, Pennsylvania and Jew Jersey. Paid off September, 1946.

Civilian occupation radio and TV repair. Joined Connecticut Air Guard. 1950, Air Force career for 33 1/2 years. Flying status, airborne electronics, intelligence operations, worldwide. Southeast Asia, 1954 to 1972. Various wars, Indonesia rebellion, Quemoy/Matsu conflict Straight of Taiwan, Malaysia insurgency, Indo-China-Laos-Cambodia flight operations. Burma operations. Chad in French Equatorial Africa. Simply put, nearly any place that my French language could be utilized.

1982 retirement from USAF. Third career as consultant in air operation area for civilian and DOD.

W1WJC, operated maritime mobile, many hours aeronautical mobile using KWM-1 with knife switch and homemade match box to aircraft antenna.

Member of GIRA and glad to have found membership in organization.

KINSEL, LORIN D., R-21

"Voices of D-Day"

My name is Lorin D. Kinsel. I am a Doctor of Chiropractic, and also a Christian counselor. My identity with WWII was a Radio Officer with the United States Merchant Marine. I made an Ensign rating in the Marine Communications from the Maritime Service and given "an honorable discharge from the United States Coast Guard on the 15th day of August, 1945 as a testimony of honest and faithful service. This was issued pursuant to P.L. 95-202 for service in the American Merchant Marine in Ocean going Service during the Period of Armed conflict, from December 7, 1941, to August 15, 1945." This honorable discharge was issued by law in the year 1992 in final recognition of the role of the American Merchant Marine in the Second World War.

I enrolled in the Maritime Service in 1942 and was assigned to the Gallups Island Radio School class R-21 operated by the United States Coast Guard on Gallups Island, located three miles outside of Boston Harbor at Boston, MA. I graduated 10 months later in 1943. Upon graduation, I went to the War Shipping Administration and requested the Pacific theater instead of the Atlantic. I had heard too much about the Atlantic and thought it better to take my chances in the other direction. My assignment was San Francisco.

My first ship was the *SS Alcoa Pennant* on which I went to Milne Bay, New Guinea with men and officers plus equipment for the SeaBees who were building a dock for an advance naval base. The only scare we had on that trip was a message from a ship torpedoed just 24 hours behind us off the Figi Islands. This area was considered the most treacherous spot on the trip. That theory proved correct for the ship following us. That radio distress message was my first SSSS (ship torpedoed) and when I reported it to the Captain, he was very apprehensive for our safety. We practiced gunnery very often. I got to fire the 20 mm guns and also had to load the huge shell cannisters. The Navy gun crew also fired their 4-inch rifle on the stern. The sailors passed along stories of what Japs did to seamen, especially the Navy men that managed to get into life boats.

On December 9, 1943 I signed off the *Pennant* in San Pedro, CA. I went home for a month and then to New York to the ROU. On February I signed on as the only radio operator on the *SS West Cheswald*, an old pre-WWI Hog Islander.

We moved from New York to Boston to load for the European theater and somewhere around the beginning of March, we set out for Halifax, Nova Scotia. We all got to go ashore for a couple of hours, but there was nothing to do in Halifax and nothing to see. The weather was foggy and cold. After leaving Halifax, we were in convoy traveling 5 knots an hour.

Our convoy consisted of 30 ships and we were told there about 30 Naval vessels which were destroyers and destroyer escorts. If I remember right, our 30 ships were in about five lines of six ships each. I often watched the little destroyer escorts which were not as big as some of the waves they had to ride and could imagine the fatigue in the men as the ships rolled and pitched constantly. Never a smooth ride even for a moment. Yet those ships were our safety network and they did their job well.

It took us 30 days to get to Liverpool with our cargo, which sure was a lot of time to go a couple thousand miles. With our zigzagging, we probably traveled a lot more.

We arrived in Liverpool, and our cargo was unloaded. Then we were sent to another dock where workmen came aboard and cut holes in our bulkheads along each side of the ship the size of a doorway. This procedure was repeated on every deck from fore to aft. After that was completed, they filled the bottom hold with sand and then welded wire baskets onto the side of each hold which later contained a large charge of TNT. As crew members, we were not informed of what was going on or any reasons for it but we knew we were ready to sink. As usual, everything was secret. Each charge was wired to a detonator which was in charge of the Navy Armed Guard Lieutenant.

One of the most amazing scenes I ever saw in England was their "Tea Time." Believe it or not, the war stopped for tea. When the whistle blew, everything came to a halt. I saw what I believed to be a 30-ton tank being unloaded for tea and their lunch.

After our ship was ready to be sunk, we moved out of Liverpool and went into one of the Fjords of Scotland and were joined by a lot of other ships plus some of the British Navy. Some officials came aboard and told us that our ship was going to be involved in an invasion and asked us to volunteer our services. We all volunteered. Then they also told us that as civilians we were not obliged to go, but would have to go ashore and spend the time in prison until after the invasion was over to ensure secrecy. My radio shack was to be demolished right after our ship was abandoned.

After about a month we finally set out in convoy. For some reason, we were delayed for one day enroute. About the time we got to the English Channel, there were ships everywhere. I never saw so many ships in my life, and never will again. We had to follow another ship through the mine fields and finally came to our beachhead and that it was D-Day +2. We had missed D1, the first day of the invasion. The water was littered with everything imaginable. Men that came aboard told us that 5,000 men had been lost yesterday.

We also found out that our ship was going to be sunk parallel to the shore in order to provide a breakwater for landing craft since there was a 21-foot tide at Normandy.

Our seamen fished a number of life belts out of the water which were long rubber tubes with two gas bottles in them. They fit around the waist and, supposedly, if you were in trouble, you could squeeze the levers and pop the seals on the CO2 bottles and you would float. The men going ashore were loaded with a heavy back pack and arms and ammunitions. When they were wounded or dropped off into water that was too deep to get a footing, they used their life belts. This became a serious problem because the top-heavy weight distribution would cause them to pitch forward and float face down and drown. This problem was the original idea for our for our ships to be sunk as block ships providing shelter against the wind and high tide for the landing craft.

However, the first wave of men did not have this protection yet. This landing hazard was foreseen but the life belt danger must not have been predicted. I often think of those 5,000 men. They didn't get to fight and every one was needed.

The order of sinking for the ships was to be one ship perpendicular to shore and six ships parallel to shore. This plan was quickly amended to one perpendicular and three parallel. The beachhead was so large that we were unable to really get a good view of everything going on and really knew very little of the battle plan.

We were at anchor in Normandy for five days and nights. I will never forget it. We saw all kinds of action and everyone was jittery. There was a constant bombardment of the shore. Three ships fired continually and blasted the hillside and yet fire was returned, even on our last day there.

A destroyer was anchored parallel to the shore and used all of their guns bombarding the shore. Outside of the destroyer was a cruiser parallel to the destroyer. The cruiser was firing over the tops of the destroyer. Outside of the cruiser was a battleship firing over the top of the cruiser and destroyer. They literally pulverized the bank of the Normandy beachhead. How any shore gun batteries could survive that bombardment is amazing, but some did.

We saw aircraft by the thousands going on raids. One day while watching a squadron of B-17s on a bombing run (they couldn't have been too far away as we could see them dropping their bombs), there was a tremendous explosion that must have rocked even the planes. They hit an ammunition dump. There was such a huge mushroom cloud that we could watch the sound wave come right at us. When it hit us we almost lost our footing and were nearly knocked down. I believe today that my deafness is partially due to that sound wave. I learned later, back in England, that people were instructed to lie down on the ground facing the sound wave and pull a coat over their head to keep their ear drums from breaking. Somebody forgot to warn us about that.

Whenever a fighter plane would swoop down over the beachhead, every ship would start firing. There were at least three thousand ships there throwing everything from 30 caliber to 5-inch shells. When this would happen at night, every shell had traces and you could see thousands upon thousands of rounds of ammunition going up and exploding. One night while observing the show, I was watching our own gunners on the 20 mm shooting straight up. Our tracers would go out, but a 20 mm will not explode except on contact. I wondered just where these would land and I looked for shelter. Every morning I would look the ship over for shrapnel but never found any.

Some planes were shot down every day. There were dog fights around that we could not see, and one great day, our ships shot down 12 planes. We later found out that seven were our own planes. We were instructed to be careful. I remember one pilot bailed out and was descending toward the water and one of our gunners on a 30 caliber machine gun was shooting at him. I yelled at him to cease fire and he did. Every one was keyed up to battle. It seemed that when one ship fired, 3,000 ships fired everything they had.

Some nights we would have cloud cover and when firing would start, we could watch the tracers go through the clouds and explode and this would light up the clouds. Once or twice there were three layers of clouds and the display was gorgeous as well as deadly. With all of that, I have a hard time enjoying a small 4th of July at home. I take my grandchildren so they can see and understand a little bit of war.

There were always small craft moving around among us which were minesweepers. One night a minesweeper was following a floating mine right past our ship. We never did hear what became of it. However, I watched their activity a lot. I saw five ships hit mines: two minesweepers and three LST's loaded with men, wounded or dead, going back to England. When a minesweeper hit a mine, there was a huge explosion and a huge geyser of water. When the water settled nothing was visible. The ship just disappeared and settled with the spray. It was no more. The LST's made a terrific shudder and settled to the bottom.

On the second day of our watching the battle, D-Day +3, I was sitting up on the flying bridge with my feet up on a railing watching all the proceedings and acting as content as I could be. If you have ever seen a picture of an old Hog Islander, they had two deck houses. The Captain and Deck officers had their rooms in the forward deckhouse with the officers mess, navigation room, navigation bridge and radio room. I had a room in the after deckhouse with the Engine officers, crew and stewards.

My room was on the starboard side above the engine room. The smoke stack was just forward of the after deckhouse. Suddenly, my reverie was broken by a shell screaming between me and the smoke stack and it exploded in the water off the port side of the ship. I didn't know what to make of that until a couple of minutes later another shell exploded in the water right in front of me. I knew enough about shooting to know that the next shell would be right on target. I got out of the way. The third shell never came.

This shelling, one or two shots a day, went on for four days and somebody got a little nervous about it. On our fifth day, we were under fire again and before we left the ship, a shell struck the water just at the waterline below my stateroom. The shell did not puncture the hull, but the #3 generator which was located near the side of the ship exploded and there was wire all over the engine room. I was told there was shrapnel on deck just outside my room. I didn't investigate to get any souvenirs. If those gun sights had been raised a fraction, that shell would have gone through my room. I was very happy to be on the other side of the ship.

Our crew was divided in half—those needed to maneuver the ship and those not needed for that purpose. The Second mate was in charge of abandoning. He was so nervous that he screamed at the men to get down the rope netting into the landing craft. I could understand his tension as he had been torpedoed twice on previous ships and the second time he was in charge of launching a lifeboat. He had given a command to loose the stern line which would have allowed the bow line to tow the lifeboat along while other men came down the ladder and got into the boat safely. However, the man on the bow was so nervous that he loosed his line and the bow caught a wave and the lifeboat promptly sank. Four men were in the lifeboat and the Second mate watched each one of his buddies lose their grip on the lifeboat and sink. He was never the same after that experience. He was the only one of the four who was saved.

Our half of the crew was taken to a large, class 3 cargo ship and put aboard. We found out that this was General Eisenhower's command ship. We had supper on this ship and, if I remember right, it was chipped beef on toast. We went to bed and were told the next morning that two bombs exploded in the water about 75 feet off the stern which is where we were sleeping. None of us heard it. What was one more explosion?

After my half of the crew left the ship, the other half maneuvered it into position to be sunk. We later learned that our ship was under bombardment the entire time the maneuvering process took place. Our Chief Officer was an old man at least seventy years old and one time when a shell came close he dived under a canvas cot for protection. It took him a while to live that down. When I say old, most of our officers were old men the government found somewhere to have enough experienced men to man all the ships need for the War.

We lost so many ships and men before we ever got into the War that it was a disgrace. I have since learned that during all the years between 1939 and December 7, 1941 (before Pearl Harbor), President Roosevelt could not politically declare war on Germany, and that our nation was not able to defend our own ships that were supplying England in their war effort. We lost 20 out of every 25 ships that left our ports. Dead bodies would wash up on our shores and no one could do anything about it. Subs would wait outside the harbors and shoot them like ducks. We lost thousands of sailors and hundreds of ships and cargoes. We wound up with a lot of 90-day wonders for officers and captains. Normally it took 15 years to work up to a Captain's license.

Only by the grace of God did we win that war and I still have my questions about it all. Psalms 91 and 121 were very helpful for comfort.

Our crew left the Normandy scene and were taken to Bournmouth, England for a two week vacation which we enjoyed very much. We had good food which the English people did not get to enjoy. At the end of the two weeks, we were taken to a train station and put on board without food or water and spent the next 24 hours traveling. We arrived in Glasgow, Scotland and were greeted by the Red Cross people who gave us coffee and doughnuts. Then we were put in a small craft and taken out to the *Queen Elizabeth* to go home.

We were assigned quarters and since I was an officer, I got officer's quarters. We were nine men to a room which had 21 bunks in it for a crossing from the United Kingdom to the United States. It took us four days at sea to reach New York. I could never figure out just why we should have to spend 30 days at sea from New York to the United Kingdom and be exposed to subs when we could have made it a little sooner.

Getting back to New York was an experience I shall never forget. It was about as dramatic as Normandy. I was used to the poverty of the people in England and the stress they were under. Back home I walked into a grocery store and the shelves were loaded. England had no fruit; one egg a week; 1/4 pound of butter a week; one pound of bacon; and one pound of beef. Our clothing was colorful, and their clothes were either black or dark brown. Depressing. Here we had everything. No one seemed even aware there was a war and millions of people were being killed just a couple thousand miles away. It reminded me of my trip home to California after coming back from New Guinea. I tried to tell my experiences to my friends and they were not in the least interested, but could only talk about their kids and jobs.

Before going home from New York, I was downtown and went to the Radio City Music Hall. I was so traumatized from what little war I had seen that I couldn't be comfortable watching the movie and the Rockettes so I finally got up and left the theater. I did have an opportunity to have dinner with Captain Eide one night and we had lobster ala Newberg. My favorite. We had a very good time discussing our experiences and then parted never to see each other again. I went home for a month leaving the serious things that were going on in the world behind. Then I went back to New York to ship out again and this time it was to Russia.

ST. ANDREWS, EDWARD, R-21

I went to Gallups Island on August 28, 1942, on my father's birthday. I had seen an ad in the Boston newspapers to join the United States Maritime Service. Being a naive young man, I thought I was joining a branch of the service like the Army, the Navy or the Coast Guard. It wasn't until I got out to the island and called my mother one night and she told me that my draft notice had arrived. So I went up to the administration office and told them: 'What's this about being drafted? Am I not in the service?'

"Hell, no," they said. "You're in the Maritime Service. Do you want to stay there and go to school?" they asked, and I said "yeah."

"Then get on the next boat, go to Boston and join the Naval Reserve."

So I dug out my old clothes—they were all balled up, wrinkled and dirty, waiting to be taken home on my first leave—and went into the Federal building in Boston, joined the Naval Reserve, came back on the next boat and stayed there through my whole school term.

I graduated in April of '43 and got my license from the FCC. Upon graduation, I was sent to New York to the War Shipping Administration to be on stand-by for a ship. After about two weeks, a call went out looking for a couple of fellows to go back to Boston. They didn't have anybody on the rolls in Boston. So, Dick (Richard) Callahan and I volunteered and came back to Boston where I spent the summer on stand-by waiting for a ship.

The first ship offered to me was the oldest ship in the world: an old Panamanian caldare. I went down there and looked at her. I didn't even know how to turn on the radio, so they excused me from taking that ship.

A little while later I was asked to relieve the man on duty on the Nantucket ferry. So I went down there one night and they showed me the radio room, went ashore and we sailed for Nantucket. Stayed there all night and came back the next morning. I was then told to go to Portland, ME, there was a ship waiting to go out. I was put up in a motel in Portland for two to three weeks before the ship sailed.

I made three trips on my first ship. The first one to London, the second one to Liverpool and the third one to the Mediterranean where we did shuttle service out of Naples, Italy, to North Africa, to Iran and back to Naples. After that I went to France and up the Rhone River. We also went to Paris right after it was liberated.

Came back, and on another ship went to Scotland and stood by at Loch Obin and went across to Omaha beachhead where we were a floating ammunition depot for about two months. Then we came home, and on my next ship I went to Antwerp. We were at the Battle of the Bulge. We had steamed up, waiting to get out of Antwerp, but the Germans managed to crash through.

I went back to Antwerp a second time, then to Poland with the United Nations Grain Relief and was in the worst storm ever seen in the North Atlantic. We came close to being swamped. We turned the ship around and headed back towards Boston into the wind. Going full speed ahead to Boston, we were driven back 12 miles toward Europe over the next 24 hours.

After that trip, I signed on a coastline tanker. We got rammed while in Galveston Bay, so we stayed in Texas for a couple of months until they decided what to do with the ship. They took it to Mobile, AL up the St. James River.

My last ship was a tanker that went from Boston down to the Caribbean, then to Buenos Aires, South America, over to Liverpool, England, back to South America and home. I finally got off that and returned to civilian life.

WHITEMAN, STANLEY, R-21

Before the events of December 7, 1941, I was interested in and studied U.S. Naval history. I had it in mind on entering the submarine service, hopefully as a radio operator because I had held a radio amateur license since 1939. Following the attack on Pearl Harbor I immediately attempted to enlist in the navy but I failed the"Stillings Eye" test. So I transferred my hopes to the Merchant Marine. My grandfather and great uncle were seamen and that's where I was inclined.

Having received a favorable response, I left Watertown, CT by bus to Waterbury. From there I went by train to Boston. Early the next morning I left from a Boston dock for Gallups Island by a large converted sailboat or fishing vessel.

At Gallups Island I fitted right in and soon became assistant code instructor because I had entered the school already testing at the graduating code proficiency speed.

The school faculty had recently become alarmed at the high rate of failures in code in all classes. A navy psychologist was assigned to the school. A fellow student made the remark that he thought that he could understand the code in his head but could not get it typed. On the basis of that statement I informally tested the class's proficiency with and without the use of the typewriter. It was the typing speed that didn't keep up with the code. A situation easily rectified and immediately incorporated into the testing program.

On leaving the school I left behind a paper of my study of RCA's emergency transmitter then being used as on merchant vessels, a complex unit which by law had to be able to transmit continually for three hours powered by battery. My study, which was a graph, concluded that the unit would be inoperable in one and a half hours. By then the battery would be completely discharged. Shipboard operators had been reluctant to test the emergency equipment suspecting the problems themselves, fearing FCC inspectors unannounced inspection would catch them with a dead battery or a defective emergency transmitter. RCA later redesigned the unit making it acceptable to the FCC.

My training at Gallups Island and

experiences encountered was most interesting and beneficial during the remainder of the war and afterward. Stanley Whiteman's Account of his Murmansk Run officially designated convoy number JW 61 from Baltimore to Murmansk and convoy JW 62 on return. It occurred between September 20, 1944, and January 11, 1945, on board the *SS William Pepper*.

The ship was built by Bethlehem Steel. A Liberty Ship, it had extra steel plates welded to reinforce the structure. Worthy of remark is that during the war the Plimsoll Line, a painted line that indicates how deep in the water a vessel should be loaded, was moved upward allowing ships to carry heavier loads—which explains how the deck could hold several locomotives, steel girders as well as the cargo hold of vital food supplies and other heavy items. The *SS William Pepper* looked to me more like a submarine, being more below the water line than above. We were loaded beyond a safe level as events would demonstrate.

We left Baltimore on September 20,1944 and then Lock Ewe, Scotland the 20th of October. The convoy included thirty multi-national cargo ships and many naval escorts, including six Russian small (SC) sub chaser craft. The enemy Battleship *Tirpitz* was damaged 15 September, thereby eliminating a serious menace to convoys. It was totally destroyed by RAF bombers by 12 November—to the relief of everyone. Leaving Scotland the sea was very rough, the skies stormy leaving us with poor visibility. Vessels, trying to keep station (their assigned position) in the convoy, placed heavy burdens on deck and engine crews. All the while enemy mine-laying submarines kept putting mines in our paths. Our escorts gallantly forced subs away and many depth charges were felt by the enemy.

Needless to say, it was a welcome relief to arrive in Murmansk on October 28th. The crews stayed aboard the ships during the layover with limited forays into Murmansk. It was a dreary place with little opportunity to have contact with the locals. Authorities were suspicious of all foreigners and people were hesitant to show any friendliness.

Meanwhile, the Germans were assembling U boats near Kola Inlet (outer reaches of Murmansk) to lie in wait for the departure of the convoy. The departure date was December 10. We were almost immediately in the midst of a "wolf pack" and *U387* was sunk by the corvette *Bamborough Castle*. The next day, December 11, *U365* torpedoed the destroyer *Cassandra,* an allied ship. She lost her bow in the explosion and had to return to Murmansk. The following day, the 12th, the Norwegian corvette *Tunsberg Castle* struck a mine sinking the ship and killing 240 men. The convoy then began sailing close to the Norwegian shore forcing the German Command change tactics. They sent nine torpedo bombers to attack us that very evening. I happened to be tuning a short wave receiver and picked up the German Command base as they were conversing with their pilots. Not speaking German, the ship's first engineer, a former German national was summoned and he very emotionally translated for us in the radio room. Our guns began blasting and our rear deck cannon, a five inch thirty-eight, roared, sucking in what felt like all the surrounding air after the blast of each shell. One of our aircraft was shot down by friendly fire because the pilot came inside the convoy perimeter during the attack—orders were to stay away until the all clear signal was given. One German torpedo plane attempted to draw fire away so theirs could come in for the kill. It looked to gunners like a plane coming in to land on the aircraft as they could only see landing (wing lights) on. The gunners fired at the center of the lights but no visible plane. The aircraft radar gunner noted that the radar didn't show an object at the vicinity of the lights so moving ahead he found a radar target, fired, and thereby downing the bomber. The plane had been towing a target sleeve fitted with landing lights. The German trick had failed. It soon became quiet, the guns began to cool off and the heavens put on a spectacular show of northern lights.

The following day our gunnery officer stated that in his view the German plane had also dropped bombs containing mustard gas, because of discolored paint on the Flying Bridge, his claim was that chemicals were put in the paint which would indicate the presence of gas. The trip back to Baltimore was just as trying as far as I was concerned. My bunk was within inches of the ceiling which was just below the Flying Bridge deck. Because fog and the general visibility were so bad, the ship's Captain manually blew our station number (11?)—that's a lot of very long touts! I vibrated to my very soul.

All too often the captain didn't bother to listen for other ships to answer. Getting near Newfoundland, radio direction finding became the primed navigation for the convoy. Monitoring the signals became my responsibility and it was necessary to it every two hours around the clock. There was concern because the wind and shore currents were constantly pulling the stern inward toward the shore. The other radio operator had a hearing problem and the captain didn't dare trust anyone else.

On reaching Baltimore I was many pounds lighter and exhausted, glad that I could say "I have done my share." It was a volunteer trip. My union didn't have a contract with the company and in addition I had agreed to sail as second operator. Always before and after I had sailed as chief operator.

Platoon R-22

DAVIS, JACKSON L., R-22

Reginald F. Brandon, Melvin H. Singleton and I met at the U.S. Navy recruitment office on September 4, 1942. None of the three of us had any idea of what we were volunteering for. The recruitment officer thought we were going to the Galapagos Island off the cost of South America. No one there had ever heard of Gallups Island.

The three of us were sent to Boston with two meal tickets for the train ride. The orders were for us to report to the pier where the *Roderick* was berthed. The night was dark and dank with nothing but an old rust bucket docked with no lights or other identification. No one at South Station including the travelers aide had heard of Gallups Island or knew where it was. It looked like something from the adventures of Fu Man Chu from Lime Street in London, England. The three of us spent the night at the Summer Street Police Station as we knew nothing of the Milk St. Seaman's Club.

The next morning we were transported to Gallups Island. After walking up the dock and presenting our orders we were told in no uncertain terms, "The cook and bakers school is closed." We replied that we were here for the radio school. This set off a big to-do from Boston all the way down the East coast to Washington, D.C. Later that day we met other members of R-22.

Monday morning arrived, we were issued clothing allowance, stenciled same and were assigned work details under Chief Gay. If I remember correctly, we were digging for sidewalks from the Administration Building to the barracks.

We settled down to routine after the aptitude test and began training. OHMS law, some algebra and code practice under Navy Chief Bradley. The routine was three hours of theory and three hours of code practice; practice with the mill (typewriter) and the stick (pencil). Some seamanship was thrown in under the guidance of a Navy Warrant Officer.

After the first four weeks we were assigned to administrative duties for two weeks: kitchen duties, fire watch and more digging for those sidewalks. At last liberty came to R-22. Since we were a junior platoon we were afforded liberty from Saturday afternoon until late Sunday evening. Some of R-22 tried to make New York City and even Washington, D.C. All of us arrived back late as the "owl," the last train from New York City was late. We learned about this and did not try this again.

R-22 plodded on with our routine, losing a few here and there because of code or theory problems. Singleton and I had some difficulty with mixed code at 12 wpm and dropped back to a code intensive class which threw us a class back. But we finally overcame the code problem. It was a sad day to go down to the dock and see the class graduate and leave without me. But because of the intensive code class I was able to attain code proficiency to copy 30 plus wpm. Also, since we did ship board lab together I got a double dose of ship equipment with the next class and zipped right through the exam with flying colors. I took the FCC exam and passed my Second Class Telegraph and Telephone on May 27, 1943. I met the instructor, Bob Clough, and gave him the good news which made him the instructor with highest number of 2nd Class Licenses in recent times.

I left the island for New York City on June 4, 1943. I stayed on the shipping list during the summer waiting for one particular ship, the *SS Robert L. Vann*, named after the publisher of the Pittsburgh Courier. The *Vann* was launched at South Portland, ME. She came down to Boston, MA to complete outfitting. She went to Castle Island to load explosives for the United Kingdom. Then she anchored in the explosive anchorage where we used to see those ships and wondered why they were placed outside the nets south of Gallups Island. Such is life. I could look over to Gallups with binoculars to see the activity. The seas got so rough out there that we had to take a tug boat to get to the convoy conference.

The ship sailed shortly after the convoy conference and after losing the convoy twice because of lost code books we arrived at Plymouth, England. The ship had anchor chain hanging over the side to pull us away from the dock during air raids since we were loaded with explosives. After two collisions at the anchorage at Milford Haven, the *Vann* returned to New York City.

I stayed with the ship for another voyage to the United Kingdom where we became *U.K. 525*. Then we were in on the invasion of the French Coast. We shuttled back and forth to Omaha and Utah Beaches finally returning to New York City on Christmas Eve, 1944. The next trip for the *Vann* was a disaster as she hit a mine and sank. I had already moved on to the *SS William A. Graham* which had been converted to carry troops and prisoners. Singleton became my 2nd operator on this ship. We stayed on the *Graham* until everyone was released from the Naval Reserve and discharged. We were not subject to the draft as the emergency was over.

Brandon and I formed a radio and TV repair company in Washington. The three of us returned to sea various times in the years that followed. I returned to sea on the *Clyde Austine Dunning* and the *Oregonian*, both nice, clean Liberty ships. Brandon went to work for the Boeing Aircraft Company working on electronic aircraft and aircraft carrier landing systems. Singleton returned to Chicago, in the postal service. Both have become silent keys.

My postal career began in 1947, but there were interruptions when I went back to sea. In 1955-56 the Postal Service began a program of expansion into electronics and automation. I became an electronic technician, foreman of automation and then superintendent of maintenance for mechanical equipment. All possible because of the basic training at Gallups Island and the FCC licenses along with shipboard experiences. Reggie and I also taught radio and TV repair in adult education during our careers.

As the years passed, I went to the University of Oklahoma to teach Postal Maintenance for Management in the Postal Service. I finally retired after 37 years of combined federal service. This included the War Department, War Shipping Administration and the Postal Service. In 1981, after retirement, the Postal Service sent me to Chicago as a consultant for the Postal Service in Maintenance Management and Electronics; again because of the training at Gallups Island.

HAROLD, JOE, R-22

I enrolled in the Maritime Commission at Minneapolis, MN on September 9, 1942 and was released from Gallups Island on May 15, 1942. I believe I was in R-22 and I can recall a few names of members of that group, Tom Hickey, O'Leary, (first name?), a Mr. Pegors from Minnesota, Walter Heinen, and that's about all I can remember for that was a long time ago. I suppose there is a roster somewhere of the members of this group. I remember that we were on the Island making sidewalks for a couple of weeks as the school was growing quite rapidly and they were phasing out any wooden sidewalks there. They must have been wonderful sidewalks because they used sea sand, not a good idea. *(Editor's note: They're still there.)*

I remember they had some interesting instructors, one in particular who was asked the last week what to do about repairs if the set did not work. I do not remember his name but his advice was practical. He said, "If you turn on the switch of your transmitter and nothing happens, do the following. First, check the fuses. If they are OK, see if you have current going into the set. Then check tubes. If they are OK—START WORRYING."

If I remember correctly, the first week of the course we had to review all the math we had in our schooling, including trigonometry. They threw it at you so fast that the story was that one man left the classroom to go to the bathroom. When he got back he found out that he had missed trigonometry.

The reason I had taken the course at Gallups Island for a radio license was to avoid the draft, on the theory that I would prefer being shot for $50 a week instead of about $50 a month as a GI in the army.

And being a hick from Minnesota, I wanted to see as much of the world as I could while the war was on. I had never been farther east than Chicago before the war.

My first ship out of Seattle was the *S.S. Delazon Smith*, a new Liberty ship. I have been on many vessels since that time but none have ever rolled as much as these Liberties but they got you in and out of port effectively. First year running to the Aleutian Islands until the Japanese were run out of there.

I remember Christmas Eve in December 1943, on Attu Island, which had been taken over the Japanese shortly before. When the war was being phased out in the Alaskan/Aleutian area I went to Los Angeles, got a ship to New Guinea. Returned to Los Angeles and shipped out of New York to the Mediterranean area, Africa, Italy, France, etc.

Last trip during the war was out of Los Angeles to Manila arriving about 10 days after port had been taken over from Japanese. From Manila went to Calcutta on

Platoon R-22

a Victory ship and from there to Charleston, S.C. as the war was over and shipping was slowing down.

Tried selling insurance, not successfully. After a while returned to sea on tankers and saw most of the rest of the world, something I had never expected to do as a kid growing up in Minnesota.

Do I miss shipping now? No. Only thing I miss is some of the great poker games we had on many of those vessels. Kids gambled more in those days from what I can see around here now. My grandkids seem to have no interest in gambling. Probably a good thing.

KUCALA, NORBERT L., R-22

Graduated Gallups Island as member of R-22, receiving 1st class radiotelegraph and 1st class radio-telephone. Served as a radio officer in the Merchant Marine from October, 1943 to January, 1960.

After my years in the service, worked in the shipping industry and as a radio technician-mechanic with Pan American World Airways.

RIEU, DANIEL P., R-22

I reached 20-years of age in June of 1942. I had just finished my first year attending night college at New York University, where I was studying to become an accountant, and was employed at AT&T headquarters as a junior clerk. In other words, hoping to start a career that was now becoming more and more overshadowed by a rapidly escalating consciousness of what was developing on the war fronts. I had not yet been classified for the draft and was seriously occupied with more personal preoccupations.

New York City, as you can imagine, was already a hub of war time activity. 90 Church St., was already a center of military activity, was within view of where I worked, and the presence of the military in the area was fast becoming more noticeable. Because of the vast amount of shipping that normally passed through the port, the news about the mounting losses in the Merchant Marine was becoming shockingly more known. Already, patriotic enlisting posters requesting men to serve in the Merchant Marine, were appearing in public areas.

The main office of the U.S. Maritime Service was also located only a few blocks from where I worked. I passed it frequently during my lunch hours. One day in August, after having given more thought to serving in the Merchant Marine, I stopped in to inquire about enlisting and, much to my surprise, because I appeared to have the necessary educational requirements, I was told that I could be a candidate for radio school at Gallups Island. After only a few days of deliberation, I returned to announce my decision to accept. On September 15, 1942, about five other candidates, and I were sworn in to serve. Later that day, with our documents in hand, we were on the train to Boston with orders to report to the launch that would take us to the island at 6 a.m.

During all my younger years, I had not had much experience being away from home environment. Nevertheless, I quickly adjusted to the routines of camp life. In many respects, life had suddenly become more comfortable and I had never before been so amply fed. Because of the initial vigorous activity of helping to build sidewalks on the base, the wholesome food, and the enforcement of 10 p.m. "lights out," I soon reached a weight that I have never since achieved. The thought of having the opportunity to learn radio theory also inspired me.

The incoming class was assigned platoon number R-22. Approximately 85 enrollees made up this class. They came from every section of the United States and could easily be distinguished by the differences in their speech. Like myself, it seemed that almost all had no previous knowledge of radio. By the time we reached the end of our training, more than half the original number who started dropped out for various reasons.

To write the details of the daily life on Gallups Island that followed during the almost eight months we were there seems like an endless task. However, I can recall a few memorable occasions that took place during my stay there. One involved an R-22 classmate and newly gained friend, John Mezy, who was involved in a confrontation with a cocky, arrogant Navy chief petty officer stationed there. To settle their differences, it was agreed that the two would meet for a boxing bout in the recreation hall. Soon the word of the event spread around the base.

For the members of R-22, this was an opportunity to show off our pride and we were eager to cheer encouragement to our hero. By the scheduled time of the bout, almost everyone on the island must have crowded into the rec hall. I was delayed and still on my way down the hill when I saw the mob already pouring out. The chief petty officer lay knocked out cold on the floor and I learned that John threw just one right hand jab which brought the fight to a conclusive end. As you can imagine, R-22 had plenty to celebrate that night.

The end of our scheduled months of training eventually arrived and the excitement of being able to successfully pass the exam to obtain a Second Class Operators license was the uppermost thought on my mind. Even before sitting for the examination, however, a class graduation party was arranged in a Boston restaurant. It was attended by the high ranking school administrators and many of the staff. After the conclusion of the official formalities and congratulation speeches, the celebrating became intense and the party became somewhat wild and unruly.

I was among the first group of candidates selected to appear at the FCC field office in Boston to be examined. The date was set for April 30, 1943. I found it to be a tense situation. You can imagine my elation, when at the end of the day, I was notified that I had passed and was granted a three month provisional Second Class Radiotelegraph license pending the issuance of the formal license. I could breath a sigh of relief.

It isn't easy for me to forget the friendships gained from having been so closely associated for so long a time. My departure from the island left me with many fond memories and, as events turned out, it was to be the last time that I would set foot on Gallups Island.

The new destination was my hometown, New York City. I had orders to report to the Maritime Service Office, the same location from where I had departed eight months earlier. I was now much more mature, however, and felt an eager sense of anticipation as I looked forward to being assigned to shipboard. As I was soon to find out, I had not long to wait!

Platoon R-23

LINDER, GARETH, R-23

I was recruited in Minneapolis. I was looking down around in the recruiting offices of the various services and right next to the Coast Guard recruiting office was the radio recruiting office. I went in and they told me what I could do if I went to Gallups Island and trained, and so I went home and thought it over for about an hour. The next day I went down and signed up.

We left Minneapolis and went out to Boston, and there were so many of us that they divided the platoon into R-23 and R-24.

We came into South Station at night. We were supposed to be met, but nobody was there. We went down to India Wharf after somebody in South Station told us that's where the boats came in. I took one sniff of the air and said, "This is the sea air?" It was low tide.

I graduated at the end of May, 1943, went to Seattle and shipped out on the *SS John P. Caines*. On its second voyage it broke in two. When it broke, all the antennas were carried away, and it was necessary for me to make an antenna that would work. I had tried some unsuccessful attempts. In rough seas, I finally discovered that the main antenna line that led inside a box was broken, so I had to attach to another one. I ran a wire back to the after-mast and brought it forward again in a V-shape to a 20 mm gun position with an insulator on the end. That became the antenna that silenced half the Pacific while we did our SOS traffic. We were fortunate that we were rescued.

We came back, and after that I served in the South and Central Pacific. After that I crossed the continent and did an Atlantic sea coast for the balance of the war. After the war I took my wife with me to Illinois where I finished my college, came back to Boston and started to work at what is now known as WBZ, Channel 4. Then, it was just radio. The television station opened up and I spent a career there. I retired at the age of 62 and 3/4 and have enjoyed it ever since.

MARKOWITZ, MARTIN, R-23

I really don't know how I got to Gallups. Somebody said: "Why don't you go to radio school?" I didn't have the faintest idea what I was doing. I never knew anything about radio. Somebody said go there, and I went there. I spent six months there. Unfortunately, I was there when they had the Coconut Grove fire.

I was a little smart kid up there. They put one of these official posters up about short hair—and I was a smart aleck. I changed the whole poster around with somebody else.

Lo and behold, they found out who it was and threatened to send me to prison and everything else. But since my grades were so good they couldn't do it to me.

Platoon R-25

BENKO, GEORGE, R-25

I turned 18 in December, 1941 just one week after Pearl Harbor while still in high school in Pueblo, CO and in January of the next year we registered for the draft. They began calling up some of us that summer so I thought I had better find some outfit to get into or I might end up in the Infantry like my two older brothers had and I didn't want that!

I was inside our main post office one day and saw a poster they had put up wanting men for the Merchant Marine. There was an address given in Denver to report to and since it sounded pretty good to me I took a bus up there and heard about the different schools they had and picked out a book on the radio school in Boston. I took the application home and went up to Denver on October 5, to be sworn in with two other young men from other cities in Colorado. I was small for my age then, and at 129 pounds, I was afraid I might not pass the physical, but I got in OK. We had to enlist in the Naval Reserve first, as Apprentice Seamen, Inactive, then into the U.S. Maritime Service. That evening they put us on a train for the East Coast.

We arrived at South Station about 8 p.m. a couple days later. At that time Boston was being "browned-out" so we stumbled around outside the station not knowing how to get down to the dock and finally checked into a cheap hotel nearby. The next morning we made it down Atlantic Ave. and came to Rowe's Wharf and got on board the *SS Calvert*.

Arriving at Gallups Island we were put up in one of the barracks and issued some dungarees and other work clothing. We expected to start radio class right away, but they had other ideas. It seemed some new concrete sidewalks were needed around some of the barracks and since all of the men had not yet arrived for our platoon R-25 we shoveled sand and cement for three weeks. Some radio school this was! Afterward we got the new petty officers uniforms and our books. We looked pretty sharp on the parade ground with all our gold buttons.

We completed all our radio training in June and went up to the FCC office in the old Liberty Building to take the 2nd class radiotelegraph exam. I passed it on the first try, but not by a grade I could write home about. The code part was easy for me, though.

After that I went down to New York and got a place to stay on the West Side and then went down to the WSA-RMO office at Trinity Place to show up every Monday morning to collect my $30 pay for standing by.

At the end of August, I was assigned by the ACA to a Liberty ship loading cargo at the Army Base in Staten Island. My assistant was from the Navy Armed Guard. He was a year younger than me and had just gotten out of the Navy Radio School. Neither of us had been to sea before so it was a pure case of the "blind leading the blind."

The ship had a RMCA 3U unit along with a RMCA H/F transmitter on a separate bulkhead and a M/F radiotelephone receiver put on in Australia probably for harbor control use—none of which was similar to the MacKay stuff at school. We got a little break, though. We did not sail right after being loaded but went up to anchor under the George Washington Bridge with some other ships to await convoy. We dug out the instruction manuals for everything and managed to fire up the transmitters into the dummy antenna and copy some Navy fox from NSS on 115 Kc, so we had a little handle on the equipment in the shack before we left.

We sailed then next day to Halifax where we joined other ships to make a 84 ship convoy to the U.K. It took us about 16 days to get there as we were routed on a far north course, near Greenland and Iceland, then into the Irish Sea; all at about 10 knots.

When we returned to the U.S. we loaded Russian cargo at Philadelphia for the Persian Gulf. On the third trip we went from New York to Brazil with military cargo and came back with coffee for the US Army.

My next ship was a T-2 tanker. I was Chief with another R/O and a Navy Armed Guard enlisted radio man. We shuttled for ten months from the Caribbean to the South Pacific carrying fuel oil to ship of Task Force 58. In the spring of 1945 I took a new C-2 out of the shipyard in Wilmington, NC, as sole operator. We carried bombs for the 15th Air Force from Texas to Guam. On V-J Day we were at the San Jacinto Ordnance Depot getting a second load and then they took everything off.

Not having enough points to get off the ships I had to stay on until early 1946. I then returned to Pueblo to enroll at the junior college there.

I made the 1990 reunion of the GIRA in Boston and took the tour of the old Island. Along with the others was Bob Gisslow from San Francisco who was also in our platoon, R-25. We walked around the old base and were amazed by the changes 47 years had made. Gone were the barracks and other buildings; they had been torn down and carted off years before and only a few of the outlines remained.

Vegetation of all kinds had taken over, some of it as high as my head. There were pathways through it and signs had been put up showing the locations of the former buildings and where the parade ground was. We made our way down to the location of the old canteen and walked along the beach back to the landing. The Island is now primarily used as a tourist destination along with some of the other outer islands. A few of the picnicers were enjoying the day there as we passed by and I could not help noticing they were not much older then we were then—47 years ago.

I'm happily retired in sunny Florida and like keeping in touch through the GIRA.

Platoon R-26

McKEEHAN, JACK W., R-26

Three buddies from Broken Arrow, OK went to Kansas City, MO and enrolled with the U.S. Maritime Service, October 6, 1942. We were sent to Gallups Island where we became part of R-26. One of us could not make the grade and since we were Naval Reserve he became active duty. My best friend, with whom I grew up, was graduated and was lost during enemy action in the Philippines. I came to California with a lot of our friends and soon sailed on a Liberty tanker and later T-2 tankers.

From my enlistment date until November 6, 1945 we served in the Mediterranean, Atlantic and Pacific war zones, we were lucky and saw very little action.

Over the years I have wondered about Gallups Island and just what the conditions were—Had it become a residential development? Had it sunk into the bay? Was it a private estate or what? I went to the library to find addresses of some government agency to write for information. Some of my letters were returned for improper addresses—finally got it right. Miss Simmons sent me some information and a picture and said she would contact you.

I am retired and no longer associated with radio. I am of course very interested in hearing about the reunions and glad there are many who can attend.

Platoon R-28

CLAUSEN, JIM, R-28

One regret that keeps haunting me is that I did not keep in touch with all of my good buddies from R-28. I wish if anyone knows the whereabouts of John Haradon that they would pass this information on to me. I have been in contact with a couple other good buddies though, Charley Kalson in Pittsburgh and Bill Greer in Yucca Valley, CA.

Now for the most harrowing experience of my whole lifetime. . . . My youngest brother was very seriously wounded in France shortly after the big invasion. I found out that eventually he was transferred to an army hospital in England.

Platoon R-24

Platoon R-25

Platoon R-26

I wanted to see him very much, since word that my Dad received was that he might not make it. So I really wanted to see him which might be the last time. Well there was one sure way of getting to England. Sign on a ship loaded with bombs. The *SS James McCosh* fulfilled this wish and we convoyed to London, spent 2 days, formed another convoy of 16 ships headed up the coast to Hull, England. During the night several "E" boats from the mainland attacked our convoy. Needless to say, not many of our ships reached Hull, but the good Lord must have been riding with us that night, because we did make it and delivered our load of bombs. Immediately upon docking, I disembarked and headed via train to the little town about 40 miles Northwest of London where my brother did survive and is still with us today in spite of his wounds. So this trip was really to no avail, but it did teach me to stay away from ships loaded with bombs.

Another experience that kept us all on edge for most of the trip was shipping out of San Francisco on a good ol' Liberty towing a huge section of dry dock and loaded to the plimsoll with other war material. Well you all know the hair raising speed of a liberty, but can you imagine what the speed was like towing this huge square section of dry dock about a quarter mile behind would be. Yes, we were moving, but only at 4-5 knots which made us a sitting duck. What a relief when we finally arrived in Milne Bay, New Guinea and got rid of this dry dock section. Then we proceeded to Langemak Bay to discharge the rest of our cargo. It was a great relief to leave this area and the bombing raids from New Britain and head back Stateside.

I also spent considerable time of T-2 tankers hauling avgas to the many Pacific Islands, mainly Guam, Saipan and Tinian for the B-29's that were bombing Japan. Also time on Victory ships hauling dry cargo and several trips on a newly converted Victory ship carrying troops back from France and Italy.

After accumulating the necessary points for discharge in December, 1945, I took leave of the Merchant Marine and was anxious to get started in civilian endeavors. My first lesson in the brutal world of competitive enterprises was a juke box business in Los Angeles. From there to a hardware store in Julesburg, CO, and then I started a butane-propane business which I enjoyed for 28 years, then the construction business building post offices all over Colorado and Arizona.

Then I retired again for about a month and wound up in the RV business with three lots in Colorado, one each in Denver, Ft. Collins, and Colorado Springs. That lasted 17 years before retiring again. This time it has worked. I learned that you have to have something to do to occupy your time and your mind, especially if you have no hobbies.

So I bought a Greyhound bus and converted it into a motor home which has really kept me busy and now for several years we have spent considerable time traveling and seeing the sights from Alaska to Newfoundland to Guadalajara and all points between. We have spent winters in Florida, Texas, Arizona, and California. Last year we made a trip to Denmark and visited relatives, then spent a month touring the rest of Europe. We also had a family reunion at our cabin in the mountains for a week and had 65 family members from seven states present. What a grand and glorious week that was.

Now I am working on our family genealogy which my uncle spent years working on and traced our ancestry back to the year 850. Yes, you read it right, year 850. He brought the genealogy up to the year 1960. Now I am compiling all of the family data from 1960 to the present. If any of you need something to do, this will keep you busy and out of trouble.

In closing I have to say that my life has been full of interesting and exciting episodes. We have all lived through some very traumatic times and survived. We have witnessed many, many changes, many new inventions, very dramatic advancements in medicine, health and sciences and tremendous strides in technology.

I must confess, and I truly believe, that the training and knowledge acquired at Gallups Island and from the Merchant Marine experience, that a firm foundation was laid, from which to proceed and build upon and enjoy a very fulfilling life.

Platoon R-29

HARONIAN, NAZARET HARRY "HANK", R-29

I first heard about Gallups Island through a newspaper advertisement in a journal. I had finished a radio repair course in Maine and figured it was worth following up, so I went down and applied. I met the requirements, got accepted, and blood, sweat and tears later, I got through!

LA PATA, FRANK R. R-29

I served in Gallups Island in Platoon R-29 during the "Big War," as we say. Enjoyed it.

It was a good school. When it was getting close to the end and it was time for us to get out of there, we had to go down to Maryland on the Navigator. It was a training ship and we had to put in so much time on it."

After school I managed to get on two or three vessels. . . I really can't be sure of the names of them all. One that comes to mind is William F. Cody. Another is George P. Garrison, Duncan Fletcher and the fourth one I have trouble recalling. But we did sail from Massachusetts to Hog Island, hauling ammunition and everything overseas to the Big War.

We pretty near got torpedoed a couple times—pretty near broke up in a storm a couple times, but it was actually a pretty good education. I got more of an education from the times when I contributed to the cause, although it was a necessary thing.

I ended up first in New York where I married my wife and went into a company called MicroMoldwell making radio capacitors, which were not for me. I putzed there for a while until they realized maybe I could do a little better, and I got into some different phases of micro-mold.

Right about then we decided we'd go back to the midwest to Minnesota where I originally came from. I got out of radio there for a long time and went into office machine equipment for a big insurance company that had about eighth or nine hundred girls. I did that until we moved back to Florida and started to get back into ham radio a little bit and met a bunch of nice buddies. We had two or three clubs—and they did well until we got a few who wanted to tell you how to run it all, but wouldn't help you run it all. You know, just typical organizations—too many chiefs and not enough Indians.

I hung on to the license but haven't done much in the past few years. I got involved with this machine business on my own and time consumption just ate me up. I could stay on the air but I couldn't get out and really help people like I really wanted to do. But we had radio clubs and talked novice, etc.

All in all it was a good school. I enjoyed it and made a lot of good friends. I wished many times that I could have backtracked to find out where they all went to. I ran into one or two of them overseas while we were in France or Belgium or Holland. It was a good cause. I just sometimes felt that if we'd had a little more backing, I think we might have had a better organization—even from the beginning. But it isn't that we didn't give, because we gave.

RAFF, ROBERT V., R-29

We knew the stories. The Spanish American, the World War, only one till then. The little radio behind the bar demanded attention. The Japanese had attacked Pearl Harbor. Not one of us knew Pearl Harbor. Not one of us. Most of us just out of high school . . . The news sinks in. First the hush. Then, the rush to anger, unbelief, curses! Before we broke for home, a pledge . . . all, everyone, bound for the enlistment places Monday. Everybody had the same idea.

Air Force had too many everybodies, Navy, too—go home and wait, get a letter from your mother—wait. Weeks and months, then a full page ad in a Minneapolis paper. . . . The Merchant Marine needs men.

I go down to the U. S. Court house—raise your hand to be sworn in. Not old enough. Get a letter. My widowed mother signs the paper.

I'm sent to Coast Guard and sworn in. Sent to Navy, sworn in. Sworn in the Maritime Service, sent home to wait. Finally, they decided I'm radio material . . . orders to report . . . think there were five of us, on the train to Boston assigned to R-28. They count

Platoon R-27

Platoon R-28

Platoon R-29

heads, too many—I was one of those put into R-29. That meant two extra weeks going to Boston on the *Roderick*, loading 94 pound sacks of cement, cement blocks. Bringing our cargo to Gallups. Unloading via gang plank, sacks on shoulders, stacking the one flatbed truck, unloading, mixing cement. Put . . . put . . . put. Building the buildings for Islanders to come.

School . . . tests, shore leave, Seamans Club, Scollay Square, the Crawford House, beer, rum and Coca-Cola, the Coconut Grove. . . . On report—no leave, the weekend of the fire. Confined to Island and extra duty.

Finally, that long day in Boston, out on the little balcony between elements of the FCC. A cigarette and some serious prayer and pleas to God.

The examiner shook my hand, said an inch is as good as a mile. I had my ticket, my Second—no TLT. Great. . . . Off to New York City. Done with Gallups, but I got assigned to the USMS training ship, *SS American Seaman*—two weeks steaming Long Island sound.

While on the *SS Lee S. Overman*, our British captain and I talked on the wing of the bridge, watched the troops going over the side. The captain says, "Sparks, how would you like to go ashore?" I thought he must be kidding. Thought he might be trying to get rid of me 'cause he caught me bluffing him on a big round of poker. When I raked in my winnings after he folded, he reached across the table and turned my face up. Thought he was going to have a stroke, red as a beet. He wanted French cheese, all I could carry back. He gave me French invasion money, American and British and I dropped into a duck with khaki pants, shirt and gold braid hat. On landing saw bodies here and there—some still in water. Went into a pill box, must have had a direct hit. Three dead Germans inside.

I hitched a ride with a Canadian jeep to a small crossroad bar, asked for any cheese I might buy. The owner was so happy to see an American I got hugged and hugged. The few people in the ball were all excited and happy. I had to have a drink, then another. People came up with about a dozen wheels of cheese in wooden circles about a foot across, two inches deep. Different kinds, some with the worms like the captain told me about. The owner went to his cellar, dug in the dirt floor, brought up fourteen bottles of Piper-Heidsich and other fine champagne. He had these hidden from the Germans all those years. He wrapped them in paper, straw, a gunny sack. The cheese into another gunny sack. His wine, a show of joy. This happy, liberated country merchant with his gift, as the Maji, the wise men of another time.

Juno beach, Mike Red sector. Big waves coming in, nobody wanted to take me out, wait till morning. Not good weather for ducks. I slept some on a beached LST. Captain happy with his boxes of cheese. *Prince of Wales* fires overhead, we're back to London for another load of troops, guns, trucks. Seven round trips to beach. Total seven times shelled by bigs guns at Calais. We would see a small spotter plane, like a Piper Cub, then the guns would fire.

The Navy Gunnery officer and I were standing on flying bridge, passing Dover, and they started shelling. We could see the shells coming if we looked towards France. Looked about eight feet long, 21 inches across. Biggest guns in the world, mounted on railroad cars so they could be retracted into the bunkers or caves to avoid air attacks.

We had been told that if we suffered a direct hit the concussion would kill everyone aboard. One blew not too far above the *Overman*. A large piece hit our stack, glanced to the cement topside deck. The gunnery officer ran and picked it up. His palm and fingers puffed from the burn. He had his piece of the war though. On that two pounds of steel was the German eagle, the swastika in its claws and the foundry number below. I got a piece the size of a bean in my shin—the pants slowed it.

I got on the *SS W.S. Jennings*, another Liberty and a top secret trip. We load . . . leave . . . pass Gibraltar and on to anchor off North Africa for only a few hours awaiting orders. We run alone north to a spot off Monaco, anchor for a short while. All a no-escort run.

Then we are off to Marseilles. The invasion of Southern France. We anchor, take a lifeboat, the one with the motor, second mate, myself, about half dozen other. We tie up the lifeboat, find command post by landing, get dockage information, go alongside, begin discharging support supplies. Had a couple hours for important duties before we went back with the lifeboat. Our quay had several shoreside cranes on tracks alongside tracks for the little European open freight cars.

The Germans are still in the city. A few days later hundreds surrendered and were in a big fenced area right by the quay we were tied to.

Back to the States and another secret run, but with all the ship prep in Brooklyn by Navy yard everyone knew we were Russia bound. This ship was the *John Sharp Williams* KXCO.

The Russian trip not as bad as the earlier months of the war when we suffered huge losses.

Last trip—late 1951 aboard the *SS John Chester Kendall*, first ship to India with load of wheat under a lend lease type of deal. Madam Pandit and lots of dignitaries down to the ship at Philly before sailing. Newsreel cameras, Voice of America, etc., etc. We were to be the first ship to India with life saving wheat.

We get half way across the Atlantic and we got trouble. The ship had been laid up in Bone Yard somewhere before being taken out for the India wheat trips. At some time salt water had got into engine room. Rust had gotten a start under the shiny surface of one of the large crank pins. After about a week steaming, the outer surface crumbled in spots leaving surface like wormwood. The sharp rusted surface wiped the bearing. Carried one spare. Replaced the two halves, wiped those. About four days total floating around. Good weather helped. The engineers disconnected one piston and tied off the one rod to one leg of the engine. The up and down job goes on two.

The *Kendall* not a happy ship. . . . The mutiny, attempted murders, fights, the race problem, the morale, the morals . . . Whites the minority, less than half. The second mate and I were elected to go to the American consulate at Alexandria to try to bring law and order aboard. Some of the ongoing moral offences by crew members were hanging offenses. Think many a British captain would have made short work taking care of that crew.

Gallups Island had been good though. Wouldn't take a million for the experiences Gallups led me to.

Platoon R-31

NEWBOLD, CHARLES G., R-31

I lived in New Hampshire at the time, so it was no problem to get to Boston—I hitchhiked—in those days you could hitchhike.

I enjoyed my stay over at Gallups, especially the weekend liberty. I would get the train up to New Hampshire, spend the weekend there, which wasn't too bad. Eventually we graduated in September—no, maybe it was August of 1943, and I got my license. It did take quite a long while before I got a ship, but I eventually got one and I wound up on five different Liberty ships during the war, mostly out in the North Atlantic.

Of course, the Murmansk Run was memorable. We started with 28 ships and wound up with 13. Yours truly didn't suffer at all, not even with a cold.

I made a lot of side trips to Scotland—my mother was born and brought up in Scotland. She had 12 brothers and sisters there; of course, I met them all. You might say, through the influence of one of my uncles, I literally owned Glasgow. I took back to the ship what I wanted to and so did several members aboard ship.

After the war was over in Europe, I went to the West Coast and spent a year out in the Pacific. I was midway between Manila and New Caledonia when the war was over. We were eventually scheduled to go to Suez which I didn't mind at all because I'd never been through the Suez Canal—it might be my first trip around the world. Well, after 48 hours, they re-routed us again and we were back through the Panama Canal, eventually into Baltimore and I got out—it seems to me it was about February of '46—and was more than glad to get out. Back into the hills of New Hampshire, which, at that time, I enjoyed very much.

I had knowledge of Gallups before I even got there—but I didn't know a heckuva lot about radio and still don't for that matter.

Platoon R-30

Platoon R-31

Platoon R-32

Platoon R-32

SCHUUR, CHARLES , R-32

After my graduation from Gallups Island in October 1943, my first two trips on the *Stephen B. Elkins*, a Liberty, were to Swansea, Wales, back to Halifax and then back across the Atlantic to an unremembered location about thirty miles or so from Glasgow, Scotland.

After a return to New York I changed ships to the Florence Martus, another Liberty, in order to become the Chief Operator. Subsequent trips were made to Liverpool/Manchester (twice), Cardiff, Wales, Antwerp and Cherbourg. The North Atlantic seemed to become sort of a "milk run" for me, although somewhat more exciting at times.

With the war then being over in the ETO I agreed to just one more trip with a load of grain for somewhere in Europe. We shuttled from NYC to Galveston to load up the grain but found that the *Martus's* holds were a little too leaky to allow the load of grain. Got sent to New Orleans where we loaded up heavy equipment for the Philippines. About halfway across the Pacific the war with Japan ended and consequently when we got to the Philippines nobody needed or wanted our cargo.

After a couple of weeks shuttling between Batangas and Tacloban, we were relieved of our cargo and ordered to Calcutta with a couple of days stopover in Singapore en route.

When the *Martus* got to Calcutta we were one of three ships assigned to take ammunition and bombs stored in Army depots, load up our holds to capacity and then dump the loads two hundred miles out in the Bay of Bengal. One hand giveth and the other hand taketh away? Anyway, we did this for five months, each loading and unloading taking about a month each time.

Having three operators aboard really left me little to do and it was quite a relaxing five months in India. We got on good terms with the personnel of the Seaman's Club and they even let us use their Jeep or station wagon when it was available, which let us get farther out beyond central Calcutta. And Calcutta was the only port I found where the USMS officers were permitted the use of the Army Officers Club. Come to think of it, I still have some drink chits left over.

(Met a fellow at the AOC who was in the process of closing up the area radio station and he gave me over eight 17" transcription records and quite a number of the 12" V-DISCS, many of which were recordings of tunes that had never been commercially recorded. Being a "Big Band" buff, I have put the salvageable tunes on tape and they keep me company a good part of the day.)

With the dumping project over we returned to New York via the Suez Canal and past that well-known life insurance symbol, the Rock of Gibraltar. Signed off the ship on April 16, 1946 and headed for home, which at that time was Cleveland. The expected two-month trip turned out to be ten months instead. Around the world in 300, not 80 days.

Won't bore you with life after my sailing days, but I'm still plugging along with a few of the assorted ailments which usually come with advancing age. Left Cleveland in 1954 and have been residing in a country area about 35 miles southeast of Cleveland since that move.

Retired in 1982 and my time now is taken up with reading, doing stained glass work which I started five years ago, taking care of a couple of rental properties, a couple of trips to Arizona each year, as well as taking care of myself and the homestead, as my wife passed away seven years ago.

The GIRA sure is doing a fine job and I'm sure us old timers all appreciate the work done by the editor and all the other contributors.

Platoon R-33

CANADY, LOUIS L., R-33

I left for Gallups Island in 1943, and was assigned to R-33. Since I worked for Sinclair Pipe Line Co as a telegrapher, code was no problem; however got to playing lots of blackjack with another R-33 friend by the name of Brezinski, and looking forward to weekend liberty too much, am afraid I neglected classes. Consequently, about one week before graduation, another friend by name of Lomansey and I were asked to resign.

To make a long story short we went to New York to take our FCC test. We couldn't get a hotel room that night as all rooms were taken. We ended up sleeping on the steps of some apartment building. Sure was one long night. The next morning we took our test, and we both messed up on element one. As a result we ended up with Temporary 2nd class licenses. I then went to New Orleans to find a ship, checked in with the Union; however every time my turn came up for a ship, a regular union member would show up and I was back on waiting list.

About this time I ran out of money, found a job at the Lone Star Cement Co., ate one meal a day, which consisted of a poor boy sandwich, 15 cents if I remember right. About two weeks later got paycheck, also checked in with United Fruit Co. and they had a new ship for me. Went to Panama City, FL. Signed aboard the *SS Frederick C Howe* went back to New Orleans, then to Puerto Rico where we picked up load of sugar. Puerto Rico was quite an experience for a small time Kansas boy. Next we went to New York where we traded sugar for gasoline, ammo, and planes. Joined a large convoy and headed for the Mediterranean. When we passed Straits of Gibraltar the Germans had a fix on us. I think we were the first convoy of over 100 ships through the Mediterranean, had destroyers and I think one cruiser with us.

Every night the German planes would come over and drop magnesium flares and then the bombing and shooting would commence. It's amazing how beautiful the sky looked with the flares, tracers, bombs and fires. Copied several SOS, which we couldn't do anything about, saw quite a few of their planes go down from gunfire. When we got to Suez it was all over, stopped at Aden, then on to Karachi, India. Unloaded there. I also ran into a flier there from my home town of about 150 people, small world. We then proceeded to Perth, Australia and then back to Seattle, Washington where I signed off June 28, 1944. Signed on next on *Fredrick C Howe*.

There I met and was a fellow Radio Officer with Barney Reynolds, on either my first or second voyage. Stayed with *Fredrick C Howe* until November 1, 1945. Took my last trip on the *Henry G Costin*, took load of coal to Belgium. Discharged February 5, 1946; then I went back to work for Sinclair Pipe Line Co. pumping crude oil.

CLAUSE, OTTO, R-33

I was very active in the high school amateur radio club, and always wanted to get into radio—I guess electronics was not quite the right word for it at that time. So I enlisted in the Maritime Service in late '41 or early '42 and was originally assigned to class R-31 at Gallups Island.

We had a work detail up there—a whole gang of us was moving a lifeboat davit, and there was an accident. I had my foot smashed, and so I spent a couple of months in the maritime hospital up in Brighton, MA. When I got back to the island I fell back a class to R-33.

So I graduated from R-33 and picked up an assignment on the *Quincy*, which was operated by Black Diamond Lines, and carried all types of war material throughout the Mediterranean to India.

"We then picked up other cargo to take down to South Africa and around to Brazil and then back up the coast. We made a couple of trips into the Mediterranean, offloaded some light military equipment at Anzio during the problems over there at the beach.

That ship was put up into mothballs in the James River, and then we moved over to the *James Pedigru*, which was operated by Moore-McCormick. We made a couple of runs up and down the coast of South America, which were practically the same ol' routes to Argentina and Brazil I made during peace time.

I never took any time off during sailing until a friend of mine had an opportunity to take a cabin cruiser down the Inland Waterway from Baltimore to Florida. He was a transmitter engineer at a radio station—WITH in Baltimore. The chief wouldn't let him take time off unless he could find a qualified licensed replacement. In those days, a telegraph license would suffice for a radio transmitter. So he talked me into staying ashore to relieve him so that he could take the trip, which I did.

At that time the station was just expanding. FM had just started on the old 43

Platoon R-33

Platoon R-34

Platoon R-35

megacycle band prior to moving up to its present 88-188 megacycles. We converted the transmitter to that present frequency band, doing a lot of construction—which I liked. I just sort of got hooked on radio.

When television first reared its head in Baltimore, I went to work for WMAR Channel Two to help put them on the air. I stayed with them about a year and a half then moved over to WBAL Television, Channel 11. After about nine years I became chief engineer, and just recently retired from WBAL TV as chief engineer. So I've basically been in broadcast, or some kind of radio, for forty-some years.

At the present I have a little company where I do a little consulting work just to sort of keep my hand in. We've also got a farm up in the country, and we're looking forward to backing off and taking life easy.

STILES, GILBERT G., R-33

I arrived at Gallups Island in February of 1943. I think I was assigned to platoon R-33. Masalon Wilson was the platoon leader. After graduation, I sailed for Esso (Standard Oil of New Jersey) and came home for good in April, 1946.

Platoon R-34

PEARSON, EVERETT, R-34

I was in platoon R-34 and graduated from Gallups Island in 1943. I ended up on a Liberty ship, the *James B. Weaver*, as 2nd RO, out of New York City, in convoy bound for England. When all cargo had been discharged in Wales, 500 bunks were installed in the #4 hold to carry troops to the Normandy beachhead.

The ship was given the number MT 243 for military use and sent to Scotland to wait. On a minute' s notice we sailed to Falmouth, England where we loaded troops and their equipment and went into the Normandy Invasion. Made many trips between to Southampton and the beachhead, to load and deliver troops and equipment.

Later we made trips up the rivers in Belgium and France carrying military supplies. We were attacked by German planes on more than one occasion; shelled by shore guns, mined—but the ship never suffered any direct hits. It was a little over a year before arriving back in port of New York. Went into drydock.

Left for vacation, then sailed on the *James B. Pratt*, a Liberty, out of the port of New York for the Pacific. Anchored off the Northern Philippines and was scheduled for the invasion of Japan when the war ended. Had discharged all the heavy deck cargo and cargo in the hatches before going up to the northern Philippines.

I sailed for awhile and quit the sea in 1948. Sailed on a victory ship during the Viet Nam war. Spent 50 days that trip.

Platoon R-35

MORRISON, WILLIAM "BILL" A., R-35

I went to boot camp at St. Petersburg, FL and from there to Huntington, Long Island. I was in the second bunch that was ever there. I started out in R-35 at Gallups Island in June of '43. Then I got a ruptured appendix and got put back in R-45. I left with R-45 on December 31, 1943.

I got on a ship, went to the Persian Gulf with a load of sugar from Havana and then I went to the West Coast in '45 and spent a year out there on a tanker running from Panama back to the fleet wherever they might be, generally in the Ulithis and Marianas Groups.

I came home in December, '45.

MORSE, ARTHUR S., R-35

Enlisted in the USMS 14 December '42 and was trained on Staten Island to be a seaman. Managed to escape to be in the second class at the Otto Kahn Estate in Huntington, Long Island for eight weeks. Then on to Gallups Island for the class of R-35 which graduated in October of '43. Somehow got on the USMSTS *American Navigator* for a two-week cruise in Chesapeake Bay. Don't remember too much except it was quite jammed with people.

After that I stayed at the 34th St. YMCA while waiting for a ship. In January, '44 two of us were offered a job as sonar ranging operators on the *SS Kernstown*, a new T2 tanker, being built at the Sun shipyard in Chester , PA. We were sent up to Boston for 40 hours of operation and maintenance of the sonar equipment—we did learn how to turn it off and on.

We signed on the ship 11 February '44 and found out that: a) all four of us were jammed into a stateroom aft designed for two; b) our console was in a closet on the bridge—about 3'x3' and the door had to be closed because of the "pinging" noise and; c) the transmitter/receiver was located at the bottom of a 6'x6' shaft from the deck to the keel. There was a hatch cover over the shaft and steel rungs to climb up and down. A pulley arrangement lowered maintenance equipment. We worked shifts of three hours on and nine hours off. Our instructions were to range out 3000 yards to the port side and inform the watch officer if we received a sonar contact.

The first voyage of the *Kernstown* started on 13 February '44 when we set sail for Norfolk, VA to get depermed (degaussed) and from there to Baytown, TX. Around February 14th or 15th, the seas became very rough—lost five lifeboats and oil drums smashed into the gun turrets and the remaining lifeboats. It was a rough trip and all we did was stay in our bunks. Only one man was injured and he was put ashore in Miami Beach. On 21 February we arrived in Houston and had the lifeboats replaced and other repairs made. Finally loaded the aviation gas and sailed for New York City on 26 February. Again we had rough seas for three days.

In New York we loaded a deck cargo of planes and on 10 March we set sail for the UK. On the morning of 18 March, the convoy was attacked by a sub and the General Alarm was sounded. We were operating as we were instructed. Then the word came over the TBY that we should shut down ranging and go switch into a passive listening mode until directed otherwise. We did this and were never directed to resume active ranging again. We docked in Avonmouth and discharged our cargo. Returned to Chester, PA, with no major incidents.

Cargo was loaded on 12 April at Marcus Hook and we set out for New York City where we joined the convoy for the UK. With very foggy conditions the convoy was quite loose and two collisions occurred. We were able to listen to the other ships' propellers, count their revolutions per minute and match our speed to the ship ahead so that we would not ram them.

On 16 April the General Alarm was sounded; the *Pan Pennsylvania* was hit amidships—800 yards dead ahead and was dead in the water. We were able to avoid ramming her but not by much. Convoy resumed and all went well until the next morning when the General Alarm was again sounded and the gun crew saw a "periscope" and began firing. After a while the firing stopped when the object was identified as a floating spar buoy. Finally arrived in Bristol Channel after a couple more General Alarms—but no attacks.

Since nobody had told us not to range in this convoy, we started again but soon found that nothing was happening (i.e., no answers). We were able to get a maintenance man from some radar outfit in England who looked at it for about 15 minutes, replaced a few tubes and pronounced it "OK." He told us the hardest part of the job was climbing up and down the ladder. Sailed home to New York City with few incidents.

The last voyage of the ranging operators started on 14 May with only three of us left so we worked a regular shift, four on and eight off instead of three on and nine off. This voyage was uneventful except for the ending. The ship was scheduled to go into drydock and that meant we had to raise the housing of our sonar into the small hull along the keel in order to prevent damage to the transmitter.

So we pushed the button, but the light stayed on. Then we climbed down into the hole and cranked it up by hand—the light still stayed on indicating that the housing was still in the down position. Two hours had passed—the drydock was waiting—time was mounting and chaos was imminent. In the end the Old Man said to take them in and and the hell with it. So they did and lo and behold—there was no housing there. It had fallen out and is still sitting on the ocean bottom in the Irish Sea. End of the ranging operators on the SS *Kernstown* .

PALLAZOLA, DOMINIC, R-35

I was born and raised in Gloucester, MA just 100 yards from the ocean. I've always loved the sea. After graduating from high school, I was working in the defense industry in Detroit when the draft board began breathing down my neck. I had read about Gallups Island in Boston Harbor and thought that joining the Maritime Service was the quickest way to get there.

I enrolled in the Maritime Service on December 2, 1942 and was sent to Sheepshead Bay for "boot camp" training. While there I took a qualifying test for radio school and was sent to the Otto Kahn estate in Huntington, Long Island, for pre-radio training school.

From there we were sent to Gallups Island for the grueling training to get our R/O ticket. The code was easy but the theory was a nightmare. I lived, breathed, ate, and slept "theory" with the old "Q&A" as my bible. On weekends I returned to my family in Gloucester and fell in love with my present wife. She always encouraged me when I doubted my ability to learn enough theory to get my ticket.

One of the highlights of my Gallups Island sojourn was when our platoon participated in a bond rally, marching through Boston with many Hollywood movie stars. Some of them were Jimmy Cagney, Betty Hutton, Judy Garland, Harpo Marx, Greer Garson, Fred Astaire, Mickey Rooney and others.

We finally went to the Custom House in Boston to take our exam and to my great relief I secured my second class radio telegraph license. I chose to ship out of New York and was on stand-by for several months. My first ship as second R/O was the Liberty *William H. Wilmer*. The chief radioman was also a Gallups Islander, Bob Richelson. We got along great. The *Wilmer* was in a convoy going to Liverpool, England. The trip over and back was uneventful except for the constant depth charges sounding off—a reminder that the enemy was out there.

My second trip on the *Wilmer* was to the Mediterranean. We went to Oran, North Africa, then to Catania, Sicily, and back to Bari, Italy to discharge cargo for the British Army. We returned to North Africa to load up with Free French forces. After a brief stop at Ajaccio, Corsica, we invaded Southern France, with the beachhead at St. Tropez, on the Riviera.

We then loaded up with Italian laborers to take them back to North Africa. I was used as an interpreter to question these laborers with the aim of trying to ascertain if they had any venereal diseases. I wasn't too successful because there were so many different dialects that it was hard to communicate. We unloaded the laborers in North Africa and loaded up with POW's who were members of Rommel's Afrika Corps—bound for Norfolk, VA. They impressed me as being fine, physical specimens and their morale was quite high. Also, sadly, in the bow of our ship were five GI felons being returned to the states for major crimes. They were highly guarded.

In Baltimore, preparations indicated a trip to a cold climate—and so it was. Our northward trip to Russia with the wintry blasts blowing, the "ash cans" forever sounding, and the rough seas was a memorable experience. After an exciting trip we entered the White Sea and docked in Archangel. We got frozen in and I remember walking across the ice into town. Two icebreakers, the *Lenin* and the *Stalin*, made a pathway further into the White Sea and we loaded up with manganese ore at Molotovsk. During this time, the war news was good. The Germans lost the Battle of the Bulge and victory was in sight.

Our ship left the convoy on its return to Scotland to repair a damaged propeller. While there I sent a cable home to my girlfriend to set the wedding date on our return to the states. We were part of a very slow convoy, battling "growlers" (loose ice) all the way back home. Finally arrived in Baltimore where I got to send a message in plain language to the home office. Upon docking, I made a beeline for the nearest telephone. Left the *Wilmer* and my shipmates. I'll never forget Bob Richelson, Captain Jubb or Timothy J. Ring, Chief Engineer—all fine men.

Returned home to get married on March 11. After an all too short honeymoon, I shipped out on the Liberty *Robert E. Peary*. We made two trip to Antwerp, Belgium. On the homeward convoy I remember V-E Day with BAMS coming out in plain language "SPLICE THE MAIN BRACE," which any seaman knows is a celebration toast. Although there were also messages to all German subs to return to port, ships were being sunk by subs up until the end of the hostilities—even off the East Coast of the United States.

The Maritime Commission sent me a Christmas present—my discharge papers in the latter part of December, 1945. After the holidays I took my bride back to Detroit to the war industries job that I had left. That was winding down so I joined the Post Office where I spent the next 30 years.

Of my impressions on discharge . . . the almost "shame" I felt for being in the Merchant Marine—some of people called us draft dodgers and even worse. I refused to discuss any of my war experiences with my co-workers. It wasn't until we got our veteran status in 1988 and since I joined GIRA that I had a change of heart, as far as my war experiences were concerned.

I started reading books about the wartime Merchant Marine and am now proud to say I belonged to that group. My medal from the Russian Embassy and my letter from Gorbachev are proudly displayed in my den. The men I associated with during those years exemplified the finest that American had to offer and I am proud to be counted as one.

RUTTER, VERNON HAROLD, R-35

Born December 12, 1929, I served in the United States Coast Guard Merchant Marine aboard the *John L. Elliott* from March 1, 1944 to August 15, 1945. During that time, I served as 1st Radio Operator and Chief Radio Operator.

I served during four voyages on the *Elliott*, traveling from New York to Liverpool, Southampton, Omaha, Utah and Sword beaches, Cherbourg, Antwerp, Ulithi, the Lingayen Gulf and other locations, transporting ammo, personnel, vehicles and other cargo.

Some of my most notable experiences while serving aboard the *Elliott* include being at the dock in Antwerp during V-1 and V-2 attacks and participating in the Battle of the Bulge. Also while on the *Elliott*, the ship voyaged from New York to Apari to transport 2,000 Japanese POWs.

The awards I received include: Merchant Marine Emblem, Atlantic War Zone Bar, Pacific War Zone Bar, and the Mediterranean-Middle East War Zone Bar.

As of February, 1990, my home is in Riverwoods, IL.

Platoon R-37

CARLSEN, FRANKLIN, R-37

I was working as an apprentice instrument tech at ARMA Corp. of Brooklyn, NY up to early 1943 and many of my friends were going in the services, so coming from a seafaring family I went to the Maritime Service office and the next day I was a "boot" in the Sheepshead Bay Training Station.

While there I was given the "code recognition test" and was told that I (along with others) qualified to train for radio operators school at the Otto Khan Estate. We were there approximately 10 weeks before we went to Gallups Island.

We finished our training around November, 1943 and went to different ports for ship assignments. My first voyage was on the *SS Joseph Story* to Cardiff, Wales. I had two new operators, both Gallups men, on my next voyage on the *Story*. We sailed on March 23, 1943 to Liverpool, England and after discharging our cargo were retrofitted as a troop/vehicle transport and sent up to Oban, Scotland waiting for D-Day.

We took part in landing troops and tanks onto Rhino barges for landing on Omaha Beach and shuttling between Southampton and the beachheads until they opened the ports of LeHavre, Cherbourg and Rouen. We finally terminated our voyage in New York on March 15, 1945—just 10 days short of one year.

Marion and I were married on April 13, 1945 and I made my last voyage on the *Story* to Antwerp, Belgium. We were in convoy at sea when the war was over in Europe.

When we returned to New York I signed off and after a short leave signed on the tanker *SS Phoenix*, for two trips.

My next ship was the *SS Waterbury Victory*, a troopship used to ferry replacements to Europe and GI's home to New York for separation. On my last trip on the *Victory* we brought home the 442nd Battalion of Neisi troops from Hawaii, who had distinguished themselves in battle in Italy. We brought them from New York through the Panama Canal and on to Honolulu where they were given a rousing welcome. The local people made us feel very welcome for bringing these heroes home.

From there we took replacement

troops to Yokohama, Japan and brought troops home for discharge or reassignment. Signed off in San Francisco and flew home.

About this time the Army Transport Service was signing up Commercial Radio Operators, so I signed on the *USAT General Ballou* for a trip to Leghorn, Italy. After that I sailed as Chief Operator on the following ships: *USAT General Taylor* (troops), *USAT Robert Burns* (coffins to Europe and war dead returned to US), *USAT Jose Valdez* (troops/ supplies for the Antilles Command), *USAT Jonah Kelley* (troops/supplies for Northern Arctic Command), and the *USAT Zebulon Vance* (hospital ship). Was then reassigned to the *USAT General Taylor* for my last voyage from New York to Naples (picked up approximately 1200 D.P.s), then through the Suez Canal and on to Columbo, Ceylon to Sydney, Melbourne, and Freemantle, Australia, back to Columbo, through the Canal to Bremer Haven and picked up another load of D.P.'s for New York where I signed off and quit sailing in September, 1949.

I worked at Kollsman Aircraft Inst. Corp. as an electro-mechanical tech for approximately 11 years. Then worked as a lab supervisor and engineer in R&D, with a small group in the development of a microfilm process and equipment for Republic Aviation Corp., a division of Fairchild Industries, for eight and a half years. A money group in California purchased this system from Republic Aviation and called the new company Microform Data Systems. They made attractive offers to persuade five of the seven of us to relocate to Santa Clara Valley, CA—I was one of them. Later the company name was changed to ICOT Corp. due to product changes, etc.

When we first started I was doing hands-on engineering but soon became the engineering administrator responsible for project controls and budget. After 18 years I was administrative assistant to the president and corporate budget analyst. I retired in 1987 and moved back to New Jersey to a retirement community.

Marion and I enjoy traveling—just visiting our three children and grandchildren takes us to Rochester, NY, Atlanta, GA and Mounds View, MN. We enjoy the reunions.

Until then—73's.

CROSBY, RICHARD H. (DICK), R-37

I signed up with the U.S. Maritime Service at Boston in January, 1943 and reported to Sheepshead Bay in February, 1943 for boot camp. While there I learned about the opportunity for Radio School at Gallups Island. I had always been interested in radio/ electronics so I immediately applied and was lucky enough to be accepted. At the end of boot camp in March, I went to the Phase I Radio School at Huntington, Long Island. This was at what had been the Otto Kahn estate, which included a mansion that had approximately 100 rooms with extensive and beautiful gardens and grounds. It was an excellent spot for a school and very conducive to study. Classes were in basic radio theory, code, math, typing, etc. After about two months we moved on to Phase II at Gallups Island for seven more months of schooling. Gallups was also a nice place to study. You could get off by yourself with your books, looking out across the harbor, or past Boston Light to the open ocean. A great place for a Maritime Service Training Station.

The instructors at Gallups did a great job in teaching us and preparing us for the FCC license exams. Our training was comprehensive—not aimed strictly at passing the exams—but the instructors, towards the end of the course, helped prepare us for the special type of multiple choice questions and ambiguous questions that we could expect.

Our platoon graduated in November, 1943 and I then went to New York to get a ship. Warren "Woody" Woods (R-37), Bill Archambeau (R-43) and I had some pleasant times in New York while waiting to ship out.

After a couple of months I signed on the *SS James Smith* as Chief Operator. She was a Liberty and a good ship with a good captain: Captain Homer L. Howser—I stayed with her till I left the Service in March of 1946.

Our first trip was to Murmansk and Archangel by way of Glasgow and Loch Ewe, Scotland. The next trip took us to Omaha Beach right after D-Day and then back to England/Wales for several shuttle trips from Barry Dock/Cardiff, Falmouth, Plymouth, London, etc., to Omaha Beach, Cherbourg and LeHavre; we were among the first ships into Antwerp—at the time of the Battle of the Bulge. We returned to the States in January, 1945 and then made two more trips to LeHavre and Antwerp.

After VE Day we went through the Panama Canal to New Guinea and unloaded our cargo at Hollandia. We then went to New Castle, Australia, and picked up a load of steel for pontoon bridges. From Australia we headed for the Philippines and arrived just after VJ Day. After the war no one knew what to do with the pontoon bridge steel, so we sat at anchor at Samar until early February, 1946 when we headed back to Baltimore. After having swung at anchor for six months we had about a foot of grass on the hull so our top speed in still water was about six knots. We played the two or three knot currents along the equator coming home for all the extra speed we could get. Arrived in Baltimore in March, where I signed off the ship, and left the Service.

For all of the time I was on the *James Smith* after the first trip, I was fortunate to have W.A. Barker of R-62/63 as my fellow Radio Operator. We became good friends and after a few months home in Massachusetts I went to Texas to visit him. I liked Texas, went to college, met my wife and stayed. W.A. was best man at our wedding. My degrees were BS and MS in Physics, a natural follow-on to the Radio/Electronic education at Gallups.

My working life after college was as an Electronics Engineer designing aircraft electronic systems with Chance Vought Aircraft in Dallas in the 1950's, Range instrumentation systems at Cape Canaveral in the 1960s and then back to Dallas, spending 23 years with Texas Instruments before retiring in 1990.

In January, 1991, I was fortunate to be among a group of fellow Merchant Marine Vets at the Soviet Embassy in Washington where we were awarded the Russian Medal for our trip(s) to Murmansk/Archangel. It was nice to have them remember our help in WWII. My memories of the friendliness of the Russian people of Archangel and of the Russian sailors we brought back to England, together with mental images of the hard work and sacrifice of the Russian people that I saw on that trip, have always been in my thoughts during the Cold War.

The last couple of years I've helped organize and hold a Merchant Marine Veterans Reunion here in Dallas. Last year (1992) over 100 veterans, plus their wives and/or guests attended. We all had a great time—great camaraderie. We hope to do it again.

PATTERSON, WILLIAM B., R-37

Got into the Maritime Service Radio Training Station at Gallups Island in the early part of 1943 and graduated in November, 1943. Went aboard my first merchant vessel, the *SS Noonday*, which was an ammunition carrying vessel on its maiden voyage. We went to Liverpool, England after loading at New York.

I sailed (on) cargo ships, tankers, T2s, other larger type tankers—both chemical and sulfur—and the container box carriers which had been converted from T2s to a combination container and chemical container.

I was subjected to the German Buzz Bombs, V1 and V2 types, while in Antwerp, Belgium and a torpedo was fired into our convoy, but the vessel I was on was not hit. I was very lucky during the War and did not suffer any torpedo attacks or injury in any war during my career due to enemy action.

I spent 42 1/2 years at sea. Retired in November of 1986. Was sailing almost constantly during that period of time and I have to thank the U.S. Maritime Radio Training Station for my training as a radio officer—I made it a career. I ended it in 1986 and I'm enjoying every minute of my retirement.

My wife's name is Donna and we live in Richardson, TX.

PIACENTE, P.J., R-37

The first time I saw Southern France was during the invasion, which was something of a letdown, since we had been all keyed up, expecting a lot of action. Our ship, the *Mercy Warren*, a Liberty, had been loaded with men and material—we even had a doctor and three medical corpsmen aboard. A small section of the bow had been designated as a hospital, and there was a lot of conversation about whether the Captain would beach the ship if it were hit, to save the cargo, or whether he would run for the open sea, etc.

Platoon R-36

Platoon R-37

Platoon R-39

We were in a convoy with two rows of ten Liberty ships each, that had followed the invasions force, and the *Mercy Warren* was the lead ship in one of the rows. We arrived about five miles off the coast on the evening of D-Day, and we could see flashes and tracers in the distance. My watch was scheduled for 3 a.m., so I went below to get some sleep, after telling the Third Radio Operator to be sure to wake me.

I awoke the next morning to find we were at anchor. It was a bright sunny day, and on deck the men were all busy with their daily routine. An LST had been beached and lay tilted on its side about 50 yards away. I heard a distant explosion and saw a puff of smoke on a hillside, so I trained a longlass on the location and observed three soldiers who were obviously exploding land mines. Between two hills, I could see a church spire. There was a road running parallel to the beach, lined with graceful Lombardy Poplars. I noticed some movement on the road and when I trained the longlass on it, lo and behold, it was a girl riding a bicycle.

The next morning, there were about 300 ships behind us that had arrived in the night. The morning after that, there must have been 1000. That evening, the Germans sent a reconnaissance plane over the harbor. We had orders not to fire our guns unless we were attacked, but any time the shore batteries started firing, our gunners and the gunners on every other ship opened fire. It was an exercise in futility, since the plane was too high for our 20 mm to reach it. However, about seven of the ships shot down their own barrage balloons that were tethered to afterdecks.

STRONG, MARVIN L., R-37

I signed up for the Maritime Service in Wichita, KS while I was working for Boeing Aircraft Company and I guess if I'd have been real smart I'd have kept my occupational deferment. But I was mad-to-get for adventure and I always thought I'd like the Merchant Marine.

I've always been interested in radio operation and I don't have any regrets at all about my Merchant Marine service. I met a lot of good people and I didn't ship any after the war.

I did a little stint in radio in Galveston, TX for about a year and then I married and went back to my home state of Oklahoma and began working for one of the oil companies in the communications business.

Platoon R-41

BRIGLIA, WILLIAM J. SR, R-41

In about February, 1943 I thought I was going to Sheepshead Bay for training (being a New York cliff hanger), but I was sent to St. Petersburg, FL instead. Anyway, I wanted to get back to New York. Heard of a radio school up there. Took a test from a record, qualified and was off to Huntington, Long Island, NY to the Otto Kahn Estate—which was a castle that had been disassembled in Europe and reassembled in Huntington. We would muster in the cobblestone courtyard and sweat the hours learning the code. Food was scarce and we were sending home for some. If you went in line for seconds you would get extra duty. (The portions were extra small.) The canteen was mobbed every night. One fella got extra duty when he turned to the fella behind him in the chow line and said, "Here take mine, there's no sense in both of us starving to death."

Some of the things I remember about Gallups are: If you didn't pass the code test you didn't get off the rock; what lead-acid batteries could do to dungarees when you moved them; don't use an overboard suit without a towel around your neck—it fills up. Your life preserver keeps you from sinking—and from swimming. The times I used to catch the last train from Boston to the Pennsylvania station on lucky weekends; discussions over whether the barracks windows should be open or closed—the resulting broken windows and another restricted weekend. I remember most of us passed our exam for our 2nd class license—even a guy who never studied—just the "Q&A" —and he got a restricted license.

I remember my first ship, the *Henry H. Richardson*. Left New York in convoy and ran into the worst storm I ever sailed into. The old Liberty rolled like a pig. We had locomotives on deck too. All the boom lines had an inch of ice on them, the lifeboats were torn loose and battered against the supports and davits, the covers were torn loose in the gun tubs. My bed (a day bed) had torn loose and was sliding back and forth hitting the bulkheads on each side of the room. When I asked the bedroom steward if going to sea was always like this, he replied "Yes, Sparky, but you'll get used to it." Five days later we pulled into Halifax.

We had two pumpmen, a very large fella named George and a very small fella I'll call Pete. They went ashore since we had finished pumping avgas. Next evening an army jeep came down the cement pier, Pete was with the driver and big George was draped like a shot deer over the hood. We called our purser down to look George over. He had them put a mattress under him the checked his eyes with a light, took his pulse, and pronounced him dead, saying he had given him the mirror test. The mate decided to leave him there for the coroner in the morning as it was now very late. Next morning the mattress was empty and George was gone. He was up eating breakfast with the rest of the crew.

DAVIES, DAVID R., R-41

"Let us keep watch"

After graduation from Gallups Island and receiving my 2nd class Radiotelegraph License, I encountered fairly routine experiences while at sea during World War II. Most of my time was spent in the Pacific, along with many others, some of the finest people I have ever known. I recall one or two air attacks (Kamikaze), torpedo wakes and some tense nights in convoy. However, it was shortly after the War when an incident occurred that left me with a permanent respect for good communication systems.

My last trip was out of Seattle bound for the outer Aleutian Islands. The small freighter *M/V Crown Reefer* was carrying food, dry cargo, and a group of U.S. Army nurses on their way to duties on the islands. I had sailed as "Sparks" long enough to feel confident in operating a shipboard station and being comfortable with life at sea. However, this last voyage put past experience to a demanding test.

We had made port calls at Kodiak and Adak, proceeding west in very rough weather. At one point the ship rolled and pitched enough to flip the typewriter out of its well onto the deck. All hands seemed to be seasick or trying to convince themselves they were not. About 0400 one morning I was asleep with elbows braced against the bunk rails when the ship rammed head on into rocky reefs on the south shore of Amchitka Island.

Our speed was probably about thirteen knots when she went aground. As I was trying to get my balance and put on a parka, the Chief Mate was shouting at me as he stumbled through the radio room on his way to my cabin. Obviously he was calling for me to get an SOS out. The ship was grinding up on the rocks and had no power. Since he had a very heavy Dutch accent, especially when excited, I had trouble getting certain information from him, such as the ship's estimated position. The need for more efficient communication came to mind. I assured the mate that an SOS would be sent, and I followed him to the bridge. The call for help would not be worth much until I could report our position.

On arrival at the bridge, I saw the Captain braced against a high wind. One of his hands gripped the bridge rail and the other had the pilot's arm in a vise grip. Some of the crew had told me the Captain had survived some treacherous runs to Murmansk. He was of Russian ancestry, and his speech reflected that fact. I don't believe the pilot was trying to escape, but he was receiving some high level lecturing from the Captain in a mixture of English and Russian. When the Captain looked my way, he simply said SOS, which was fine by me, but the need for our position remained. The Second Mate was in the chart room. He realized what was needed and handed me a piece of paper with our position on it.

I returned to the radio room. It was lonely, cold, dark and very quiet, except for the grinding noises from below. As most of us probably remember, the primary reason for "Sparks" to be aboard was to transmit a distress message when called for. Good communication became very important very quickly, just as they told us at Gallups Island.

So the process began. The emergency battery switch gave me a dim light over the equipment. I was glad that I had kept the charge up and was relieved when the main transmitter tuned up on 500 KC. I started by sending the auto alarm signal followed by SOS and statement of our situation. At that moment we did not know if the ship would sink, roll over, or grind to pieces.

Instantly, upon completion of my transmission, a surprising reply was received. Two voyages previous, I was Chief Operator on the *SS Typhoon*, KMXW in the South Pacific. Unbelievably, that same ship was the first to answer me and offer assistance. Of all the ships at sea in those days, roaming the oceans in all directions, how could it be that the first reply to the first SOS that I had ever sent was from my first ship? As it developed, the *Typhoon* was many miles away and agreed to stand by while I took other responses to my SOS.

Complications began to arise. The U.S. Navy at Adak acknowledged my message, and as I recall nearly 50 years later, ordered a sea-going tug to our rescue. Direct signals with the U.S. Army stationed on the north of the island were needed, but apparently they could not cover my frequency. I thought again about how vital good communication could be. It was necessary to have the Navy relay messages to and from the Army, which then sent a rescue party across the island in snow and high wind conditions.

We were grateful to see U.S. soldiers appear high on the cliff across the heavy surf. It took them several hours of hard going to reach our location. Attempts at firing a Lyle Gun line to them failed, too much wind, too much distance. Using a blinker light on the bridge, powered by jury rigged wet cells, I was able to send basic information to the Army. My Morse code by light in a freezing storm was probably not too clear, and it seemed they were not accustomed to blinker light messages. Appreciation for ease and efficiency in talking to others came to mind once more.

To conclude the rescue story, as I recall the events, we put our passengers down cargo nets onto a flat bottom barge which was able to clear the surf. Crew and officers abandoned ship the next morning. The life boats were near to being swamped, all of us in heavy clothing. Can't explain how I got down the cargo net with the heavy emergency transmitter. We reached the rescue tug that was standing off in deep water behind a fog bank. All turned out well, thanks to a lot of help and the ability to communicate.

I would like to make a point. We should always be aware that the ability to communicate clearly is a vital part of our society. I hope we all keep watch and protect our ability to freely communicate. One of the important factors in good communication is the listener. I remember there were some good listeners keeping watch while the above events took place. All hands were glad of that.

Platoon R-42

GONZALEZ, VINCENT L. JR, R-42

Born in Havana, Cuba on October 10, 1918 of American parents. Moved to New York City about 1930 and went to school without knowing a single word of English.

Graduated high school and went to work in New York as a runner delivering documents to the steamship companies and consulates. Got married in 1940.

When I was classified 1A, I entered the United States Maritime Service. Spent my time at Sheepshead Bay, Brooklyn, NY then the Kahn Estate in Long Island and Gallups Island, Boston, MA. Got my second class telegraph license from the FCC early December, 1943. Graduated from Gallups Island on December 19, 1943 and obtained certificate #1785 .

Went on training ship for a week and while on duty at the radio room, we got a message from the Skipper to transmit to shore requesting aid from the Coast Guard to transport a ship mate to shore since he was very sick. Called WSL and transmitted text as fast as I could with the Bug since subs were around the East Coast of the USA.

During my last week at Gallups Island, I was called to the office and was interviewed by a Navy Lt. in Spanish. He wanted to see how my Spanish was. The Skipper then told me to report to the main office of the USMS in New York in the next few days. I was interviewed by a gentlemen who asked me if I would be interested in a job which could be very dangerous. My reply was "What can be more dangerous that being on board a merchant ship during the war."

I accepted his offer and was instructed to go to the Department of State, Passport Division, and apply for a passport without giving any destination and they would pick it up in Washington, D.C.

After two weeks on stand-by at the USMS office, I reported to the office and they told me to speak to a Mr. Jones. He handed me my passport and told me to report to 21st and Constitution Ave., Washington, D.C. the next day. I arrived in Washington without knowing what was at my destination or who I was to see. Upon entering the building the security was very heavy and I was taken to an officer who informed me I was now a member of the Office of Strategic Services (OSS).

I went to basic training for five weeks and one day they told me I was ready to ship out and to report to Mrs. Shipley of the Department of State in Washington. She told me I was going to the American Embassy in Madrid, Spain as a clerk. (This was my cover while in Spain.) Took 18 days to get to Lisbon, Portugal on a Portuguese freighter out of Philadelphia with a two day stop in Madeira. Took the night train from Lisbon to Madrid.

Return to USA via Lisbon where I took the Pan-American Clipper to return to New York. Released from OSS October, 19, 1945.

Amateur radio operator since December, 19, 1935.

LASSETTER, NORMAN, R-42

I enlisted in the Maritime Service in Dallas early in 1943, and went to St. Petersburg, FL for seaman's training; from there to Huntington, Long Island and Gallups Island, Class R-42.

As I recall, there was only one pay telephone on the island and weekend liberty only—a scenario for a much-overworked phone. The coin box would fill up and then the phone wouldn't work. I tried it one time, and in disgust hit the phone with my hand and also hit the jackpot——it disgorged several dollars in coins, and then worked fine!

Sea duty began early in 1944 in San Francisco as 2nd operator on the Liberty *Frank H. Evers*, operated by American President Lines, a voyage which lasted 14 months, from March, 1944 through May, 1945. In May, 1944, while in port at Hollandia, New Guinea, our Chief Radio Operator, Spence Leather, passed away on board ship as a result of a heart attack and it was our sad duty to lay him to rest there.

Other destinations on this long trip were several times to Australia, the Philippines (including Lingayen Gulf), and, as I recall, about half the islands in the Pacific.

Other voyages, all aboard Liberty ships, were from San Francisco on the *SS Hinton R. Helper*. My first knowledge of the atomic bomb was via Scott radio about two or three days before getting to Manila; some of the crew refused to believe this until they heard it themselves on the radio.

In 1946, I made trips from Galveston to the Mediterranean on the *SS William L. Yancey*, and from Port Arthur again to French, Italian, Greek and Algerian ports on the *SS Paul Hamilton Haynes*.

After leaving the Merchant Marine in July, 1946, I worked for broadcast radio stations in Brady and Del Rio, TX, in South Carolina, and then built and operated KALG in Alamogordo, NM before moving back to Fort Worth. For the past 30 years I have operated Emerald Hills Memorial Park and Funeral Home. I was married to my wife, Velma, in 1947, and we have three children.

MCCULLOUGH, JIM, R-42

My path from high school to Gallups Island had a zig-zag course. I was released from high school at the advent of Pearl Harbor to attend a mechanic learner radio school preparatory to working at Hill Army Airfield in Utah. I used this training to get a job as an electrician in the Kaiser Shipyards in Richmond, CA when I turned 18.

I couldn't stand watching the Merchant Marine sailors taking those ships without me being aboard, so I joined the Maritime Service with a promise of a radio operators job. Boot camp at Catalina Island, then Huntington and finally Gallups Island in platoon R-42.

I did barbering for those who didn't like the island barber. Great time hitchhiking the East Coast on some weekends. Hotel Touraine on others but mostly studying. I earned my Radio Telegraph and Radio Telephone licenses. R-42 was a great platoon as we studied and worked together and played together, making it all fun—including the occasional extra duty we qualified for!

Doyle Catron and I went to San Francisco to ship out. There was a long waiting list at the union hall and we were on standby with the USMS. Because of my previous experience we were able to go to work in the shipyards on a 30-hour week. We were wiring 5" 38 gun turrets; the Navy inspectors told us they would get us released from the USMS and the draft if we would stay and work with them. We said, "Sorry."

I shipped as 2nd radio officer on the *SS Robert C. Grier*, a Liberty ship, in March, 1944. If my memory serves me right we spent 55 days on the hook at Kwajalein and had some air action. We were also in Pearl Harbor both coming and going, I think.

Late in July I shipped as chief radio officer on the Liberty, *SS Charles Paddock*. This ship had a fantastic skipper, crew and armed guard. Almost all stayed for two trips, lasting about a year. We hit Pearl, Eniwetok and then followed the invasion of Guam with hospital supplies. I fearfully remember going in the harbor along the cliffs. Japs were holed up in the caves in the cliffs and occasionally took pot shots at us. We had some high explosive deck cargo and a change of shorts was called for when we were in and safely anchored.

On one of the shuttles to Pearl we were alone so they gave us a minesweeper escort with his sonar inoperative (who was escorting whom?). Between Pearl and the States we had a sub chase us on the surface. The chief engineer (Klein) tied the boiler safety down and handled it manually. We assumed the sub couldn't overtake us underwater and we lobbed shells at him when he was on the surface if he tried to overtake us. Fortunately he gave up before we blew the boiler. We dropped anchor in San Francisco on New Years Eve. Sparks made it ashore on the pilot boat so that he could patrol Market Street.

We departed San Francisco January 19, 1945 to Hueneme for cargo and then to Freemantle (Perth), Australia. There is a lot of water between Los Angeles and Perth over a 35-day period. We had hurricane winds off the southern coast of Australia for four days. We lost rafts, lifeboats and antennas and some plates just aft of the deck house buckled. I had to climb the mast in that wind and string an emergency antenna. I don't know if I was afraid of heights before that, but I have been ever since.

Shortly, I had a chance to try my new emergency antenna. Just at sunset a sub surfaced ahead of us in the sun. We opened fire and sent an SSSS message. The sub dived and a bit later we received a message that it was friendly and for us to proceed normally. Had to change britches again.

Found a lot of pretty girls and warm beer in Australia. We went commercial at this time. We sailed north up the Hoogly River to Calcutta then Ceylon, Mombasa, Zanzibar, Beira. (Was it Mombasa or Beira where Ma Gorman had her little dancing establishment complete with all the current popular records, good scotch, girls, etc?) VE Day while in Beira. Then on to Port Elizabeth (almost married the Dutch Mayor's daughter but the skipper wouldn't let me take her home with me), Capetown, Trinidad and New York on June 17, 1945.

With VJ Day occurring during my time ashore, I went to work for United Airlines as a radio operator but soon found out that the Army was still looking for employees. Fortunately I was on good terms with the Union in San Francisco and they put me on the *SS Sapulpa Victory* one day before I was to report for induction.

Signed on the *Sapulpa Victory* on October 4, 1945 and went to Hueneme for cargo then Okinawa, Saipan, Guam and Nuka Hiva. At Okinawa there was a tanker at anchor that had discharged its cargo sometime before but still had a crash boat aboard as deck cargo. One night it took a terrible list and the crash boat fell overboard. We commandeered this boat and used it for awhile around the area with the help of the SeaBee fuel dump. The chief engineer and another crew member were planning on returning and trying to drive it back to the States. I've often wondered if they made it.

When southeast of Pearl on our return trip I picked up an auto alarm distress call for the *SS Augustus Merriman*. Their boilers had been out for several days and their food was bad. We raced a French ship to get a line aboard her first, which we did. We towed her to Nuka Hiva, a French mandate island some 500 miles off South America. I would like to know how that crew made out with all those bronze-skinned native gals. And no, we didn't get full salvage. Our company negotiated with their company and we each got $300 for settlement.

That's enough wartime. I worked for United Airlines for 11 years ending up as a supervisor of communications in San Francisco. I wanted to get back into the Salt Lake area so I went to work for the Sperry Rand Corp. when they opened a missile plant in Salt Lake City. I worked in the Engineering Lab, Manufacturing Engineering and Quality Control. I was working as the test and inspection manager when I retired in August, 1981.

Since then I have sold real estate, raised Rex rabbits and have and still am running a small horse boarding stable in Sandy, Utah.

My wife, Jackine, and I have five children, 15 grandchildren and one great-grandchild and love every one and every day. We travel a bit, ride a dirt bike and compete in Mountain Man Rendezvous.

It was great to meet everybody at Long Beach in 1992.

MITCHELL, BOB, R-42

I wound wire on an oatmeal box and fired up a UX-201A with discarded telephone batteries when I was 12 and had a class B ham license at 13, class A at 15. At 16 I enrolled in Massachusetts Radio and Telegraph School in Boston. There I learned, as did others, from September to June, sufficient electronics to get our first phone and second telegraph.

Anyway—six weeks? Do you remember how QST said if you had a class A ham license the army would make you a staff sergeant on enlistment, or the navy a third class, or the marines a three stripe sergeant? Well, they wouldn't me—I was 6'4" tall and

Platoon R-41

weighed 119 pounds. I nearly had to wear snowshoes to take a shower. So, I was walking down State Street in Boston and here is a window "U. S. Maritime Service." What's this?? In I went, and in about 10 minutes I was enlisted. Off to Sheepshead, where I had to compete with everyone else for radio school. (I was just a kid from Kennebunkport who didn't know I could have shipped out direct, having the license already.)

But I made it, and on arriving at Huntington, you remember the query, "Anyone with previous radio stand by." Six of us did.... the first was in CCC, the second a class B ham, one had worked in his father's radio shop. When he got to me, I'll absolutely never forget the question: "Well, Sonny, what can you do?" "Sir," I answered, "I have a first telephone, second telegraph, class A ham, I can type 75 wpm, send and receive international Morse at 45 wpm and Continental Morse at 25 wpm, as well as run a Teletype." "Oh, yeah, can you prove it? Be in my office at one o'clock." (He must have been a 90-day wonder, he didn't say 1300.)

So the next day I went to Gallups, reported, and was put in R-34 for the last six weeks, shipboard training. To keep me busy, they put me in a code class, where I promptly ticked off the instructor as I sat looking around, apparently not copying the code. When he started to charge up the aisle toward me, I quickly put it all down on paper, so he couldn't say a word.

Another thing I remember about Gallups is what happened to me on Saturday morning inspection. The big wheel (excuse me, I should have said Big Wheel) walked past me and asked me howcum my pants didn't match my jacket. The junior officer with him leaned over and told him I was the kid that had my licenses, and they could only find a jacket, but no pants my size. "Restricted to the island." Same thing next week, only this time I was asked if I had civvies. "It isn't fair to restrict you because we can't supply pants." So I went ashore in civvies.

I'm sorry to relate that I was at Gallups such a short time that I don't remember much about it or any of the personnel there. However, I sure enjoy meeting you guys now, on the air and/or in person. Oh yes, now there wouldn't be any trouble finding pants, I've gained 100 pounds in 47 years.

Platoon R-43

BLACK, JOHN, R-43

Class R-43, Gallups Island. Coincidentally, that was in 1943 that we started that class and we finished in early '44.

There was not a need for radio operators at that time... so they were waiting disposition of how to handle that because we had all these graduates who were trained to operate. So, they changed the setup about that time.

When I arrived in New York I didn't have a ship and they came around and wanted a class for underwater sound. So I was in sonar—took a class at Bell Labs. Went on a tanker. It wasn't feasible to run tankers in a convoy with underwater sound, so that program was discontinued.

When I came back I was on the Liberty ship *Martha Sinclair*. Made several trips to Europe, the war was over, V-E day, came back with the 500 lb bombs, went through the Panama Canal and went over to the Philippine Islands. Had orders to go over to Okinawa and there we were in time for a hurricane or typhoon.

We were sent to sea during that storm and then after it quieted down we went up to Yokohama, then back to the West Coast. That time was late 1945 and that was my service.

Platoon R-44

DINSMORE, CHARLES H., R-44

I had been quite interested in communications since my days in the Boy Scouts, and had set up telegraph keys and buzzers in a couple of rooms at home. When I graduated from high school in 1942, I went to Massachusetts Radio School in Boston, and was in the same class with Bob Mitchell (R-42). We took our FCC exams and got our radiotelegraph licenses, which of course gave us a leg up to become Merchant Marine radio operators. I took boot training at Sheepshead Bay so as to learn some of the ropes of life at sea, then spent a few weeks at Huntington, Long Island radio school, then to Gallups for about six weeks to get some experience with shipboard equipment and wartime radio procedures, graduating in December of 1943 with R-44. I took until March of '44 til I got my first ship, a WW1 Hog Islander, the *SS West Honaker*. Although I continued sailing for seven years, and each ship had its own story to tell, I would say that more things happened on that first trip than any other!

We left Boston and joined a 130 ship convoy at Halifax, Nova Scotia. This was a slow convoy, 5 knots, and unknown to us, was scheduled to make a one-way trip to Europe! We were supposed to have two aircraft carriers and two destroyer escorts for protection, but the carriers didn't show up, so we left without them! We took a quite northerly route across the Atlantic, with waves running 50 feet at times, so as to avoid subs. One night, at about 2:00 a.m. while I was on duty (second operator of three), about half way across the "puddle", the ship just in front of us got torpedoed, then the ships immediately on each side of us got hit and lit up the sky as they were tankers, then the ship just behind us got it! That left us as a sitting duck and I can assure you everyone was nervous. Just at that time we overheard the destroyers closing in and they managed to chase the sub(s) away without further incident. A couple of days later we saw someone from a ship in the distance hanging onto an iceberg (don't know whether they got rescued or not).

Eventually we got to Liverpool, England and discharged our cargo. We also hit a couple of English ports and experienced night air raids by the Nazis, but no damage to us. We weren't allowed to go ashore so we kept moving around to different anchorages, and this lasted for about a month, until about June 3, 1944. By this time we realized we were going to be involved in the Invasion of

Platoon R-43

Platoon R-44

Platoon R-45

Platoon R-46

Europe! We also learned that our ship was scheduled to be sunk to make a breakwater for the portable landing docks that were to be towed over from England. This also explained the mystery of several ships constructed of concrete which had been in our convoy—imagine how far in advance the invasion had to be planned in order to allow the design and construction of these special ships.

Just before June 6th we received a message that the 2nd and 3rd radio operators were to be removed and repatriated back to the U.S., so we got off in Oban, Scotland, a seaside resort, and were repatriated in an old German ship, the *SS Kungsholm*. The ship I was on went to Bari, in Southwest Italy, where we were the first ship to arrive after their liberation. We had to zig-zag around all the ships the Nazis had sunk in the harbor to make it difficult for us. They had also destroyed all the dock facilities, so the bulk cargo of wheat we had aboard had to be unloaded manually by the workers carrying burlap bags down into the holds and bringing them up individually. There were still Nazis hanging around killing people, so we were cautioned to only go ashore in pairs and just to go to the Officers' Club a couple of blocks away. It took almost a month to unload the ship.

My next ship was the first one to bring supplies to Rotterdam, Holland. The country had just been liberated, and the people were just getting outside and enjoying the feeling of freedom again. As we were going along the 50 miles or so of the (I believe) Scheviningen River, people along the banks kept playing their instruments and singing the "St. Louis Blues." That and the other indications of their gratitude were things that I'll never forget.

I continued sailing until 1951, delivering Marshall Plan cargoes to Europe, and also sailed with the Military Sea Transportation Service. All these trips were adventurous but nothing like the previous.

I started in broadcast engineering at a couple of smaller stations, then one night at a union meeting in Boston, Gareth Linder (R-23) asked me if I would be interested in working at WBZ, a 50 KW clear channel station in Boston. In those days you had to wait for someone to die to get a job like that, so of course I jumped at the chance! Some years later I learned that the Chief Engineer and Engineering Manager, Willard Hauser, was a graduate of Gallups (I don't know his class)! When I learned this I wondered if he had hired me because of the Gallups connection, but there may have been another reason, too—in those days we had a synchronized operation with WBZA in Springfield (about 90 miles away) and any communication between the two stations for engineering purposes was carried out by Morse code on a special Western Union Line, so everyone was required to know the code! I spent the next 31 years there, always finding it to be a challenging job as it seemed to be company policy to keep one jump ahead of the competition in both engineering and programming.

I finally had to call it quits in 1983 when it was discovered that I had Lou Gehrig's Disease (one of the Muscular Dystrophy family of illnesses). I have been quite lucky compared to most people with this, as I'm still able to get about fairly well.

Platoon R-45

SEALE, JACK K., R-45

I learned of the Maritime Service Radio Training Station on Gallups Island, MA through an article in QST while I was still in high school. I had to wait another year after graduation in 1942 because of my age. At that time the minimum age for enrollment was 18. I traveled from Longview, TX to Washington, D.C. and spent the next year working in a defense plant in suburban Maryland helping build special electronic test equipment for the Naval Research Lab.

On enrollment in April, 1943 I was sent to Sheepshead Bay for boot camp. After seven weeks I was sent to the U.S.M.S. station at Huntington, Long Island to begin radio training. I met several of my future Gallups Island classmates at Sheepshead and Huntington. After eight weeks we were sent on to Gallups. Our class finished up in December of that year. On the whole, I enjoyed my training at all the schools, especially at Huntington. However, at times the food at Gallups and the winter weather was not the best.

After graduation we were given two weeks leave and told to report for assignment in New York, New Orleans, or San Francisco. I chose New Orleans. What a ball we had in that wonderful city! There were so many of us of all grades, they had no place to put us and nothing for us to do. It took about 30 minutes of our week on Friday afternoon to report in and pick up our pay and per diem allowance. Needless to say, we explored the French Quarter and many other parts of the city.

In March, on very short notice, we were sent to New York for shipping assignment. Dick McChord and I drew the *George G. Crawford*, a Liberty. We took a load of material to the U.K. After discharging the cargo in Newport, Monmouthshire we laid in Loch Long, Scotland awaiting orders. After a while things began to move. We loaded equipment and a few special troops, including some Free French, and made the Normandy invasion. What a show that was! We continued to shuttle between the beachheads and Southampton and Belfast. On August 23 we were paid off in New York.

Looking forward to some relaxation, we checked into the McAlpin Hotel. It was then mid afternoon. I made the mistake of going down to the ROU hall to see what was going on. Nobody was there but Fred Howe. He was looking for somebody to take a pierhead jump way up the creek near Perth Amboy. If you ever tried to argue with Fred, you will know that I joined the ship as soon as I could get there. It was a Liberty tanker, the *Carleton Ellis*. It was built like a regular Liberty except the holds were made into two tanks each and the deck gear was fake. I stayed on her six months while she ran coastwise, refueled convoys to the Med and hit the troop ship George Washington near Guantanamo Bay in a blackout. Nobody was hurt, but it sure made a mess of the other ship's port side plating.

In April I joined the *Alexander Martin*, another Liberty, in Baltimore. We loaded there and at the new Naval Stores facility on the James River in Virginia. We took all that out to Honolulu, loaded some more at Pearl Harbor, then island hopped in the Pacific for a while. We discharged the last of our cargo at Okinawa shortly after it was "secured." We went to anchor off Saipan to await orders to prepare for the invasion of the Japanese homeland. While there the war ended, much to the relief to all. Shortly afterwards we proceeded to San Francisco and payoff.

I continued to sail Liberties and T-2's through 1950, when I went ashore. I worked in mobile communications for two years then got into the newspaper business. I spent 16 years at the Dallas Morning News, leaving there as the Plant Engineer to go back to sea.

I was back at sea during '68, '69, and '70. I first took a worn out pre-WW II tanker, the daddy of the T-2s, from Houston to East Pakistan. We lost the plant four times crossing the Indian Ocean. On arrival at Chittagong the ship was reflagged Panamanian and we were flown back to New York. I then took a relief trip out to West Africa. Then a permanent job on the *American Corsair*, a nice new ship. She averaged 21 knots round trip from New York to Viet Nam. I stayed with her a while, then came back ashore in 1970.

I re-entered the newspaper business, working for a couple of equipment manufacturers in various engineering positions.

In 1981 I started taking an occasional relief job on ships. In late 1985 I started shipping regularly through the ARA. I've had a coastwise tanker, a ship taking relief food out West and East Africa, and the last five years I have been on Lykes Bros. container ships running from the Gulf to North Europe. I expect to continue for another two or three years.

Several years ago we bought a home and a few acres in the green rolling hills of Northeast Texas. There is nothing exciting about it, that's why we like it. If you are in the neighborhood, drop by and visit with us.

STEIN, MAX, R-45

SS William M. Loring. Date of Shipment: Jan. 17, 1945 . Date of Discharge: April 3, 1945. Trip to Cuba and back to the U.S.A.. *SS George R. Poole*. Date of Shipment: April 27, 1945. Date of Discharge: Aug. 24, 1945

Sailing in the Chesapeake Bay, on the way to Newport News, VA, we passed a Liberty ship, outward bound for the Atlantic. I notified the captain, who was with the pilot, that the Liberty was just torpedoed and I would send out a distress signal for help. In minutes, planes were dropping depth charges. We went for our repairs and then left for New Orleans to pick up cargo. We entered a convoy and went through submarine attacks all the way to Dover, England. Anchored off Dover, watching V-1 Rockets flying overhead, not knowing when they would come down. Joined convoy for crossing English Channel and went into Antwerp, Belgium. Ship ordered to stand guard over guns as the Battle of the Bulge was going on. V-2 Rockets were coming down in Antwerp, so it wasn't pleasant. I came out of a Canadian movie theater and a half hour later the theater was hit by a V-2 Rocket and all were killed. After unloading our cargo, sailed for Naples, Italy. Going through Straits of Gibraltar in convoy, ran into submarine attacks and one day afterwards, ran into air attacks south of Spain. Picked up high explosives, 7,000 tons of it, and sailed for the Panama Canal. Rerouted and went to Maine, U.S.A.

SS Cape Sandy. Date of Shipment: Oct. 24, 1945. Date of Discharge: Jan. 20, 1946

Ship destination: China. Sailed through the Panama Canal and off Japan, danced around mines and went into China. My radio transmitter stopped working and could not answer E.T.A. message. Sailed into bay before going into Shanghai. Captain notified me that Admiral Halsey wanted to see me and was sending his personal boat to take me to his carrier. Was escorted aboard the carrier and met his aide. Was given complete respects and they sent the radio repair ship to repair the transmitter. They could not find the fault in the transmitter as the fault was in the motor-generator. They could not find the fault in the motor generator and then left. I told the captain and he had the Maritime send out two Chinese electricians. They took the motor generator off the ship, came back three days later, connected up the cables, I set up the dummy antenna, pressed the key and the transmitter was on the air. Unloaded cargo, sailed back to the U.S., San Francisco.

SS Denison Victory. Date of Shipment: Feb. 24, 1946—March 4, 1946. Date of Shipment: March 5, 1946—May 7, 1946

Ship destination, the Baltic Sea. Sweden, Denmark and Finland. Around Ireland, going through 55 foot seas and anchoring in Scotland. Captain picked up special orders and we headed into the North Sea, then into the Baltic Sea. The first American ship to enter this mine-laden body of water. The Captain told me he didn't have any charts about mines and this body of water had all the mines in the world. The English mined it by planes and the Germans mined it to protect their submarines coming and going out of their pens in Germany. I was sending out mine reports at least seven times a day for five days. The war was over, and there we were fighting for our lives against German mines. Mine reports and dense fog made it worse. Who said the war was over for the Merchant Marine? We had no minesweepers or radar, just guts. We sailed into Gotenburg, Sweden, then Stockholm and went to Malmo. Then, we sailed to Copenhagen, Denmark. Then we sailed for Finland. I radioed for an ice-breaker and a Russian ice-breaker took us into Finland, all through solid ice. It was like sailing through land. Amazing. We then sailed back through the Baltic Sea, sending out mine sighting reports every hour. Finally got out of the Baltic Sea and into the North Sea. Off Greenland, we ran into some smoke coming out of the ocean. I told the Captain that I would call the Coast Guard weather ship and describe the condition. They replied, "under sea icebergs" and told us to slow down and they would escort us through, as they had radar. They arrived in nine hours and took us through the iceberg field. Then we sailed back to the States. Long live the Merchant Marine. I had my honorable discharge from USMS and went home.

Platoon R-46

COX, HERBERT C., R-46

They said I was crazy! I could not prove otherwise! Fourteen ships sunk off the Atlantic coast and in the Gulf of Mexico the day I signed on with USMS in Nashville, TN.

I was a Tennessee "red neck land-lubber." I had a deferment as I was working for US-TVA as a Safety Officer. I was young, married, with a pregnant wife. I couldn't see the infantry at $75 a month, so when I saw an ad in the paper "Go to Sea - 6 to 9 months - Training - $275/month" I thought it was an answer to my prayers.

I was sent from Nashville to St. Petersburg, FL "boot camp". The personnel office kept putting me off about sending me to Radio School. I marched into his office one day and informed him that I had joined USMS to go to Radio School, and if he wasn't going to send me, I was going back to Tennessee and take my chances with the draft board. Later that week I was ordered to go to Huntington, L.I. Radio School.

I have many pleasant memories of three months at Huntington. There was lots of hard work and lots of good times with lots of swell fellows. We went to New York City's "Hell Kitchen" neighborhood one Saturday to attend the Chief's wedding (I can't remember his name).

We had some of the best instructors in the business at Huntington. I particularly remember the instructor in Basic Radio Theory. He was a former Columbia University Professor. He would lecture about 55 minutes then ask for questions. If there were several who had the same question, he would try to clarify it. If you were the only one with a question, he would not waste the class' time.

I know some of you will remember the officer who was newly in charge of base who, when barking orders at dress parade (Saturday morning Ritual before weekend liberty), spat out his false teeth and caught them in his hand!!

I left Huntington for Gallup's Island not knowing what to expect. Boy, was I in for a surprise—a pile of windswept rock seven miles out of Boston Harbor. There are lots of fond memories, though, and lots of hard work. Liberty weekends were spent in Boston. I missed Monday morning Liberty Boat; Captain's Mast extra duty for two weeks. In November—on my birthday—I was called to the phone and my wife told me that we had a beautiful red-headed baby daughter. (She is now a real estate broker in Las Vegas.) I graduated Gallup's Island in January, 1944. Afterwards, I traveled to Tennessee for a few days of well deserved (I thought) leave and visit with the new daughter and wife.

Then it was on to New Orleans and my first ship. Shipping orders were slow with about 40-50 radio operators on a waiting list. On March 22, 1944, I was sent to Mobile, AL drydock to board a new T-2 tanker, the *S/S Autossee*. As I lugged my gear down the dock she looked like the *Queen Mary* to me. She had no ballast and the gangplank would not reach the dock, so I had to climb a cargo net to get aboard. I made several trips on the *Autossee* hauling 180,000 barrels of aviation gas, various deck cargoes, planes, trucks, etc. to the European Theater. I signed off the *Autossee* in Marcus Hook, PA, on June 9, 1944.

On July 6, 1944, I signed on the *SS Earl Layman* in New York City with Captain A. Debozy. This was my longest period (of time) outside States. We shuttled various cargoes from England to LeHavre, Cherbourg, and Normandy Beach. The GI's unloading on Normandy furnished us with five-gallon Jerri-cans of "Calvados" (apple brandy). They used it for antifreeze, lighter fluid, etc. We drank it! Then it was back to the States where we signed off *Layman* on January 9, 1945.

Next, I signed on the *SS Leonardo L. Romero* on February 10, 1945 with Captain Edward R. Downing in Houston, TX. She was a new ship, so I was on board for her trial run. I remember the skipper and crew as pleasant. Upon loading in Baltimore, one trip, Skipper brought a Plymouth sedan down from New York. We drove downtown to a liquor store and loaded the little sedan with liquor and beer. He drove it down to the dock and loaded it aboard, lashed down atop No. 2 hatch. Needless to say, when we got to Europe, we were very popular around port!

As best I can remember, this was the ship that on VE Day was in the convoy in the English Channel headed for Belgium. We met a convoy going in the opposite direction. Somebody lost their cool, and there were ten ships involved in five collisions. We wound up with a ship's bow in #2 hold, starboard side. We headed for an English port for a patch job, then proceeded to Antwerp drydocks for a six-week repair job. When we returned to the States, I signed off the *Romero* on September 4, 1945.

I signed on the *SS John G. Carlisle* on October 25, 1945, with Captain James J. Galviani for an uneventful trip to the United

Kingdom. I signed off on December 27, 1945.

I signed on the *MS Gulf of Venezuela* on January 14, 1946 with Skipper G. W. Kemp. This was a World War I—I repeat, World War I—relic! The radio room looked like it came out of Noah's Ark. This was to be a Gulf Oil coastwise run from Port Arthur to Providence, RI; however, the auto-pilot steering mechanism was very erratic, and we spent the better part of one day going in circles off Key West. There were so many holes between tanks they could only carry one cargo. Still I survived one month before I signed OFF February 14, 1946. This was my wartime sailing career.

In 1952, I signed on the *SS Mercy Warren* with Captain A. Alks to help deliver "Marshall Plan" wheat for Europe. I signed off my last trip on March 10, 1952.

I then went back to US-TVA for several years. Afterwards, I went to work in Oak Ridge, TN at the Oak Ridge National Laboratory where I retired after 30 years in 1982. I have since (and presently) worked part-time as a utilities construction inspector.

Platoon R-47

JONES, SHERMAN, R-47

While at Gallups Island, Knable, Tushin and I were usually together and at times Czesnowski, Eisinbach, Rosenfield and a couple of others were with us. After graduation until the early 1950's I met some classmates when I came to New York City to ship out, but never got chance to sail with any of them.

In late January, 1944 I shipped out of New York on the *R.J. Reynolds*. They had just started putting three RO's aboard that trip, and the three of us were first timers. The master picked me for chief but had to get a waiver because I didn't have 6 months sea service on my license. We sailed to Europe and back in convoy. The second trip going over in one of the largest convoys, just a few days from England, we came under a submarine attack in the late afternoon which scattered ships and convoy over miles of the Atlantic, as far as you could see. The attack only lasted a few hours and a few ships were hit, but we didn't see any go down. It wasn't until the next morning some time before the ships and escort vessels were back in position. Return voyage was uneventful. I got off in New York and decided not to sail in slow convoys.

After a few weeks at home, I signed on the *Mill Spring* and a couple of days later the chief got sick or something and the other RO had not sailed as chief. So I sailed as chief in the fast convoys to England. The *Mill Spring* was rated at 16 1/2 knots. Had a pretty good thing going because the trip was about four weeks, then we were in New York for at least one week before leaving again. Then the second or third trip it was over because every trip at about mid-ocean the convoy would slow down to 10 knots to allow the D.E. escort vessels to refuel which took about one day. When this happened the master would break radio silence and claim to have engine problems or something; requesting to proceed alone. The last time this happened, we received orders to proceed to the Panama Canal.

After transitting the Canal Zone we ran alone in the Pacific for about eight months with one trip to the Indian Ocean and up to the Persian Gulf. We made a couple of ports in Australia and returned to Panama Canal about three times before returning to New York about a week or so before Christmas, 1944, for drydocking. Naturally, I got off. Got a couple of nice letters from the company and phone calls trying to get me to return after drydocking. I politely declined because I was pretty sure I could get something that wasn't crossing oceans.

From about 1947 until 1988 I either worked ashore or I shipped out; almost evenly distributed. Most of the time I worked ashore as an electronic technician. The company's card said service engineer but we were servicing, installing and repairing just about any make or type of electronic equipment aboard U.S. and foreign vessels, plus some shore side stations.

I've had a first class telegraph license since about 1946 and my current license is still valid, but never applied for a Ham's license. I moved from Pennsylvania to California in 1960 and have been here ever since. Took me almost five years to get climatized and think of here as home. Have been married and divorced a couple of times. Am still married. Have two grown sons and neither ever got into any trouble growing up.

Regards and 73's om.

SLABOTSKY, KENT E., R-47

I got to Gallups Island like most of us did—through the basic training at Huntington, Long Island. After the basic camp at Sheepshead Bay in Brooklyn. After graduating in January of 1944, I went to San Francisco to attain my first ship.

It took about five weeks. My first ship was the Liberty *Arthur A. Penn*. We went to the Solomon Islands with a cargo of 500-lb. bombs and a cargo of planes, jeeps, tanks, trucks, etc. From the Solomons we went to the Hebrides and down to North Island, and South Island, New Zealand. We loaded fresh potatoes and went back to the Solomons where we discharged quantities of the fresh potatoes to New Georgia, the Treasury Islands and also to Bougainville. We observed some mortar fire from the Japanese who were hidden up in the hills behind the smoke which hung over the forest from the volcano at Bougainville. From there I went back to the states.

I shipped on the *SS Emperial* which was a C2 class cargo ship—very nice ship. I stayed with that ship through four trips to the Pacific. Pearl Harbor, Eniwetok, . . . Saipan, Kenyon, Guam and we were making round trips from the West Coast carrying B-29 engines over to the Marianas for the bombing runs to Japan. Returning with used B-29 engines. Round trip took approximately eight weeks.

After that trip and a short time ashore, I shipped again on the *SS Brunnel Victory* and V-J day was announced. After the two day national holiday we continued to San Francisco, loaded war material and took it to Okinawa even though the war had ended. At Okinawa, we were in some horrendous typhoons and they would not allow us to leave a 33-ship convoy. Subsequently seven ships directly behind us and a destroyer escort sank in that storm.

Platoon R-47

We finally departed Okinawa in the middle of the November after a lot of trouble discharging cargo—and returned to the states in December to the port of Portland. At this time I had thought I was through with my merchant service. However, as it turned out four months later, I had to ship again because I lacked three days service to be relieved of my military obligation.

I returned from my home in Kansas City to San Francisco and signed on the SS *Santa Cruz* owned by Grace Line which was being transferred to the Philadelphia port through the Panama Canal. However, within three weeks the vessel was sold to the Alaska Shipping company and we were sent to Seattle to load 1,500 Filipino laborers and ship them to Alaska for the salmon run of the summer of 1946. At our first port of call we discharged 50 laborers and an amount of small stores to sustain them through the period of the run.

Went farther down the Alaskan peninsula and at that point, through a malfunction in the engine room, a severe leak developed in the machinery and the vessel filled with water. The ship was beached by two tugs in 30 feet of water. We stayed there for eight days until we were rescued by the *SS Baronoff* on its regular run between Nome and Seattle. We returned to Seattle as survivors.

I returned to my home in Kansas City and began my 41-year career in the automobile business, ending up as the home service manager for the Oldsmobile Division of General Motors in Oklahoma City. I retired over two and a half years ago.

TRIMBLE, GLEN, R-47

Enlisted in the Maritime Service at Topeka, KS, in February, 1943. After a brief time in inactive duty (boot camps were full), attended Sheepshead Bay, Huntington Radio School in Long Island and Gallups Island in Boston Harbor.

Spent a year on first ship, a C-1 carrying 500 pound and 1,000 pound bombs on trips through the Solomons, New Guinea and with the first convoy in Leyte, Philippines, October 22, 1944. We were used as a warehouse for two months with 114 direct attacks on our section of anchorage. Had a 30' and 75' near miss and a Kamikaze crash on #3 hold boom where he exploded. Shrapnel and burns around the bridge area but no fatalities.

Next ship was a new Victory out of the Richmond, CA yards, same cargo, and this time the Okinawa invasion. Japs had learned a plane plus pilot for a ship was a good deal. We were only at Hagushi Beach two weeks and the closest Kamikaze missed us by a 100 foot or so. Back to San Francisco and left just before the surrender, but instead of returning to the States we were routed to Ulithi, 700 miles east of the Philippines, and sat there from the last of August to December. We went to Saipan and unloaded the bombs and came back to the States. Received Philippine Liberation medal and Mariners Medal in 1946.

We had been sent back to Pearl Harbor for repairs after Leyte and then back to San Francisco. I took 30 days leave, came home, married my high school sweetheart and we will celebrate our 47th anniversary in February. Spent the first years after the war with the Highway Patrol and then joined Boeing on Electronic and Missile systems. Spent 20 years from Florida to Seattle and in 1970 bought a ranch in Missouri.

A horse bent over backwards with me and put me out of the cattle business. Spent the next eight years running a Motorola business radio service station in the center third of Kansas. Retired in 1981, but started in real estate after about six months of retirement and worked till spring of 1987 when congestive heart failure took me out of action for a year. That was the start of real retirement!

Started organizing Kansas Merchant Marine vets in the fall of 1989 with lists from USMMV WWII and Gallups Island. First reunion last spring had 20 attendees plus spouses. Secured a Coast Guard printout of Kansas vets last summer and will have about 65 vets plus spouses here in Wichita the last weekend in April, 1992.

Platoon R-49

BERGER, ROBERT L., R-49

I shipped out to Sheepshead Bay at the end of May. On to Huntington, R-49, and then Gallups Island. Received my 2nd telegraph, 1st phone and ham ticket in January and shipped out on the *Mission Santa Yenz* from San Francisco in early March of 1944. We went to Honolulu and then through the Canal and over to Curacao. We made two trips to Esperito Santos and two to Ulithi before our economizers went out and we had to return to Los Angeles.

The chief op by this time was Ev Soule, I think in R-47 or R-48. After some shore leave, we teamed up in his home in Minneapolis. Ev knew everyone in town and we were treated to real Scandinavian hospitality. We were poured aboard the train and slowly made our way to Seattle. The union assigned us to the *Westminster Victory*. After shake downs up and down the Columbia River, we loaded lumber up to the wheel house and headed for Honolulu.

When they unloaded us, it was unbelievable how much lumber we had. They then loaded us with war scrap and sent us to Baltimore. My R-49 friend, Bill Becker, had just come back from two trips to Murmansk and Archangel so he welcomed my suggestion of the South Pacific. We hitchhiked to Los Angeles by way of Boston (old girl friends, of course), Omaha, Oregon (that's another story) finally making it to Los Angeles.

The Liberty ship, *Allen C. Balch*, was just finishing repairs from a fire fight at Guadalcanal. The crew said they got two enemy planes and painted them on the stack to prove it. We sailed in late July bound for the Philippines and managed to just miss August 15 as we crossed the International date line. Did pick up the VJ from Guam around midnight. Diverted to Subic Bay where we made up a convoy for Buccner Bay arriving in time for the big typhoon around October 6-10 (850 small craft destroyed). The convoy, joined by the 3rd Marine Division, then went on to Sasebo. We were the first merchant ship to enter the beautiful Sasebo harbor since 1930. We were a month unloading due to another typhoon.

We were then ordered home to San Francisco with 55 very happy guys who started with the 3rd Marine Division in 1942. Arrived at the Golden Gates at 6 a.m. What a shout those guys made. By the time we docked there was the whole works to greet them, brass bands and all. Union hall said there was a ship in the harbor that had been waiting five days for a radio operator, so Bill went back to New York and I sailed on the *Hinton R. Helper*, another Liberty ship, bound for New Orleans. We were diverted to Galveston where we loaded grain for Italian relief. The ship made two trips to Italy before it was sent to be moth balled in the James River. I then shipped aboard the *Attleboro Victory* with a load of horses from Galveston to Greece. She was laid up as well.

It was now September of 1946 and I decided it was time to start college. None of the schools that had accepted me in 1943 had room for a civilian but by good luck my mother had been able to arrange for me to go to Colorado A&M, now Colorado State University. It was a great place to be for many reasons including my being able to start an amateur radio club, 10 meters. It still operates. I also worked at a new local radio station as engineer and disk spinner.

Ran out of money and took the *Charles A Dana*, yet another Liberty ship, out of Galveston with a load of sulfur for Brazil and Argentina. We loaded grain for India, by way of Durbin, but couldn't enter Bombay because Gandhi was killed that day. The Indian government sent us to Kundala, an artificial port between Bombay and Karachi. It turned out that we were their first ship. We were unloaded by hand, the grain being transferred to small coastal sailing ships upon which the whole family lived. We left half our load there and the rest at Cochin. We were then sent on to Newcastle, Australia where we loaded grain for Marseille. We went back to Norfolk and mothballs and I went back to college.

Graduated in physics in the summer of 1950 and also got married. First job teaching at small college in Missouri for a year and then on to graduate school at Penn State (Nittany Lions). Received a Ph.D. in physics in January, 1956 and spent the next two years in Cambridge, England as a post-doc going over as a passenger with my wife and two kids on the *Ryndam*, Holland-American. Was fun, as was coming back with by then three kids.

Five years on the faculty of Utah State University and then a permanent move to the National Heart Institute here at NIH in Bethesda, MD where I have been since 1962. A research program in Milan gets me to Italy once a year plus short trips here, there and yon, so that my seafaring wanderlust days are still partially satisfied. Spend most of my life on computers these days as we automate all of our instruments for the blood work we do. If you get to Washington, do come see us, my laboratory is one block from the Medical Center Metro stop on the Red line. My wife is the NIH historian and curator of the Medical Research Instruments Museum, so we'll be glad to show you around.

HARTMANN, FRED D., R-49

I met a friend of mine while waiting to be drafted and he knew I was interested in radio, and he told me about this terrific deal that would enable me to get a commercial license and also do something for the war effort. So the next day I went over to lower Broadway and enlisted into the Maritime Service in the summer of '43. I went over to Sheepshead Bay and was fortunate enough to be chosen for radio training.

I went to Huntington—beautiful place—for two months, and with pardonable pride I can say I was the honor graduate from Section 49. I originally started with R-47, but halfway through training I was sick for two weeks. When I got back they put me in R-49.

In September of '43 we went up to Gallups Island and were there for the winter of '43-'44. That was the coldest winter I have ever seen in Boston. But it was a wonderful experience that prepared me for my life's work: 18 years in the electronics industry. I later got a job teaching electronics in high school.

KUBANY, ALBERT J., R-49

I was 17 and just out of high school in 1943 when I enlisted in the U.S. Maritime Service as an Apprentice Seaman. After basic training at Sheepshead Bay, I passed the qualifying tests for radio school. I graduated in January, 1944 with Platoon R-49. Shipped out of Brooklyn, NY on a Liberty ship, the *SS James McHenry* in a 106-ship convoy bound for the Mediterranean. Just 18 years old, I was the youngest officer on the ship.

From Gibraltar, German spies could count the number of ships from Franco's Spain. Nothing could hide coming through that narrow strait. High level German bombers hit the convoy. Two planes were shot down. One ship was seen burning out of control and floated away from the convoy.

After shuttling cargo from Oran and Algiers, North Africa, and Naples, Italy, we carried troops and cargo to the invasion beachhead of Southern France. 1944 was an interesting year in the Mediterranean geology. We observed Vesuvius erupting as well as Mt. Etna and really flaming and boiling, Stromboli. These were awesome sights. We sat at anchor in the Bay of Naples for over a month while ships were unloaded and assembled. During that time, a severe storm arose and some ships were blown around dragging anchor. Ships were helpless since their boilers were down. One ship kept coming closer, slid along our starboard and ripped away the lifeboats on that side. During wartime, lifeboats were fixed outside the hull ready to drop to the sea. We spent the rest of the voyage with only half our lifeboats.

We returned to New York in September, 1944, and after a few weeks on the beach, I took a ship, the *SS Cyrus T. Brady*, out of Philadelphia. After assembling at Hampton Roads, VA, 94 ships sailed in convoy toward Gibraltar. We carried supplies for the Russians via the Persian Gulf. We were placed in the coffin corner of the convoy. Our cargo was high explosives and toxics. If we were torpedoed and exploded, we were expected to float away without endangering the rest of the convoy.

While going through the Suez Canal we had an experience that caused some official anxiety. Our ship lost steering while the First Engineer was being relieved for dinner by the 3rd Engineer. It was due to a faulty steam reducing valve. We went aground and glanced off the side of the canal. We were soon righted, but the canal pilot was suspicious. We were suspected of attempting to sabotage the canal. This was critical since Rommel's Afrika Corps wished to do the same thing but they failed.

When we reached the southern end of the canal at the city of Suez, our officers had to answer for the incident. It is somewhat reminiscent of another suspicion that we radio men experienced. Our union, the American Communications Association (ACA), was thought to be Communist dominated. I don't know how true this was except that I encountered a hint of it when I went through the Naval Armed Guard reports I obtained through the Freedom of Information Act from the National Archives. Included in every report by the Armed Guard Commander was a statement "The commercial radio operators carried out war-time instructions to Merchant Vessels and kept all messages confidential." Evidently the government ordered the Navy to keep an eye on us.

After delivering supplies to the Russians, we were routed to Calcutta, India in January, 1945. At 0257 local time, while sailing west of Bombay, the forward lookout reported sighting a submarine. General alarm was sounded and all guns were manned. The Captain ordered a hard right to bring our stern to bear on the submarine. We were running away from it, and , besides, our biggest gun (4") was aft. Visibility was poor with patchy fog. We opened fire at 3,500 yards but the results were undetermined. That was it. When we arrived in Calcutta, our Captain reported the incident to the authorities. When they checked time and location, we learned that a British submarine reported being fired upon at that location. As everyone knew in wartime, any submarine is an enemy submarine.

We were diverted across the Indian Ocean to western Australia. Crossing the Equator, all rookies had to undergo King Neptune's initiation. We sailed underneath Australia to Melbourne, Sydney and Brisbane. We carried war supplies to New Guinea, Dutch East Indies, Philippines, etc. Nine months after leaving the East Coast, we had circumnavigated the earth and returned to San Francisco. It was quite an experience for a young man who had not yet turned 20.

Ironically, the war ended when I had just signed on a ship in New York with a load of bombs bound for the Far East. The bureaucratic war wheels grind down slowly and we left heading for the Panama canal. After about a week, they caught up with us and ordered us to Galveston, TX.

After the war I could have used the G.I. Bill which we never got for college. I went to the University of Pittsburgh and obtained my B.S., M.S. and Ph.D. in psychology. During the Korean War, I joined a research group and flew to Korea as a civilian and traveled along the MLR (Main Line of Resistance) working with infantrymen, studying critical incidents of combat effectiveness and later working with Chinese and North Korean POWs on a psychological warfare study. I practiced psychology at the General Motors Institute from which I retired at 65 in 1990.

RIST, EDGAR , R-49

How I got to Gallups Island was a mistake: I was in the Army and was discharged after 89 days because they thought I had an irregular heartbeat. So I tried to get into the Coast Guard and the Navy, but they wouldn't have me. The only people who would take me was the Maritime Service. And that's how I got in.

Platoon R-50

CRUSE, TOM, R-50

Left Gallups Island during the time when radio operators were waiting for ships. Sent to New Orleans to wait. Got a job as an electrician in the shipyard in Algiers. We reported to the Maritime Service for per-diem once a week. All at once, we were notified to start reporting daily. They started putting two operators per ship. Gallups was still turning them out faster than they could get ships built. Then we had to report twice a day (no more ship yard work, reporting times in middle of both shifts.) Three Ops per ship. Then the shipping halls emptied out fast. I ended up taking a train ride to Wilmington, NC for a new C-1, the *Talisman*, built by Mississippi Shipping Co., to England.

Next a Liberty from New York to Murmansk, second trip back to Hull, England then for Invasion of France, signed off in New York.

Went back to New Orleans for ship, took *SS Republic* to Mobile for rework in ship yard. Got a T-2 from Mobile, to North Africa with aircraft on deck, then to Italy to unload avgas. Repeated run to Italy, but went to Curacao for reload then through the Panama Canal to Guam to Panama to Ulithi.

War ended on way to Panama where I got off. From Brooklyn, the *SS David C. Shanks* to Italy to bring home troops. Discharged troops in Brooklyn, then to the Philippines for more troops. Went to San

Francisco, converted troop ship to dependent ship, and loaded dependents for Hawaii. Then on to Auckland and Sydney where we loaded GI brides and babies for San Francisco.

Got off and went to Pittsburgh, PA for a cousin's wedding, met my future bride and went back to San Francisco. *Shanks* had left without me, so I took a job in shore station at Fort Mason as an operator. Sent for bride-to-be to fly out from Pittsburgh. We got married at the Fort Mason chapel. Got RIF notice in pay check while at Russian River on honeymoon.

Next life started as Flight Radio Officer for PAA. About 14 months later, received transfer to Bangkok, Siam. Wife discovered she was pregnant with first of four while we were waiting for visas in Pittsburgh. Next life started selling Quaker Oats in Pittsburgh. I got notice of 1-A from draft board and joined the Pennsylvania National Guard to avoid the draft, but got Federalized for next life...23 years later, retired from U.S. Army as Master Sgt, Nuclear Power Plant Superintendent, General.

Began next life as a manufacturer's rep selling electronic components to industry. New life just starting as Computer Components Rep.

DONLEY, CHARLES P., R-50

By the time I was 16, way back in '42, I was "chasing the war" and tried to convince my father to let me join the Marines. His sage response was "as soon as you graduate from high school." I was pretty restless until May, '43; finally went to Marine enlistment center and failed physical due to suspicion of heart murmur. Much to my chagrin, was also turned down by Navy and Army pilot training due to less than perfect eyesight. My father, back to sea again when war broke out as a Chief Engineer, came to my rescue by suggesting that I apply for Kings Point. With a six-month waiting list in store for me, I was convinced to put my name on the list and join the Maritime Service, so off to Sheepshead Bay in late May, 1943.

Heard about openings to become a radio officer, took various tests and was sent to Huntington in July and on to Gallups (R-50) where I studied all week, on most weekends and graduated in January, 1944.

There are many positive and pleasant memories regarding Gallups, but there is little I can add regarding how it was on the Island that hasn't already been eloquently and graphically recounted by other GIRA members. Somehow, athletics seem to have dominated my Gallups memories. Like intercepting a pass during a snow storm and returning it for the winning touchdown, or out-boxing the Gallups boxing instructor (two previous years in the New York City Golden Gloves helped), or not making the starting five on our platoon basketball team.

Following graduation, I picked New Orleans as my port to ship out from. After a month of strenuous training at Pat O'Brien's Hurricane Emporium I finally shipped out. February, 1944 to August, 1945 was a blur of Liberty ships and tankers; general alarms, depth charges, air raids, mines, ship collisions and atrocious weather conditions as I saw service in the North Atlantic, Mediterranean Middle East and Pacific theaters.

Interestingly, the big moment in my GIRA career came shortly after the war. I was on the *SS Joseph Hooker*, a Liberty ship that our National Secretary/Treasurer, Ray Anderson, had spent considerable time aboard during the war. I came aboard the ship in July, 1945. On the second trip, coming back from Europe and off the coast of Newfoundland, loaded down with troops, we ran into rugged weather, started taking water in #3 hold and I had to send out an SOS.

Fortunately, the Navy sent out tugs and a cruiser, but even more fortunately, we somehow managed to stay afloat, keep the pumps working and didn't have to abandon ship. That would have been an unmitigated disaster as hundreds of American troops who had just survived combat in the E.T.O would have drowned. As it was peace time, Walter Winchell used this as his lead story on WOR radio that night while we were still floundering at sea and the outcome very much in doubt. I lived in New York City at the time and my family had quite a start at the dinner table when Winchell's voice came crackling out over the old reliable Sears and Roebuck Silvertone radio, featuring the plight of the *SS Joseph Hooker*.

I left the Maritime Service in January of 1948 and, after an abortive attempt at operating my own business, joined the Army in April, 1950 with the intent of gong to OCS and making it a career. Korea erupted in July, I volunteered for that show (still not cured of chasing wars) and spent, or I should say, survived 16 months over there. When I got back, I was in no physical condition to qualify for OCS so I got discharged in May, 1952 and then spent ten years with the NSA and CIA organizations.

Left that para-military world in May "62" to enter the business world, make money and have fun. I was in a variety of business enterprises but over the last ten years I have worked with two former Navy Seals and we have our own firm that provides investigative and personal security services worldwide.

In recapping my modest odyssey, I would be remiss if I didn't mentioned my wonder wife Diane. We have been married for 25 years and are proud parents of a great son, Chuck, who just graduated this summer in the Navy ROTC program with a BS in Physics from Georgetown and a Master's Degree in Mechanical Engineering from Catholic University. He was commissioned at the Marine Corps barracks in Washington, DC and is currently stationed at the Pentagon in Nuclear Research.

As I reflect on all of this, I have no doubt that my exposure to the Gallups Island Program at a very formative period in my life had a significant influence on whatever I have been able to achieve subsequent to that. I learned how to focus on a single goal (graduating), to sacrifice (weekends in Boston) and to discipline myself (never get discouraged). Those attributes, defined and developed at Gallups Island, have stood me in good stead in all the endeavors that I have been engaged in and I am very thankful for what the Gallups Island experience has meant to me.

Platoon R-51

SHELLENGER, CHARLES, R-51

Entered TV broadcasting in 1947. Employed by Triangle Publications, RCA, ABC, and CBS. Joined Sports Network in 1960, which became Hughes TV Network, as chief engineer. Spent 18 years at Hughes. Three years as engineer in charge of approximately 500 sports originations at Madison Square Garden.

ANDERSON, ROBERT W., R-51

In the early stages of World War II, with enough high school credits after three and a half years, I officially graduated by mail.

I opted for the Merchant Marine via the War Shipping Administration rather than wait for the draft. A Portland, OR teenager, I arrived at Gallups Island via Sheepshead Bay and Huntington, Long Island, NY United States Maritime Radio Training Station. Huntington was an excellent transition from the regimentation of Sheepshead Bay boot camp to the concentrated study that awaited us at Gallups Island.

As I recall, the Huntington estate of Otto Kahn had been donated to New York City and converted to the training school. It provided a very comfortable setting in which to adjust and prepare for the rigors of the coming winter on bleak and confining Gallups Island.

While Gallups was a compact island installation, I learned to appreciate this when I had the early morning security patrol with time clock punching at various check points. This included the onerous chore of giving wake up calls before sunrise to our physical fitness instructor who delighted in leading us in pre-dawn exercises on the chilly, wind swept drill field. I became very well known the morning I was unable to rouse him from an exceedingly heavy sleep following a late night party.

I can't say enough about the high caliber of our instructors. The officers and crew of one of my early T-2 tankers might not be here today if it were not for the extra effort of one of my favorite instructors. We had passed our FCC exams and were getting the finishing touches before shipping out. He cautioned us, "If anyone reveals what I'm about to show you, I'll deny it and you may end up in the brig."

He went on to show how the code for the classified BAMS (broadcast to allied merchant ships) could be decoded. This was strictly under the jurisdiction of the Naval Gunnery Officer who had charge of the gunnery crews we carried to protect us since

Platoon R-50

Platoon R-51

Platoon R-52

we traveled without convoy.

I had copied a top priority, coded message to our vessel and given it to the captain for delivery to the gunnery officer for decoding. Hours later the captain got me out of my bunk, asking if I had any idea how to do the decoding because the Gunnery Officer couldn't do it.

I informed him that I'd never done this and that while it wasn't part of our training perhaps we could try to work something out? Then I recalled the clues our Gallups Island instructor had given us under secrecy and at some personal risk to his career. He had anticipated just this type of situation.

My captain provided the necessary code books and together we decoded our orders to alter our course to a revised destination. We broke radio silence as ordered and I radioed our coded confirmation of compliance. Our Gunnery Officer was replaced when we reached Australia from the war zone.

I will always carry the warm feeling of pride that the officers and men of Gallups Island instilled in me and the men of platoon R-51.

Shortly after the war I formed my own export business which was moderately successful. I soon realized my need to get started on my college education. Since I was not entitled to veteran's benefits, even though I held the rank of Lt. (j.g.) USMS reserve, I returned to sea to earn my tuition and expenses. Here again I'm very thankful because my Gallups Island training made earning college money relatively easy and provided interesting and educational post-war travel.

Another plus for the Gallups Island training is that without it I probably would not be getting the enjoyment I now have in amateur radio (WA2OPF). My wife, Irene, and I are both retired, travel extensively between our home in New York and our dot on the map on the Oregon coast and find ham radio certainly amplifies our joy in traveling by motorhome.

We both enjoyed the Gallups Island reunion in early August, 1990 and celebrated the great occasion by touring Boston harbor area and visiting Gallups Island with platoon R-51 classmate, Arnold Smith, whom I hadn't seen since a chance meeting when our vessels were in the same port during World War II.

As I look back from my retirement years, I'm very thankful that I had the benefit of Gallups Island training which opened the doors for me to world wide travel, lasting friendships through personal and radio contact and the curiosity to continue learning through travel.

Platoon R-52

NEESON, JACK H., R-52

After I graduated from Gallups and got my ticket, I sailed on April 6, 1944 on the *John G. Whittier* from Boston. I was very lucky to have a really experienced operator named Philip Binns who had been sailing since the early '30s as first operator. He was a trifle on the surly side at the start and began conversation by offering me a chocolate bar. When I refused he said, "Rummy eh!" But when he saw that I had a large bag of books—"Thank God I got someone who can read," he said and directed me to the top bunk. He added, again a little surly, that union regulations prohibited having to share a cabin but when you worked with the Navy you should expect a little crap.

We got along famously after my reading collection was checked out and approved and we sailed together later on the *Armstrong* and the *Vandevanter* (selected by him on the basis of the cargo they were loading—absolutely no ship carrying locomotives to Murmansk and no ships with a reinforcing plate around the gunners, or built by Kaiser Yards in about 45 minutes) until I signed off for good in 1946.

My operating skill improved very little in the next two years because our major activity was to keep the transmitter turned off—I think I only sent about eight messages all together—but out of boredom I did read the manuals and worked with Binns on repairs to the gyro and Scott ship's radio and calibrating the primitive radar, so I kept up some on theory.

This turned out to be an unexpected blessing because when I got ashore I decided to try to work in theater and I was one of the few theater people who could understand and work with the revolutionary developments in theater switch board. This was the era of the thyratron and the thyratron reactance board which made use of tubes developed during the war and the SCR (Silicon Controlled Rectifier) which led the way to electronic control rather than electrical control of stage lights.

This made getting jobs much easier and I stayed in theater until I retired as a director and drama professor in 1984. Lots of thanks to Gallups Island. . . . even if they had something else in mind.

SWEGER, MELVIN P., R-52

I was born in Perry County, in the mountainous area of central Pennsylvania. Being born and raised on a farm, I learned responsibility at an early age.

After high school, I moved to Baltimore to work in an aircraft factory and it was there I was hooked by the slogan "Heroes in Dungarees." I enlisted in the Maritime Service in the spring of 1943 and was at Sheepshead Bay for six weeks basic training. As a result of various aptitude tests, I was appointed to radio school; first at Huntington, Long Island, and in the fall of 1943 to Gallups Island. I would certainly say that the discipline and diligence that I learned at Gallups Island, especially in learning code, has greatly helped me through various difficulties in life.

Upon completion of training at Gallups, it was with eager anticipation that I was looking forward to sea duty. With a fresh 2nd class FCC license in my pocket, I checked in at the union office to acquire a shipping assignment. There was only a short wait of a few weeks to get a ship, as I believe around that time (March, 1944) a decision was made to put more than one operator on each ship.

I went to Baltimore to sign on the *SS Grenville M. Dodge*, a Liberty ship, and in a few days was in convoy eastbound for North Africa. With mixed feelings I anticipated seeing the Rock of Gibraltar, as we were told to expect enemy action there, and was not disappointed, as at dusk on the first day in the Mediterranean we were hit with aircraft.

Oran, Algeria was a staging area for forward action in the Mediterranean and many Atlantic convoys stopped there first. Our ship was designated to shuttle duty and spent the entire summer of 1944 cruising in the Mediterranean Sea. As a result of that summer, I feel a certain kinship with Odysseus of Homer's "Odyssey."

You recall that the tween decks on a Liberty were designed to be used as troop accommodations with the addition of bunks, five tiers high on metal stanchions. On various occasions we had Muslim troops aboard and, in as much as the devout ones were always prayed five times a day, a delegate had to ascertain the proper direction. Finally in October we were homeward bound with a fairly large pay-off as we had been a 100% (war) zone for most of the time.

Shortly before Thanksgiving in 1945 I signed on the *SS Louise Lykes*, a C-2 type cargo vessel, bound for Galveston and then China. We made two trips to Shanghai with cotton and flour. A cotton bale weighs about 500-600 pounds and were all loaded by machinery in the U.S., but in China when the bales were dropped on the dock two fellows with a stick on their shoulders and two rope slings would carry them at a trot into the warehouse.

It's a long haul from Panama to Shanghai (around 9,000 miles) and took 30 days one trip and 31 the next without sighting land. One trip we went to Hong Kong and then Manila where we waited about 30 days for our turn at the dock to load some heavy army equipment, tanks, howitzers, etc., returning to the states. We had been sitting in warm water so long the barnacles that grew on the hull curtailed our speed so we had to stop in Honolulu to refuel.

I stayed on the *Louise Lykes* for about eighteen months. I made three more trips to Europe, going once to Venice and Trieste. One of my more pleasant memories is of a train ride from Cherbourg to Paris for a four day excursion to see all the tourist attractions available. It was December of 1946 and with heating fuel scarce I very seldom took off my overcoat, even to sleep.

In May of 1947 I quit the sea to return home to Pennsylvania and marry a Perry County farm girl and raise a family. Shortly, I obtained a job with the local post office and later transferred to the position of rural mail carrier, a challenging job at times, especially with the snow.

I have been retired for about ten

years with hobbies of woodworking, crossword puzzles, reading, bird watching, and coin and stamp collecting. I have maintained the same address (R.D. 1, Landisburg, PA) all of my life except for my wartime wanderings. My wife, Arley, and I have been blessed with two children, a boy and a girl, who also live only about 1,000 yards from us so we see our grandchildren everyday. I feel the Lord has blessed me in my life and thank Him that I do not have any major illness, only the minor aches and pains of being 70-years old.

Platoon R-53

BURKENS, FRED, R-53

Having sailed on all Liberties and never having seen available models, except in the Smithsonian displays, I have kept the picture clear by a ships manual and pictures taken in the English Channel in April '45.

An ad I read in "Spark Gap" brought back memories of the letters received just before the Korean outbreak—the ad is very similar except the pay was $500—it was interesting I was only making $250 then. The fact of a fairly new bride and the idea that I had used up a lot of chances kept it just a tempting ad.

Only met one guy during the war, Harry Hood, in Antwerp during '44 or '45 (had three runs in there from October '44 through spring of '45—sailing down the English channel on V-E Day—bright and sunny and beautiful as well as joyous). Hood's Old Man wouldn't let him off the ship so I hobbled over. Tried to look up George Renau in Charleston, SC on a return trip; but he was at sea. Remember him arriving at Huntington with a "Health & Strength" magazine with his picture on the front—didn't put him off on his left foot.

I often think of the Gallups parade ground when I hear some Sousa march music. Also remember R-53 being picked to march in a Boston parade—right in front of a Kilte Band. marching to the drum beat was OK, but those bag pipes screwed up your step. And there was the SOS meal which put much of the Island in sickbay—three of us made out very well, we were digging a sewage ditch due to restriction and didn't get the meal.

You only remember the funny ones

LANG, JOSEPH F., R-53

I sailed for three and a half years before quitting in the fall of 1956 when I obtained the position of radio license examiner for the FCC in Detroit, MI. After a year I got on at radio station WJR in Detroit. After eight years there I began working for the automotive companies and lasted 20 years at Chevrolet Engineering at the GM Tech Center.

As for war stories, I can only report that I saw only one submarine and after close inspection it was determined that it was Canadian. I was almost mugged on the beach down in Venezuela late one night after taking on a load of native beer. One of my shipmates saw action in Oran—he was stripped of all possessions including all clothing and came hopping three feet off the hot tanker deck with only a Navy raincoat on when brought back to the ship by shore patrol.

My ship, the *SS Lookout Mountain*, was rammed by a cargo ship off the New England coast when our captain refused to use the canal and chose to put out to sea in the fog. Oh, I nearly lost Chinese laundry in New York, but that is another story.

Platoon R-54

McELHINNEY, DANIEL F., R-54

On July 8, 1943, I enlisted in the Maritime Service and after a couple of weeks of basic training, I was accepted in Radio Officers Training School at Huntington, Long Island and thence to Gallups Island, Boston. Nine months later I had a 2nd Class Telegraph license and sailed as Chief Radio Officer on the Liberty ship, *John P. Mitchell.* It was an uneventful five month voyage along the west coast of Africa and thence to Trinidad and at last, Baltimore.

In October, 1944 I set sail on the Liberty, *James Bowdoin* only to anchor outside of Scotland with no shore leave for what seemed like days and days and days. Finally, we hauled anchor for the south coast of England where we were scheduled to join the first convoy going to Antwerp. However the gyro compass was damaged and the second engineer proved to be among the missing. Scotland Yard conducted an investigation. The gyro was replaced and we joined the next convoy three days later.

That night we were attacked by German E-boats. The convoy was dispersed but managed to join up again the following morning for the trip up river to Antwerp. While in Antwerp we were under attack around the clock by V1 and V2 bombs. Eventually a V2 hit the Rex theater killing 1,100 people including two members of our gun crew.

Then came the Battle of the Bulge and they got us unloaded in a hurry and out of Antwerp. Back in the English channel, on Christmas Eve, we got word that a submarine had sunk a troopship trying to get to Antwerp. It was not a merry Christmas.

We returned to the States and in February, 1945, I sailed on the *Clarence F. Peck.* In Maine, the ship was loaded with very high explosives. Because of this cargo we were considered the most dangerous ship in the convoy and were accordingly positioned in the coffin corner. On leaving the convoy conference at Halifax, Nova Scotia, we were informed that a German wolf pack was waiting for the convoy. They followed us all the way across the Atlantic. Before entering the English Channel, we received word that another wolf pack was waiting inside the Channel for our arrival.

Fortunately, the Channel was in heavy fog and we entered without any attack by the subs. However, the *Commodore,* in an effort to elude the Germans, called for a specific turn to a new course. The fog was not favorable, so shortly after the signal of execution all hell broke loose. Apparently not all of the ships in the convoy executed the turn at the precise time of the signal. Ships all over the convoy opened up on the radio with messages of collisions with unknown objects, other ships, etc. The Commodore spot checked the convoy and managed to get the bulk of ships back in line.

We unloaded our dangerous cargo in Cherbourg, France. I went into the town, but was unhappy with the attitude of the French townspeople. I returned to the ship to stay there until we returned to the good old U.S.A.

In June of 1945 I sailed on the Liberty ship, *Nelson W. Aldrich.* We carried U.N.R.A. supplies to Yugoslavia. While there, the war finally ended when Japan surrendered. Upon arrival back home, I learned that my niece Dorothy's husband, Charles Sweeney, piloted the plane that dropped the bomb on Nagasaki. But that's another story.

RESNICK, JIM, R-54

My career in the Merchant Marine started out on a new Liberty In Panama City, FL. From there through the Canal and up and down the coast of Chile, loading nitrates; home by way of Hawaii and Oakland. Then three trips on a T-2 with high-octane across the North Atlantic to the UK from November through January.

That earned us a charter to the Admiraltys shuttling between Abadan, Iran and Oman, India, Ceylon, Burma, and Australia. We didn't get home to Boston until December, 1945, and my discharge says I ended my service on August 15. I had to continue my service until I had 32 months on active duty. I didn't get home for good until I had 33 months active duty, including boot camp and radio school.

WORTMAN, KARL KEITH RAY, R-54

Platoon R-54 graduated March 4, 1944 at Gallups. Our 30 days leave was interrupted by order to report to New York City immediately. I was taken to a ship in the harbor and put aboard as"the" radio officer. The ship was an old tanker, the *SS S.B. Hunt* with a top speed of about eight knots. It was 30 hours before I even got out of the radio room to see what the ship looked liked.

We were on the South American run until November: Aruba, Curacao, Panama, Trinidad, Rio De Janeiro and Montevideo, Uruguay. We had some very rough weather in the South Atlantic and ended up needing repairs in Uruguay—plates had to be welded up forward to stop taking on water in the dry cargo holds. One rough night our own American planes came over and lit us up with flares for identification purposes—they had been having problems with surface raiders.

After returning to the States, we left

Platoon R-53

Platoon R-54

Platoon R-55

New York City in November, 1944 and went to Panama. We were headed for Saipan, but ended up in Pearl Harbor. The *Hunt* had a new skipper and many new crew members for this trip—our new Armed Guard had enough combat bars to represent two wars. The Navy commissioned our ship into the Fleet for service at Saipan, and our merchant crew was repatriated to New York via San Francisco on the *SS Mastonia*, "Pride of the Pacific," bringing home 5,000 veterans of the Pacific war. The reception we received coming under the Golden Gate bridge was the closest thing I got to a WWII homecoming.

I was then assigned to the *SS Hubbardton*, as chief radio officer. It was the first time I had a second and third RO aboard. We ran the Blue Ribbon convoys to England until V-E Day, March 8, 1945. The war may have been winding down, but it was far from uneventful. We carried 14,000 tons of aviation gas plus eleven fighter planes: P-47s, P-38's and the Black Widow night fighter on deck.

One trip we were two days out of New York bound for England, when the escorts picked up a sub sounding and came back through the column of ships laying depth charges. One charge went off under us and it must have been shallow—it knocked me flat on my face on the radio room floor and I was sure we had been hit. The cables on my emergency battery were torn off; I crawled to the transmitter and had it ready for an SOS when word came from the bridge that it was just depth charges.

It was a jinxed trip because we had a near head-on collision in the English Channel during a dense fog near the White Cliffs of Dover. Our port side tore all the davits and lifeboats from a freighter carrying 500 pound bombs—she got the worst of it. Again, I got knocked clear across the radio room. We anchored at a refinery up the Thames River that night and during the V2 rocket exploded outside the gates to the refinery. It seemed like we were raised a foot in the water. As we left the Channel on a Sunday morning we had a special prayer service to thank the Lord we got in and out. It was well attended.

Platoon R-55

JENNE, FRANK, R-55

I enlisted into the Maritime Service in Mobile, AL in May, 1943 and was sent to Sheepshead Bay in New York and got there about the first of July in 1943. I qualified for radio school and went to Huntington, Long Island, in September of 1943. I had about six weeks there and then up to Gallups Island in R-54 in October and graduated with R-55 in March 10, 1944.

They sent me to my first port—New Orleans—and I shipped out for my first time on April 1st. I went on an old sugar boat down to Cuba.

The next ship was a new ship, a C-2 just finished being built out of Mobile, AL. I shipped out of there and went up to New York and made a trip to England before the invasion and returned to New York. About that time, things were beginning to wind down in Europe and they shipped a bunch of us by train to the West Coast. I shipped out from San Francisco on a Liberty ship and went to New Guinea, up to Leyte, Lingayen Gulf, then back to Frisco. After that it was back to the East Coast again where I made a trip West Africa and another trip to South Africa.

I made two runs on a ship out of Baltimore, the *Beth Ore*, an ore ship running from Baltimore to the west coast of South America. I pretty well finished up my time there 'till I could be released from service. That was about my extent in the Merchant Marine. April of 1946 was when I got out.

From there I went into broadcast work, radio broadcasting in West Palm Beach, FL. I moved to Florin, SC, where I've been ever since, and have been involved with radio and TV broadcasting here to this date.

MELTZER, IRVING, R-55

No one asked to be born, but it is most important to know who you are and what you are. I was born Jewish and proud of it.

I was in a convoy with Norwegian refugees, bound for Scotland. After the war, they returned to Norway; I was in Bremerhaven, Germany.

We had heard horrible stories about what the Nazis and others had done to those whose beliefs/nationalities were contrary to their own, particularly Jews. However, we tended to disregard most of this as propaganda.

I learned of a Jewish Community Center in Bremen, and while there met survivors of the Holocaust. It was there I learned of the horrors and systematic murder of innocent human beings, simply because they were born Jewish, etc. Being a ship's radio officer, I was asked by the survivors to help them return "home to Palestine," which I did.

Tempered by war and trained in its technology, we Jews and good Christians, decided WWII meant nothing if there was no "home" for the Jewish people: "No room at the Inn." We ran the British naval blockade with "illegal immigrants ships," were captured, interrogated, processed and confined to detention camps on Cyprus, only to return with other "illegal immigrant ships" and later served with the Israeli armed forces.

After almost 2,000 years, an ancient people returned to its historical homeland and we were privileged to see it.

In the 1950's I was employed by a defense contractor as an electronics inspector, after having been cleared for security. One day, the personnel manager (with a face as white as a sheet) informed me that there were two gentlemen in his office who wished to talk with me. They were from the FBI.

One agent interrogated me for almost two hours about my Russian experiences while the other took notes. This was during the Senator McCarthy period, when they were looking for Communists. After the interrogation, the notes were read back to me, for accuracy. Then I was asked to sign, which I refused to do except in a court of law.

Some time later, the Freedom of Information Act became the law of the land.

Knowing that the FBI had a file on me, I contacted a lawyer and through a request, with the proper forms, for my file. Some months later, I received a letter from the FBI stating they could find no file on me. Whereupon, with the cooperation of the office of our state senator, Abe Ribicoff, and after almost three years of correspondence, I received a copy of my file and a copy of the file on American Veterans of Israel.

Most of the information is deleted, but the sense of the files appears to be that the re-birth of the State of Israel was nothing more than a Communist plot to create turmoil in the Middle East; and we were all suspected of being Communist agents.

That about brings me up to date. (*Editor's Note: In honor of his participation in the Murmansk-Archangel convoys, Irving Meltzer was awarded the Victory in World War II medal at the Russian Embassy in Washington, DC, on December 8, 1992.)*

Platoon R-57

VIENT, BILL, R-57

I sailed on the *Ethan Allen* for U.S. Lines, in the Mediterranean between early April, 1944 to early November, 1944 between ports in North Africa and Italy, Sicily and southern France. In a trip from Bizerte, North Africa to Baniloi, Italy (between Naples and Anzio Beach) we took 500 Italian prisoners, major and up, all over 65 years old to be given medical exams. We then took them to Palermo, Sicily. We took everybody's grandfather home and they gave us a party for three days.

From Oran to Bari, Italy we carried 150 Gerka tribesmen with their horses. They cooked, ate and slept on the main deck with their horses. They never shaved, their uniform was sheet-like with balloon type pants and they carried curved knives that they were always sharpening. They served as Scouts with the British Forces.

We loaded 500 American soldiers, with their equipment in Naples and anchored in Naples Harbor with 49 other Liberties for almost a month waiting for D-Day. We landed them in August, 1944 at St. Tropez on the French Riviera on D-Day +1.

Back to Naples and took about 600 more troops with their equipment to Marseille. From Bizerte we carried 575 German POW's to New York City (close to 30,000 were brought to the USA during the war to prevent any possible rescue by their forces).

There were other trips during this period, but these show the variety of human cargo carried.

During the seven month voyage nobody but the captain and mate, and maybe a few others knew that the bottom six feet of each hold had a temporary deck which was to store cargo for the future invasion of Southern France, typically: coffee, sugar, flour, K-rations, D-rations, C-rations, 30 cal., 50 cal., 20 mm, 3" shells, 5" shells, mortar fire and other items including clothing, tents, blankets, etc. All 50 Liberties were kept in the area until the invasion; shuttling back and forth between the ports moving the trucks, tanks, and food.

On various trips on the *David B. Johnson*, of Wilmore Steamship Lines between late November, 1944 and November, 1945 we carried: group of men and women from Cardiff, Wales to New York who were members of the "Moral Rearmament Movement" and were on the way to a conference on Mackinac Island in northern Michigan in Lake Superior; and a group of English brides to New York to their respective husbands. So much for serving as a passenger ship.

On a trip from Searsport, ME to Cherbourg, France in December, 1944 we carried 10,000 tons of 250, 500 and 1,000 pound bombs to shower on Germany. It took 365 + railroad cars and 40 men seven days, working 24 hours a day, to load us at Searsport. Our return ballast cargo, to keep the screw in the water was 1,600 tons of Bells scotch, White Horse scotch, and Drambuie for Mr. Joseph Kennedy who was the U.S. Ambassador to England and exclusive importer of these items in the United States at the time.

On one trip the *Johnson* carried 9,500 tons of coal and picked up the Norwegian Consulate from a pilot boat in London, England and proceeded via the North Sea to Oslo, Norway. We were the first civilian cargo into Norway after the Germans surrendered. Half of the cargo was unloaded in Oslo and half in Drammen. This was in July, 1945 and we were in Drammen on V-J Day, August 15, 1945.

The return ballast cargo was the strangest one—never mind trying to make a Liberty a coal collier. We backed up into Frenchman Creek at Fowel (no room to turn around at the port) in Cornwall, England to load 2,000 tons of white china clay which we took to Portland, ME and unloaded for shipment to Toronto by rail. How the stevedores ever got the black coal dust out to load the white clay I'll never know because I locked the radio room and disappeared each day—the dust was terrible.

Final trip to sea was on the *Johnson* from Portland to Corner Brook, Newfoundland to load rolls of newsprint. We unloaded a third in Tampa, FL, a third at Galveston, TX and the final third at Houston. I left for Boston on Nov. 5, 1945. Goodbye to the sea.

I have tried to point out only the unusual cargoes. The most mixed were the ones to norther Russia where holds were full with 144 two and a half ton trucks, each loaded and railroad engines; PT boats on the deck—all this with four icebreakers—through ice 14 feet thick, down into the White Sea to ports of Arkhangelsk.

Platoon R-58

FARRELL, JAMES "PAT", R-58

When we all left Boston in '44 we had a great "sing along" on the train to New York. I think I spent the night at the home of Art Natvig in Brooklyn. A day or so later, Jim Biersach and I stood on the train from New York to Norfolk. I lost track of Jim, since I was assigned to a ship and had to leave with my ride to sign on... Basis for tale one!

SS Andrew Moore. William Gillispie, Master. When I arrived, signing on was in progress and my turn was next. I was signed on as Second Operator, as first was already occupied by James Parce (R-13). Third operator was a Navy operator from the West Coast.

The *Moore* was a troop/cargo ship. We transported 600 troops and the accompanying armed guard and an Army transportation personnel group assigned full time.

We left Norfolk and had the usual scares from there to the Mediterranean. Our destination was Naples, and the first real excitement was going through the Straits of Messina when we lost our steam. We kept drifting back, the convoy commodore kept signaling to get back in line, and Captain Gillispie kept swearing at him and the chief engineer. I soon learned what a sailor's language consisted of (being a young 18-year-old from the mountains)!

For the next eight months we ran shuttle service in the Mediterranean. Took a shipload of ammo to Anzio just before the break out. Saw battleships and cruisers shell German lines. Sailed between Oran, Bizerte, Sicily, Sardinia, and dropped anchor four hours after H hour on the invasion of Souther France.

We took supplies and equipment from Italy to France from August to November, 1944. Then came back to New York, where I took leave of the ship to make a trip to western Maryland, where I hadn't been for a year.

As an add-on, many happy events happened on the *Andrew Moore*, such as Jim Parch explaining Pompeii and insisting I go with him to see the ruins. Sad things also, as when we had Coast Guard trials for seamen accused of mutiny. Don't remember how that turned out.

JACKSON, HOWARD R., R-58

After graduating from Gallups Island I spent 1944 and 1945 carrying 100 octane aviation gas to England preparing for the invasion. Aboard the *Esso Hartford*, two other radio operators, Harvey Long, Stan Jennings, and I made some seven trips to ports in Blighty as well as to Naples, North Africa, Venezuela, Aruba, Curacao, etc, Texas, too.

With the war's end, S.O., like all American companies, no longer was no longer required to have three operators. I signed on as Radio Operator on the *Esso Montpelier*. Our new captain had been a German Navy veteran, submarine service, during WWI. He was not allowed to sail as an American captain until the war was over.

We didn't get on well. He took a dislike to me. There was some trouble as we left Corpus Christi. A man was thrown overboard. I sent word by signal light, and the tug came around, took him up. Then there was an inquiry; I spent the trip to New York taking depositions on the typewriter. We got lost off Chesapeake Bay. The Old Man put me on the DF to get a finding. He told me I would be responsible. With triangulation, I managed to locate our position and we found the buoy.

On another trip, 300 miles south of Buenos Aires, we were leaving Bahia Blanca, when the captain told me, "Sparks, there's a new regulation. We have to get a port report to New Jersey within 24 hours." What to do? New York was more than 5,000 miles away. I wracked my brain, then remembered the Gallups Island teachings of the Kennelly Heavyside layer and the bounce. I had the Bridge call me an hour before sun up the next morning. I cranked up on 22 megs and called WCC in Tuckerton, NJ. :WSID DE WCC K" I couldn't believe it! Loud and clear, a nine or ten. I banged out the, maybe, 150 word message and got acknowledgement. Just enough time. Soon after, he was gone.

With some pleasure I told the Old Man his message was sent and received in New Jersey. Back to Trinidad for fuel. Against all advice, the captain let the crew go ashore. More depositions. FBI in New York again. In July, 1946, Howard Jackson, RO, signed off. Quit.

JENNINGS, STAN, R-58

Remember back in the olden days at Gallups Island how they pushed radio theory, electronics, higher math at us so steadily we thought our craniums would crack? To prevent such a fate, we had to simultaneously absorb Mr. Morse's bell-like euphony while our wrists bounced to return the music as fast as it was coming in. Rules of the Road. International Q signals. Transmitters. DF's. Logs. Lights. Radio in convoy. Auto-alarms. Emergency antennas. From all those months of mauling our metabolism we emerged from the FCC examining room, proudly clutching our hard-won Second Class Radiotelegraph License, and sailed over salty seas to places we just couldn't believe. Places so far and Sinbads so young.

A couple of months later some of us returned to that source of all things good, the FCC, to get our Second Class Radiotelegraph License while our heads still retained all that scientific knowledge.

Then, after a year of brass-pounding practice and likewise receiving, up to the phenomenal speed of 25 words per minute, going and coming, Mother FCC checked our dazzling performance, noted our required year of ship service, and granted a Radiotelegraph License—FIRST CLASS>

Forty years come and go. Other areas of endeavor come. The licenses go. They went after five years.

Our little daughter meanwhile

braved the FCC and came away with a General Radiotelegraphic Operator License. When it was renewed after five years, she received another one. Then she received a little card, a license—good for life.

After the war I worked in graphic art for several newspapers and magazines before going to National Geographic. Geographic hired me partly because of my world travels. They put a stop to those travels and permitted me to design many publications, magazine pages, the masthead and the gold rectangle Geographic uses on every book, TV program, etc. Their ID, their logo. Thanks, NGS. Geographic gave me the opportunity to work with thousands of great photographs and begin my own photo and graphics business—Jennings Publications.

Jennings Publications, with 25 years of Washington newsphotography and newsletters at the National Press Building, now continues with office and darkroom at home.

I have two daughters. Phyllis is a California counselor in social work. Shelley, a musician, is also a sound engineer and licensed Boston electrician. Married to Jane, my editor, who helps with the Gallups Islander—and also with this GIRA book.

Platoon R-59

CURTIS, ROY R-59

I made a trip to "Inja" on the *Nathaniel B. Palmer* in 1950 that still remains pretty vivid amongst the blur of my other seafaring days.

We went to Karachi, Pakistan with coal, unloading half there and the remainder to either Vishakapatnam or Chittagong. It was Karachi where I learned the hard way about local liquor entrepreneurs using a hypodermic needle to remove the "good stuff" from Black Label and Johnny Walker, replacing it with something that is unique to them. I was bad sick from that.

After coal delivery was completed, we went to Old Goa (Portuguese colony on the coast of India). While there I went to a local "dentist" who removed an impacted tooth I had with a rusty hook. It became quite an involved process—using whisky for both deadening and rinsing. The "dentist" also cleaned my teeth with a foot-powered drill. Its wheel kept coming off in my mouth and raced around, creating quite a physical sensation.

Then the ship's steward brought some local produce aboard that gave most of us, dumb enough to eat it, dysentery.

Oh, the good ol' days! I reside in Phoenix, AZ with my wife, Irene.

KYLER, JAMES R R-59

Retired from Reynolds Metals Company, Marine Division as radio operator in 1975, after 13 years with them. Spent seven years, between '44 and '51 with various SS companies. Wish I could do it all over again and remember it all.

Best of 73's.

SCHAFER, WILLIAM H. R-59

I enlisted in the U.S. Maritime Service in late July or early August of 1942 and arrived at the Sheepshead Bay boot camp on August 5, 1942. I was let out of quarantine on my 19th birthday, August 17, 1943, to register for the draft. During boot camp I qualified for radio school and accepted with pleasure. From boot camp I was sent directly to radio school at Huntington, Long Island and from there directly to Gallups Island where about 60 students formed class R-59. Training was intense with little free time.

I signed on the *SS Kernstown*, a T-2 tanker, on June 25th and stayed with the ship for five crossings until November 27, 1944. On each east bound crossing we carried aviation gasoline and a deck cargo of fighter (P-47's and P-51's) airplanes. We detoured south on each return crossing to Aruba or Curacao to take on aviation gasoline. The first voyage we stopped at Casablanca and Oran and unloaded at Mers-el-Kebir. On the return crossing for the third or fourth voyage we dodged a major hurricane all the way from south of Puerto Rico to New York and sent many weather radio messages as we were often in the hurricane and once in the center.

As we turned west (from east of the Ambrose lightship) the hurricane went across Long Island, unstepped our main mast, contaminated a tank of fuel, caused much damage ashore, while turning about 75% revolutions ahead, and as an added thrill we drove right through the middle of an outbound convoy.

The last crossing was to Southampton, England. This was nearly fatal. We completed unloading in late afternoon and it was too dark to go to sea and we were full of gasoline fumes. There was much harbor traffic and we were sent to an anchorage shaped like an isosceles triangle. We were told to anchor at the apex of the triangle. We did. A strong tide was running as we anchored (and for some time afterward) and the tide was running directly towards the apex of the anchorage.

At that time we went to dinner and while eating the ship jumped straight up. The captain was almost the last man out of the saloon. We had been struck by a British motor vessel troopship, loaded with troops standing at the rail. The MV was outbound with the tide and didn't realize how hard the tide was running. The MV struck the *Kernstown* a few feet aft of the bow on the starboard side, skipped forward and scraped her entire port side across our bow. After the MV had passed we could see the hole in our bow—a large hole with cherry red edges and big enough for a good sized vessel (to drive through). The hole was closed with plywood, timbers and concrete.

On return to New York it was decided to send us out again without going to the shipyard for repair. During the return crossing we had observed that several degrees of rudder was required to keep us on course. I opted to sign off.

Several ships and experiences later, I signed on my fourth ship, the *SS Beloit Victory*. I made two voyages on this ship. One at the end of the war with Japan and later my first post-war voyage. The SS company kept me on the payroll between the voyages.

The first voyage was from July 25, 1945 to December 11, 1945. I was the chief operator, Milton V. Bitner of Salt Lake City was the First Assistant and Walter A. Woolson of Syracuse, NY was the Second Assistant. The captain was John Silvia. We were loaded for the invasion of Japan. There were amphibious tanks on deck fueled and ready to go. They even had crystals in their radios. The holds were in layers; the first layer of each was 3/4 ton trucks on pallets—

Platoon R-59

all fueled and ready to go. The second layer was perennial tents and C-rations. Below that were boot socks and winter underclothing and so on until the very bottom where they had loaded spare army artillery barrels.

On August 6th we dropped the first atomic bomb. There wasn't much ado or comment among the crew. On August 9th we dropped the second atomic bomb. There was quite a bit of talk on the ship now and a lot more information on the news. (I wondered what peace would be like). At 5 a.m. local time, Friday August 19,1945, I heard on the Armed Forces radio that Japan had officially accepted unconditional surrender. We started a party on the ship. I believe that my life and the lives of many American servicemen, as well as Japanese people, were actually saved by our dropping the atom bombs.

We arrived at Wakanoura Wan the harbor for Wakayama, Japan in late morning October 15, 1945. We off-loaded the tents, food and clothing for the occupation troops and returned everything else to the States. We departed for home the last week of November, 1945. We fed many Americans there and brought home 210 servicemen of all ranks and services. We were short on rations and glad to reach San Francisco about December 10, 1945.

On December 30, 1948 I was drafted into the U.S. Army. I served for a year and a few days and was released in early January, 1950 "at the convenience of the government." In early January, 1951 I was recalled to active duty in the U.S. Army and released in late November, 1951 "at the convenience of the government." Therefore, I became a veteran as a function of the Korean War.

After my last military service I returned to my home in Columbus, OH and applied to North American Aviation, Inc. I was hired on December 5, 1951 as a technical writer, because of my radio training, and retired from the company, now Rockwell International, Inc., on August 30, 1985. And though I upgraded my electronics training three times, my career was a direct result of my U.S. Maritime Service training at Huntington and Gallups Island.

My first wife and I had five children; three girls and two boys. After being widowed I married a widow who had two children. My second wife, Joyce, and I have a total of seven grandchildren.

All of my post seagoing employment involved naval military aircraft, the Apollo space program or the Space Shuttle program.

Platoon R-60

AVERA, A.W. R-60

Summary of WWII sailing history: Served aboard the *SS P.T. Barnum*, August-December, 1944. I went aboard in New York while the ship was underway, in convoy with about 500 ships, to the UK. We were escorted by Canadian corvettes and British carriers. Scuttlebutt was that we lost several ships at the tail-end of the convoy, and one trailing rescue vessel was torpedoed while picking up survivors.

Our first stop was Oban, Scotland. From there ships left daily for different parts of England and we finally went up the Thames River to London and Antwerp, which turned out to be "Buzz Bomb alley." Most of them (German aircraft) were shot down by antiaircraft fire from flack towers built on stilts in the channel, near where we were anchored. I saw several (Buzz bombs) explode after being hit by Spit fire fighters, and a few that got through to explode on the mainland. The most I saw was 32 in one day. We had to stay under cover during attacks to prevent being hit by shrapnel. We saw only three explosions of the V-2 rockets—just the explosions, not the rockets because they were too high or too fast. It was interesting to watch the American P-17s in their perfect box formation going out to and returning from the continent.

We returned without convoy to New York. We didn't encounter the enemy but did run into an arctic storm that didn't sink us but did scare the hell out of us. After that trip I got permission to leave until sailing time and went to North Carolina where my wife met me. (While on my way to the railroad station in New York City, in very dense fog, I heard a low-flying plane passing over. I stopped to listen and heard it crash. read in the next day's paper that a B-25 had flown into the Empire State Building.)

Later I returned to the *Barnum* traveling from New York to Baltimore to Italy. The deck cargo was crated aircraft or parts, and railroad steam engines which had LIMA stamped on them. While in Naples we got acquainted with several of the Air Force pilots stationed there. They ate a few meals with us, took baths and got soap and towels, etc. In turn, they put a couple of Jeeps and personnel carriers at our dock for us to use for sightseeing.

The order came to level all cargo and prepare to move the Canadian Fifth Army from Italy to France, where the Canadian First needed assistance due to heavy German attack—now known as the Battle of the Bulge. While in convoy to Toulon, we went through the channel between Sardinia and Corsica, when the Germans made an attempt to attack us with their E-boats (motor torpedo) but were sunk by the escort ships. The German air attack was held off by the American pilots we had met in Naples.

After we returned to Naples and from Toulon, we went to Gibraltar, North Africa, back to Gibraltar and back to New Orleans, before I left the *Barnum.*

I then signed on the *SS Am-Mer-Mar* (American Merchant Marine). We went directly to Georgetown, British Guiana. From Georgetown to Port of Spain, Trinidad then on to New Orleans, where I left my two pet monkeys with the customs agents, and to Miami for R&R.

From New York I was assigned to a troop train to San Francisco where I got a T-2 tanker. We went to Pearl Harbor and after several days layover went to Ulithi and then to Okinawa. After riding out (at sea) my first typhoon, we went to Yokohama then to Tokyo and Wakayama in the Japanese Inland Sea where we got our second typhoon. We anchored (in the middle of a mine field) and rode out this typhoon with two anchors and kept sea watch. When floating mines came toward us, the captain had the ship turned so they missed us. The Navy signaled us to wait for escort where we were, but when they didn't send an escort ship, the captain used some 1914 charts and we went to Wakayama. From there we went to Nogoya and back to Yokohama and finally unloaded the last of our cargo of fuel

We left for San Francisco and about two days out of Pearl Harbor we got orders to go to Pearl for bunkers, through the Canal to Curacao and Surinam. After one more stop, we traveled across the Atlantic to Texas City where I left the ship for the last time.

Once back in Miami, I went to work for Delta Air Lines for several months, then to Eastern for 30 years and retired.

CAMPGNOLO, F.J. "NINO" R-60

In 1942 I worked at a local shipyard as maintenance electrician. In 1943 I decided to leave the yard and sail on the ships that we made. So, on Memorial Day, 1943 I signed with the Maritime Service and was sent to Catalina Island, CA.

After a few weeks I was accepted as a radio cadet. Several men in our group boarded a train for New York. At Huntington, Long Island we learned basic electrical theory, typing, code and math. This place was like a palace, lavishly constructed. We were then sent to Gallups Island. There we stayed in barracks. I soon learned that the bed next to the front door would be snowed under whenever the door was opened in the winter time. It seems like it was constantly opened. But I got used to the snow on my bed.

The idea in class was to hang on tight to your pencil because if you dropped it you would lose three weeks of normal instruction while you picked it up. We were certainly given a lot of material to absorb. Luckily, I teamed up with Jim Woelfec, we studied all the time. We even took our books and notes with us on weekends to the servicemen's club. The girls could not understand why we would rather study than dance. We were very determined to pass the final tests and get our license. Although after a couple hours of studying, the music sounded so good and so many girls wanted to dance, we gave in and had a heck of a good time.

Leaving Gallups, I got my first ship out of New York, the *John L. Sullivan*. I sailed to England (twice), France, Italy (twice), Holland and Ireland. In the Pacific we visited ports in the Philippines, Japan, New guinea (Biak), New Caledonia (Noumea) and Eniwetok.

On the way to Noumea we were sailing alone. Our ship, the *SS Cape Archway*, was a C-1 transport loaded with bombs and various ammunition. One day near Eniwetok while having coffee with the captain and first mate on the port wing, I spotted something in the water. I mentioned to the captain I though that we were being followed by a periscope. With that all hell broke loose—the alarm went off loud and clear—I made it to the radio shack in about six giant hops where I was ordered to send "SSS SSS SSS." Eniwetok answered.

Meanwhile, our 5" forward gun and our aft 3" gun was having a field day. It was very exciting. Full steam ahead or go direct to heaven with a bang. All I could think of was all those bombs (in the hold) below. I wondered how high I was going to be blown. Somehow, with luck, we got clear. I thought we might have damaged the submarine but found out later that one of our destroyers did the sub in.

The ships I sailed on were the: *John L. Sullivan* (Liberty), *Brander Matthews* (Liberty), *William H. Claget* (Liberty), *Emory Victory, Cape Archway* (C-1), *American Forwarder* (C-2), *American Ranger* (C-2), and *Explorer* (C-3).

The *Explorer* was a beautiful ship; my quarters were the front corner of the starboard side. It was a large room with a good sized table and the radio shack was behind me. We carried several thousand troops from Nagola, Japan to New York via Panama. It was December 17th when we went through the Canal. Many soldiers slept on deck all the way—which was fine until we got a little north of Florida. I looked out one morning and saw piles of snow all over the deck—those poor guys sure got a surprise.

My last trip was on the *Emory Victory* from New York through the Panama Canal to San Francisco. I went to work as a salesman and buyer for a hardware store. Then at a general electric supply as counter salesman. After that, I worked for 13 years at Hewlett Packard Co. in Palo Alto. Got my real estate license and sold homes, gave lectures on real estate, got my brokers license, managed a real estate office and finally retired and moved to Santa Cruz, CA.

I must say I still have an active radiotelegraph license. I feel that my time at Catalina, Huntington and especially Gallups, and the ships I sailed on were the most cherished years in my life. I feel proud to have served as I did.

Yes, you bet, I still long for the sea.

Platoon R-61

CROSS, GOLDEN H., JR. R-61

After enlisting in the U.S. Maritime Service in June, 1943, I requested to sent to Radio Officers School because I had been discharged from the U.S. Army Signal Corp school in North Carolina, having completed the course.

I was sent to St. Petersburg, FL for basic training. During training I was promoted to Petty Officer and given a company of men to see through basic training. After just a few weeks in late August, 1943 I was sent to Radio Officer's School at Huntington, Long Island, at the Kahn Estate. At this school we were given basic theory, electronics, receiving and sending Morse code, typing, etc.

After graduating from Huntington we were sent to Gallups in November, 1943. One funny memory to me at this time was we were assigned to different bunks the first night and also given Fire Watch and Guard Duty. It was our responsibility to find where the person who was relieving us slept, in order to awaken him. The person I was relieving forgot to do this, therefore he went from bunk to bunk awakening people, asking them "are you Cross?" . . . One fellow told him, "No, but I'm mad as hell." This item made the Islands paper, the "Spark Gap."

Our platoon, R-61 graduated in May, 1944. During our leave in January, I had married Edith Hollen of Greenville, SC. If you remember we had been promised 30 days leave and 30 days when we graduated. We had been put on "Hurry Up" training and never received either. We were ordered to New York where we were assigned to ships.

I signed on the *S/S Robert Jordan*, Captain Atwell G. Collins, skipper. I was the second radio officer, with Phillip Karent, chief . . . he was a great guy to teach me a lot of practical tips. The Jordan was sent to Belfast, Ireland, in convoy. Upon arrival we anchored and spent about three weeks—no leave, no going ashore! We saw many other ships, all types, arriving. About June , we were ordered to proceed around Northern Scotland to go to Southampton. When we arrived, we were ordered to anchor and await orders. . . . we found out we were going alone through the English Channel to the Isle of Wight and then to the Invasion of Normandy! We were on duty 12 hours and 12 hours off. We were the first ship to go into Cherbourg . . . after the minesweepers and were to make sure all the mines had been gotten.

I was also in on the Invasion of Southern France and after the war was over in Europe, I was sent on ship to the Pacific Theater of War.

I served on the following ships: *SS Robert Jordan* KXGC; *SS Rachel Jackson* KKTA; *Harry Leon Wilson* KUOP; *Lyman Abbott* KOJT; *SS Cape Lilibeo* KWOU; and *SS Dr. Lykes* AOPI.

While on board the *SS Cape Lilibeo*, headed in a Super Convoy to invade Japan, it was announced the Japanese had surrendered. We were a very thankful crew.

LONIS, MAYNARD J. R-61

I arrived at Gallups Island by a different route than some of the other recruits.

I was sworn into the Maritime Service in August, 1943 at Raleigh, NC. I had been an Air Force civilian instructor of air-

Platoon R-60

craft engine mechanics at Seymour Johnson Field, Goldsboro, NC. Arriving at boot camp in St. Petersburg, FL, I was subjected to the usual brush cut, and shore leave was restricted for the next five weeks while we studied maritime subjects, had daily practice on the drill field, swimming lessons, etc. I recall swabbing the deck during kitchen duty at the Vinoy Park Hotel, which was taken over by the Maritime Service.

Learning that there was a chance for some apprentice seamen to receive radio training, I resolved to try for that. I sent home for my algebra book and studied every available night that they gave me spare time to be ready for an exam to qualify.

Some of our seamanship lessons were conducted on a square-rigger, the *Joseph Conrad*. I remember I was polishing brass near the captain's quarters when the notification came of my acceptance to go to radio school at Huntington, Long Island. (The square-rigger is now 108 years old, and is tied up at dock in the Mystic Seaport Museum, Mystic, CT. The caretaker skipper brought out a log book which he was allowing visitors to sign, if they had previously served aboard the training ship.)

Being sent to the maritime school in Huntington was a pleasant eight weeks. (It was formerly the banker Otto Kahn's estate and last year the TV program "Lifestyles of the Rich and Famous" showed some of the estate, including the famous staircase. It was tradition that the underclassmen only used one side of the staircase while studying there until they were ready to graduate.) Then it was on to Gallups, where I graduated in April, 1944.

No doubt most GIRA grads have many memories of Sheepshead Bay, which is where I went in 1948 for a radar course.

MAYHEW, BOB R-61

Forty-nine years ago tonight, at about 2200, I boarded a launch at Lewis Wharf to join the *SS Washington Allston* at anchor in Boston Harbor. When I awoke the next morning, we were underway. Louis J. McCourt (R-12) was chief and I was the only Indian.

The first two trips to North Europe were relatively uneventful. We spent long periods at anchor in various ports around the British Isles waiting for the discharge port to be named. The second trip included quite a while at Omaha Beach—weeks after the invasion and finally discharging at Cherbourg.

Trip number three had much of what the first two missed. Sailed from New York in January, 1945 with Oran as the first stop. From there to Alexandria, Egypt, through the Canal, Red Sea, across the Arabian Gulf and Bay of Bengal to Calcutta. After a few weeks in Calcutta we went to Madras and from there to Sydney, Australia.

Sydney was heaven for the young and virile, especially if you weren't hindered by a sense of loyalty for someone back home. The Gunnery Officer and I spent two wonderful weeks in an apartment on "Bondi Beach" with two of Sydney's finest—a very enjoyable interlude.

We were holding down the apartment when the Germans surrendered and the city went nuts. One of the especially exciting events was watching the ship's engineer go riding past on top of a tram car.

As always happens, when everything is going real well, sailing time came and we were due to leave for the Philippine Invasion. At sailing time, the skipper didn't like the way the Aussies had secured the deck cargo so we undocked and anchored a short way off, where the deck gang spent the night securing deck cargo to the captain's specs.

During the night I had severe stomach cramps, got the purser up early in the morning and he diagnosed acute appendicitis. Within a couple of hours my gear was on a launch and I was lowered over the side in a basket. To this day, I remember the entire crew lined up along the rail with comments like, "You lucky bastard" and "How I wish I was in your place."

My friend the Gunnery Officer rode ashore with me and phoned our two lady friends so I would have company in the U.S. Navy hospital. As soon as I was able they visited me and when I could go on liberty, one came and got me.

After about three weeks, I was discharged from the hospital "fit for duty" and reported to the WSA/RMO. The man in charge greeted me, told me I was the only RO on the beach and said he had a ship going to the States. I said I'd take it. He said, "You don't have to." I said I'd take it. He said, "It's flying the Panama flag." Once more, I said I'd take it and he'd take me to the ship.

The ship was the *Isolator*, a coal burner with hand-fired scotch boilers, built in Germany in 1896, 1500 hp; with a three man U.S. Navy gun crew, who, with the steward and I, were the only Americans aboard. It was a true "League of Nations" crew. The radio room was truly a "shack;" added on probably 25 years after the ship was built; with the head down two decks and a big bathtub for my exclusive use. The "Sait" receiver had plug-in coils to change bands and the break-in relay was an alligator clip.

The third night out, the captain woke me and handed me a message to send at once. The message was "cancel So-and-So's allotment—missed ship Sydney." I reminded him that there was a war going on and we were operating under radio silence. He informed that his provisions included a provision that we could operate radio within 300 miles of the coast and we were still within 300 miles. "But we've been under way for three days, surely we've gone farther than that." I then found out that under the best of conditions our top speed was 5 knots (and we had nearly 7,000 miles to go).

The following day brought the big disaster. The ship's reefer system failed and the engineers were unable to repair it. We ate like kings for a few days and then all the spoiled food had to be thrown overboard. It was almost like someone had known it would happen because it developed that we had emergency supplies ready.

At breakfast we had a choice—vienna sausage, canned Australian mutton or canned stew. At lunch it was canned stew, canned Australian mutton or Vienna sausage, and so on almost without end. Within a few weeks we were out of coffee, canned milk was next—the only thing that was constant was the three choices at mealtime. What saved me was the gun crew's foresight, at some port they had visited before my item, where they had appropriated large amounts of D-rations—very hard, very filling lumps of chocolate. I ate it for appetizer, main course, dessert, and when the coffee was gone, as a beverage.

The most upsetting thing during this period was to watch the Giant Albatross, gliding along, passing us with no effort, as if we were going full speed astern.

Next, as we were approaching a position abeam of Hawaii, the chief engineer broke the news to the captain that we didn't have enough coal to reach any port on the West Coast. The captain got me in his office and we composed and encoded a message to the authorities that we were diverting ourselves into Honolulu/Pearl Harbor due to insufficient bunkers.

On arrival at Pearl Harbor, the bunker barge came alongside only to get the word that we couldn't use what they had, we needed coal. The next word was that there was no coal and hadn't been for years. Then, some genius in the WSA realized that they had a big, sea-going tug leaving for San Francisco, so one of our anchors was detached and the chain became part of a giant towline. To this day I don't remember if we got stores or not—the important thing was that we were now going to be moving.

Even then, problems developed. The tug couldn't go full speed for fear that the anchor windlass would pull off the deck, but we made 10 knots. The steering wheel had a hub with a couple turns of cable on it leading out to both bulwarks and astern to the steering engine. The cable would break and two black balls (day) or two red lights (night) would be hoisted while the cable was spliced.

Eventually, we reached San Francisco and for the only time, before or since, I believe everyone went ashore to eat.

Two days later I took the train back to Boston and later discovered the final irony—when I turned in my pension time this trip didn't count as it was not the U.S. flag.

Guess how many times during this two month period I said, "If only I'd stayed in Sydney with full pay, full subsistence and a fine bed-warmer to keep me happy. . . . " (

Editors Note: Mayhew wrote this while aboard the *SS Energy Independence* WBJF, arriving Salem, MA from Maracaibo, Venezuela, 24 May 1993.)

MYERS, LEON GLENN R-61

Gallups Island was definitely an

Platoon R-63

Platoon R-65

Platoon R-69

impressive part of my life for six months during the latter part of 1943 and early 1944.

I was born in the deep South (Macon, GA) 85 years ago. When I was a child, talk of the Civil War was much alive, and I was taught that Yankees were awful people. Let me say that I met two of my dearest friends at Gallups Island Radio School. They are Maynard J. Lonis and Barnard Dorsey, both from New York, who are men of the strictest honor and virtue. They would put some of my southern gentlemen friends to shame. I still correspond with them both, and Lonis and his wife have visited me twice.

I got back home from serving in the military in April, 1946, with 24 months of actual sea duty.

In June, 1990 I was honored for 54 years of continuous membership in the Macon Exchange Club.

Although I did not make a career of radio after the service, I worked as an attorney, once as a partner in a law firm, and retired in 1974 from the Bibb Company. I have also served over 60 years on the administrative board of the Vineville United Methodist Church. I am a member emeritus of the Macon and Georgia Bar Associations.

Platoon R-63

BRZOZOWSKI, VICTOR R-63

I'm a graduate of R-63. Prior to that, went through Sheepshead Bay and...stayed at Huntington.

"Most of my duty was (in the) North Atlantic, Mediterranean and the Indian Ocean. We were happy to get our discharge from Uncle Sam under 45 years.

"I want to thank the Gallups Island Radio Association, and I want to thank God we're still alive."

CORCORAN, DESMOND P. R-63

Honoring your request for info of ex-Gallups Islanders, I respectfully submit my "DIT DAHS" worth. WWII came along when I was living in Los Angeles and working for Lockheed Aircraft Corp. After waiting over a year for call-up, having enlisted in the US Air Corps, I was granted permission to enlist in the USMS. I enlisted in July 1943 and was sent to Catalina Island for basic training.

A group of us was sent to Huntington, Long Island, NY for some kind of radio school. From there I was sent to Gallups Island in November of 1943 and started with R-63. Halfway through training I ended up in the Marine Hospital in Brighton, MA for three weeks having suffered an injury during training. After hospital release I was placed in R-64 and graduated in July, 1944 with that group.

Was assigned to the *Horace H. Harvey*, a Liberty tanker, making several North and South Atlantic crossings to the Northern point of the Orknees to Africa. When hostilities ceased in Europe I was sent to the Pacific where I sailed North and South Pacific from Iwo Jima to New Zealand and all points between. Was discharged in San Francisco, December, 1946.

I decided to continue the electronic training I received at Gallups and decided to enroll in an electronic trade school in Omaha, NB. When I went to enroll I found out they needed an electronic instructor, said I was capable of teaching, thanks to the instructors at Gallups Island, and was hired as an instructor. I kept two hours ahead of my students.

While engaged in this type work I obtained a first class phone licenses along with an amateur license, WOEEH. Quit the teaching business and went to work with Philco Corp. as Tech-Rep ending up in Fort Bliss, TX where I taught Radar and Guided Missile systems. After 10 years with Philco I was made corporate supervisor so decided to quit to enter my own business. The business was an electronics school. This was very successful and branched into two other electronics type businesses.

I received a U.S. Patent for one of my electronics training aides. Sold out everything in 1969 and at 48 years of age decided to retire. During my retirement I've kept busy as a SCORE member (Service Corp of Retired Executives) and am now serving my second term as mayor of Horizon City, TX.

I'd say I've had an interesting and successful career with the pivotal points being the electronics training I received at Gallups Island. Although the winter there was terrible, the beautiful memories of the wonderful instructors, members of R-63 and R-64 will be warm and everlasting.

KILGORE, JACK R-63

When I signed off my last ship in the summer of 1946, I took a good last look at the rusty hull and decided to change back to my old profession of Commercial Radio and Television Broadcasting.

In 1952 a friend convinced me to enter the Life and Health Insurance field, where I have been ever since. In 1968 my wife, Kay, and I began providing Life and Health Insurance coverage for the clients of Independent Insurance Agents. When we get the business, the client and agent are well served.

We have no plans to discontinue our business operations, but hope to make every GIRA meeting as they come up.

WHITEIS, ULYSSES E. R-63

After graduation on May 12, 1944, I was assigned to be chief operator on the *Joshua Alexander* at dock in Boston. Before embarking, Helen Luby and I were married at Gallups Island. The marriage came close to being aborted because a fog settled on the ferry, so heavy that we had to dock just inside the submarine nets. When the fog disappeared as suddenly as it came, we proceeded to the Island.

When the *Alexander* departed Boston, it went to Halifax, Nova Scotia, undergoing navy target practice. We dropped anchor at Milford Haven, England on the night of June 5, 1944. Before sunrise the ship moved into open waters and, under the cover of darkness, joined other ships. When daylight occurred we could see several thousand ships all moving in the same direction: To the east and south. This could only mean one thing, so I marked an "X" as usual on my calendar and wrote the word: INVASION.

As we approached the beach, the naval craft, the minesweepers and the landing crafts moved towards the sands. The supply ships moved closer to shore and dropped anchor. Soon the small craft got into action. The Higgins boats and DUKW's went back and forth from ships to sand—carrying men and supplies. All of this occurred under fire from German strongholds in the bluffs which overlooked the sands.

On the night of June 6th none of us could sleep, with the thunderous noises of aircraft overhead and our gunners with tracers passing over the bluffs.

None of the crew knew the location of the invasion until we returned to the States. We learned from news reports that we had been at Omaha Beach, on the Normandy coast.

In another convoy of 15 ships, my ship went the North Atlantic Run to Murmansk, Russia. We were there about six weeks. When we were about to leave the German submarines torpedoed five of our convoy. The remaining 30 ships went around the cape and lost one more ship which was sunk off Norway.

In still other convoys, I was on the *SS Kearney*. Our ports of destination were Antwerp, Liverpool, Swansea and Marseilles.

After the war I returned to school at Harvard, earned the doctorate degree and became a professor of psychology at what is now Central Connecticut State University. I retired in 1978. Next month (April, 1993), I will be 82 years of age.

Platoon R-65

OZBURN, PAUL R-65

I joined the Maritime Service in Kansas City, MO after getting a leave of absence from TWA. Trained in St. Petersburg, November-December, 1943. Went to Gallups in January, 1944. Graduated with 28 men in August, 1945 (R-65).

First ship was the *SS Harpers Ferry*, a T-2, in August, 1945 to Swansea, Wales and Liverpool, England. Back to New York then to Aruba to New Guinea where we fueled destroyers in Invasion of Philippines. Back to Panama to Eniwetok to Ulithi.

While we were there a Jap sub broke through the nets and for the first time they broke voice silence and told us to get the hell out of there—which we did. Back to Balboa, Cartagena, New Jersey.

Next ship was a Liberty, the *SS Roger Griswold*, New York to Sicily to Bari, Italy to Onan, Africa. Back to New York to Naples to Leghorn to Oran to New York

The *SS Santa Clara*, a T-2, was my next ship out of San Francisco to Yokohama, Japan to Balboa to Curacao to Casablanca to Algiers to Curacao to Liverpool to Port of Spain, Trinidad to Mobile, AL.

Went to Pittsburgh, KS (where I grew up), married my girl on April 11, 1946. Went to New York. No ships. Went to New Orleans, signed on an Army ship until July 4th and then signed off and went back to work for TWA in Kansas City as a radio mechanic. Retired after 41 years in 1983.

My wife and I had three girls and we now have five grandsons and a granddaughter. I had a first phone when I went into the Maritime and I got my second telegraph at Gallups.

The education I received at Gallups didn't really help in my peace time job, but I did enjoy shipping and had many interesting experiences and met many interesting people.

Platoon R-66

ANDREW, HAROLD B. R-66

My father Harold B. Andrews died February 19, 1990. I am so happy that he was a member of GIRA, as he was very proud of his service during WWII. He worked for the Department of Defense until 1974 when he retired.

Currently, I am writing a memoir of him. If any of you remember my dear father, please contact me.

(Editors Note: Andrews' daughter, Maureena Andrews Renner, submitted this information on July 15, 1991. She is a resident of Westerville, OH.).

JORDAN, LEONARD R-66

In late 1990, the Russian embassy notified some maritime organization, that sailors who made the Murmansk/Archangel run in WWII were eligible to receive medals that had been especially struck for that event. The British and Australian sailors had received theirs in 1988.

During the course of the war 840 ships were involved, of which 390 sailed under the American flag. Of these, 97 were sunk (11 1/2%).

I was the "Sparks" on an ESSO tanker that was carrying a deck cargo of "duck" and 55,000 barrels of alcohol (for munitions manufacturing) to Archangel in the White Sea.

Most ships carried a crew of 52 plus 12 in the Armed Guard supplied by the U.S. Navy. We departed Philadelphia early November, 1944 and returned March 28, 1945. Because our cargo was quite explosive in nature, we were assigned to the coffin corner of the convoy. (Thus, if we were hit, only three other ships would be endanged.) We also shared this dubious honor on the return trip as the Russians couldn't take all of the alcohol, so we had 12,000 barrels to bring back!

When the necessary papers had been processed, we received a formal invitation to the Russian Embassy in Washington, D.C. on January 31. We could bring one guest and it was suggested that we dress in a business suit, or a blazer, (and tie) and wear whatever medals we had earned.

The number two man at the embassy gave us an earnest and congratulatory speech, then gave each of us the medal and a certificate (with our names in Russian) signed by Gorbachev. We then repaired to the main reception room—a very grand place indeed. Here we were served champagne by three maids—followed later by a gala luncheon at the nearby Hilton.

Only 51 were there and another 50 had their medals mailed to them. Of the 20,000 sailors in the Murmansk run, it seemed a small turnout, even though it (Murmansk Run) did take place 45 years ago.

Platoon R-70

PAFUMI, JIM, R-70

I had numerous battle experiences during the war, but an incident that stands out in memory occurred after the war was over.

I was on board the *SS Dusquesne*, a T-2 tanker under Captain A.N. Sole. Japan had surrendered on August 14, 1945. We were in Buckner Bay in Okinawa, my thoughts were of home and the future.

In that month a typhoon hit Buckner Bay and wrought terrible destruction on the ships in the Bay, as well as installations ashore. All hands were called through the night and we pulled many sailors out of the drink. Captain Sole did a great job and we came through the storm without any damage.

We finally departed for the states about a month later. I began receiving reports of mines floating all over the Pacific. The storm had apparently broken them loose from their moorings. Captain Sole plotted the reports on his charts, alerting the crew to keep a sharp lookouot.

I was leaning on the starboard railing on a beautiful day. The sea was calm, the war was over, we were homeward bound. Suddenly I saw a contact mine two points off the starboard bow, and I shouted a warning to Captain Sole who was on the bridge. He took evasive action immediately, ordering hard left rudder, and at the critical moment, hard right to prevent the stern from striking the mine. Captain Sole saved the ship from sustaining serious damage. He was one of the BEST.

The mine slipped astern and it came within a couple of feet of blowing a hole in the ship. The captain then had me send a message on 500 KC, warning everyone of the presence of the mine in that position. I guess it didn't have our number on it.

Platoon R-71

GERCKEN, CHAUNCY GENE, R-71

I joined the Merchant Marine in 1943, and went to Sheepshead for basic training, and got into Gallups Island in the spring of 1944 and graduated in October or November of 1944.

We shipped out immediately and sailed for 18 months pretty consistently until the war was over in the summer of 1945.

I went to school and then in 1950, when the Korean War started, I was just a couple of months too young to be exempt from the draft. So, I got drafted in the first draft of the Korean War and as one of the first

Platoon R-70

replacements to go into Korea, and did tour of duty with the 2nd Infantry Division.

I came back from there. I didn't go back to school. I was married and had a family, so I went to work and spent my life working , until I retired, in the printing business in Kansas City.

I've been to two reunions: The one down here in St. Louis and in 1978, the Gallups Island Reunion.

KING, JAMES GORDON, R-71

I was somewhat dismayed to find that only three R-71 class were listed in the roster. I didn't keep in touch with many—only with a couple, and now that I'm retired I reminisce about the Gallups Island days. Thinking back to the old R-71 days makes me proud that I was a member. During my stay at Gallups, platoon R-71 was obviously one of the best of the school. We reaped the dubious honor of being chosen to march on October 12, 1944 in the Boston Columbus day parade. I say dubious because when we marched through one section of Bean Town we were booed. I couldn't figure that one out. Must have been some of Michners' family!

My days on the island were filled with fear and trepidation that I would not pass the algebra so I didn't wait, but attended night classes in algebra during the whole time I was at school. It worked admirably. I tried to get into all sorts of trouble—I practiced radio at an early time in the schooling, as some of us stayed on the island to study one week-end and at 5:30 a.m. "Oh, What A Beautiful Morning" came bellowing over the loud speaker over my head. As I was cognizant with radio I disconnected one of the speaker wires and blissful silence wafted through the barracks. Shortly, the officer on duty appeared at my bunk and asked who disconnected the speaker. Knowing I was doomed by circumstantial evidence I confessed and was assigned duty detail for the next week.

My fond memory of R-71 occurred in the spring of '44 when the island organized a whaleboat race from Deer Island to Gallups—I would be remiss if I did not report that the winner was the R-71 boat! (Somewhere I have a picture of some of the boats being towed over to Deer Island for the race back.) While on KP I learned first hand how they made those miserable gelatin like eggs! Happier though, I learned that there was a shower room next to the mess hall and I discovered that it was empty while everyone was on the drill field for calisthenics and since I didn't eat breakfast I could spend a leisurely hour having a shower and shave. It worked good until three weeks before the end of school. Yep, mast and a week of work detail.

I was very happy to have been able to play with the very fine dance band that was organized by Roach. I can't remember his first name but he was from Pittsburgh and played bassoon in the Pittsburgh symphony. I remember the dance we had at a hotel in Boston and it was a wild party. One of our saxophone players was such a nice kid that everyone that came by the band wanted to buy him a drink and by the time we played the first tune he was in the john with his head hanging over a bowl with a very amorous grip around the porcelain.

My last brush with the powers-that-be on the island occurred a day before we were supposed to ship out. For eight months I had stretched out on top of my bunk for a catnap before class, only this day Chief came down and chewed me out and I didn't have my glasses on so I sassed back and here came the Navy Lieutenant who was in charge of discipline. He informed me that I wold be lucky to get out of there in two weeks instead of the next day as I had impolitely informed the Chief! He was right!

I'll always be indebted to R-71 member Paul Gatts for his efforts to keep me out of trouble. He invited me out to visit friends in Everitt, MA which got me involved in attending church with them. The hospitality shown by his friends and the church members left a lasting sense of indebtedness and gratitude.

The last ship I sailed was in 22 different countries in 11 months, T-2 Tanker!

This is all for now. . .

Oh, yes. I finally got my amateur license in January . . . KD4vtjA.

SMITH, JOSEPH R., R-71

It was December 7, 1942. I was in the "Big Apple" on business for my employer, Ingersoll Rand. I was walking down Broadway when I noticed a building with posters in the window for signing up in the United States Maritime Service. I went in and that afternoon I enlisted in the USMS.

I was very excited since I had trying to join the Army, Navy, Marines and Air Force. I had been turned down since I was sole support for my widowed mother.

I was put on inactive—pending recall—and returned to my home in Painted Post, NY.

It was not long before I had my orders. It was December 24, 1942 when I reported to New York and Hoffman Island for training. When I had completed my training at Hoffman, I was sent to Gallups Island for radio school.

It was then on to New York to the RMO office where I received orders to report to Norfolk. At Norfolk I was assigned to the SS Charles C. Jones, loaded with 500 pound bombs. After several days in Norfolk, we sailed for Bari, Italy with our cargo.

We had difficulty getting into the Bari harbor since the Germans had bombed the harbor previous to our arrival. We finally got off loaded. Then some of us had shore leave.

We found a hotel called the Mirmar. The hotel sold only one drink called a "special"—and it was! It tasted like gasoline.

When we returned to the ship, a British "Sparker" with a bottle of rum came on board to my quarters. Our Armed Guard, Ensign J. McCarty and our chief engineer joined us and helped us dispose of the rum.

Upon leaving Bari we returned to Newport News, VA where we took on another load of bombs for Marseille, France.

On this trip we encountered very rough seas. We had a tank on each hatch cover and each time we rolled, the tanks would go under the water and back again. Sometimes, as we rolled, the bridge rolled and touched the water. When I came off watch, my port was open and my bunk was soaked.

While we were at the pier in Marseille, an M.P. came aboard and tried to sell me a German 38 pistol. While we were dickering over the price, the damn thing went off. It just missed hitting me. Hearing this, the captain called the Armed Guard. All that I could see were red lights. The purser came and gave me an aspirin and I didn't buy the gun after all.

From France, we went to Charleston, SC and picked up another load. We went to Colon to go through the Panama Canal. However, we were held up and could not go through the locks as a flat top was having trouble welding on fenders.

The next morning we finally started to go through the locks, but had a delay as one of the mules had a problem. On top of that, the *Jones* had engine trouble. We finally got through and headed out into the Pacific.

Once in the Pacific, we stopped at various islands, many of which I didn't know.

Eventually, we stopped at Saipan. I went ashore and found a Japanese POW who had an outside barber shop. He would cut your hair but was not to be paid. He cut my hair and then started to shave me. He had a straight razor—when I looked up and saw his almond eyes, I handed him two bucks and was out of there in a flash.

One evening, my friend the pilot, our Armed Guard and I spent most of the night at the Officer's Club. When we left, we took the wrong jeep, drove it down a narrow road, missed a curve, overturned and ended up on some coral beach. Our Armed Guard received some injuries. I woke up in the 369th hospital in a big white room and it was then I learned my left eyelid had been cut and was just hanging on.

We later learned the jeep belonged to a Lt. Colonel. There was an investigation but nothing happened.

We were hospitalized about a week and then returned to the *Jones*. Shortly after this, we started for Seattle. Arriving there, we received our pay and were discharged.

I returned to Painted Post and my job, retiring in 1979. My wife of 43 years and I are still living in Painted Post. I have my Ham license, KA2VOX, and am really enjoying it.

All in all, it was a great experience on the *Jones*, but I'm not sure I'd relish doing it over again.

Platoon R-72

ANDERSON, WILLIAM A., Jr, R-72

In September of 1943 I left the old

farm in Kensington, NH to join the Merchant Marine. A friend of mine had told me that they needed radio operators because the submarines were sinking so many ships. I was going to try for radio training in the navy but when I heard about the Merchant Marine I thought it sounded more like what I was looking for.

My basic training began at Sheepshead Bay, New York from October, 1943 to March, 1944. I started engine training while I was waiting for radio school. I completed engine training and was scheduled for a training ship the following Sunday. That Friday afternoon I found a tag on my locker that I had been accepted for radio school. I worked two weeks in the power house at Sheepshead Bay and after that I was off for radio school on Gallups Island in Boston harbor. I arrived on Gallups Island in March, 1944 with the class of R-72. Completed radio training September, 1944.

After graduation from Gallups Island Radio School I went to Philadelphia for ten days waiting to sign aboard the Liberty ship *SS John M. Harlan* (Moore McCormack line). We sailed from Philadelphia, in a split convoy of 102 ships, to the Rock of Gibraltar where the convoy split up.

My ship *SS John M. Harlan* continued on to Oran and then on to Khoramshahr where we unloaded a cargo of war supplies for the Russians. We moved out into midstream and on loaded a cargo of barley for Catania and Palermo.

We went empty to Beni Saf, North Africa to pick up a cargo of ore for Baltimore.

Second trip was from Norfolk, through the Panama Canal to Pearl Harbor with a load of supplies for the naval yard. Then we headed back to Washington state and Oregon to pick up a load of lumber for Manila. While in the Manila area we carried 1,500 army troops.

We were ready to return to the states but were held up for two weeks because we were going to be part of the invasion of Japan. It was at the end of that two-week period that the "A"-BOMB was dropped on Japan and thankfully that ended the war.

With the ending of the war we headed back to San Francisco. I waited four days for a ride by train to Boston. I had to sit up four nights and five days to get home, but who cared, at least we were able to go home.

After thirty days leave I signed on the *SS Berkshire*, a collier for Sprague Coal Company out of Boston. I sailed on the *Berkshire* for one year and ended my sailing days in November of 1946. Getting home I worked for a building contractor for a few years and then went out on my own and started my own building business. I worked at my own building business for thirty years and retired at age sixty-two.

Thirty seven years passed before I renewed my acquaintance with radio. I earned my ham ticket in 1982 and went on the air with the Gallups Island Radio Group.

In July, 1984 a suggestion was made to form a Gallups Island alumnae associaton. Pro tem officers were David Bulkley, President and Ralph Albers, Secretary and Treasurer. The name Gallups Island Radio Association was suggested by Ralph Albers and all agreed. In 1986 Ed Hayden became pro tem Vice President, Dave Bulkley, Ralph Albers and Ed Hayden were the first editors of the two GIRA news letters, the "GALLUPS ISLANDER" and "GALLUPS NEWS QUARTERLY." In January of 1990 a thriving GIRA was turned over (approx. 611 members) to the newly elected President, James Kinkle and his officers.

GUNTER, URBAN, A (BUD), R-72

I entered the Maritime Service and went to Sheepshead Bay for basic training. I had background in electronics, or radio as it was called in those days, having worked for the Bendix Corporation and already had a great deal of theory and code when I went through Sheepshead Bay.

I was selected for Gallups Island and went up there and got in in September, 1944. Sailed out of New York on tankers to England and Europe and came back, got a tanker and went out to the South Pacific for eight months. Then back to the states, then on a Liberty for Europe and spent about one year on that Liberty, the *Wallace M. Tyler*.

MOFFITT, LEONARD E., R-72

Enlisted in USMS January, 1944. Sent to Sheepshead Bay Training Station for Engine Room Training. Saw notice of Radio Candidates and applied for training as I was fair at Morse Code from scouting experience.

After night school at Sheepshead Bay was sent to Gallups Island and assigned to R-72. First class of speeded-up education. Graduated in December, 1944. Went to the Philadelphia WSA and was put on ship same day. We sailed that night on *SS George Weems*, a Liberty ship (Moore McCormick Lines).

I was only radio man aboard as they were holding ship for a radio man. First trip—never on a ship before—and alone! Did not even know where battery room was or how to get to the engine room to check on the generators.

Convoy to Cherburg and return—made three trips on the *Weems* and two trips on the tanker *America Sun*. *"Sun"* all to South America (Ship Doxford Diesel). Right up and down torpedo alley along U.S. East Coast.

Education: High school, two years college and Gallups Island.

Had a family, two kids, house and mortgage. When war was over left the sea for good. Next 30 years in construction machinery business: Cranes, shovels, backhoes, etc. Retired in '82 to Ocean City, NJ. Now I golf and fish.

Platoon R-73

CHEATHAM, BOB R-73

Shortly before I was to join the Navy's V-5 program, my brother-in-law talked me into joining the Maritime Service. He was a deck officer on a merchant ship that had just been torpedoed and sunk off the coast of Sydney and was on leave in Los Angeles. He knew I was strongly interested in radio and electronics so he strongly recommended trying to get into the training program at Gallups Island. I signed up for the basic seaman's training program on Catalina Island with only the assurance that I would be given an opportunity to take an exam for the USMSRTS at the end of the basic training period.... Following the exam six of us were selected. We were given railroad and meal tickets to cross the U.S. to Boston and on to Gallups Island.

Following graduation we had a choice of ports of embarkation. I chose San Francisco and again was given a railroad ticket along with a ship assignment.... My ship was the *SS Dunham Wright*. Being so young, daily experiences were new and exciting. Of course, the radio room was familiar because of radio school, but the ship seemed so very big—endless rooms, offices, quarters and hold facilities. But it didn't really come into focus until some years later that we were mostly a bunch of young kids running this vessel and being responsible for its operation. The captain was 28, the first mate was 24, the second mate, 23, the purser, 21 and I was just 20.

We got our orders, headed out under the Golden Gate bridge, loaded with invasion equipment for Okinawa. We were part of a small convoy, which I think split up somewhere near Eniwetok Atoll, where we anchored inside for what we thought would be four or five days. The stay extended to 110 days. Either our orders were misplaced or this was a planned delay because so many ships with so much invasion equipment were being sent toward Okinawa.

We were also denied the request to go ashore for recreation. Nevertheless, we became very innovative and because of the many activities we created, there was not one fight among the crew or the Navy gun personnel during that long stay. We prayed a lot too and that was crucial. The captain allowed us to drop a motorized life boat into the water each day to explore the nearby uninhabited islands making up part of the atoll. Some of us also learned to dive from very high areas of the ship, including the railing and the bridge.

We left Eniwetok and headed for Okinawa; but instead docked at Guam for a couple of days, which gave the radio staff an opportunity to go ashore for several hours each day. Naive as we were we collected undetonated bazooka shells and other live ammunitions, even several grenades, from the brush outside the port city. We learned later that there were quite a number of Japanese soldiers still holding out in the area.

During this trip which lasted about seven months, our ship was not subject to direct attack by aircraft or submarine. We did receive many SSSS and RRRR (or was it AAAA?) warnings, but other than a few bogees, nothing of consequence happened. Our Navy gun crew even earned a citation for shooting down a target drone during one of the alert drills.

The war was over and in 1946 I

enrolled in the University of California at Santa Barbara. I majored in industrial arts and obtained a BA degree and a teaching credential for drafting and electronics. I was very appreciated of my Maritime service for it directly led to my college experience. Furthermore it provided, as did all the services, an important period in which youngsters grew up. I wouldn't be a bad idea for every kid today to have some kind of service experience.

With the exception of a three year period as principal of a small Beverly Hill private school, I worked for the Los Angeles Unified School District during my entire professional career as a teacher and administrator. My last assignment was Director of the Information Technology branch.

I've held a general amateur radio operator's license (WTTMM) since 1950, but have been active only for a few short periods. I thought that I would have more time to devote to the airwaves when I retired by my wife had found new avenues of responsibility which keep me occupied. Some day. . . .

HICKS, ROSS, R-73

Arriving in New York City, after a final farewell to Gallups Island, I found myself in lower Manhattan where shipping selections were made. A giant, ruddy-faced man pointed to me and said, " I want this one." He introduced himself as Captain H.V. Rasmussen of the Sinclair tanker *William Boyce Thompson*. After arrangements were made I picked up my sea bag and followed him out to hail a taxi. We boarded a launch out to the ship's anchorage not far from the Statue of Liberty. I was shown my quarters and given a paper showing meal hours. We were to depart that night for Little Creek, VA, where we would anchor, awaiting the formation of a southbound coastal convoy to the Florida Straits.

The old ship was by no means a thing of beauty, being built in the early 1920's. The staterooms were nothing in comparison to the outside; all were paneled and well cared for. The crew proved to be very congenial, the officers were long-time Sinclair men and apparently serious about their duties. The captain was a father image to all, his Norwegian manners gave me confidence—as he was most patient in dealings with me. How little did I know then how this would pay off later on.

Three days later we dropped anchor at Little Creek, which was primarily a Marine training base. The captain and I went to the Navy convoy conference. When eight to 10 ships were assembled to form the convoy, we departed, again in darkness.

Numerous "SOS SSSS" reports were logged—Gulf, Atlantic and Sun tankers being the main targets. Once, when we sighted a glare over the Eastern horizon, we could only imagine what some ship just like us was going through. Each dawn found the old *Wm. Boyce* trailing, the convoy being sighted on the horizon ahead. When they left us, proceeding toward Gulf ports, we were then alone, to sail on at around 8 knots to our planned destination, Panama. We knew too well that German subs could easily change our plans.

Friday, May 22, 1942 dawned the same as always. We proceeded southbound past Cuba into the Caribbean. Rasmussen came into the radio room and said, "Sparks, maybe you'd better make up a distress message, leave the position blank, just in case." That chilled me, in spite of the warm day, but I told him it would be done, hoping never to need it.

The day passed slowly as we headed for Panama. After ending an afternoon watch, I was outside the radio room when the sea erupted all around us with a deafening explosion. I was thrown into a stanchion and felt a sharp pain in my left knee. Confusion reigned among the crew, one of whom released a life raft located on the port bow. The ship stopped; we were beginning to reverse course. I found my way to the radio room, where I saw the H/F transmitter sitting at an angle but the main seemed to look normal. Our position was called to me by the third mate. The message I had prepared at the captain's suggestion proved to be a wise move as I put in the coordinates and went to work. Fortunately, the antenna was not damaged.

QRN (static) was high and it seemed eons later when NAW, Guantanamo Bay, Cuba advised receipt and they would send a plane out over our ship, only to cancel it as darkness came. The captain was going to the bridge when we were hit. He was thrown to the deck, bruising some ribs and an arm. I noticed that my left knee was swollen and turning blue. We were now enroute to Montego Bay, Jamaica, dodging reefs and any possible torpedoes remaining in the area. Before dawn the next day we anchored off shore there and it was a badly scared crew that strained to see the small boat coming toward us. Customs and a doctor came on board and the injuries were looked at and given treatment. For me, it was cold packs and aspirin!

Rasmussen went back with them for more treatment of his injuries and to contact the Navy as to where they wanted our ship. He returned with news that we were to go over to Gitmo (Guantanamo) for a temporary patch for the large hole in our starboard side. The tanks having been empty and steam cleaned back in New York saved us for another day. I went down into the damaged mid-ship tanks with some others and still have a fragment of the torpedo casing that I found there, along with the original fateful message and reply from NAW.

Still later on, we sailed into Galveston and Todd shipyard for drydock. If ever a sad farewell was bid, I did just that, leaving a great crew and a wonderful Norwegian captain. My next stop was San Francisco, (via a short visit to my Oklahoma home) to see how the war in the Pacific was going.

As a foot note, when I was R/O on the *Sinclair H-C* in 1945, who should come on as relief captain in Chester, PA but Rasmussen, who was to be with us for two or three trips to Puerto la Cruz, Venzuela and back to Chester. We were on a shuttle, bringing crude oil to the refinery. With no blackouts and a much newer tanker, it made me feel like I really was on "top of the heap." The captain took me with him on one trip to New York and I believe it was 500 Fifth Avenue, where I met Mr. Harry P. Sinclair at lunch. Yes, one could think, "Those were the days, my friend."

Platoon R-74

McPHEE, HAROLD, J., R-74

After graduating from radio school at Gallups Island as radio officer 2nd class, I was sent to the union in New York. I shipped out on tankers and Liberty ships to Liverpool and Southampton, England; Oran and Odessa and last run to Italy and France from Texas. I was discharged in New York in 1945.

I returned to my home in Canfield, OH where I joined my wife and eldest son. Our family eventually consisted of two sons and one daughter. Our youngest son died in 1976 at age 30.

My brother and I bought a Ford-Mercury dealership which we had for 30 years in Canfield. I retired in 1978 at age 58. Since retiring, I play a lot of golf and work from April until November at a local sod farm. My wife and I spend February and March in Florida. We will be married 50 years this June. We still live in Canfield.

Platoon R-75

FRICK, RICHARD J. "DICK", R-75

Graduated Gallups Island in October, 1944, with fellow R-75 classmates, including Ken Watson, Martin Stebbins, Jack Simons and others. Frick, Stebbins and Simons then spent a week or so in a tiny hotel in the garment district of New York City while awaiting their first ship.

Frick and Simons then shipped out as 3rd and 2nd operators, respectively, on the S/S Bear Paw, a T-2 tanker, headed for the Pacific ocean.

In the mid-'50s, Frick worked for a radio condenser. He also was employed in industry, working for several years with Ray Maurstad. Frick currently resides in Quakertown, Pennsylvania.

STEBBINS, MARTY, R-75

We left Gallups Island on October 13, 1944. Jack, Richard and I were sent to New York. Jack and Richard shipped out on the same ship and I was sent on to Baltimore, MD and signed on the *Leo J. Duster*, a Liberty ship as second radio operator. The first Radio Operator was Bill Archambeau from Class R-43. (By the way, Bill and I had breakfast together at the Boston reunion.) We left Baltimore on November 11, 1944, stopped at Oran, Marseilles and on the way

Platoon R-73

Platoon R-75

Platoon R-76, Graduation Dinner.

home at Beni SAF, North Africa for ballast.

This stop I will never forget. The third cook and myself stayed ashore too long. When we came back to the pier the ship was moving out. Fortunately for us it dropped anchor in the harbor. The third cook negotiated a deal to have us rowed out to the ship. This was one time I can definitely say I was glad to be aboard. The captain requested my presence and I was given a fatherly lecture and made to feel like a dumb 18 year old, which I was.

Our return home to New York was during the Christmas holiday period. My most memorable part of the meals on this trip was the Thanksgiving and Christmas dinners. The Standard Fruit Company certainly had the cooks put on elaborate menu with roast beef, pork, turkey and all the trimmings. Fantastic!

After reaching New York they began to prepare the *Leo J. Duster* for a trip to Murmansk. Living at the time in Pennsauken, NJ across the river from Philadelphia, I returned home. I visited the FCC office in the Philadelphia Custom House and obtained my second class radio license. I then proceeded to the three big oil companies in the area; Atlantic, Gulf and Sun.

February, 1945 Sun Oil signed me on the *Sabine Sun* tanker for a Pacific run. The first operator was John Wiley, I believe of Class R-16, who was from Texas. Third operator was John Kelly, a Navy operator from Washington, D.C. The *Sabine Sun* took us to Venezuela, Hawaii, Manus Island, San Pedro, Suracao, Ulithi Atoll and home after V-J Day. I continued to sail with Sun Oil on the *MS Sun* and the *Southern Sun*—a trip to Le Havre, France, Antwerp, Belgium and several trips along the eastern seaboard.

January 1947, I left Sun Oil to join Western Electric Co. to install central office telephone equipment in the Philadelphia area. Of course, my Gallups Island training and sea experience provided me the qualifications for this employment. I spent 17 years in installation, 12 in engineering and five in the National Training Organization in Columbus, OH.

Retiring in 1984 after 34 years service at the age of 58. Spent my enjoyable seven years in Daytona Beach, FL and after a departure of 24 years returned to New Jersey, my home state, just twenty miles from my birth place.

This is the epitome of my 48 years since Gallups Island, which now, looking back on this period, seems as short as this tale.

WATSON, KENNETH, R-75

Our group graduated October 18, 1944—there were 50 of us; two or three had dropped out before finishing the course.

I sailed on four or five Liberty ships and one T-2 tanker until November, 1949—saw a lot of the world. Then I went to work for Kraft Food Company at their big manufacturing plant in Garland, TX in February, 1950.

I was drafted into the Army in September, 1950 when the Korean War broke out. (My Merchant Marine time didn't count as military service.) Served 33 months in the Army and ended up in Japan and Korea.

Am a lifetime member of the American Legion and Veterans of Foreign Wars. Worked for Kraft over 20 years; over 10 years for U.S. Steel Corp (now USX) and presently work in a machine shop for Automatic Products Corp. in Garland. Am retiring and plan to start enjoying my retirement by doing some traveling around while I am able to do it and enjoy it as well.

Platoon R-77

MIONE, PHILIP, R-77

On April 17th of 1989, I received from the United States Coast Guard an armed forces discharge for honorable service during World War II from December 7, 1941 to August 15, 1945. It is signed by F.J. Grady, captain, U.S. Coast Guard, by direction of the commandant.

For the same service, I have another certificate issued by the War Shipping Administration, stating that I served in the U.S. Merchant Marine from April 6, 1944, to August 30, 1946. It was issued for continuous and substantial service at sea and was signed by Granville Conway, administrator, War Shipping Administration, and dated August 30, 1946.

I personally was content sometime in 1946 to put the matter of not being designated a veteran to rest as I began to pursue my peacetime interests. I attended college for four years, leaning heavily on my merchant marine skills to earn enough money for my tuition and other expenses. I sailed each summer for one or another of the various shipping companies as an FCC licensed wireless operator, substituting for a vacationing regular on each occasion.

When I graduated from N.Y.U. in the class of June, 1950, I was soon surprised to learn of my new 1-A draft status, and was subsequently drafted into the Army of the U.S. on September 8,1950. Our troops were already fighting in Korea.

I spent my first six weeks at Fort Dix with a rather odd assortment of young people. At least 20 percent of the inductees in my barracks were former merchant seamen, all unmarried and under 26 years of age. Some were refugees from Europe.

It took all of my six weeks of rifle training to unlearn seagoing nomenclature for the new G.I. language. And after eight or nine more weeks of engineer training at Fort Belvoir, VA, I was returned to Fort Dix, NJ, to join a combat engineer battalion, made up of mostly active reservists who were WWII veterans with combat experience.

Surprisingly, we were not sent to Korea as I had expected, but we did go overseas to West Germany as part of the Seventh Army. Eventually, we were deployed along the Czechoslovakian and Austrian borders, ready to perform demolition duties if the Russians should attack. Essentially that was our mission.

In August of 1952 after almost 20 months of overseas duty, I was separated from the Army to be placed in a reserve, and was to receive my honorable discharge about four years later in November, 1956.

Notwithstanding all of bureaucratic nonsense I have had to put up with over the years, including my 25 years of teaching in the New York City public school system, many fond memories persist in my mind, especially my years at sea, for which I would never exchange for any paper documents. But thanks anyway, commandant.

(Editor's Note: Philip Mione originally wrote this article for the "New England Senior Citizen Journal" in 1989. Due to space restriction it is not printed in its entirety.)

Platoon R-78

HILTON, GEORGE (JOE), R-78

In 1944, I had been a teller in a Brooklyn bank for 11 years trying my best for a transfer to a sit-down job that would relieve the pain in my knees. I was married with a 2-year old, but the draft was coming closer—first taking married men, then married with children—and I knew if it caught me I would be marching with the infantry. So into the Merchant Marine I went. Sheepshead Bay was an experience—in five minutes after walking through the gates I was reduced to a cypher—what the hell had I done? After weeks of painting anything that didn't move, it was on to Gallups Island. Not that I knew anything at all about radio, I just wanted to learn something new. By that time training had been cut to five months and I never saw the inside of a radio. I never studied so hard in my life, and wound up with a 2nd class license and signed on the Liberty ship *Judah P. Benjamin* as 2nd operator.

When I was a kid and my mother took me shopping downtown every week, I'd get sick on the trolley. So I knew I'd spend my seagoing career hanging over the side. Sure enough, the night we sailed, while still tied up at the dock on Staten Island, it happened. But in the next two years, with two round the world trips, I didn't even feel queasy.

We had some close calls on the *Judah P.*, not necessarily because of enemy action. The skipper was an 80-year-old graybeard who subsequently acquired the title "Lighthouse Eddie." He earned it by identifying a light in the middle of the Pacific as "probably a lighthouse." On another occasion, ignoring the exhortations of the mates, he insisted on maintaining a course which took us at night into the middle of a coral atoll. The mate's call of "breakers ahead!" was all that saved us and we spent the rest of the night thrashing around trying to stay off the reefs.

On one ship where I sailed as chief operator ("Don't worry about it—just double talk the captain." But he knew more about radio than I did!), I managed to be a bit of a hero to the crew when, instead of dismantling the emergency entertainment receiver antenna when reaching port, I rigged a permanent antenna around the boat deck. And again, when the entertainment receiver went on the blink (how did I know what was wrong?) I bought a complete set of 23 tubes at the next Navy port and eureka! it worked. How did we manage to win the war?

Although the Liberties had a bad name, with stories told of breaking in half, etc., you can't prove it to me. In spite of the pitching and rolling and praying that when it did roll over (as you knew it would) on the starboard side, you'd be on the port, it was a sturdy, reliable ship.

I'm 75 now, never went back to sea, although I still miss it; retired ten years ago as executive vice president of the same bank I started in—really a very daring person. Have been doing volunteer work with SCORE (Service Corps of Retired Executives), trying to advise and assist embryo small business people. Hope they know more about their business than I did about radio!

73's.

LARSON, LES, R-78

As for my war experiences, I spent all my time in the Pacific theater. My first ship was the *SS Nicaragua Victory*. Also sailed on the *SS Smoky Hill*, and the *SS Alice H. Rice*. We went to the Philippines on the Rice and ran aground on a reef just north of Subic Bay. We spent 23 days on that reef while the Army came out with LST's and removed enough of our cargo to re-float us. The bottom had been damaged, but we all stayed aboard and sailed her back to the states.

My last ship was a coast-wise tanker called the *SS Comet*. After getting off the *Comet*, I went back to civilian life, leaving the radio career for a life as an electrician. I worked at 20th Century Fox Film Corporation for a number of years as a maintenance electrician before leaving there and going to work for the City of Los Angeles and the Department of Water and Power. I worked as a lineman, line patrolman and line mechanic supervisor. I retired from the Department of Water and Power the first of July in 1982. Moved to the state of Washington in 1987 and am enjoying retirement.

BRIGLIA, SALVATORE, R-78

The year was 1944, the month was October or November, the cargo ship was the *SS Steel Mariner*, the company Isbrandtson Lines, NY, and we were in a 130-ship convoy on our way to the Isle of Wight, the English Channel and Antwerp Ghent, Belgium.

The cargo was various war materials including food supplies ("C" and "K" rations), alcoholic beverages labeled "embalming fluid," some small arms and ammunition. I was one of the two radio officers aboard. Note: To go from the radio shack to the bridge one had to go outside as one door was warped. In fact, one night the captain shouted at me, "Sparks, don't you know better than to smoke a cigar in the black-out?" "I am not smoking any cigar." It turned out to be the radium dial on my wrist-watch, which was quite bright after a time spent in the darkness.

The night of the incident was a moonlit night and I was out on deck getting some fresh air after standing a radio watch. I was alternately looking over the side, watching the glow of the barnacles. phosphorous as the ship sliced through the smooth sea and the silhouette of the ships in the next column on our port side. In the next moment—and I couldn't believe my eyes—a periscope appeared, a conning tower (so close that I could see the German cross on it) and part of the sub's deck.

I tried to shout but for some reason my mouth became so dry that it seemed to seal my lips for the moment. To this day, I don't know if it was me shouting, "SUB, SUB—PORT SIDE!" or if it was the others shouting that I heard. By the time the gun crew attained the proper angle to fire at the sub without endangering other ships close to us, the sub began to submerge, and slid back past our stern. An American D.E. (destroyer escort) and a Canadian corvette rushed to the subs' vicinity and started depth-charging. The usual debris such as clothing, some fuel oil, etc., rose to the surface and nobody was sure if we hit the sub or not.

The rumor was that the sub was of vintage design, out of torpedoes and the type that had to surface in order to re-charge its batteries. In an emergency situation such as this what safer place could there be than to surface between two columns of ships, thus reducing the danger of ships firing at them?

When we finally got to a dock in Antwerp, known as "Buzz-bomb Alley," another incident occurred. We were sitting around this huge, heavy table in the officers' mess, playing card to get our minds off the Buzz-bombs overhead. "Axis Sally" was broadcasting, "Antwerp will be a city without locks or locks without a city, or both." We got a kick out of "Axis Sally" and were enjoying the music, when a Buzz-bomb sputtered directly over the ship—or so I thought.

We knew that when they sputtered in that manner, it meant they were out of fuel, (the V-1) and it was just before they crash-dived. The complete silence that followed seemed very odd, to say the least. I looked around at all the expressionless faces surrounding the table, and thought to myself (in a second or less) "If I dive under the table, possibly avoiding some bomb fragments I'll probably look silly, embarrassed, and frightened, but then again—if I don't dive under the table, thereby showing my inexperience, I may be lucky and not be hit."

I decided to dive under the table and believe me I found myself, at very close quarters, staring into a lot of silly faces. Everybody was probably thinking along the same track that I was. . . later, of course, nobody was frightened, just cautious—oh yeah!

Platoon R-79

BERCOS, GEORGE R-79

After leaving Gallups Island, I went to sea for a couple of years and qualified for the Russian Medal with trips Murmansk and Archangel

I got a job with Allied Radio as a trouble shooter, repair expert. During the job interview I was asked if I had a license. I did, so got the job. After some time, Allied wanted to know what my call sign was. I explained that I had a commercial license. They told me to get an amateur license so I could properly deal with their customers. It was that or find a new job. I got a ham license and was allowed to use all the new Halicrafter gear as it was introduced. After 20 years, Allied was bought out by Tandy Corporation and turned into Radio Shacks. I went job hunting.

Next job was with Universal Oil Products doing various tasks with the process instruments group. I found myself in Boreno working with liquid gas. The local sultan has gas gushing out of the ground. They installed a pipe line out to deep water to be able to load ships. This some three miles of pipeline had to be refrigerated to keep the gas in the liquid state. *(And I thought I had a touchy job working in Nuclear Power plants.)*

I did fairly well for myself. I put two children through college +. A son is a lawyer and a daughter is a CPA.

(Editor's Note: This excerpt was written from a telephone interview with Bercos.)

CANTRELL, NED P. R-79

I joined the maritime service in Detroit and did my boot camp in Sheepshead Bay. At Sheepshead Bay, I made application for radio school and was fortunate enough to be chosen for that. And after completion of training there I went to the Pacific area and sailed in the Pacific.

Had the privilege of serving as the radio operator on the first victory ship ever built—the *SS United Victory*—and we carried ammunition into the battle area on a couple of trips. Then I had to make one peacetime voyage, for the experience of it, in 1946, after having a short leave of absence at home. When I got to Shanghai in 1946, I was in port there and I received a message that our first son had been born. That was the last convoy I made.

When I got back to the states I went back to Illinois and I've been there ever since.

CLARK, HENRY R-79

I got to Gallups because I'd been in the U.S. Army Infantry in a training unit and they found they couldn't use me in officers training school, which is why I went into the Army. They sent me back home and decided they wanted me back, but I decided I didn't want to go back in the Army. I decided I'd go in the Merchant Marine as a radio officer.

Since I'd already been beckoned from Uncle Sam, they said "You can't join the Merchant Marine. You've got to go back to the Army, " and I wrote my congressman; he passed the letter on to Commander Land of the U.S. Merchant Marine. Along comes a few days and I received this long telegram saying "I am directing that you report to Maritime Service—as soon as possible—at your convenience."

I went over there and they were ready for me, except they looked at me and said, "You're wearing glasses. You'll have to sign a waiver." I said, '"All right—does it keep me from going anywhere?" They said, "No, you can't be a captain or an engineer." I said, "I don't want to be an engineer, I want to be a radioman." They said, "Well, you're aboard."

They shipped me out in three days. They were shipping everybody out—two in a bunk on the pullman cars—out of St. Louis to Sheepshead Bay. I ate in the diner. When I got to New York the station wagon met me and the honeymoon was over. From there I went to Sheepshead Bay . . . through the grinder. . . and from there on to Gallups Island.

I have more to the story. At Sheepshead Bay they had a little duty called Work Week and every now and then they'd say, "Everybody here on eye-waiver raise your hand." Some of these people raised their hands and they disappeared and you never saw them again. Turned out they were in cook and baker school. I didn't want that so I just wouldn't raise my hand.

Then came up Work Week; I was in dress uniform, I was captain's messenger. Had the run of the place—could go anywhere—even in the file room and read my file, found my waiver, tore it up and flushed it down the toilet and never heard anymore about cook and baker school.

Went on to Gallups; had a little fun there. I could do 13 words per minute. When I got there I couldn't break the barrier . . . having a helluva time on code . . . I got on the telephone, called the FCC down in Washington and asked could we test there. "Yes . . . We'll let you sit for it." I went to Clark and told him and he said: "Pay your own way."

We went to Washington, missed it (the test), and had to go to Philadelphia. All four of us passed.

Got passed on to a ship as 3rd operator. "When you get to Manila, you can take the test." . . . Out there for a year on an ammo ship. . . New Guinea . . . By that time they dropped the bomb. Spent 140 days at anchor in Batangas Harbor—sitting there while they decided what to do with the ammo.

The captain would run up signal flags showing the number of days we'd been there. They started sending these ducks out, started unloading this ammo—these 1,000 lb. bombs, and they found these parachute flairs. Found they could cut the parachutes off and sell them for about 50 to 60 pesos. Some guy pulled it too far one morning and the parachute flair went off with flames as high as a king pin. The alarm went off, people running around. . . After three hours we got the fire out. After 140 days we got out of there and took the great circle route out of the Philippines.

DOYLE, PAUL J., R-79

Graduating from maritime basic training at the Sheepshead Bay station in Brooklyn, NY, I was assigned as ship's company to Gallups Island. Eighteen years of age—wondered what it was all about as took first boat ride to the Island aboard the *SS Roderick* . . . on a cold, grey, choppy day. Poor PJ!

I believe I was the first person to go from ships company into the radio-cadet program. Working aboard the Roderick for a month I was admittedly envious of the cadets as they marched down the Gallups Island pier, resplendent in their blue uniforms and officer-type hats. "That," I thought, "is what I want to be." Took written and oral exams (details not recalled) and I was aboard!

In a fit of an 18-year-old's fantasy I bragged to our platoon leader, Ned Cantrell, of how my Dad was the Chief of Police in Salem. Could not realize that Ned and his wife (also from Xenia, OH, I believe) would walk into the Salem Station and ask to see "Chief Doyle" "Chief WHO???" came the laughing reply. But later Ned was most gracious about it all. . . just smiling the whole episode off. Fears of being publicly drummed out of Platoon R-79 never materialized.

I recall the efficiency with which they taught us on the Island how to use the RDF. With the ear phones on you turned the ring-barrel contrivance until you picked up the two strongest radio signals coming from shore. Where those drawn lines intersected . . . why that's where your ship was located at sea . . . right?

Aboard one ship the CRO (Chief Radio Officer) warned me that the captain evaluated the new radio men by their accuracy of the use of the RDF. Ah ha!. . . I practiced faithfully for days in the chart room, always when the captain was elsewhere. Then it happened. . . the "old man" so casually asked me one day (in his pronounced Southern drawl) if I wouldn't give him an RDF bearing. "Why, of course, Sir," came my cadet-like reply, "I'd be glad to." I called out the first degree of bearing. Drawing a line on the chart before him the good captain said: "That looks good, Sparks. Now let's have the other bearing when you get it." Rapidly rotating the RDF I confidently and quickly called out the second bearing. The captain drew another line. Silence. He stepped back; scratched the back of his head; looked at the chart again, and said (also in a pronounced Southern drawl): "My, my, Sparks. I've never had this ship in downtown Hoboken before."

Short version: Took test to be Purser at the Coast Guard quarters in NYC. Passed. Obtained shipping assignment aboard the *SS Francis B. Ogden* and away we sailed. Radio operator not too friendly. So did not associate with him nor even tell him that I was also a "Sparks." There was just no occasion to do so. On the return trip I met him halfway down (up?) the ladder (a ship's stairway). He was unusually glum. Most depressed. "What's the matter, Sparks?" I inquired.

Our glum chum responded, essentially, that for hours he had not been able to get the ship's radio transmitter working; that he'd tried everything he could think of—nothing worked—and it was critical that it

Platoon R-77

did work because the captain's wife was to have a baby at anytime soon, and while he could receive that message far at sea (and, mind you, after the war open communications were again allowed) he would somehow have to tell the captain that he, the captain, could not send any congratulatory responses because the transmitter was inoperative. Befuddled, bewildered and beside himself, Sparks was going below to the Officer's Mess (restaurant) to have some coffee, settle down and rethink the whole process. Curious, I went to the radio shack (room) and, sure enough, the transmitter was totally inoperative. Nothing. "What," I thought, "would they have told us to do at Gallups Island in a situation like this?" Admittedly, I was temporarily stymied. What? What? Light bulb goes on! "Check the fuses first," said the voice of a distant, unseen Island instructor.

Upon opening the fuse box it was glaringly evident that one fuse had blown as evidenced by its blackened color. Replace-yes. Fire up transmitter-yes. Upon his return, one astonished—but very happy—radio operator asked, "How in the hell did you get it working?"

That trip aboard the *Lee* was my baptism under fire. First, at night off the shallow Grand Banks my ship, although fully loaded, was tossed from one side to the other, always shuttering, at its furthest pitch, to regain its equilibrium. Time after time I rose from the mattress I had jammed against the bulkhead (wall) to put toiletries and medicines back into the medicine cabinet and to rehang clothes in the closet—to no avail. The earthquake motion of the oceans would not be denied. Not denied to the extent that the davits were truly bent and twisted by the ferocity of the seas; and both lifeboats, on the upper boat-deck on the starboard side were swept off into a broiling black ocean. Also, that next day, calm, clear, serene, I could see that all of the gliders and their wooden crating had been figuratively pounded into splinters. Only ghostly 2' x 4' uprights marked the areas the congested crates had occupied. Just two of the gliders remained intact and they were immediately behind the main housing of the ship.

It was the trip in which we later lost the whole convoy . . . some seventy ships . . . in the middle of the night. Exposed, vulnerable, and extremely explosive we sailed on toward our destination—toward the battlefields of Europe. "What happens now," I asked our good skipper. He replied that while the convoy would zig zag (to avoid torpedo hits), we would sail a straight course; this, because we knew, from the voyage plans, approximately where the convoy should be at certain hours of the day . . . zig-zag and all.

By mid-afternoon we could see the convoy, in mid-Atlantic, sailing as fast as its slowest ship would go. Out came a Canadian corvette (like a small destroyer) to challenge us. It must have been proceeding at "flank" speed (all out speed) as it was awesome in its speed and maneuverability. (Later a lot of the guys said they were damn glad that ship was on our side.) Blinker-signal recognition (via lights) and we proceeded back to our "coffin corner" placement within the convoy. So-called "coffin-corner" because we were the last ship on the outer row of many rows of ships; thus, if we were hit we'd likely not take any other ships down to a watery grave with us; as we surely might if we were in or near the center of the convoy. Ammunition ships exploding, incidentally, give the appearance of an atom-bomb erupting.

But when we were in the northern latitudes of the North Atlantic in the middle of winter, with salt water spray literally freezing upon all ships and the seas (waves) so massive and powerful I actually began to think more kindly of my first-ship assigner, William Barr of Boston. That is, if the Liberty ship the *SS William G. Lee*, was hit I'd never know the difference. No damnable efforts of trying to abandon ship under these conditions. We were likely to freeze solid in a lifeboat or surrender to the icy waves. However, one hit on our ship and none of us would ever know the difference.

Arriving without any enemy encounter by the time we got to England the convoy then split up. The *Lee*, with her intrepid Chief Radio Officer, Ralph S. Rovsek (Mr. Cool). and five other ships were to proceed up the English Channel to discharge their cargo in Hull, England (about half way up the Eastern Coast of England above London). Our captain, I am told, asked the Admiralty Office what kind of an escort we would have, being just off the coast of embattled armies and with the high-arching vapor trails of V-2 rockets almost constantly heading from German-held territory to England on our immediate left. "None," came the reply. "You'll just have to do the best you can."

Two abreast we headed north up the (English) channel, a little uptight, perhaps, and a little skittish. But we'd made it so far and our final destination was now less than one hundred miles distant. Evening. Chow time. All watches set. All was well until the loud, clanking gongs of a General-Quarters alarm blared throughout the ship. The mess hall was emptied in seconds; everyone scrambled to their pre-assigned emergency destinations, mine being on the starboard wing of the bridge, battle helmet and all. My assignment was to orally notify various parts of the ship of the captain's orders in case internal communications was tendered inoperative.

Striving to gain the reality of it all, I saw a huge, dark submarine emerging right off the starboard bow. My heart in my hand, I felt they were not going to waste a torpedo on us. Firing the submarine's deck guns would be just as effective and far less costly. I stared at the Navy Armed Guard below; these are the men assigned to fire the merchant ships guns and to try to defend the ship. They were trying gallantly. Still, it seemed they were in slow motion as they tried to rip the heavy canvas tarpaulin off the guns; all this as the large sub was now fully gaining the surface of the sea!

The sub's hatch flew open! Sailors scrambled onto its deck racing toward the after gun which all submarines carry. I remember trying desperately to cheer the Armed Guard on—to get them moving faster—as if it were some kind of a sporting event, rather than a possible life-and-death encounter. But my eyes could not deny that our side was going to lose!

Past the after gun the sub's crew went and, just as hurriedly, they hoisted a British Union Jack flag from the aft flagpole! A sigh of relief? Not quite yet. How could we know (for sure) this menacing ship, so close and so unexpected, was truly a friend? Secret signals were exchanged by our blinking lights. There was no doubt that she was a friend. She was having battery trouble and, learning of our destination (Hull, England) the sub's skipper indicated he would proceed with us, on the surface, into Hull. Which, of course, was a lot better than proceeding into Hell . . . as I had earlier thought was a near certainty.

Platoon R-79

JONES, JAMES M., R-80

I came to Gallups Island from St. Petersburg in 1944, I think it was in May. I graduated in November of '44 and went to New Orleans to serve aboard ship until the war was over. We were in the Pacific most of that time with not a whole lot of action, but we were in convoys that did see some action.

I got out in '45 when the war was over and went home to Birmingham, AL. From there I was transferred to Tennessee as a store manager for Goodyear Tires and Rubber Company. From Murfreesboro, TN, I went to Lebanon, TN, where I bought a dealership and was owner of a Goodyear store until about 1972 when I retired.

GUEST, PETER, R-84

My first ship after graduating in February or March of '45 was a T-2 tanker, the *SS Ponca City*. It was built in Sausalito where I picked it up direct from the shipyard. After a shakedown up and down the coast, then Hawaii we embarked for points west with a load of Aviation gas and a deck load of aircraft—P-51's and P-38's. I think we were supposed to land these in the Philippines. We off-loaded the piling in Batangas then went on to Manila. The whole area was a beehive of activity in preparation for the invasion of Japan. We were in Manila for a good long time—including August '45 when the two bombs were dropped and the truce happened.

We hastily reloaded and set sail for Japan where we lay on the hook for three days until the peace ceremony. We were anchored very close to the *Missouri* and had a ring side seat. The happenings of that day included a sky filled with aircraft from horizon to horizon. It has to be one of the most memorable events of my life.

The day after the ceremony we docked in Yokohama. I believe we were the first, but if not, then the second ship to anchor there. Things were interesting and chaotic. Army brass made our ship their headquarters for chow and showers. A couple of generals were there regularly and a Colonel Shelby and Captain Davis were chowing down regularly. (These two I later discovered from newsreels were in charge of the Japanese POWs.)

On the way home we experienced a very big storm which sunk several ships in the Okinawa area with substantial losses of life. I think a couple of destroyers with about 800 souls went down.

While in Japan we did much unauthorized exploration, visiting Tokyo before it was occupied and other unknown areas when we boarded the wrong train attempting to get back to Yokohama. Wherever we went that day we were definitely the first Americans to arrive at that spot. In retrospect, we were very foolish and lucky that the Japanese were such disciplined people. If circumstances were reversed and it was this country, I'm sure the chances of being bushwacked would have been excellent.

This trip concluded in San Francisco. Transportation back to the East Coast was all booked so I signed on the Liberty *SS Benjamin Fischer* for a trip to the bone yard in New Orleans with the intention of making my way home for Christmas. It worked out that we went to Mobile and picked up a load of coal for delivery to Savona, Italy. Since we had signed foreign articles and it was close to Christmas nobody volunteered to replace me, so off I went. We stopped in Oran, entered the harbor on Christmas Day, '45 and hung for a couple of weeks, then up the coast for iron ore/ballast and back to Baltimore.

In order to accumulate 32 months of continuous service so as to avoid the draft I shipped out, after a brief vacation, on a C-1, the *SS Cape Borda* which made a regular run stopping at Havana, Barranquila and Cartagena, Colombia and Panama.

In spite of that I was drafted into the Marine Corps in 1951 which is another whole story. I plead eloquently before my local draft board that a) I'd done my bit, and b) if they needed me so badly why not let me do what I was trained for? While I was at Paris Island I received an urgent letter stating that 24 ships were tied up in New York harbor for lack of a radio operator. I sent this to the head of the draft board but never heard from him.

HEIMBACK, ALBERT, R-84

I took my basic training down in St. Petersburg, FL. On the bulletin board was a sign-up sheet for anyone interested in radio school. I signed up, and to my amazement, I was one of the few that were chosen.

I got to Gallups Island in '44 sometime and graduated in early '45. I was only 17 at the time. I went into the Merchant Marine because my father wouldn't sign for me to go into the Navy. He said it was too dangerous.

I put in 30 years of state service, working 30 years with the state penitentiary, retired, and now live in Florida and in Pennsylvania.

ASHTON, LEON, R-85

Even though we have all gone our separate ways over the years there was still a camaraderie among those of us while at Gallups Island.

COLLINS, RAYMOND B., R-86

The ship was the *Montclair Victory* and the 2nd Operator was me. Imagine my surprise when I returned from ashore to be told by the chief operator that the new 3rd was aboard and that his name was Bradley and that he had been an instructor on Gallups. I knew immediately who he was since my class had graduated form Gallups about six months previously. I remembered him as quite aloof and of course the two stripes were very impressive to a then 16-year old.

We soon met in the Radio Room and although he didn't remember me from classes on Gallups it was evident we would get along OK and he didn't resent the fact that I, his former student was 2nd operator while he was 3rd. Actually, we got be real shipboard buddies and I must admit I relished the tips I was able to pass on. Since this was my third vessel since leaving Gallups I felt like an "old salt" and entitled to teach the new man.

As I remember Bradley, he was a born teacher. He posted a notice in the crew's mess offering to teach a course in Morse code and typing and actually had a couple of kids take him up on it.

I remember we went ashore in Southampton, England to a place called the Royal Oak. Had such a good time we tried to think of a fake message we could give to the skipper ordering us back to Southampton from Le Havre, France, but nothing came of it.

The period I just referred to was late summer/early fall, 1945 and I stayed with the *Montclair Victory* until the spring of 1946 when I went ashore for good.

Subsequently, I joined the Army and spent 32 months in Japan with the Army of Occupation as Communications Chief and High Speed Operator. Upon discharge, I joined the Air Force Reserve and was called to active duty during the Korean War. At the time of the recall I was a student at the Massachusetts Radio School in Boston trying to renew my licenses. One of the instructors there had been an instructor on Gallups, don't remember his name.

Adding the Coast Guard discharge to the U.S. Army and U.S. Air Force discharges, and the U.S. Maritime Service release from active duty sure means something, but I'll be damned if I know what. (Mis-spent youth maybe?)

DREHER, DAL, R-86

In the fall of 1943, a 16-year-old, lean, lanky, East Texas farm boy, standing over 6', weighing (soaking wet) not over 135 pounds, too young for military service, but large enough to elicit, "Boy, why ain't you in the service?" from every pot-bellied, yellow toothed traveling salesman that came through the town of Singleton, TX. Disgusted, too young for the military, and the ridicule of every 4f "ner-do-well" that came by.

Some strong badgering produced parental consent to join the Maritime Service. So I boarded the "Red-Headed-Dinkie," a three car diesel powered train with a red nose, for Houston. Two days later I was on my way to St. Petersburg and boot camp. In the fourth week of boot camp I took the exam for Radio Officer's School. Boarded the four masted *Joseph Conrad* for two weeks seamanship training. When the *Conrad* returned to St. Pete, I boarded a coal burning train for Boston, Gallups Island, and Class R-86.

Damn! it was cold from Texas to Florida. Then I saw the dirtiest snow banks in the world; having never seen snow I thought it should be white. Not so; five minutes after a snow-fall coal dust and cinders were everywhere.

After the FCC exam in Boston and graduation, I went to New Orleans, and assignment to the Liberty ship *Pio Pico*. After outfitting we proceeded to Biloxi, MS, took on a cargo of commissary stores and "Red Fox" ale (never heard of before or since). Next we loaded construction explosives in Mobile. Down to Panama through the canal. The authorities sealed the radio room and placed a Marine Guard on the door while in the Canal zone. We joined a convoy forming in the Gulf of Panama, then started the long eight-knot zig-zag to the Pacific, but not for long, two days out the convoy commander directed us to leave the convoy and proceed solo to Palau. As we neared the Carolina Island group, there were four submarine contacts over a two day period. Ash cans were used on one contact with a resulting oil slick. During this time there was always debris: mattresses, life rafts and other flotsam. Our course took us within sight of Truk Island. This island had been by-passed, but still had a Japanese contingent on board, a scout plane came out to look us over but stayed out of range.

When we dropped anchor at Palau, an armed guard was posted on the anchor chain each night, the Japanese on the island would swim out at night, climb the chain and steal food, etc. On April 12, 1945 President Roosevelt died, BAMS advised that the ensign would be flown at half staff for 30 days.

From mid-April until July we island hopped to Yap, Ulithi, Ifalik, back to Talaud, discharging small amounts of cargo at each port. In mid-July we arrived at Leyte Gulf and discharged our final commissary cargo and all of the explosives we had loaded in Mobile. This was our first shore leave since leaving Mobile on March 12th.

On August 1, 1945 we left Leyte Gulf bound "State-side." Five days later, August 6, 1945 the Enola Gay dropped the first atomic bomb on Hiroshima, Japan. This would bring down the curtain on hostilities for WWII. The S/S Pio Pico came about. When we got to San Francisco the papers still had photos of the wild parties on the streets VJ Day. We missed it all!

I returned to Texas and signed on the S/S Coquille, this T-2 ran between Baytown, Bayonne, Guantanamo Bay, Lake Charles, and one trip to Hull, England. When she returned from this trip she went into drydock at New Orleans.

My next ship was the *SS George W. Barnes*, a WWI Hog Islander, fitted for crude molasses. The WWI radio gear was still on board. An RMCA "Coke" machine had been assembled on the opposite side of the radio room, my quarters was a small alcove behind the "Coke" machine with a bunk, closet, and toilet facilities, you had to go on deck to the shower room. The *Barnes* was mothballed after my fourth trip hauling crude molasses for the Dominican Republic.

My next ship was the *USAT General William Wiegel*; I signed on as chief operator while she was being refitted in Brooklyn, NY. After the shakedown off the east coast it was back through the Canal zone, to the Pacific and Fort Mason in San Francisco Bay. For the next 18 months we ran cargo, troops, dependents and civil service employees to Honolulu, Midway, Manila, Inchon and Yokohama, and back state-side. After 18 months the *Wiegel* was decommissioned in San Francisco.

I hired on with the Flying Tigers in San Francisco, and went to Tokyo as a communication contract worker under contract to the U.S. Air Force. The contract was for a year and when it expired I signed on the *USAT Wall and Crown*, a reefer running out of Yokohama, to Inchon, Korea. We loaded perishables in Yokohama from state-side ships, then went to Inchon and the tidal Basin where the larger ships could not enter.

In 1949 I "swallowed the anchor," and returned state-side, went to work for Northwest Airlines as a ground radio operator, in Miles City, MT. After six months I was caught in an employee reduction; needing a job I went to work for the John Deere-Caterpillar dealer in the parts department.

Being a naval reservist during the Korean "Police Action," I served 18 months at the communications center in 11th naval district headquarters. After Korea I returned to the farm and construction equipment business. I have been in that business ever since. My wife Mabel and I have a farm machinery dealership in Roans, Prairie, TX, just five miles from where I was raised. We have nine children and 19 grandchildren.

It's been a good life.

NOWLAND, HAROLD C, R-86

I joined the Merchant Marine in 1944 and was sent to Sheepshead Bay where I briefly held the position of a math instructor for seamen wishing to apply for radio training at Gallups Island Radio Training School. Then one day I said to myself—"What the heck are you doing here?" So I signed up to go to Gallups Island.

I was an experienced radio amateur and had no problem getting through the course in record time. I served as a section leader of R-86. Upon graduating I was sent to New York City and I sailed for Wetinc. My first ship was the *SS Champions Hill.* I served as third radio officer. On all succeeding ships I served as chief radio officer.

I served on the following ships: *SS Elks Hill; SS Hall of Montezuma; SS Camp Charlotte,* recommissioned the *SS Stoney Creek*; and last ship was the *SS Albert G. Brown*, which I remained aboard until late 1950. All the ships were tankers. I served in the Mediterranean, North Atlantic, North Sea and Pacific Theaters of war. I spent VE Day in London, England and VJ Day in the Philippines loaded with aviation gas for aircraft carriers of the 7th Fleet. As you know the war ended at that point.

The companies I sailed for after the war were Bernuth Lempke Inc., Tidewater and American Trading and Production Corp. At the close of the war I opened up the amateur radio station W8ZXH as one of the first maritime mobile amateur radio station after the war and had a ball on the ten meter band.

During my time in the Merchant Marine I traveled to 38 different countries and logged thousands and thousands of miles. I enjoyed my time sailing in the finest position aboard ship—namely the radio operator. I would do it over again if I was given the opportunity.

After leaving the Merchant Marine service I returned home and went to work at McLouth Steel Corp. in Trenton, MI. Found an attractive young lady named Betty L. Wales and married her. Raised two lovely daughters, Debra and Dorene. We have been happily married for over 40 years. If I had it to do over again I would even marry the same woman.

While working for McLouth Steel, I started a mail order company in 1968, "Haltronix;" since incorporated and family owned. Please note the wife is the boss in the company as I made her the owner when I had to take a medical disability retirement from the mill at the age of 56 in 1980 due to a heart condition. In 1982 I had a quad bypass at the Cleveland clinic in Ohio. Went along fine for a good many years until November of 1990, when I had two more heart attacks and got that under control and since then I have been in good health. We bought a store front and moved the mail order business into it and I operate it six hours a day. How's that for bankers hours, second retirement. Got something to do every day and I love it.

Managed to get to the Boston, Minnesota and Long Beach reunions and I'm signed up for the New Orleans. If the good Lord is good to me I expect to see a few more of them.

JOHNSON, JAY L., R-87

On December 7, 1941, I was 13 and living in a small town in southeastern Montana. On that day I began dreaming of joining the Navy. The following year my step-father and brother both went into the service leaving me at home with my mother and a younger sister.

Shortly before my 17th birthday, my mother agreed to sign the enlistment consent for me to join the Navy. I was sent to Helena for my physical. It was determined that I was color blind and I was rejected. When I arrived home, the recruiter gave me another eye test which I passed and he said he would send me back again the following month. I told him not to bother—I had decided to join the Merchant Marine.

I was inducted into the Merchant Marine at Denver, CO on September 10, 1944 and sent to Catalina Island for basic training. On the day before I was to ship out as an Ordinary Seaman I was told I had passed the exam for Radio School and off I went to Gallups Island to become a part of platoon R-87.

After a few weeks in Seattle I was assigned to the *Wellesley Victory* which was still in the final stages of completion in Vancouver, WA. We were on our way down the Columbia River when we heard that the President had died. A sad beginning for my first voyage and the following week was even sadder. Upon crossing the bar at the mouth of the Columbia, I was stricken with seasickness and spent my first week with a paper bag close at hand and the rest of the time in my bunk.

Although the ship was brand new, we were experiencing one bad problem. The fresh water had been contaminated—the tanks had evidently not been flushed out so all our fresh water tasted like creosote. While showering, a blob of it (creosote) would hit you now and then. Even the coffee tasted like creosote.

We were carrying a cargo which consisted of about 50% ammunition and 50% canteen supplies and a deckload of belly tanks for B-29's. The crew was quite upset as the Army Cargo Security Officer would not allow us to buy any Coke or beer and yet he was known to have his own personal supply.

We arrived in Guam and were eventually off-loaded there. Arrived in San Francisco as the United Nations was meeting for the first time. Our Army Cargo Security Officer didn't wait for the ship to dock. He left the ship very quietly with the pilot as he had evidently heard some of the crew were talking about having a party for him in San Francisco.

I was sent to Norfolk and signed on the *United States Victory* which had been converted to a troop transport. We sailed for Le Havre the day after VJ day with a load of German POWs. After discharging them at Le Havre we went up to Bremerhaven for a load of troops and took them back to Boston. Spent several months doing this ferry service. Occasionally we would lose a POW or two enroute. Whether it was because they didn't want to return to Germany or whether someone else didn't want them to was never determined.

Those of you who were running in the North Atlantic that winter know how rough it was, especially for those of you in Liberties. Our only claim to real excitement was having a floating mine skim by us as we were picking up a channel pilot off Dover. A most beautiful sight was the fireworks along the coast of Belgium on New Years Eve.

We had little chance to get ashore as we were on a quick turn around schedule. From what little I did get ashore I was so disgusted with the filth of Le Havre and apparent lack of effort to clean up, impressed with the industriousness of the Germans and the efforts they were making to clean up all the debris from the bombings.

Signed off the *United States Victory* for the last time February 26, 1946 and went to Connecticut where I married Patricia Sellew on March 9th.

Went to work for a marine hardware manufacturer and about a year and a half later while working at the New York Motor Boat Show I became ill and found I had tuberculosis. Entered a sanatorium at Meriden, CT in March of 1948 and the following month my wife received a letter from my doctor stating that they did not expect me to last for more than a month. The next morning our first son was born just a bit prematurely. I was fortunate to be chosen as one to receive streptomycin in its experimental stage. It worked well but it was determined that I should still have surgery to collapse my right lung. This was accomplished by removing a number of sections of ribs from my back on the right side.

After 18 months in the sanatorium I was discharged and along with my wife and son returned to my parents' home in Montana for a period of recuperation.

Spent a couple of years learning watch and jewelry repair but found that sitting and leaning over a bench was too hard on my back, so I worked in the hotel business for about five years managing a couple of small hotels and then took a job as field officer manager for a Chicago construction firm. We were going to go to Libya for two years when things got hot over there and the Suez was blocked. That canceled that job so the company said they could use me in Little Rock. At that time Little Rock was in the middle of school integration and I didn't want to take my family into that, so we packed up and went back to Connecticut.

Retired after 21 years with the Department of Corrections at the age of 53.

On looking back, I consider being turned down by the Navy as one of the most fortunate things that happened to me as I was so proud to serve with the Merchant Marine and am proud to still be a part of such a great organization. As we've traveled, I've tried to locate some of my classmates from R-87, but have been unsuccessful. Would sure like to hear from them.

RILEY, CARL G. R-87

After graduating from radio school I was sent to San Francisco, where in April, 1945, I sailed as 3rd radio officer on a Liberty ship to the Philippines.

I was able to obtain verification from Norfolk, VA on a subsequent voyage to Italy (I was in Trieste when the war ended) and received an honorable discharge from the Coast Guard. However, the people at the National Archives in San Bruno, CA cannot verify my trip to the Philippines because they say all the records are in cardboard boxes filed by name of ship. And to save my life, I couldn't recall the ship's name. Neither the Coast Guard nor my U.S. Senator can help. Now my only hope for this information, which will complete my wartime record, is to see if anyone with the GIRA remembers the voyage and the name of the ship. Here is what I remember of my time aboard.

I signed on at San Francisco in April, 1945. Went down the coast to Port Hueneme (Long Beach) and took on a load of lumber. We also carried jeeps and trucks topside. Then we sailed to the tip of Samar, near Leyte. FDR died while we were midway across the Pacific.

Once, our Navy gun crew tried to blow up a mine with gunfire but missed. We broke radio silence to report its location. The 1st radio officer was an older man, about 40 at the time. The 2nd officer was my age, about 17. Our mess boy was a cartoonist and used to leave cartoons in the officers mess. One was pretty raunchy showing the 2nd officer in the lower bunk, me in the upper bunk. I was playing a tune on an Ocarina ("Sweet Potato"). That's all I should describe of that.

I remember the bos'n's mate was a big man with a full beard and a scar on his right arm that went all the way to the bone. He built a small rowboat in the chain locker, and once when we stopped at an atoll for a day or so, he put it over and went exploring.

When we were in the Philippines, we could see Japs on a nearby island, busily trying to rebuild their airstrip. They had be "hopscotched" and left to live off their supplies. Anyway, we had to maintain a watch to prevent their trying to board our ship, maybe at night.

We returned to San Francisco in July, 1945.

If anyone can remember that voyage and tell me the name of the ship, I'd be forever grateful. I would be listed on the ship's log as the 3rd radio officer.

ANDERHEGGEN, GEORGE C., R-88

One of the members of my family asked me, "Why after 45 years, are you so excited about getting the "Gallups Quarterly Newsletter?" Why, after 45 years, were a bunch of old seamen bothering to try to be counted? One possible answer—something called "self-esteem."

On a rainy Sunday morning in July of 1988, a boat load of men who once attended the proud Gallups Island Radio School descended on the Island. Each was no doubt looking for something he had lost. Or perhaps it was a stroll down memory lane.

As I listened to the old salts talk of old friends, of ports of call, of ships sunk, I began to become aware of one common theme. Most of these men were just kids when they were on the Island. Some had been rejected by the services for health reasons, or were too young. Some didn't need to be there at all. What a crew.

After the war, most who had gone to sea to serve God and country returned to take up where they had left off. No longer did the public hear radio spots asking desperately for merchant seamen to carry the supplies of war. Instead, they were accused of having had comfortable jobs and being highly paid. This attitude cut into their self-esteem. They had gone off proudly to serve their country, and wound up being ridiculed for having given that service.

Today, 45 years later, the facts come out. The pay scale was researched and found to be "essentially the same." The ratio of killed to wounded was ten times that of the military, and the Defense Department's own data show the US Merchant Marine as hav-

Platoon R-80

Platoon R-84

Platoon R-88

ing a higher rate of loss than all the Armed Forces, with the exception of the Marine Corps. What does this do for the old salts who survived the war but who were shot down with verbal abuse when they were just kids? Perhaps it boots up their self-esteem a bit. It is the least that can be done for them now.

So in answer to the member of my family who wondered why I was so excited over the "Gallups Newsletter," I answered, "It is because it gave me back a bit of the self-esteem I didn't know I had lost by volunteering to serve my country in the the United States Merchant Marine when I was fifteen years old." So, I would like to thank you as Editor and all the other old salts who over the years have taken the time and trouble to get at the truth.

HENRY, FRANK, R-88

After a week in New York City, Frank Grabski and I were ordered by the RMO (Recruiting and Manning Organization) to Norfolk, where, after four days, we were assigned to our first ships. Mine was the *MV Farallon*, a V4 seagoing tug (50 were built during WWII, ten of which were lost).

I served as second radio operator for two voyages under William L. Smith. While on board the *Farallon*, I made ports in Panama, New Orleans, Mobile, Gulfport, Panama again, and Kwajalein (from where we were to join the forces in the Philippines for the planned November,1945 invasion of Japan's home islands).

When Japan surrendered, we were ordered to Pearl Harbor, and enroute diverted toward Wake Island to take a battleship in tow. However, the order was canceled and we continued on to Pearl Harbor. From there we went to San Francisco. I had the midwatch prior to arrival and stayed up after my watch ended at 0400. We passed the Farallon Islands at dawn and at 0800 passed under the Golden Gate Bridge, an experience I have never forgotten.

A couple of weeks later, the V4's cut back to one radio operator, and I was transferred to the *MV Sands Point*. Just prior to that, Admiral Halsey's Third Fleet came into San Francisco, and the *Farallon*, with dignitaries aboard, was the first ship inside the Golden Gate, giving us an excellent view of the ships as the Fleet filed in. Unfortunately there was some fog over San Francisco, slightly marring the viewing conditions from the city.

My third voyage was aboard the *Sands Point*, to New Orleans via the Panama Canal. The night before we reached Panama, we received the message, sent worldwide, which ordered merchant ships in designated areas and appropriate water depths to jettison ammunition. The Captain, Joseph D. Strohm, the remaining Armed Guard Gunner's Mate, and I turned to, heaving the ship's allowance of 3" 50 caliber and 20 mm ammo into the sea.

Other interesting experiences included seeing a very brave Boatswain's Mate First Class, Guido DeAngelis of Torrington, CT, who was in charge of the *Farallon's* Armed Guard gun crew, disarm a berserk crew member wielding a fire axe, in port at New Orleans, prior to our departure for the Pacific, on my second voyage. . . .An engine breakdown in the Caribbean, near Swan Island, on my first voyage and the temporary loss of two barges of our three barge tow when a towing line parted during the night in the same general area of our second voyage. . . . The ship being challenged by blinker light in mid-Pacific. . . . Traveling blacked out, we were invisible except to a ship carrying a radar, and conversely could not see our challenger which identified itself as a U.S. battleship, on whose surface radar we must have looked like a column of four destroyers making all of six knots.

After leave at home in November 1945, I made inquiry at the Navy Recruiting Station in Providence. Thanks to my USMS/USMM training/experience, I was given the "Eddy Test" and offered Seaman First Class and Radio Technician training, for which I enlisted four years. Regular Navy. Not only did the Navy send me to 20 weeks Elementary Electricity and Radio Material School at Great Lakes (following the requisite boot training there, not easy after have been through it previously at Sheepshead Bay), but also 28 weeks Radio Material School at Treasure Island (learning the Navy's radio, radar, sonar, IFF, and Loran equipments), and because four-year men were scarce among RT's (subsequently changed to ET-Electronics Technician), to GCA (Ground Controlled Approach) to Maintenance Phase School at Olathe, KS (search and precision radar systems and aircraft type radio equipment).

I completed my four year hitch with GCA Units, eventually becoming leading Electronics Technician, and in 1948 volunteered to go to Berlin with a GCA Unit which had been ordered to Tempelhof Field to aid the Berlin Airlift and would have gone but for the refusal of administrative authority to undo the orders of a technician who was willing to swap duty with me.

In 1949-50 I prepared to enter college in September, 1950. However, the invasion of South Korea in June, 1950 led to my recall to active duty from the Naval Reserve. Instead of reporting for college freshman Week on Sunday, 10 September, I instead boarded the 0700 train at Providence's Union Station, in uniform and with full seabag, for Boston. There followed two weeks of processing before I received orders to a destroyer-minesweeper already operating in North Korea.

Two weeks after departing Boston—traveling by train to San Francisco, by air to Japan, and by troop train from Yokohama to Sasebo—I boarded the *USS Comstock* LSD-19 and we headed up to Inchon. After several weeks, during which the *Comstock* participated in the landing of Marines, tanks, and landing craft at Wonsan, as part of a 35-ship force, I finally joined *USS Endicott* DMS-35, back in Sasebo, and eventually became the ship's leading electronics technician. I served aboard her during two tours, much of the time along the northwest coast of Korea. We patrolled as far north as Chongjin, just short of the Korean-Manchurian-Siberian triple border and Vladivostok. Besides escort, blockade, shore bombardment, and minesweep operations the *Endicott* destroyed trains and rail lines, opened the siege of Wonsan (reportedly the longest in Naval history), and led the way out of Hungnam during the evacuation of Hungnam on Christmas Eve, 1950.

After 18 months, I was released from active duty and returned home to work and once again prepare to enter college.

I entered Brown University in September, 1952, eight years after graduating from high school. I did well with my (electrical/electronics) engineering studies, thanks greatly to knowledge, study habits, and determination acquired in the Maritime Service and the Navy. Upon graduation in 1956, I embarked on a thirty year U.S. Government career, spent mostly with the Department of the Navy. By the Grace of God I muddled through, strengthened often by my service technical training and experience.

Finally, I cannot help but wonder what course my life would have taken, fatherless and facing the world at sixteen, had I not been given a jump-start by the U.S. Maritime Service and the U.S. Merchant Marine.

HESTER, JIM, R-88

I graduated in early 1945 with R-88 and was sent to San Pedro, CA for assignment to sea duty. I was two months past my 17th birthday when the RMO assigned me as chief radio operator in the T2 tanker *SS Pocket Canyon*. The *Pocket Canyon* had just returned from the Leyte Gulf invasion where they had encountered heavy kamikaze attacks.

The captain had refused to promote either the second or third radio operator (who were old enough to be my father) to chief, stating that they lacked sufficient experience. To say that he was less than enthusiastic about my assignment is an understatement. He didn't have apoplexy, but two days later he got off the ship saying he was not going back out there with an inexperienced crew.

His replacement was an elderly gentleman called out of retirement who was much more tolerant of my tender years. He just stated that we had something in common: He was too old to be out there and I was too young. But some of the crew were not as tolerant, and I found that the boxing expertise I picked up in Catalina sometimes proved to be more valuable than the training I received at Gallups.

So off to Ulithi with a load of bunkers we went. We returned to San Pedro for a second load and this time we went on an odyssey of several months. We visited Ulithi, Inchon, Manila, Singapore, Abadan, Hobart, Melbourne, Sydney, Panama, Aruba, Curacao and probably a couple of places I do not remember. The war was over before we got back.

I was paid off in Newport News, VA and went home for a month or so. But the thirst for adventure was still in me, and I went back to sea for another eight years. I picked ships that would ensure my travel to ports of interest throughout the world. It was a wonderful life and I would probably still be pursuing that career were it not for a young lady in San Francisco who led me to the altar.

My wife convinced me I should get a job ashore, so I applied at what is now GTE. They hired me because of my first class radiotelegraph license, which incidentally I still have, but I never used the license in my 35-year engineering career with them. I worked with switching equipment and was involved in the conversions from electromechanical to electronic to digital.

This is just a brief overview of my seagoing career. In nine years I had so many interesting experiences that I could tell sea stories for days. It was a rich and rewarding experience for a young man, and something to cherish in my later years.

HIEKEN, CHARLES, R-88

I sailed from San Francisco on the *Escanaba Victory* in April, 1945 to Okinawa. We then went to Guadalcanal, then through the Panama Canal to Norfolk and then to South Africa and Mozambique. My last freighter to Africa later wound up stranded in the Suez Canal.

I sailed on the maiden post-war voyage of the *America* for a year before attending MIT. I worked at WCC/WIM during Christmas vacation and the following summer. The next summer I was supposed to sail on the maiden voyage of the *United States*; however, I wound up substituting on the *America*.

I was drafted into the Army after MIT, returned to MIT for additional graduate courses and then attended Harvard Law School while working as a patent assistant at Laboratory for Electronics. I graduated in 1957, went to Chicago for three months and then returned to Boston where I have been ever since. The north windows of my office face the Custom House Tower where we took our FCC exams in 1945.

I have been practicing intellectual property law since 1954. Since 1988 I have been a partner in Fish and Richardson, a firm started by a patent lawyer who represented Alexander Graham Bell and the largest intellectual property firm in Boston. We also have offices in Washington, Houston and Silicon Valley.

My amateur call is W1TKS. I still renew my first-class radiotelegraph license and Merchant Marine Officer's License. If I

Platoons R-90-R-99

MAURSTAD, RAYMOND B., R-93

Born April 4, 1928, St. Paul, MN. To Puget Sound Navy Shipyard at Bremerton Washington at age 15, Riggers Helper then Electricians Helper, wiring guns on the Mississippi, West Virginia and Yorktown. Tried Navy/Nil. At age 16 trained at the USMS School on Catalina Island and graduated from the USMS Radio Training School on Gallups Island, Boston Harbor, April 12, 1945. (AWOL from R-92; graduated with R-93.)

First ship, the Liberty *SS David Lubin*, as Chief Radio Operator. Sailed independently from Puget Sound to the Marshall Islands and on to the Caroline Islands where we joined a 40-ship convoy backing up the invasion of Okinawa. Thirty days on the hook at Hagushi anchorage near Naha under Kamikazi attacks. Under sniper fire during convoy conference ashore. Arrived at San Francisco on VJ day.

Other ships: 1945, *SS North Sea*, passengers, Seattle to Alaska; 1946, *SS Topila*, oil, California to Oregon; *SS Typhoon*, troops, Okinawa and Philippines to U.S.; 1946 Pan Am on Wake Island Air-Ground and Pt/Pt Comm; 1946-47, Northwest Airlines, Minneapolis and Grand Forks, ND; 1947, *MV American Packer*, dry cargo and passengers, U.S. to Europe; *SS Marore*, ore, Chile to U.S.; *SS Ampac*, Los Angeles, dry cargo, Europe to U.S.; 1948, Western Union, St. Paul, MN cable office; 1948-49, *SS Marine Marlin*, passengers, Europe, South America to U.S.; *SS Tydol Bayonne*, oil, Texas to New Jersey; 1949-50 with U.S. Mission to Korea with six Baltic Coaster vessels and wife Marion, who was pregnant, evacuated women and children to Japan; first son born in Tokyo, August, 1950.

As amateur HL1CE along with HL1CD sent first word to McArthur's Tokyo HQ that Korea was under attack. Turned ham stations over to U.S. Army when they landed.

1950, *Ordinary Seaman Ore Boat* on Great Lakes. 1951, *M/V Tydol #2* oil, Gulf to East Coast; *SS R E. Wilson* oil, Gulf to East Coast; *SS Lyons Creek*, oil, Gulf to East Coast; *SS Chiwawa* oil, Gulf to East Coast; *SS Frank Haskell*, oil, Gulf to East Coast; *SS Strathport*, dry cargo, Europe to U.S.; *SS Wm. F. Humphrey*, oil, Gulf to East Coast; *SS San Mateo Victory*, supply Narsarsuak, Sondestrom and Thule bases in Arctic; 1955-56, *African Dawn*, dry cargo and passengers, Africa to U.S.. 1956, *SS Robert E. Hopkins*, oil, Gulf to East Coast; first Radio Officer to respond to SOS from *SS. Andrea Doria* and *SS. Stockholm* collision, the fourth vessel alongside the *Andrea Doria*. 1956-57, *SS. African Dawn*, 1957-1960 Press Wireless Inc., New York. 1960 CBS TV, Channel 10, Phoenix, AZ, KOOL/TV and disc-jockey KBUZ Radio Mesa. 1960-89, Remington Rand Univac, Univac, Sperry Univac, SPERRY, started as Production Engineer and retired as a Director. 1989-92, Administrative Electronics Officer, Texaco Marine Services (shipboard Radio Officer). *SS. Texaco Massachusetts, SS. Star Rhode Island, SS. Star Georgia, SS. Star Mississippi.*

Reside with artist/real estate wife, Marion, in Coon Rapids, MN with four children and five grandchildren as close neighbors. Retired December 22, 1992 to ham radio computers, fishing, coins and stamps, writing, gardening and trying to remember what has happened since Gallups Island.

Platoon R-89

HARPER, C.E. GENE, R-98

R-98 was away down the line! About six of us arrived in snowy winter of 1944—goofy youngsters from the South! Raymor-Play-More, Buddies Club, Stagedoor Canteen .. up to Salem, etc., and sailing from Sun Shipyards, Philadelphia, in June... down to Aruba, carried navy bunkers through the canal, up the west coast, and emptied at Seattle ... *SS Hubbardton*. Light, going down to San Pedro for a load destined to support the invasion fleet of Japan proper.

We heard the bomb had fallen. We crossed the Pacific with that load, Ulithi Atoll, to Buckner Bay, then watched the proud warships from several nations as they gathered between typhoons. Our attention-getter was "Cavalcade" and we all sailed into the China Sea to miss the worst of one typhoon!

The deck trainee from the academy had mis-figured our fuel! We came home in "slow motion" and had about a day's run left when reaching Panama!

When we crossed the Marianna Trench we were told to throw away our shells. We did!! Too deep to ever go back for them!

On the next voyage to England we dodged aviation mines in the water right where the radio told us they were to be found. Coming home, hit a small iceberg, tore a big hole in the bow, laid this ship, the *SS Kettleman Hills*, up at Norfolk, VA. Later, I read in a "Life" magazine where some female had bought it and sold it to the Red Chinese!

Sailed coastwise a bit, came home, used the dot-dash to convert to click-clack and worked for the railroad. Then, used the typing training to work for retail credit company (EquiFax) doing reporting. Later, packing house (meats) and left that to enter the insurance business. Sold the agency last year, but am still specialized estate planning and Mutual funds.

PALMER, KEN, R-98

With the Persian Gulf-Kuwait crisis heating up in the fall of 1990 and a critical need for Radio Officers, it was decided (with my wife's permission) to help out. My 1st Telegraph was renewed in October, but my Coast Guard license took longer to renew because of coronary and other problems. A Federal Drug test was also taken.

Finally, in March 1991 the ARA quickly assigned me to the *MV Green Wave* KRHL-NADN (Military Sealift Command) at Oakland Army Base. After a few hours break-in the previous R/O left and we departed for the Pacific Island bases. Quite a big learning experience after "swallowing the anchor" over 44 years ago.

I finally mastered the TELEX-INMARSAT and Weather FAX equipment. The main CW rigs were ITT-Mackay. A neat 1 kw CW, SSB rig was part of the gear also. There was a SITOR unit, but it was never used. All Navy and commercial traffic sent and received was via satellite via Santa Paula Coast Earth Station. A total of 591 messages were sent and received. I was busy! An 8-hour per day watch on 500 kHz was a requirement. A few contacts were made on CW but no traffic.

Our first stop after loading some lumber and poles at Tacoma, was Pearl Harbor Naval Base. One day was spent swimming at Waikiki; plus shopping and a bus ride (60 cents) around the island.

Second stop was a remote Air Force Base on Wake Island atoll. Shore leave was via LCM. Was able to operate ashore as K2FJ/KH9, plus lagoon swimming. Only seven Air Force personnel and 170 contractor and Thai workers here. Also a NOAA Station.

Next was Guam Naval Base. Setting foot ashore here was a great feeling. Last time here was 47 years ago—June, 1945. We docked near the *USS San Jose*, just returned from Desert Storm.

From Guam it was off to Okinawa, Japan. (Another 1945 port). The high-light here was a $3.75 haircut and massage in Camp Kinser Marine Corps base!

With Super-Typhoon WALT moving our way, we moved out, along the coast of Taiwan and arrived safely at Subic Bay Naval Station in the Philippines. Quite an installation. A good time was had snorkeling at Grande Island, racing a Go-Kart, mini-golf and Chapel. A couple of weeks after we departed Subic, Mt. Pinatuba blew its top!

Throughout the "cruise" I operated K2FJ/MM (trap vertical) and by a real coincidence had exactly 591 logged contacts: (Same as the ship traffic total).

The ARA had called me for other assignments, but paying social security back $1.00 for every $2.00 I earned over $7K was highly objectionable. So, it's back to the rocking chair in East Aurora, NY.

EDWARDS, RUSSELL D., R-99

I graduated with Platoon R-99 on June 1, 1945 and was transported by train to home in Missouri for a few days and then on to Los Angeles, CA for assignment to my first ship. After sitting around and waiting several days I was sent to Oakland, CA and signed aboard the *SS Eagle Wing* as 3rd Radio Operator on June 23, 1945. (Note: the *Eagle Wing* was a sleek dry cargo vessel, type C-11, with a cruise speed of 18 or 19 knots.)

We were loading general cargo in No. 1 and 5 holds and I was elated to have as my first trip a fast dry cargo ship loading general cargo. I had dreaded the thought of drawing a tanker.

After loading No. 1 and 5 holds we moved upriver to an ammunition depot at Port Chicago near Pittsburgh, CA and proceeded to fill No. 2 and No. 4 holds with mines, small arms, ammo, etc. No. 3 hold was topped off with approximately 92 large wooden boxes about 10 to 12 foot square.

During the loading of these boxes the ship and dock were under armed guard and photos were taken of each box as it was hoisted aboard. When we asked about the contents and the excitement caused, we were told it was a mystery, that the size and shape of the boxes, as well as the codes on the boxes, were all new and a first for the depot.

On July 3, 1945, we steamed out of San Francisco Bay and when the pilot boat came out to pick up the pilot, a second Army cargo security officer boarded our ship. The story later unfolded was that he had been flown directly from the Pentagon with the manifests for the big boxes in the No. 3 hold. I believe his name was Hitchcock and he supposedly was a colonel, but he was not wearing a colonel's insignia.

During the 21-day trip to Tinian Islands, "Mr. Hitchcock" first showed obvious irritation that the *Eagle Wing* was not escorted, and when told one ship didn't justify escort service, he stated we had the hottest cargo ever to leave the Unite States. Later, during and after bridge games with Captain Woody Pugh and other ship's officers, he offered to bet anyone that the War would be over by Christmas, 1945. He later said that the "Big Boys" in the Pentagon were betting it would be over by Thanksgiving so he felt safe to bet on Christmas.

We arrived at Tinian Island, I believe on July 24, 1945. While the mysterious boxes were being discharged the *USS Indianapolis* steamed in and dropped anchor in the so-called harbor. During her short stay of a few hours a landing craft went alongside, and we were told she discharged a small amount of cargo pertaining to the mysterious boxes which contained atomic explosives. (Note: The book, <u>All The Drowned Sailors</u> placed the *Indianapolis* at Tinian Island on July 26, 1945.) After discharge of our mysterious boxes, Mr. Hitchcock left and we were told he was flying back to Washington, D.C. The original Army security officer, a 2nd lieutenant, remained on board until we arrived back in the States.

We departed Tinian Island on or about July 31, 1945. The *Indianapolis* was sunk just after midnight, July 30, 1945, after five days in the water. Only 316 of 1,196 men survived.

Hiroshima was bombed August 6th, and Nagasaki on August 9th. We arrived at Los Angeles, August 19, 1945.

During my short stay in Los Angeles, I had the opportunity to have dinner with my cousin, Captain Edwin I. Mann, who was a flight engineer aboard a B-29 on leave from Tinian Island. He told me they had a number of B-29's with bomb-bay's modified to accommodate the atomic bombs.

The *SS Eagle Wing* was scheduled to take a full cargo of 55 gallon drums of gasoline into Japan with the invasion troops, but since the War was over we loaded the gasoline and departed on August 26, 1945 for the occupation of Japan. Another rather uneventful trip except for riding out two typhoons which beached about 60 ships in Okinawa.

Needless to say, Harry S. Truman has been my hero for making the decision to drop the bombs.

Platoons R-100-R-123

ASHLEY, DALE, R-100

I graduated from Gallups Island in May of 1945 with my platoon R-100 several weeks before my 18th birthday. Soon afterwards I shipped out of New Orleans on the *F.J. Lukenbach* with a full load of pregnant mares bound for Greece.

Some 36 vessels later I retired in 1983 having spent all but about five years of that time at sea. I never had much luck landing steady jobs but I did manage to hit every dog hole from Point Barrow to South Africa and Singapore to Leningrad. I wonder if that number of ships is unusual?

McCOY, GENE, R-100

After leaving Gallups in June, 1944, I sailed from Seattle on Liberty, *SS Louis Weule* as third operator for points unknown in the Pacific with a cargo of beer and commissary supplies. Stopped at Eniwetok on the way to Guam where we unloaded. Beer tally was only 25% short. (There were some strange vehicles in the transporting line). Then to Manus Island to load gasoline barrels for the Invasion of Japan. The big bombs were dropped while we were there ending the war.

The ship then spent several months in the Philippines unloading the gasoline barrels, a few at at time. In December they converted the ship to haul Japanese prisoners home and sent us to Japan to deliver it to a Japanese crew. By then we had no food left except for canned mutton and weevil bread.

On the way north we hit a typhoon and with only water in the hold the ship was quite unstable—rolling more than the book said was possible and still rolling back. The screws were out of the water and the deck plates were popping.

I was the only operator on my next ship, the *SS Walter Luchenbach*, a WWI ship which had been chartered to the Navy and just turned back to the shipping line. We were in drydock in Seattle for a month, keeping the batteries up and trying to figure out the old equipment before we headed out to Puget Sound. Just a few miles out the shafts started vibrating and I sent the only message of the trip, a call for a tug to bring us back to Seattle. The crew was discharged upon arrival.

Went back to Liberties after that. Took the *SS Richard Maczkoski*, on a trip to Kobe, Japan with a load of flour. The trip was made memorable by a banquet for the ship's officers put on by the local government in gratitude for bringing the flour.

At that point a career on the ships didn't look too great to me so I went back to school, first to a community college and then to Washington State. Started in electrical engineering but soon found my talents best used in civil engineering.

Made trips on Liberties in the summers of 1947 and 1948 to help finance schooling. In 1947 went to the Philippines and China on the *SS Henry Stephen* and in 1948 made an intercoastal trip on the *SS Pere Marquette*.

Shortly after graduating from WSU I was drafted and fought the Korean War in Germany as a topographic computer. After discharge I went to work for the Corps of Engineers as a geotechnical engineer, working 33 years and retiring as chief, Geotechnical Branch of the North Pacific Division in 1985.

I still do some consulting work, along with voluntary, working a 50 acre tree farm with a large garden, hiking and skiing and being an active grandfather.

PEDERSON, C.W., R-101

I sailed until March, 1948. Paid off the *MV Gulfwing* in Philadelphia, and in April, 1949 started operating heavy equipment on a road job in North Dakota. Am still involved with earth moving on a part time basis but not operating any more due to some physical problems.

I was originally scheduled for graduation from Gallups in April of 1945, but used nearly a month in sick bay with scarlet fever and this made it (graduation) in May.

ENGEL, PETER, R-102

Got caught up in Desert Storm and got back home in spring, 1991. I have kept my FCC license and Coast Guard license up-to-date since retiring from the sea over 17 years ago. Heard there was a shortage of radio officers and called the ARA. They grabbed me fast and put me on a ship called the *MV Green Ridge*, owned by Central Gulf Lines in New Orleans and under MSC charter.

During Desert Storm we took a load of Marine Corps equipment from the west coast to Saudi Arabia and Bahrain. MSC then sent us to England for a load of bombs. Loaded 8,000 tons, mostly 2,000 pounders, and took them to the Persian Gulf. Got there just as the war ended, so they put us out in the middle of the Gulf for a week until they decided what to do with the bombs. Finally they told us to take them back to England. I was relieved there and flown home after being gone about 5 1/2 months.

It was a most interesting trip and I enjoyed catching up on the latest innovations in maritime communications, i.e., Satcom and Sitor, neither of which existed when I quit the sea. I was somewhat surprised to still hear quite a bit of CW on the air, mostly from ships of Third World countries. But a rep from Magnavox told me most of their sales of Satcom are going to Third World country ships now. I did very little CW, although MSC in Bahrain did send us a couple of messages by CW through Bahrain Radio. Why, I don't know, maybe just to keep me on my toes. And Suez Radio still sends out messages on CW to ships transiting the Suez Canal.

KING, RAY, R-103

I can see almost all of Gallups Island from the front window of my home on Great Hill in Weymouth, Massachusetts. I am probably the only graduate of Gallups Island Radio School that sees the island every day.

I joined the U.S. Maritime Service in 1944 immediately after I attained the age of 18 and after having been rejected for military service because of an eye problem. During my period of service as a radio officer I served on 25 different merchant and army ships, believe it or not.

I stopped going to sea full time in 1948 but continued to go to sea summers until 1954 while working my way through college and law school with my summer earnings. I graduated from Harvard College in 1952 and from Harvard Law School in 1955. The reason I served on so many ships is because most of my summer jobs were short term jobs in which I relieved radio officers taking vacations.

During the course of my law school training I made the acquaintance of Charles Hieken. He was in R-88 and is, I believe, the only other Gallups Islander who is a graduate of Harvard Law School. Chuck is also a graduate of Massachusetts Institute of Technology and is a distinguished patent attorney. Like myself, he now practices law as a partner in a large Boston law firm.

Like all Gallups Island graduates I had many exciting adventures at sea and could tell many a tale. One of my more interesting voyages was to China in the fall of 1946 on *USAT LST 916* operating out of Manila. Our mission was to deliver food carried in refrigerated truck on our main deck to a meeting being held in Nanking, China, about 150 miles up the Yangtze River from Shanghai and the sea. This meeting was sponsored by the U.S. It was held in the midst of the Chinese Civil War and was attended by the Chinese Communist Nationalists leaders, including Mao Tse-tung and Chou En-lai on the Communist side and Chiang Kai-shek on the Nationalist side.

The U.S. Delegation to the meeting was headed by General George C. Marshall. The object of the U.S. in sponsoring the meeting, was to try to end the Chinese Civil War by encouraging the formation of a coalition government of the Communists and Nationalists. That effort failed and the Communists won the war by 1949.

From Nanking, my ship sailed to northern China with a load of spare aircraft engines for a U.S. marine fighter squadron stationed near the city which is now known as Beijing. I was able to get to Beijing and tour the city and the surrounding countryside. I have a number of photographs of myself in historic places in the city and its surroundings in backdrops almost identical to those in photographs of ex-president Nixon on his famous trip to China many years later.

As a product of institutions of higher education I can confidently assert to all Gallups Islanders that the most important educational experience I have had in my lifetime was my sea time which I share with all of you.

KINKEL, JAMES, R-103

I went to Catalina in October of

Platoon R-101

Platoon R-105

Platoon R-107

Platoon R-111

Platoon R-112

Platoon R-116

1944. In my high school I had taken some extracurricular classes with an instructor who was also a ham radio operator. He also taught us CW, so I was able to copy at 8 to 12 wpm. It went on my high school transcripts, and apparently the maritime picked up on it while I was at Catalina.

We ended up in Boston in the first part of January (1945) in about 18 inches of snow and were were in dress whites. Our peacoats were in our sea bags and our sea bags were taken from us, and we nearly froze our fannies off. One of the Coast Guard boats took us up to the Island.

Some of the guys ended up shoveling snow off the grinder and the rest of us were put on galley duty. That's where I wound up, on galley duty and fire watch and then on anchor watch during my first two weeks.

I really didn't understand at the time what I was scheduled for, because when I left Catalina I was going to go to cook and baker school. I figured the galley work was the beginning of the training. Next thing I know, I'm in there taking code and wound up being a radio operator, so it was a strange circumstance which led to my becoming a radioman.

We got out of the Island in the first part of July. We went to San Francisco and caught a ship in August after the war was over. It was a leased ship called the *Arizonian*, commissioned in 1917. She could do a good six knots if you had a wind to the stern and had the safety belt tied down. We took a cruise around the Pacific to Hawaii, Tokyo Bay, came out of there and went to Guam, Saipan, down towards the Philippines and wandered around out there while they figured what they were going to do with us.

I got home after that excursion in April of '46 and off-loaded basically the same cargo we took on. I went home and, being the war was over, decided to get on with my life.

Coming back from the ship, I picked up my 1st class radio phone ticket at the same time I got my ham ticket. After the first part of the '50s, when they started changing the ham ticket from general class to extra class to advanced class, I updated my ticket to an advanced class ticket.

Although I tried a number of times to get away from electronics, for some reason or other it kept coming back into my life. Finally, in the first part of 1950, I was actually locked into it permanently by taking a job with the telephone company. Prior to that time I had worked with Western Electric, worked with a power company and a bunch of other little jobs trying to figure out what to do. But invariably I kept coming back to electronics.

I started out as a grunt on a line crew and went on up to district manager here in New Mexico.

MANBY, WALTER, R-107

I was born November 21, 1926 in Norfolk, VA. Joined the USMS November, 1944 and entered Sheepshead Bay Training Station December, 1944. Transferred to Gallups Island, spring of 1945. Graduated with 2nd class radiotelegraph operators license, July 27, 1945.

During the next three years I acquired my first class and served as radio operator on the following ships: *MV Sea Serpent, SS Colby Victory, SS Alexander E. Brown, SS Alexander S. Clay, SS Tarleton Brown, SS George W. McCrary, SS Michael DeKovats* and *USNT Cedar Creek*.

Inducted into the U.S. Army, September 26, 1950. Served as radio operator with the 426th Field Artillery Battalion, 7th Army, European Command. Honorably discharged after two years mandatory service and 5 years inactive reserve.

Currently retired from the Baltimore Gas and Electric Company after 34 years as a marketing specialist.

Have held an amateur operator's license for over 40 years. Current call sign is W3VJX.

Never saw actual combat anywhere. Came close to a couple of floating mines while in the merchant marine, but that was all.

NESMAN, ED, R-109

I finished my time at sea, spent a lot of time in Latin America after graduating from college, and I've been a professor at the University of Southern Florida for 25 years.

BRAINARD, ROBERT K., R-110

On December 6, 1944 I enrolled at the USMS recruiting office on 844 North Rush St. in Chicago, IL. I reported for apprentice seaman training December 29, 1944, at Sheepshead Bay, Brooklyn, NY. April 1st I was assigned and transferred to the USMS radio training station in Gallups Island, Boston, MA.

I was released from active duty in the USMS and placed on inactive status on August 17, 1945 after having completed radio training at Gallups Island. Transported from Boston, MA to Seattle, WA and reported to the USMS Recruiting and Manning office for stand-by. On September 5th I reported to the Master on board the *SS Cape San Martin* at Winslow, WA. We signed on for a foreign voyage to the Far East (Okinawa), September 25th. Signed off March 4, 1946.

Was issued a Certificate of Substantially Continuous Service on April 5, 1946. No berths were open and military cargo was winding down so I returned home and reported to the local draft board.

BULTEMA, LARRY, R-112

As often happens to midwest farmboys, I was fascinated by sea stories, ships, sailors and the like. When it became apparent that something military was in my future, I hied myself to Chicago into a recruiting office for what was called the USMS; this was not the military and we became draft deferred.

One fateful day, I reported to the recruiting station and having previously passed a physical exam of sorts was sent along with a group of fellow sad jacks to the railroad station.

There was a USMS official who went along to segregate us from the rest of the passengers on the train. Now that we were sworn, I guess they feared we would contaminate the other passengers.

In due time, we arrived in New York and can remember all the speculation of the groups as to what our fate would be on arrival. Onto a bus, we went off to the USMS training facility at Sheepshead Bay.

Our training began. I'm sure each of us has some special memories of those days of training. We were young and full of vim and vigor. Remember climbing the ropes in the gym at P.T. time. What about that huge swimming pool? We had to jump off a 25-foot tower and swim out of the pool in various ways. No problem if you were a swimmer, but pure terror if you weren't.

The training was tough and demanding but nothing that fazed the kid from the midwest farm and the intrepid fellow member of section 25. In the astoundingly quick time of six weeks our training was over and we were waiting for new worlds to conquer.

Now what comes next? What becomes of this group of civilians who so fearlessly started training six weeks ago? In my case, I was fascinated by an announcement that appeared on the bulletin board some weeks previous. "Interested in being a radio man? Enroll in math class, pass the test and go to Gallups Island for a twenty-two week radio school."

I did—passed and was interviewed by a man recruiting people for that infamous radio school. His answer to me. "Bultema, I'm sorry. Tests show that you have no natural aptitude for Morse code. You'd be a square peg in a round hole." And sent me on my merry way. This depressed me to no end so each day of my work week, I appeared in this good placement officer's office at 8 a.m. to beg him to reconsider.

His reply on seeing me sit there was invariably, "Get outa here!" Nevertheless, I persisted and on Friday he said, "Bultema, go back to your barracks and pack your sea bag—in two hours a group will be transported to Gallups Island—including you. I'm tired of seeing your face."

So I departed for Gallups Island. The only member of section 25 to go. You know something? That placement officer was right. I had no natural aptitude for Morse code and had to work my a— off to master it.

RAETZ, EUGENE H. JR., R-112

Attended Gallups Island Radio School, I am not sure of the date our platoon graduated, but it was R-112. After graduation I sailed as a radio operator on a Liberty ship, *SS Sam Adams*, transporting troops. We ran into a storm and the main deck cracked due to rough sea.

I signed off, by this time the war was over, and there was a surplus of radio officers as they cut back to one on each ship.

I took a test sail as electrician and luckily passed. I then sailed as chief electrician on a new Victory ship, the *SS Pass Christian*. First trip was to Gdynia, Poland. We carried a load of horses and lost hundreds on the way over due to poor ventilation in the holds. When we arrived there were Russian soldiers were every where. I feel sure the horses went straight to Russia and not to the Polish farmers they were intended for.

Served my required time, got my honorable discharge and left the Maritime Service.

While at Gallups I buddy'd with a fellow student, Matt Rini. Have not seen or talked to him since graduation. If anyone knows his whereabouts I would appreciate his address.

Nothing very exciting in my Merchant Marine experience to talk about. I did enjoy the time spent and would do it all over again.

LIGGINS, EDWARD, R-112

The year was 1945. I was in a class—I think it was R-103. I caught pneumonia and when I got out of the hospital I was in R-112.

There were tests for people who wanted to be radio operators. I took the test and passed it. After I got out of school I went to sea for about four years.

Then after that I went to the University of Hawaii. From there I became a police officer for 25 years before becoming an attorney.

I retired as an attorney and moved up here to Maine. I couldn't handle retirement and went back into the practice of law.

THOMAS, CHARLES E., R-112

I'm a civil service retiree who trained at Sheepshead Bay and graduated from Gallups Island Radio School, R-112, August 31, 1945. My friends knew me as a farm boy from South Dakota.

I am hoping I can find someone who can tell me what happened to my good friend and high school buddy Isaac Olson (or just Ike). He trained at Sheepshead Bay and graduated from Gallups sometime in 1943 or 1944. He received his radio license and shipped out of San Francisco on the *Belgium Victory* (owned, I believe, by Olson & Tuntland). Something happened to him during the shuttle runs in the Philippines, and even though he returned to San Francisco and left ship, he was a mental and physical wreck and was never able to tell family or friends what happened. I am hoping that someone who reads this can tell me something about it. I now live at 17844 Glen Oak Ave., Lansing, IL, 60438. Telephone (708) 895-3299.

After graduation, Larry Bultema and I signed with Olson & Tuntland in San Francisco. I remember a buddy, France, was there too but not with us. We lived in the York Hotel for over three months waiting to ship out. Ike's ship came in while we were there, but because docking times were still classified, we did not know the exact time he would come in. We missed him by about half an hour. Someone said they saw him leave the ship, running west on Market Street. We searched the town but did not find him.

Two girls I knew from high school called and said they had seen him briefly in Seattle and that he was "not himself." From there he disappeared and somehow signed with one of the U.S. armed services. A short time later he was given a medical discharge and sent to the VA hospital in Sturgis, SD. From there his family sent him to private hospitals where he received shock treatments and therapy but he never regained his mental or physical healthy and finally lived the rest of his life in a nursing home. Ike died in 1989 in a nursing home in his home town (and mine) Centerville, SD.

Larry and I gave up on shipping out from California and at Christmas time bought tickets for New York. I stopped off at Centerville and Larry stopped off at his home in Lansing. After New Year's I joined the Army Air Force and Larry joined the Army. I was home on leave when I heard Ike was in the VA hospital. His dad and I went to Sturgis to see him.

Things were so bad (at the hospital) that we checked him out against all advice. I was instructed on what to do when he was violent and kept charge of him during my leave. I was the best one to communicate with him in putting together the pieces. It sounded like his ship was in a 100% war zone and he was blamed for their being there. Once he said, "They tried to kill me." Other than that, information just wasn't there.

As for me, I benefited from my Sheepshead Bay and Gallups Island training. Basic was a breeze for a farm boy who had already been through boot camp. It also boosted my IQ test score so high I qualified for any school that I wanted to attend. I chose weather school and was sent to Chanute Air Force Base.

After graduating from weather school they kept sending me to special schools and finally George Vlay (my Air Force buddy from Buffalo, NY) and I spent the last year and a half of our three-year hitch, visiting Air Force bases to maintain and teach weather personnel. I got married in 1948 and George did the same shortly after. We were both were discharged in January, 1949.

Thanks to my maritime and Air Force training I spent the rest of my career working as a civilian for the Air Force and retired at the age of 53 in 1978. Lost my wife in 1986 while in Georgia. Came to Lansing where I now live and in 1987 I married Larry's sister Esther who had been widowed in 1980. (Her husband had been one of the first Marines to go into Nagasaki, Japan after the atomic bomb. He died of leukemia.)

As Es is also retired we enjoy traveling. Life is good to us. There are, of course, many tales I did not tell. But maybe at some later time.

Platoon R-114

WILLIS, ARCHIE, R-112

In January, 1945 I dropped out of Amarillo High School with only a semester to go to graduation and began my trek to Gallups Island via Sheepshead Bay, Brooklyn, NY.

My ambition was sparked by a course in high school where the final exam was the FCC license exam, we got credit for the course if we passed the test and of course most of us did. I also would have been fodder for the Army as soon as I graduated. We all expected the war with Japan to go on for two or three years with most of us young ones being the casualties. I wanted to make my life count for more than just a statistic.

My first few months out on the town for the first time were exciting but when I got to Boston and Gallups Island I was really in what my mother always called "hog heaven."

I was very fortunate to have Bill Yount (William D. Yount R-77) as an instructor. I remained in contact for a time and I had a very nice long visit with him at the 1992 Reunion. R-112 was his last graduating class in August, 1945 as the next class he had was not graduated.

When we began testing for code I used an eraser when I shouldn't have because I was anticipating and of course I failed the test for five words per minute automatically. The chief code instructor called me in and assigned me to extra code practice for the next week. I passed the next test for 14 WPM with flying colors. I was then summoned to see the chief code instructor again and thought—"What have I done now?" He asked me how did I fail 5 WPM and pass 14 WPM in a week. I told him the mistake I made with the eraser and that I had an amateur radio operator's license. He just shrugged and said, "OK, you did well."

I did not get assigned to a ship until December, 1945, a Liberty, the *Martin Johnson*, bound for Italy to bring back ammunition. We crossed the North Atlantic in January and had to watch for German submarines and mines all the way across. My next ships were in 1947-1948, my last voyage being one to Bombay and Calcutta. We left port one day before Ghandi was assassinated.

I have been an experimental machinist for many years, after trying to get into the electronic technician field. My father trained me to be a machinist and welder from an early age. I remained an Amateur Radio Operator with the call W5YA in Texas and Arkansas and change to W6LPJ in 1952 after a move to Los Angeles.

I am an active ARRL member, Quarter Century Wireless Association, Old Timers Club, Society of Wireless Pioneers and of course, a charter member of the Gallups Island Radio Association. I have had many elective and appointed positions in Amateur Radio groups and the longest lasting and most time consuming is as manager of the ARRL Sixth District (California) Incoming DX QSL Bureau. This position placed me in an almost full time job and I really cannot afford to retire yet.

WALLACE, KEITH E., R-116/119

I started in R-116 on May 4, 1945. As a result of sickness I ended up in R-119. On September 5th I was transferred to Hoffman Island when Gallups Island was closed, completed training and received 2nd Class Radio Telegraph License on September 28, 1945. I shipped out on the Liberty ship, *SS William Blount* as third radio officer.

I graduated from South Dakota State University in 1949 and taught Vo-Ag at Barnard and Groton, SD. I then attended graduate school at SDSU, married Marilyn Page from Haverhill, MA in 1953 and in 1954 received an MS in Agronomy from SDSU and accepted the position of Weed Specialist for Washington State University at Spokane, WA.

In 1976 I was promoted to Full Professor at WSU. In 1977 I took sabbatical leave in New Zealand, Australia and the Orient. In 1982 I accepted a position as agronomist with WSU at an agricultural research station in Lesotho, Southern Africa.

I retired from WSU in August, 1984, still live in Spokane and have been contracting to do research on weed control since retirement. I have three children: Dee Ann Fagan, Bruce and Loretta.

CARPENTER, ARCHIE H., R-119

I was trying to get into the U.S. Navy and couldn't pass the physical because of sugar in my urine, so I enlisted into the U.S. Maritime Service and ended up in Sheepshead Bay, Brooklyn, New York. They took applications for those who wanted to go to the radio school and gave what they called the INT test in code. I passed the code test and the algebra test and they accepted me for radio school.

I was moved up there in May of 1945, I believe it was. I was a member of class R-119 and I was taken to Sheepshead Bay March 21, 1945 and started some time in June. I was supposed to graduate October 19, 1945, but our school was closed down in August immediately after the war was over.

I was then transferred to Hoffman Island because I had passed the test and was eligible. We were in Hoffman Island for four weeks and they closed that school. Consequently, I never had the chance to take the examination.

I was transferred from there to 44th Street in New York city and was transferred again to Norfolk, VA. I sailed out of Norfolk as a mess boy on the *USS Samuel Griffin*. When we landed in Marseilles, France, we were boarded by American troops to be returned to the United States. Two of our three ship operators skipped ship, so the captain made me 2nd operator. The 1st operator and I managed to cover the radio 24 hours a day.

When we got back to New York City, I went downtown, took the examination and got my 2nd class radiotelegraph license. I shipped another tour on the Griffin, which eventually got tied up in the North River. I couldn't get another ship so I went home.

I had lost all interest in all types of radio. When my son became interested in radio code in his Boy Scout days, I became the instructor in CW. He obtained his license at that time—and that got me inspired, so I studied up and got a novice called WNB FTX. I later passed the technician test, then the general test and held the call W8 FTX until about 1975 when I became WD-4 NFK. I then moved back to Ohio from New Port Richey, FL, and picked up a new call of KA8 HOR, "ham on rye" or "horny old rooster." After passing the extra class, I transferred back to Ohio and applied for a license and ended up with NS8D—'"north/south 8 degrees."

When I left Gallups Island the commander-in-chief at that time was Lieutenant Commander Larkin. There was a junior officer under him by the name of Rippige, I don't remember his first name, but he was sentimentally called 'Rip the Bitch' because he was in charge of discipline. Ed Pleuler was our code instructor and in charge of our particular platoon. I think he was a lieutenant, j.g., or something like that.

We had a character in our class—his name was Bob Berko. I always though he would have made a tremendous comedian because every time the officers tried to discipline him he broke them up with his jokes or wit. He really had some wit.

I don't remember any of the other officers. I do remember my Horton radio book—I've still got all my radio books and the Q&A as well as all my notebooks from Gallups Island.

One of the things I do recall while I was on the island: We were called out of the sack about 3 a.m. and all ordered to report dockside of the liberty launches. Seems as though one of the fire watch on duty that night didn't pay enough attention to the liberty launches, and one of them got the bow underneath the dock. As the tide came up, the stern was sticking about ten feet in the air. They wanted all the men they could to get on board in order to float it underneath the dock to get it righted before it sunk. I recall I lost a very good Parker fountain pen that was a graduation gift from high school.

Another humorous thing: I was written up and sent to captain's mast for getting sunburned. '"Destroying government property.'"

HOOVER, ALAN, R-119

I was in R-119, went in 1945 and I was there 'til they closed it (Gallups).

I didn't have too many experiences except that we were sitting on the bank and we were eating canned chicken when Germany signed the peace. We saw a ship coming out of the bay and he was zig-zagging and blowing his horns. We knew that something had happened so we went to the auditorium and found out the war was almost over. They sent us home for the rest of the week.

We were supposed to come back on Saturday for a review but nobody showed up. We got gigged for three weeks; they canceled two of them. It was well worth it.

Weekends, we would go to the

Jewish Center and they would put you up for a quarter and give you breakfast the next morning. If you didn't have a quarter, they'd put you up anyway.

When you were on dock watch you'd have to bring the ropes in on the captain's boat, yacht. If you didn't the captain's yacht would float underneath the dock and get crushed. One morning at two o'clock we were all called out for firedrill. Somebody forgot to bring the ropes in and the yacht had floated underneath the dock and they called us all up and loaded the yacht with merchant marinemen until it was low enough that they could float it back out.

PETERS, ARTHUR T., R-122

Entered U.S. Maritime Service March 12, 1945, Sheepshead Bay, NY and selected to attend radio school at Gallups Island June, 1945 as a member of training class R-122.

When war with Japan ended in August, 1945 Gallups Island trainees who had not completed 70% of their training were shipped to Hoffman Island in New York Harbor to complete their training. I was one of these trainees. The remaining personnel were to take their FCC license tests.

At Hoffman Island there was much confusion as to consolidation, what class we should be in, etc.

In September or October Hoffman Island was to shut down with similar conditions as at Gallups.

The not fully schooled trainees were shipped to Sheepshead Bay for reassignment. After several weeks at Sheepshead I shipped out on the *William D. Pender*, Liberty ship converted to a troop ship. We sailed to Marseilles, France where we picked up about 500 Army troops and returned to Boston.

For the next several years I made various trips to Europe, South America and coastwise ports.

On shore I decided to go to radio school and obtained various FCC licenses including: first and second class radiotelephone, second class radio telegraph and my ham ticket W1RGZ.

I worked at a radio station in North Adams, MA as a broadcast engineer. The Korean War broke out and I was drafted in the Army in November 15, 1950. I attended radio school at Camp Gordon, GA and eventually shipped to SHAPE headquarters which at that time was in Paris, France.

I was discharged from the service in October, 1952 and worked for the U.S. Army Signal corps as a civilian and also in broadcasting in the Boston area. During these years I attended Northeastern University evenings and received an associate degree in electronic engineering, a bachelors in engineering and management, and a masters in business administration.

Since then I have worked in various companies in the Boston area as quality assurance engineer, product assurance manager, etc.

I am now employed as an engineering manager with a private company that does contract work for the U.S. Air Force in the field of aircraft communication, working out Hanscom Air Force Base, Bedford, MA.

TRAVOLI, WARREN K, R-123

Our group had taken basic training at Sheepshead Bay, starting in March of 1945. Completed that and was sent to Gallups Island.

When the war ended in August, 1945 we were all shipped out, mostly as messmen on troop transports.

I did not complete the course at Gallups Island, however as soon as I had my three years in and got the certificate exempting me from additional military service, I went home and started school at the Valparaison Technical Institute (formerly Dodge Institute). After completing the course I had a 1st phone and 2nd telegraph.

In 1948 Congress voided the military exemption and I was drafted in 1950 for the Korean Conflict!

When this was over I joined the Indiana State Police as a communications officer and operated both phone and CW networks for the next 20 years. At the time our CW network included all states west of Ohio and usually operated around 30 wpm.

Platoon R-119

Platoon R-58

Platoon R-21, June 1943.

Platoon R-66, 1944.

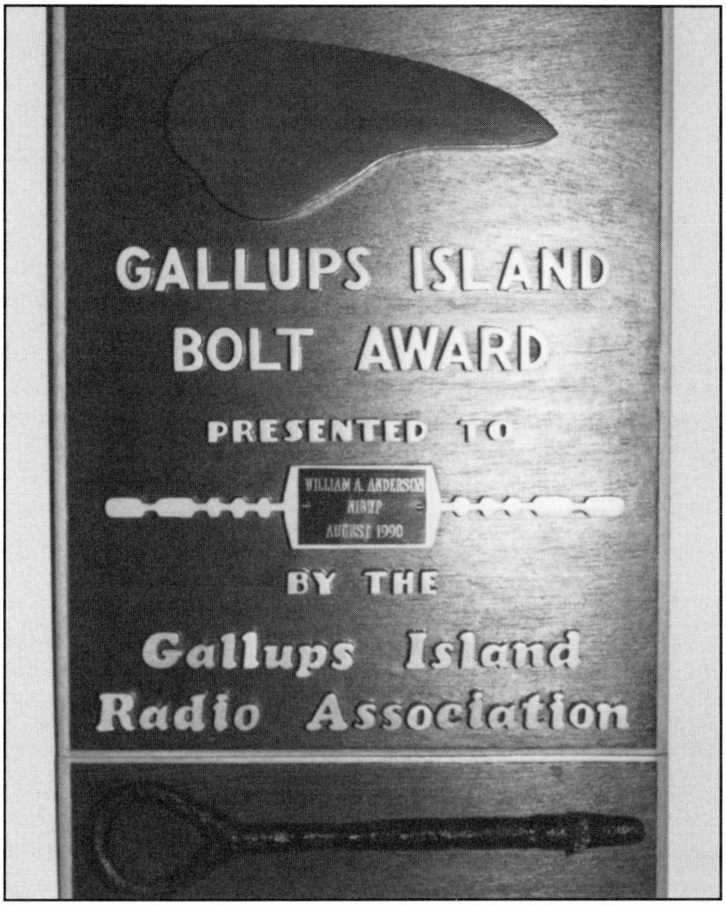

GIRA's Bolt Award, fabricated and donated by Edward F. Pleuler, Jr.

Special Stories and Memories From Gallups Grads

U.S. Merchant Marine Group marching in the Australian Merchant Navy Parade, Melbourne, Australia, April 1993

ARCTIC ADVENTURE

by John J. Ward, R-19

In 1942, by mere happenstance, while thumbing through a nickel pulp magazine, I spotted a Maritime Service ad seeking seafaring officer trainees. Perhaps because an average of five merchant vessels daily were falling victim to U-boats, within two weeks I was on a northbound train headed for Boston and Gallups Island. My companion, a fellow named Stern who had been a student at the University of Cincinnati, was pushed ahead into R-18 and I, with a more modest background, settled into R-19. My quarters, clothing, and food, not to mention a whole sixty dollars per month, were far better than I was used to in the depression hill country.

Some nine months later, clutching my brand new Second Class license and with memories of the severe Boston winter fresh, I headed to New Orleans and Houston with classmates George Wolfe and Dale Harris to wait for a ship. I never saw nor heard from either of them again. My first assignment was an ancient Panamanian-flag tanker built in Scotland circa 1908 which had been scuttled by the Italians at the beginning of the hostilities. I should have been left on the bottom. We headed for Australia then New Guinea for duty refueling the 7th Fleet, known then as McArthur's navy. However, this is not a story of WWII, the Big One, but a later war.

In 1949, needing a break from college campus life, I headed again to New Orleans for assignment to the Navy tanker *USNS Sappa Creek* that had been in the temporary reserve fleet in Algiers. The Navy had forty or so of these T-2 type tankers manned by civilian crews with the mission of resupplying military bases throughout the world. They were operated by Marine Transport Lines and three other steamship companies. About half were West Coast-built "mission" types that seemed to be plagued with more mechanical problems than the basic T-2.

The *Sappa Creek's* call sign was NDES, my favorite of all the ships I sailed on. NDES contains only two dahs and seven dits arranged in a beautiful, almost musical, rhythm. Our first trip was from the Gulf of Mexico to Germany and up the Weser River to Farge (oil port for Bremen) then to the Persian Gulf and back to Farge. During the early summer months we resupplied the Arctic air bases on the Atlantic side. In April we left Texas City for Lake Melville, Labrador, making our way through the iceberg-clogged Belle Isle Straits. Labradorian pilot Charlie Chalk took us up the river into the lake where we anchored and smaller tankers (one was *USNS Peconic*) lightered the jet fuel and avgas into Goose AFB. This method of unloading required a week or so during which we passed time by bartering with the natives, trading potatoes for salmon—imagine such an exchange today.

One day a boat circled the *Sappa Creek* and the first assistant engineer and I hailed it and went ashore with them to a tiny village called Northwest River. There we met several Ivy League students known as WOPs (without pay) who came north during summer college break to work without salary among the natives. We met two young men from Harvard and a woman from Smith. They were the forerunners and inspiration for the Kennedy Administration's Peace Corps that came a decade later.

Northwest River had a Hudson Bay Company store and a dozen or so houses with several Indian villages nearby. There were two remarkably pretty village girls, dark headed Evelyn and a ravishing beautiful blond, who had worked at the Goose AFB and apparently been unable to cope with the cultural shock (fast talking airmen). She was now very visibly pregnant. The pilot had raised the ire of the Hudson Bay Company for some critical remarks by Charlie Chalk, a character in the book <u>Checkmate of the North</u> by a visiting British author. The Hudson Bay Company wasn't exactly endeared to the locals so the pilot became sort of a folk hero and henceforth known as "Charlie Chalk."

Our next trip was to BW-1 (Narsarsuaq) in southern Greenland. Ice was everywhere in Baffin Bay and within the fjord leading to BW-1; it snapped and popped, sounding like giant hands crushing cellophane. The BW-1 officer's club claimed its drinks were chilled with million-year old ice. When he needed ice the bartender would simply go outside with an icepick and fill his bucket from a glacier only a few feet from the back door. That night at a banquet bidding civilian employees goodbye, a pretty, cocktail-fortified teacher brought down the house when she got up to thank the base people and as a parting shot, said, 'I've really been well taken care of these past two years.'

At about the 50th parallel the weather changed abruptly and off came heavy clothing and out came the Bermuda shorts. Going north was the opposite. We reloaded in Venezuela for discharge farther north along the Greenland coast at BW-8 (Sondrestrom Fjord). While moored at the underwater line the following day, an ice floe silently pushed us seaward popping the bolts on the discharge line and sending avgas gushing across the well dock. Miraculously, it didn't catch fire.

In late summer the *Sappa Creek* rejoined the reserve fleet in Algiers, and I went back to college. When they called me the next spring I eagerly rejoined for the familiar trip through Belle Straits and found pilot Charlie Chalk waiting to take us up the river. We met a different group of Harvard WOPs at Northwest River. Subsequent trips to BW-1 and BW-8 found both places little changed. Activity had diminished, and we backloaded avgas that had been in storage too long for blending at the Texas refineries.

The Truman Administration continued to cut back on military spending and our Arctic resupply mission was completed early. On June 25, 1950, as we approached the mouth of the Mississippi River headed for the reserve fleet in Algiers, the WWL radio announcer in New Orleans interrupted a musical program to announce the North Koreans had invaded the south. Within an hour we had a message from WNU to proceed to Baton Rouge and load for Fusan, South Korea. Our destination was kept secret from the crew but in the Panamanian paper we were listed in the shipping section as inbound from Baton Rouge enroute to Fusan. By the time we arrived there our forces held only a small perimeter around that southern port. That year we made two complete trips around the world, with endless shuttles between the Persian Gulf and the war zone.

That spring we were brought back to the Galveston shipyard for some heavy repairs before heading north to Lake Melville, Labrador. Because of the war, there were no WOPs in Northwest River that summer. Charlie Chalk said Evelyn was working as a nurse in Chicago. The blond girl had married an American pilot. And most of the local young men had gone south to work in the cities. Reports of submarine sighting in Lake Melville's mists caused jitters and recon sorties but nothing more.

While loading in Houston, we were equipped with cold weather gear (parkas and the like) and sailed northward to a place called Thule in northern Greenland where, in another chess play with the Russians, a new air base was under urgent construction. In Baffin Bay we met the returning freighter that took up the steel plates to construct the storage tanks into which we were to discharge. As we left darkness behind, the ice floes increased. Our radar screen looked like a bad case of measles. Among thousands of icebergs a huge one was visible at least a hundred miles away. Some time later the visibility began to decrease slowly, finally evolving into a dense fog. A short time later our radar failed. We had been traveling in a loose formation with another Navy tanker and the Coast Guard icebreaker *Eastwind*. We were positioned between the cutter and the second tanker with the *Eastwind* calling out course changes to avoid the ever-present ice. Things went well for awhile.

I was taken off radar maintenance duty to operate the R/T and relay instructions to the helmsman. The cutter instructed us to go left to avoid a growler (small berg), said we were by it and to come back on course. Suddenly the bow lookout rang the bell, pointed ahead, and sprinted aft. It was the big berg we'd spotted six hours before, looking in the fog like the white chalk cliffs of Dover. It was impossible to see either end and the top disappeared in the cloud (fog). We were going full astern for a full minute before making the contact and the bow collapsed as if made of cardboard. The bow had been recently chipped and painted with red lead and as it was hurled rearward, it looked like fire. The action of the waves on the ice had formed a mushy border almost like a fender. It saved us.

Since there were countless targets on the radar, the cutter took an hour to find us even though we blew the whistle at short intervals. We limped into Thule and discharged while CBs shored our bow with timbers for the slow return trip.

The "unsinkable" *Titanic* grazed a small berg in the north Atlantic, was cut open like a can, and sank within hours. The *Sappa Creek* hit one of the biggest bergs on record head-on, but stayed afloat without loss of life or cargo.

While we were back in a New York area shipyard for a new bow, I listened with interest to a port engineer extolling the skills of the Brooklyn Dodger's sensational Jackie Robinson. Outside the gate in a bank of telephone booths, I attempted to call an old friend, but the phone wasn't working. I tried another instrument with the same results. When a loud-talking caller abandoned the next booth, I moved to one that was functioning, but couldn't get it to work either. All else had failed so I read the instructions—"Deposit a dime before dialing." The pay telephone had gone up, by a hundred percent, for the first time since its inception, from a nickel to a dime. Inflation had arrived, never to leave.

Epilogue. My last Arctic voyages were in 1955 on the *USNS Tamalpais*. It had a dull CS of a dah-laden NKJO with 9 dahs and only 3 dits. Charlie Chalk, grayer now, said he was still working for a pittance. And as the only pilot he should be well paid, but the organization was always several jumps ahead of him. The WOPs had stopped coming, and all the young people had left Northwest River for "southern" metropolitan cities. Charlie's son, recognizing how his father had been exploited, disdained becoming a harbor pilot and went south.

Then came the trip I dreaded, a return to Thule. After relating to the *Tamalpais* skipper and mates the horrors of our previous trip, I was chagrined but delighted to find Baffin Bay calm and tranquil. Traveling alone we found nothing more exciting than the aurora borealis that barely flickered.

TWELVE FOUR-SECOND DASHES

by James A. Jolly, R-8/7

During the last few years of high school, I got my first amateur radio license. A big thrill and the result of dedicated individual study—no classes, no tutor—just a secret desire to do something different and unique.

In the summer after my first year of college, I read an article in *QST* magazine about Gallups Island and the exciting opportunity of going to sea as a radio operator on a big ship. It was too good to resist. Correspondence was swift, and within a few short weeks I found myself on a train headed to the East Coast from California. In transit, I met two other fellows headed for Gallups Island to be in the same class. Both were ham operators and both had responded to the article in the *QST* magazine. They were Andy McCloud, from Washington and Herkey Knight, from Oklahoma. We became buddies and remained close friends all through school.

School was fun and interesting. For the three of us it was quite easy, especially the theory and code, because we were hams. This was not unique at the time. In fact, most of the class members held amateur radio operator licenses. But we three were perhaps a little more competitive and a little more motivated. In any case, we were promoted from class R-8 to R-7 at mid-term.

Memories bring back some of the good and some of the bad. Our platoon was sent to help out on the fire at North Boston Pier. That was a tremendous with great clouds of smoke bellowing into the air. It required dozens of fire engines to contain the blaze and still many piers and warehouses were burned. On the Island there was the fire watch, the KP duty, the drill parade, the seamanship practice; they all fade with time into a dim but interesting memory mosaic.

I arrived at the Island in September 1941. We were not yet at war. The Coast Guard was operating the school, known then as the Coast Guard Maritime Service Training School. We had a Coast Guard Captain as our Commander in Charge. From time to time he would call all of the students together and give a lecture on whatever happened to be on his mind.

One of these lectures was important; it very well saved the three of us from what could have been a dishonorable discharge. It was shortly after the U.S. Senate had declared that the U.S.A. had entered WWII as part of the Allied Force. We had been accustomed to taking pictures, and each of us, Andy, Herkey and I, had a camera. On this particular night we thought it would be interesting to take pictures of the code classroom during the evening practice session. This we did. However, the officer in charge decided that it was a secure area and put us on report for violating security.

The next day we were summoned to the mast before the Commander. He was in a rage. We were instructed about the severity of our deed and how such irresponsible acts could cause the U.S.A. to lose the WAR and further, we could be sentenced to dishonorable discharge. What was our explanation? Our only defense was this, "You Sir, just recently in one of your briefings, advised us that if we expected a 'no' answer, then avoid the question and forge ahead. Prove that the act or idea to be carried out was useful and would be beneficial. We followed your instructions." Well, the Commander didn't like the response, but he didn't send us home. Instead, we got lots of KP duty and were grounded for a couple of weeks.

Our class R-7, was tested, received their FCC licenses and were graduated in May of 1942. Many of my classmates were assigned ships on the Murmansk run. I was informed my buddy, Andy McCloud, had gone down with his first ship while on that run. Torpedoed and lost. Some of the others in R-7 had also lost their lives. I was instructed to report to the Captain of the *MT Edmond J. Moran*, a seagoing tug, in Key West, FL.

During the World War II period there were not enough radio men trained and available to put three radio men on a ship so that there would be a 24 hour watch. Most ships had only one radio man. The solution, an automatic emergency alarm system, was installed on every ship. When the radio operator was not on duty he would activate the auto alarm. When a ship was in distress the radio operator on the sinking ship would send a series of twelve four-second dashes with one second of space in between. This would trigger the automatic alarm system on any ship that was within radio receiving range. The radio operator would then rush to the radio room and copy the location of the torpedoed ship.

This was the early part of WWII when the German submarines were sinking several freight ships each day along the eastern coast of the US. The assignment of the *MT Edmond J. Moran* was to respond to SOS calls, often received by the auto alarm, and when possible to place a tow line on the sinking freighter and tow it to the beach. By beaching the torpedoed ship it was then possible to save some of the cargo.

One might expect that these waters would be rather unsafe for a seagoing tugboat; after all, several freight ships were being torpedoed each day. We did have a Navy gun crew and 50 mm anti-aircraft guns. We also had depth charges useful to destroy a U-boat. Fortunately, though, the *Edmond J. Moran* had a defense that was perhaps unexpected. Because we had a diesel electric power plant, the screw turned at about 180 rpm under full power. Merchant ships' engines turn their screw at about 75 rpm. Navy escort ships and sub destroyers also have high speed propellers which turn in the range of 180 rpm. The German submarines had sonar listening devices and could detect the speed of the propeller of the ship that they were about to attack. They apparently did not wish to expose themselves to the possibility of depth charges and so stayed as clear as possible of the Navy escort ships and sub chasers. The propeller speed was our best protection.

Eugene F. Moran and Louis Reid, in their book <u>Tugboat: The Moran Story</u>, published in 1956, summarize the activities of the *MT Edmond J. Moran* in these words:

"Of all the ocean-going tugs that saw service in World War II, none had such a record or performed such a series of hazardous missions as the Edmond J. Moran under the command of Captain Hugo Kroll. This valiant 124-foot-336 ton Diesel-electric craft battled submarines, mine-infested waters, shellfire from enemy ships, hurricanes, and savage seas with a devil-may-care dauntlessness that won her recognition as the tugboat heroine of the war."

I will relate only a couple of adventures to illustrate the validity of the above quote. It was July 7, 1942, we were off the shore of Florida. Miami lights could be seen in the distance. We were towing the *SS*

Umtata, a British registered ship with ninety men and women on board. We had put her in tow in the port of St. Lucia. The *SS Umtata* had been torpedoed while at the dock and sank, but the water at the dock was shallow. The torpedo had struck the engine room and a crew put a cement patch over the hole and refloated the ship. It was our task to tow the *SS Umtata* to the States where her cargo of valuable ore was needed.

We were accompanied by a Navy escort. The moon was out, the sea was smooth, it was about two in the morning. The *SS Umtata* received two torpedoes. In short minutes she was listing and down at the stern. Our crew quickly cut the tow line. The *Moran*, free of its tow, streaked through the sea zig-zagging to escape the U-boat's possible shell fire.

The U-boats were very bold during this period of the war. The U-boat had fired its torpedoes from the side of the *SS Umtata* opposite the side protected by the single Navy escort. The liner was sinking. Her hull was filling rapidly. By now her lifeboats were being lowered on the main deck, and their occupants could be seen climbing into them. The Navy escort made a valiant effort to destroy the U-Boat, which had disappeared under the sea. We fished a large number of the crew out of the water. The Navy escort returned to aid in collecting the crew of the torpedoed ship.

On October 29, 1942, we were part of a single file troop transport convoy approaching Guantanamo passage between Cuba and Haiti. The convoy had been arranged with a large seagoing tug spaced between each of the large troop ships to be immediately available in case of an emergency. Circling the convoy were Navy escort ships. Everyone was a bit tense because the Guantanamo passage was known as a place where U-boats could lie on the bottom, wait for a ship to approach, then come to the surface and fire their torpedoes.

It was high noon, the sea was deep blue and smooth as silk. Suddenly the warning signal flags appeared on the halyards of the Commodore's ship. A U-boat was attacking. Moments passed. Then there in the water to our port bow a U-boat was attempting to surface. It apparently changed its plan and went into a dive passing under the bow of the *Edmond J. Moran*. The *Edmond J. Moran* struck the U-boat—there were a lot of bubbles coming to the surface. A few minutes later, a Navy escort ship made a sounding and sped over and dropped several depth charges. We saw a dark object, something like a whale's back, break the surface and disappear. We kept going, with no more trouble. Apparently the tug's hull had caved in the top of the sub's conning tower without any apparent damage to the tug.

Our days in the warm waters of the Caribbean ended when we were sent to Alaska for the Kiska and Attu action. Alaska was a challenge. Once in the Aleutians the *Moran* found herself in the very center of the fighting. Running supplies, docking and undocking warships in strange harbors with Japanese shells flying and gales blowing, pulling dozens of boats off rocks following storms. These were routine. But not so May 1943.

A vast armada of war ships signalled the invasion of Attu by the American troops. The Navy ships were constantly blasting the shore with their big guns and making bombing runs with their planes. The shore of the island was very steep, not satisfactory for normal landing craft. The technique used was to place a 100 foot by 300 foot 200 ton capacity barge on each side of a seagoing tug. (There were other tugs besides the *Edmond J. Moran.*) The tug would then come along side of a troop ship and men and equipment would be loaded on to the large barges. Once loaded the tug would head for shore. The tug and barges would strike the beach and the tug would tread water to hold the barges against the steep beach. Ramps were put down and bulldozers led the way for the mechanized equipment and troops to follow.

While the landing was in progress, there were bombers overhead and the Navy ships were continuing to shell the Japanese placements. As no surprise to anyone, the Japanese were returning the fire. The *Edmond J. Moran* was strafed during the early hours of the landing. During the first few days of the landing over fifty trips were made from the troop ships to the beach by the tugs with barges on each side. One Japanese plane dropped a bomb within 500 feet of the *Moran's* stern—she took a serious shaking, but no major damage.

I had been on the *Edmond J. Moran* for a year and three months when it was decided to replace the crew. I was sent back to the States as a passenger aboard the *SS Indian Arrow*, a Liberty ship, and arrived in Seattle, September 9, 1943.

In the war years that followed I served as Radio Officer on the *SS Samuel W. Williston*, which made voyages to the Pacific Islands and South America, the *SS Frontenac Victory*, which made voyages to Europe and the islands of the Pacific and as Chief Radio Officer on the *SS Young America*, a troop ship, which brought home Marines from the Pacific. I was involved in several incidences involving the enemy while in the South Pacific, but the exposure to the enemy while aboard the *MT Edmond J. Moran* was unique because of the uncontrolled dominance of the German submarines during the first year after the U.S.A. entered WWII.

35 GIRA MEMBERS RECEIVE SOVIET MURMANSK RUN MEDAL

Some 117 U.S. merchant marine veterans, including 35 GIRA members, who made the dangerous "Murmansk Run" during World War II were honored recently by the Soviet Union at its embassy in Washington, D.C.

Of 730 U.S. merchant ships lost during the war, 97 were sunk on the "Murmansk Run." The casualty rate for merchant seamen during the war was higher than for any branch of the Armed Forces except the Marines. The most tragic Murmansk expedition was Convoy PQ-17, which lost 23 of 34 ships.

At the presentation ceremonies, Soviet Charge d'Affairs Sergi Chetverikov read a statement of appreciation from Soviet President Mikhail Gorbachev, then presented each mariner with his medal and a certificate.

The number of merchant seamen who will receive the Soviet medal is expected to exceed 1,000 after additional award ceremonies are held.

GIRA members honored in Washington on Jan. 31 were:

John V. Brucker, R-50; Richard H. Crosby, R-37; Luther T. Cruse, Jr. R-50; William J. F. Davenport, R-47; Melvin H. Glazer, R-61; Leonard N. Jordan, R-66; Irving Meltzer, R-55; Robert H. Mitchell, R-42; Charles J. O'Leary, R-22; Edward F. Pleuler, Jr., R-3; Henry B. Rowland, R-1 (Hoffman Island); Lloyd E. Slater, R-14; Donald L. Waldrop, R-7; John J. York, R-15. Honored in absentia were: Francis H. Clark, R-19; Donald F. Davis, R-8; Chauncy G. Gercken, R-71; Bernard Grossman, R-61; Michael J. Grosso, R-21; Carl L. Hammond, R-43; David B. Henderson; Roy L. Kidman, R-62; Robert H. Lee, R-B2; Robert S. Luckenbill; Robert F. Mirvish, R-19; Roy C. Moore, R-0C1; John H. Ost, R-68; William B. Poteet, R-0A2; Robert H. Richelson, R-16; Jared S. Smith, R-0B1; Edward R. Stambolian, R-36; Arthur Stanton, R-14; Genevino Tontoni, R-1; William W. Vient, R-57; Edward J. Wilder, R-19.

from news clipping:

SOVIET MEDALS for U.S. MARINERS

by William B. Falk

Staff Writer

On the Murmansk Run, 19-year old Merchant Mariner Hugh Stephens lay in his bunk, trying to sleep, as waves of Luftwaffe bombers dove like birds of prey on his moonlit convoy of Liberty Ships. U-boats bristling with torpedoes stalked them unseen.

"They attacked us every night, all night," Stephens said yesterday of his World War II mission. The night, in the North Atlantic near the Arctic Circle, lasted 22 hours and 44 minutes. "It was mental torture."

The housepainter's son from Minneapolis repeated the Boy Scout oath to block out the drone of propellers, the thud of exploding depth charges, the mental images of another ship swallowed by the black water.

"If you didn't have a way to control your mind, you were gone," Stephens said. "A lot of guys cracked."

At the end of his 20-day run to and from Murmansk, a Soviet port on the Arctic Ocean just east of Finland and Norway, half the convoy of 30 Liberty Ships was gone. But the Soviets had received from them another load of locomotives, ammunition and food to sustain their struggle against Hitler's army. From 1941 to 1945, 800 Allied ships made the Murmansk Run.

Last month, Stephens opened a

brown paper envelope at his Port Washington home and found a bronze medal and a letter from Viktor G. Komplektov, the Soviet ambassador to the United States. On behalf of President Mikhail Gorbachev, Komplektov thanked Stephens for his "outstanding courage and personal contribution to the allied people of the Soviet Union." A 47-year-old debt had been acknowledged.

"It gives you a warm feeling inside," Stephens said. "If we didn't make those runs, the Russians wouldn't have survived, and it might have been a very different war."

Stephens was among more than 250 merchant mariners from the Murmansk Run who were honored yesterday at a ceremony at the U.S. Merchant Marine Academy in Kings Point. Six of those honored live on Long Island: Stephens; Karl Aarseth Jr. of Lynbrook; Bernard C. Flatow (R-44) of Floral Park; Gustave E. Andreassin of Sound Beach; Rosario S. Puleo of Lynbrook and George E. Kraemer of New Hyde Park.

Several years ago, amid the glow of glasnost, the Soviet Union issued commemorative medals to the British sailors who provided the military escorts to the American Liberty Ships from 1941 to 1945. Reminded of the merchant mariners' participation, the Soviet bureaucracy delayed, but finally issued medals to the mariners last month.

"We had given up, and then, there it was in the mail," Stephens said yesterday, with a laugh.

Like many other mariners at the ceremony yesterday, Stephens is now a white-haired veteran of years of service on the world's seas as a tanker, tugboat and troop-transport captain. Nothing in that time, he said, matched the intensity of the 20 days he spent on Murmansk Run aboard the Liberty Ship John Ireland.

The Ireland was part of a convoy of 30 Liberty Ships and dozens of military escorts that made the run in the winter of 1944 from U.S. bases in Iceland and Scotland. Despite constant attack by Germans, Stephens' convoy—huddled together for protection—reached Murmansk intact. But on the way back, a fierce North Atlantic storm scattered the ships like toy boats, and divided they fell under the German attack.

Stephens remembers seeing Norwegian refugees picked up on the return trip watching the flash of a bomb hit another Liberty Ship. "These women and children stood there on the deck watching the ship go down, crying. They figured that their husbands and fathers were on that ship."

"At nineteen," Stephens said, "that leaves quite an impression on you."

FIVE HUNDRED AND FIFTY

A unique event in U.S. Maritime history occurred in the fifth month of 1992; the last two operational Liberty ships in U.S. waters were under steam and sailing on the same day. These remain of the 2,751 Liberty ships built during the second World War—by a nation which had launched but two dry cargo ships from 1922 to 1937.

Both of these ships are approaching their 50th birthdays; the *John W. Brown* in September of 1992 and the *Jeremiah O'Brien* in June of 1993. Both were built on the east coast; the *O'Brien* at South Portland, ME, and the *Brown* at the Bethlehem-Fairfield yards in Baltimore, MD. The Brown now resides at her birthplace, Baltimore, while the O'Brien is home-ported in San Francisco on the west coast. Combined, they have spent a total of 97 years on the waters of the world.

A total of 97 Liberty ships would have celebrated their 50th birthdays by May of 1992. They are all gone. But their total illustrates the burgeoning volume of Liberty ships being launched for the American Merchant Marines in the early stages of the war: There were two Liberties launched in all of 1941; in January of 1942 three; in February eleven; in March nineteen; in April twenty-four; and in May thirty-eight. Of the sixteen launched through February, nine were lost; and of the thirty-five through March, sixteen—almost fifty per cent. Twenty-seven of these ninety-seven ships did not make it through the war.

Most were lost as a result of, or initiation by, a German submarine attack, only three by Japanese submarine activity. However, the first Liberty ship sunk as a result of enemy action was the *John Adams*, in May of 1942, by a Japanese sub near New Caledonia The *Albert Gallatin* was struck by three German torpedoes off Savannah—survived—then was sunk by a Japanese sub in the Arabian Sea.

A number were rendered helpless, and were sunk by friendly fire. The *J.L.M. Curry* suffered structural failure. The *Rufus King* went ashore, and broke in two. The *Stephen Hopkins* was sunk by a German surface raider, the *Stier*, off the coast of Brazil; but the *Hopkins* had inflicted such damage that in short order the *Stier* blew up and followed her beneath the waves.

In 1492, the *Nina*, the *Pinta*, and the *Santa Maria* sailed west in convoy to discover the riches of the New World and to claim these lands for Spain. Five hundred years later their replicas visit our ports. On the fiftieth anniversary of the invasion of Normandy may not three ships—the *Jeremiah O'Brien*, the *John W. Brown*, and the *Lane Victory*—sail again? This convoy would sail east, not for conquest, not for war, but in peace. The holds of the vessels filled with gifts from America for the poor and destitute of Europe and Africa. In return, we would accept, with gratitude, gifts from Europe and Africa for our own poor and destitute.

May we not dream that these ships might also proceed through the Mediterranean, the Suez, and back across the Pacific? The Great White Fleet, sailing around the world in the first decade of the twentieth century, would be followed by the Small Gray Fleet in the last decade.

The nation that built two cargo vessels between the first and second World Wars built close to five thousand ocean going vessels in a four year period five decades ago. Is it inconceivable to think that this same nation could not perform the above should it will to do so?

Robert L. Gisslow
7-4-92
KB6YYL
S.S. Jeremiah O'Brien

A TORPEDOING IN THE MEDITERRANEAN

At Christmas time in 1943, I graduated with R-46 at the tender age of 19. After a short first voyage, I joined the *SS Thomas Masaryck* in New York, and after the usual delays we sailed in convoy to Norfolk where we joined a larger convoy bound for the Mediterranean. Our eventual destination was the Persian Gulf with military cargo for the USSR including P-47s on deck.

After we passed Gibraltar things began to happen. In the evening of April 11, the convoy was attacked by high-level bombers and torpedo planes. They lit us with flares and then the torpedo planes came in from the convoy's shore side.

Since there were three ROs, I was not on watch. My duty was running ammo to the 20 mm cannon midships, so I saw the whole attack. We had a British AA cruiser in convoy and she put up so much cover it appeared she was exploding. We fired on the torpedo planes without success except that we shot down our barrage balloon. There were ship losses but we never knew who they were.

A major part of the convoy split off at Italy, and UGS-37 went on toward Suez with about 25 ships. Off the Libyan coast near Tobruk/Djerna, the convoy was attacked by submarines. Our first warning was the general alarm bells and depth charges. Our ship was in the rear of the convoy—in the "coffin corner"—a vulnerable spot, I was told. Dave Young, another RO, and I picked up our lifejackets, ran out to see what was going on and were joined by a large part of the crew. About that time the *SS Meyer London*, another Liberty ship, was torpedoed just to our port side. The torpedo hit the *London's* #5 hatch and broke her back. She went dead in the water.

We were standing at the rail, still gaping at what we had seen, when the bos'n shouted, "Get away from the rail. We'll be next!" I took a couple of steps back, when suddenly a torpedo hit us in the #3 hatch just forward of the midship housing. The hatch was filled with acetone stored in drums, which burst into flames as high as the mast and covered part of the flying bridge.

I ran to the radio room to see what the man on duty was doing and found the radio console tipped over. I peeked under it, expecting to see our man there, but he was gone. I picked up the portable transmitter and ran for the boat deck, followed by Young who had been getting some gear. When we got to the boats it was apparent that the forward lifeboat could not be launched due to the flames, but the crew along with the skipper, was launching the other boat. I took a manline into the boat, which was halfway down the side of the ship, and yelled for Dave to let the transmitter down to me on a line. About that time, the after fall jammed and the lifeboat went into the water bow first despite the captain's shouts to stop lowering away. The ship was coasting fast enough for the boat to scoop water and the plywood craft broke into a dozen pieces. We were in the water!

Behind me on the manline was a kid from the Navy gun crew who I had seen leap off the flying bridge barefooted and then onto the manline. In the water, he grabbed me around the neck. His eyes were as big as saucers as I saw the ship's prop turning slowly over his shoulder. When we surfaced, I yelled at him to let go, but he screamed, "I can't swim!" I pulled off my lifejacket and tied it on him—I didn't feel much like a hero doing it—I was just glad to get rid of him since he was choking me. I did a quick survey of the scene and saw a dozen flotation cans from the broken-up lifeboat all around us. I took one of them and had my first chance to look at the action.

The *London* was nearby, the convoy was moving off in the distance, and our ship, blazing like a torch, was still moving and had circled around and was nearing the *London*. She came very near colliding with the *London* but luckily missed. Our crew was in the water in the wake of the *Masaryck*. It was evening now, and an offshore wind was making the sea choppy.

About that time, I spotted an empty liferaft about 100 yards away—it had been blown off the ship, or a crewman had pulled the pin. I swam to it and just as I put my hands onto it I thought "SHARKS!" and flew into the raft. One of the ABs came along a few minutes later and joined me. We got the sea anchor into the water to slow our drift.

As other crewmen joined us we were shocked by the sound of cannon shells buzzing overhead and hitting the water not far off. The gun crew on the *London* had seen something in the water that appeared to be a periscope and were firing the 5-inch rifle at it. We could see what they were shooting at. It was one of those yard-long lights on a lifering which was counterweighted so it would stand in the water. They put six or eight shells near it before they had to abandon their ship. All the while, we were yelling at them to stop.

Just when our worries seemed over, we saw that the sinking *London* was drifting down onto us. If we had been trapped along her side, we would have gone down with her. We cut loose the sea anchor, broke out oars and tried to move the overloaded raft away. Progress was very slow. Everyone was scared. I remember one kid from the mess crew crying, "We're all going to die! I want my mama!" Another crewman told him to shut up and paddle with his hands.

We just cleared the *London* before she sank. We were close enough to hear the suction when she went down. The real danger was over. Now we had to wait.

Then we heard shouts from a few more crewmen who wanted to join our party. The first engineer came paddling over yelling, "My legs are gone! My legs are gone!" I got him by one arm, another man took his other arm and we pulled him aboard. I turned my head so as not to see his legless torso, but we were relieved to find that he was in one piece after all. Funny, he had on his work khakis, but had stopped to get his uniform blue overcoat with shoulder boards—I decided that he figured if Jerry picked us up he would get better treatment as an officer.

Our raft was seriously overloaded now, and shipping water with every wave. It was miserable. A couple of the escort vessels had remained behind the convoy to attack the sub and they were running around dropping depth charges—fortunately, they didn't harm the swimmers. Finally, one of them, the *HMS La Malouine*, built in Canada for the Free French but manned by a British crew, came over and picked us up.

A couple of sidelights. Our second mate had been rescued by the same ship after a torpedoing in the North Atlantic the year before and the exec on the *La Malouine* remembered him; and when I went into the wardroom, I saw a fellow R-46er, Bill Cowger from the *SS Meyer London*. I have often wondered what the odds were on two such meetings.

We landed in Alexandria, Egypt a week later and I spent five weeks waiting for repatriation. We received no pay for that time—we had been "signed off" the day of the torpedoing. However, Alex was a pleasant place to spend the time—but that is another story!

By Harold O. Hanson, R-46

A WITNESS TO THE FIRST "AIR-TO-GROUND MISSILE: AND THE FIRST SMART BOMB"

by Glen Trimble, R-47

We believed it to be another trip delivering 500 pound and 1,000 pound bombs to bases in the Solomons and New Guineas (my third trip) on a C-1, the *SS Cape Constance*, with Dorm Dahlen, R-21, as Chief Radioman.

Our first two trips had been about average: into Esperito Santo, New Caledonia, two stops at Guadalcanal, Treasury Islands, and Bouganville. During unloading at Bouganville, we were intermittently shelled for four days and nights from guns in caves up in the mountains overlooking Empress Augusta Bay.

We dropped hook at Finchhaffen, a New Guinea base just beyond Lae, and sat there for 30 days. We went up to Hollandia and found we were part of a convoy making up for, per rumor, either Formosa or the Philippines. This was the first time I had been in a convoy. It didn't take long to catch up on rumors of the up-coming Philippines invasion, especially when McArthur was rumored to have gone out to a cruiser. We left on the 18th of October and were assigned #13 in the convoy! We arrived San Pedro Bay on the night of October 22nd and waited till morning to go into the Tacloban area, off Red Beach. We were initiated with a smoke screen, Navy ships shelling inland areas, and troops going upshore in LCI units. Twenty-three planes came over about 0820, seven were shot down, with one crashing into a loaded LCI inbound, and one into a Liberty ship. This was enough to get our attention.

Due to incessant rain, typhoons, etc., getting coral down for a runway and storage areas became almost impossible. The crushed coral just kept sinking. There was no storage for bombs, so we were used as a warehouse for about two months. During this time we became aware of the "Kamikaze" program, even though it was hard to believe. We now know that there was only about 180 available planes in the Philippines on October 20th, but by the 23rd, over 800 had been moved in from Formosa and the homeland. The Japanese Navy and army both developed their Kamikaze groups due to their lack of operational carriers and experienced pilots. These were truly the first guided "air-to-ground" missiles!

During the seven plus weeks, our area had over 320 air raid alerts and 114 actual attacks. The *Cape Constance* suffered one bomb off the bow about 30 feet and one 70 feet off the stern (we later found out this sprung the prop), and one Kamikaze hit the starboard king post of #3 hold, exploding and part of the aircraft going over the bridge to the stern setting several fires on the ship. With all of this we suffered no fatalities, only some shrapnel wounds and blast burns to some of the gunners, including the Gunnery Officer. At least three of us received the Mariners Medal during this period. On the next leg up to Mindoro two of our fellow ammo ships were not so fortunate. The *SS John Burke* and the *SS Lewis Dyche* both were hit by Kamikazes and both vaporized! The ship we say hit at Tacloban included the *SS Jeremiah Daily*, that took a Kamikaze while anchored aft at our position. She was loaded with troops and about 120 were killed and over 150 were injured with burns, including the ship's crew, (including the Radio Officer, Henry N. Brown) and the Armed Guard crew.

We left Leyte in December and soon found the prop problem because of bearing overheating. it took 31 days of stop and go to get to Pearl Harbor for temporary repairs. We were in Pearl for about a week and went on to San Francisco without any problem.

We did have some Army people from New Guinea being returned to the States and a couple of days out of Pearl, one jumped overboard during the late evening. That was the first time the ship had transmitted at sea.

ANDREA DORIA COLLISION

by Raymond B. Maurstad, R-93

This is the *Andrea Doria* story by Ray Maurstad presented at the ARLL Convention in Fox Borrow, MA.

I was a 28-year-old radio operator with 12 years of shipboard experience and had gone out to sea at the end of World War II at the age of 16. ... I was on leave from my regular ship, the *SS African Dawn*—a combination passenger-cargo ship running to Africa—when I was requested to go on board the tanker *Robert E. Hopkins* as relief operator. The regular operator had become ill. I went and I took that ship for one trip, and I was no sooner on it when ... this (collision) happened.

There were two books written about the *Doria* and the *Stockholm*. The first one was written only three years after the collision by Alvin Moscow and is called Collision Course.

Saved was written 20 years later in 1979 and is a much better account of the incident. Neither book concludes which ship was guilty, and that's still not concluded today. They made an agreement in the courts to settle the lawsuits against the two steamship lines and it remains a mystery today.

Perhaps the closest anyone has come to answering the questions of how it happened and why it happened is Mr. Crothers, who is now retired in Florida. He was completely obsessed by this incident. He surmised from his calculations that there had to have been a third ship between the two ships. Both were watching it and reducing their radar ranges to keep it in sight. Much too late, at a combined closing speed of 40 knots, they increased their ranges and finally saw each other. They thought they were making the (correct) turn ... but it's a little like meeting someone walking down the hall: The two of you are zigging and zagging, and at the last minute you both decide to zag. And that's what they did.

What I'm going to talk about is the actual communications between the ships in responding to the SOS and going to the rescue. All of these communications took place on 500 kilohertz, which is the marine distress and calling frequency. It's the frequency that ships must mandatorily keep a 24-hour watch on while at sea.

The point of collision was near Nantucket Light. We were 45 miles away with the *SS Private Thomas* and the *Cape Anne*. The *Il de France* was outbound and it (the collision) happened behind the *Il de France*, which is the ship that did most of the rescue work.

What a lot of people don't know is that the *Il de France* sailed on for 30 minutes before deciding to turn around and help. Why? They had 900 passengers of their own, and the captain was a relief skipper who was uncertain what he should do. He deliberated and agonized over it, and finally decided to turn his big ship around and go back. Had he not done that, my own personal opinion is that more than 1,000 people would have gone down with the ship. As it turned out, 40 died at the point of impact, and 11 more died in the confusion of trying to reach lifeboats or falling over the side.

Let's set the scene: It was 11 p.m. on the 36th of July, and the *Doria*, which was Italy's largest and newest ship, was approaching New York at full speed with 1,700 passengers and crew members. There were very large patches of very heavy fog.

The *Stockholm*—a modern liner—had cleared New York and was east bound heading back to Sweden with 550 passengers and crew members. The ship that I was on, the tanker *Robert Hopkins*, was southbound out of Fall River, MA. We were headed for Houston with 42 on our crew and no passengers.

The two ships collided at 11:10 and for the next 12 hours I can tell you it was the most exciting thing in my life.

I was sitting at the desk next to the radio console and I should have had my auto alarm on. I wasn't worried about it because I don't have to put it on until I leave the radio shack, so I had not switched it on. I was doing some payroll work, because I was also the ship's purser. Once in a while you might have a distant station calling Tuckerton switching off to a working frequency.

Many of you are familiar with the two silent periods every hour on board a ship: from 15 minutes after the hour to 18 minutes after the hour—a three-minute period of absolute silence—and from 45 minutes after the hour to 48 after. Absolute silence. Well, you see, this happened, it happened after a silent period. I think it was WSC Tuckett Radio had come on 18 minutes after the hour and said, "Shift up to 436." (436 or 456, I forget what it was in those days.) I remember hearing that and went on with my paperwork.

It was absolute silence when I was startled by the SOS signal reeling at 15 words per minute. But what jarred me out of my chair was the strength of the signal. ... You can listen and tell when a signal is 500 miles off, 200 miles off, or next door—and I knew that thing was next door. Boy, was it next door! And that's what was so frightening about that SOS. It was so close that I was wondering if we were going to collide with whatever that SOS was coming from. It nearly blocked the receiver.

What a lot of people don't know is that the *Stockholm* sent a Triple X (XXX), which is one level down from an SOS. It means: "The next thing I transmit may be an SOS." What a lot of people didn't realize was that the SEJT—the *Stockholm*—could not hear the *Doria* because she was 'plowed' 75 feet into the side of the *Doria*. *Doria*'s SOS transmission blocked the *Stockholm*'s receiver completely because they were now one mass. Consequently, *Stockholm* didn't realize she was sending her XXX on top of *Doria*'s SOS. I could hear them both distinctly and clearly.

Two minutes later, in the dark and the fog, *Doria* announced that they had collided into a ship. I'm sitting there alone. The captain was on the bridge with a small crew, his head under the radar hood navigating the ship through fishing boats in the fog. Pandemonium had set in, and now the *Doria* was asking who had collided with her.

Positions were given for the first time and now things were getting clear. From the time when the SOS started, it was eight minutes before I transmitted—and believe me, I sat with my hand over the key ready to transmit for the whole eight minutes. Why? Because I wanted to be first (to respond). Why? Because it was my duty. Why was it my duty? Because I was right there. And I wanted that ship to know somebody had heard them.

I came down on my key the moment (the Do*ria*) finished his position message. I told him that I acknowledged and to stand by for our position. The *Doria* came back immediately and said three little letters: "R—T—U. Roger. Thank you."

I immediately went to the bridge carrying their position on a piece of paper. The captain had his under the hood and it was dark in the wheel house. I came from the lit chart room, went into the wheel house, and stood there for a few seconds to get accustomed to the darkness. I saw a silhouette over at the radar and I quietly said, "Captain, I have an SOS message." He said, "Sparks, just a minute. I'm busy with these fishing boats." I said "Captain, you'd better come now. It's right next door."

He flipped out of the curtain right away. We went into the wheel house and he said, "Let me see the message." He grabbed his parallels, struck the position right away, and his comment to me was: "My God, it's only 45 miles away."

The captain said that he would plot our position and ordered me back to the radio. The *Cape Anne*, a small ship with only four lifeboats, gave its position and it was the closest ship. A Norwegian ship, the *Laura Maresk*, was only 150 miles out and they came on.

Twenty minutes after the SOS was sent, the *Il de France* came on and said to the *Doria*, "How do you read me?" The *Doria* replied, "Roger. I need immediate assistance." The *Il de France* said at this point that she would assist. That was 33 minutes after the SOS.

The *Stockholm*—now aware of its situation 45 minutes later—made assesments and gave their position. They were still investigating their damage. They had penetrated into the *Doria* 75 feet—killing 40 sleeping people. They backed out, allowing waster to gush into the *Doria* (and) causing her to list. The *Stockholm* was standing off inspecting her damage.

We're hearing about this from the Boston Coast Guard about 45 minutes after it happened. *Cape Anne* is on its way, the *Stockholm* is badly damaged, her number one hole was filled with water. We had to straighten our position and ask the *Doria* to lower their lifeboats and row them over. We, in turn, would pick them up.

At this points, *Doria*'s screeching high level dropped to barely audible. Mind you, we're still next door to each other, but the signal is very,very weak. It's obvious that the *Doria* had weak or damaged batteries, or lost her main generator power to the radio room. From that point on, it was at times difficult to read the *Doria* clearly.

Then the famous message: "We are too bending. . . . We are listing too much. . . . It's impossible to put our lifeboats at sea—please send us your lifeboats.

Cape Anne to the *Doria*: "Your message is hazy, but we are sending our boats."

The *Doria* begs for more boats. . . . "Many passengers in the water. . . . " There couldn't have been too many who didn't get out, because like I said, only 11 after the initial 40 were victims.

The *Thomas* advised that they had eight lifeboats and asked for the *Doria* to play their lights on the boats and signal for the boats coming alongside. The *Cape Anne* was letting us know when the boats were coming back. The *Doria* (sent) some more distress calls. To the *Stockholm:* "Do you have your lifeboats in the water." "No, not yet, we're watching our number one hatch fill with water."

The *Il de France* neared the damaged *Doria* and radioed, "What can we do?" The *Doria* replied "We have immediate need!" Now this is the first message we get as to the potential size of this catastrophe: She said, "We have 1,000 passengers and 500 crew." I said to myself—and not quietly—"Oh—my—God!" And it echoed around my radio room because that's the way it hit me. Up to that point in time I knew it was a passenger ship because they were saying "Many passengers . . . " and "Passengers," but I didn't think in terms of 1,000 and 500 crew. . . .

(Editor's Note: This letter was written by Bernard C. Flatow to Tom Cruse on January 14, 1992.)

On July 7, 1991 I was speaking to Ian Millar, administrator of the Trident Archives, and he inquired if I was available and willing to permit an interview by the bureau chief of Soviet TV and Radio. The Russians were hastily trying to obtain the names of merchant seamen who had been involved in the convoys to Murmansk, Russia during World War II. They were preparing a documentary to be shown on Russian TV. I informed Ian Millar that I would be available for such an interview and had no objection whatsoever.

The very next morning, Monday July 8, I received a telephone call from Vladimir Zvyagin, bureau chief of Soviet TV and Radio, informing me that my name had been give to him by Mr. Millar and that, if possible, he would like to interview me that day. I informed him that I was not available for the next two days and an appointment was made for 10 a.m. on Wednesday, July 10.

Promptly at 10 a.m., July 10, a large, black limousine pulled up in front of my home in Floral Park, NY and the occupants proceeded to unload TV equipment in preparation for the taping of the interview. We had preliminary discussions pertaining to ground rules and the types of questions that would be asked.

Some of the questions were: What was Murmansk like when you were there? Was there much destruction? What supplies did you bring on your ship? What did the American sailors think about he Russian people? What do you think about the Russian people of today? What route did your ship take? How many ships were in the convoy? How often were you attacked by the German planes, submarines and surface ships? How many ships were sunk? Were there any survivors? Did your ship stop to pick up survivors? How long did it take to get to Russia? How long were you in Murmansk? Did the Germans attack the city while the ship was in port?

I answered all of the questions honestly and without reservation. The interview took an hour and 45 minutes and the video tape was taken directly to Kennedy Airport for immediate dispatch to Moscow.

It is my understanding that the TV program has been shown on Russian TV and it was very well received by the Russian people. I have been promised a copy, but as of this writing it has not arrived. I do not hold much hope that it will ever be sent to me since there is so much turmoil and upset in the Soviet Union at this time.

President Gorbachev authorized the issuance of a medal to me, as well as the other 250 seamen who participated in the Murmansk Run. The medal was sent by the ambassador of the Union of Soviet Socialist Republics in Washington, DC. The accompanying letter reads as follows:

Dear Mr. Flatow,

On behalf of President M. Gorbachev, I am honored and pleased to inform you that you have been awarded the Commemorative Medal "The 40th Anniversary of the Victory in the Great Patriotic War (WWII)."

This award is a token of recognition of your outstanding courage and personal contribution to the Allied support of the people of the Soviet Union who fought for freedom against Nazi Germany.

Please accept our heartfelt congratulations and wishes of good health, well-being and every success.

Viktor G. Komplektov

A formal ceremony was held at the U.S. Merchant Academy at Kings Point, NY on November 9, 1991, at which time other medal recipients were recognized.

I was a radio officer on the *SS Thomas Donaldson* and sailed to Murmansk during August-October, 1944. The *Donaldson* was sunk by the German U-boat, *U-968* (Westphalen) on the return trip to Murmansk.

I sailed in all of the war theaters and I fully agree with Ian Millar in his speech to the Merchant Marine Veterans at the Merchant Marine Academy on November 9: "The Murmansk Run was, without a doubt, the most destructive and bloodiest; where the ships were bombed, strafed and torpedoed around the clock, and where survival from sunken ships often meant a cruel death in the icy Arctic Ocean."

MEMORABLE SEA STORIES:

"I think my most memorable time is when we came back into New York Harbor and there was the Statue of Liberty. Now that was the most beautiful lady I've ever seen—other than my wife."

"The armed guard gun crew came back from liberty in Galveston, TX one evening and they'd found a puppy in the alley. It was lost, or abandoned or something. So anyway, they brought it on board.

It had the best of all worlds, certainly better than any back alley in any city. It got named Salty, grew up and learned very quickly the ways of a ship.

It was amazing to watch her climb up the ladder—this was a liberty ship—and if the ship was rolling or pitching when you were climbing the ladder you soon found yourself almost falling off. . . . Salty would wait for the pitch to go the other way and then would go up the ladder."

" . . . I remember . . . while eating on ship, the best thing we had was swiss steak—it was out of this world. So one day I'm in the kitchen and I'm looking at this hunk of meat that had moss growing on it—it was green—and I said to the chef: 'What are you making out of this?' He said, 'Swiss steak.'"

" . . . This was probably in 1944. There were a number of aircraft that had ditched during the hotter activities around Sicily. Being cooped up on the ship with nothing to do Thurston and I decided that we would take a donut raft and investigate on shore and maybe get a souvenir or two. So we paddled this raft—and it took awhile to get there.

" . . . When we started thinking about getting back, we knew it would be kind of tough, so we negotiated with some Italian fishermen to get us back to the ship. For a carton of cigarettes they would row us back to the ship.

"When we get back to the ship, off we go to get this guy his cigarettes. In the

At Sea

Clockwise from right, Merchant mariners fought both weather and war. Loading nitrate from the Maria Elena mine in Chile, July 1945, SS Frontenac Victory. C-1 type ship, SS Mormaclark. Liberty ship, SS Samuel W. Williston, 1944.

Clockwise from right, radio room, on MT Edmond J. Moran, 1942. SS Derbley. SS Pan Georgia, Cease making smoke, proceed straggler route. Graf Spee, 1945, Rio de la Plata. MT Edmond J. Moran in Alaska, 1942-1943.

meantime, the captain had noticed this foreign-looking boat alongside, and he had been warned that there might be saboteurs, so he gets this big ol' pistol out of his room, and goes out there waving it at these four Italian fishermen who are out there waiting for their payment. They were scared off, and we never got to pay that poor fellow."

"...In Augusta, you usually anchor waiting your turn to go through the straits of Messina to go up along the west coast of Italy. These native boats—we called them bum boats—one came alongside and he had a guitar with this beautiful oil painting (on it) of Mount Etna, which was an active volcano. He wanted to sell it. I said, 'How much?' He said, 'Two cartons of cigarettes.' Guess what a carton of cigarettes cost us aboard ship then? 80 cents. $1.60 for that guitar. ...

"...Well, anyway, I was determined I was going to learn to play the guitar. I had a pal I went to high school with and we used to take out girls in his brother's car after we shined it so you could see your face in it. We paid for the gasoline with our fifty-cents allowance. On top of the Blue Hills overlooking Boston Harbor 22 miles away, we would tell the girls that we were going to watch the submarine races. You couldn't see the harbor, let alone watch any submarine races. But he could play the guitar, and we had a hell of a time.

"So I was gong to learn how to play the guitar. I practiced and I practiced, and if you've ever been aboard a liberty ship you know that your quarters are on the same deck as the captain. And there's a little walk along the back called a 'captain's walk,' and that's where I was trying to play the guitar.

"Finally the captain came over and he said, 'Bill, it's you or the guitar.' I said, 'What do you mean, captain?' 'Exactly what I said! One or the other is getting off this damn ship'

"He gave me the option of giving him the guitar to put in his closet in his cabin, and when we got back to the states, he gave it back."

"... We spent the first half of the winter in Archangel, the second half . . . anchored about three miles off Murmansk.

"The most interesting thing that happened to us up there occurred while manning the guns. ... We were only 7 1/2 minutes flying time away from a Stuka German dive bomber, so we manned the guns 24 hours a day, seven days a week. They flew over very regular; on the half-hour, on the hour—you could set your clock by them.

"Our orders were if anything in the air came within range of your ship, you shoot it down. About three days before Christmas, an aircraft came down into the harbor, (which was) like a gigantic ditch about eight miles wide with mountains going 6,000 feet on each side. This plane comes right down the middle with ships anchored on each side, and four ships, including the one I was on, opened fire. When he got alongside our ship, he was hit by our weapons and went into the water. The pilot ejected and landed in the water and was picked up. The aircraft was recognized before we fired: it was a P-39 Belaire Cobra made in California. But the orders were to shoot, so everybody shot.

"The very next day, a Soviet admiral with gold right up to his elbows came out with the harbor master and presented a certificate to the captain of our ship, commending him for the fine accuracy of the gun crew for shooting him down. We asked what happened to the pilot, and they said, 'We'll shoot him tomorrow..'"

Taken from German submarine captain Stanley O. Strofford, commander, U-boats, 17.5

"...During the first part of the war, everything was hunky dory. Every time I sunk one of your ships, my name was automatically signed to it: Stanley Oriel(?) Strofford—get it? SOS.

"I had 37 SOSs notched to my periscope when my luck ran out. After we were depth-charged out of the deep and abandoned the U-boat, we were picked up by a liberty ship and turned in to the United States. There I was ... for the remainder of the war. ...

"While I was in the prisoner of war camp in No Where, GA, I was always the joke of the camp. Every morning when they served a particular dish, everyone called me by the dish's name. It drove me crazy! And when you're talking about mess hall food, you know what SOS means! And they called me SOS every time they got, and they used those same words!

"It is with great difficulty that I wish you radio officers good luck, but I will."

MY THIRD AND LAST TORPEDOING

by Lloyd "Pete" Maffett, Jr., A-2

I was interested in morse code communications, and we had a little class in Jacksonville. About four or five of us—Billy Brewer, Leon Aboud, Thomas Manning and myself—went up (to Gallups) about the same time. We got up to about 18 words a minute down here, and believe me, I never would have made it through Gallups if I didn't have that advantage. ... I (also) had two years of typing morse code in high school and that gave me a good deal of extra time to fight the math, the theories and the written stuff I was a little slow at.

But I got lucky and got all the way through. Manning didn't. He dropped out ... Leon Aboud dropped out. Billy Brewer completed and graduated from the Island, but he flunked his FCC exam. We took the FCC test on Saturday and he was drafted into the Navy on Monday. They didn't waste any time with him—he was overseas within two months.

Right now I'm interested in what the Soviet Union is doing: They are giving participation medals to anybody who was up there. I spent the entire winter of 1942 in Bogota, Archangel and Murmansk. I kind of want to get my hands on one of those medals. I think I earned it.

I was picked up and rescued out of a lifeboat by the British on two occasions: the first out of the *Stockport*, and a second time by the *Pert*.

We had been in some heavy weather on the way up from Porta La Cruise, Venezuela, with a load of crude for Imperial Oil at Dartmouth, Nova Scotia, and—crack! Cracks across the deck, and cracks from keel to keel along the bottom. We were streaming oil, and we had an oil slick all the way back to the horizon for about a third of the voyage.

And they weren't going to let us into Halifax Harbor. But due to the threat of submarines, they permitted us to come in and discharge our cargo and got us out of there in a hurry.

On April 5, 1945, we were struck by what was assumed to be an acoustic torpedo which homed in on our propeller noise. It blew the propeller off, and the rudder off, and put a hole in the stern of the ship ... large enough to drive a tractor-trailer through.

But anyway, she went down. ... I got off the SSSS right way. It was about 7:30 in the evening before the skipper put out the nightly estimated position report. It had always been my decision to never sail with a skipper who was one of those types who said, 'You are not to use that radio unless I give you orders to do so.' I was sailing with 'Captain Lemonloose' at that time.

Anyway, of course, I had everyone standing by up and down the coast before I could get out a position. After I got our position off—I think it was about 20 minutes—there were a couple of Navy fighter planes buzzing us. I had already indicated that we had four boats in the water and 42 men. Then, maybe an hour later or so, a Catalina came out—a flying boat—with its landing lights on searching for our boats. By this time it had become dark and we had drifted out of sight.

I got confirmation later on that night that we had both Coast Guard and Navy vessels picking up the men in the lifeboats.

About 10:30 that night, the destroyer *Mayrant 402* came up alongside and stayed with us all night. We were rolling back and forth in the swell with about a third of the ship under water back up to the main deck. The engine room had filled and all the gear in the engine room was crashing back and forth on each roll.

It was the five of us who remained aboard: the skipper, the chief's mate, the gunnery officer, one AB and myself. About every hour the *Mayrant* would use her blinker lights to signal over and the skipper would say to me, "Hey, Sparky, they want you again." They would inquire if we were taking on any more water and wanted to abandon, or anything like that.

Now this is another discrepancy appearing in Captain Moore's article and in his book: We were strictly volunteers—we were not asked to stay aboard by the skipper.

The next morning (the *Mayrant*) wanted to take us in tow, so they shot over a line which we secured. Then they tried to pull a towing harness over, which was just an impossibility. We couldn't get the damn thing halfway to us, so we lay there rolling in the seas for another night. The next day, a couple of Navy tugs took us in tow.

The first thing I did after I was picked up by the *Stockport* was to go up to their radio room and ask for a copy of my message, my SSSS, which the radio operator—whose name was Al Livingston—provided me a complete on a Marconi radio gram form. The company was getting copies of all the traffic that I was sending, thanks to the efforts of Raymond Pierce, who was standing watch down at Fort Mittland Marine Terminal and copied all my messages and comments, forwarding them to the marine department. I saw copies of those after I got back to Philadelphia later on.

In the meantime the company had ordered a couple of seagoing tugs to come out. They met us about break-water, took over from the Navy tugs and took us into the shipyard where the ship was repaired and eventually put back to sea, though I never sailed upon her again.

So that is my story concerning my third and last torpedoing. Incidentally, they bird-dogged that sub. We were torpedoed on April 5th and they sank the sub on April 9th.

(Editor's Note: Donald Davis, of Memphis, TN, submitted the following letter written to him by fellow Gallups Island classmate Lloyd "Pete" Maffett. In April, 1991, Davis wrote of Maffett: "I doubt if there are many other seamen (who) survived three torpedoes. It would be interesting to inquire if this record has been equalled or passed." According to Davis' note, Maffett then resided in Humelstown, Pennsylvania.)

DEAR DON,

MOTHER WROTE SAYING YOU HAD SENT AN X'MAS CARD. I ASKED FOR HER TO SEND ME YR ADDRESS, I JUST RECEIVED IT THIS MORNING.

I'M GLAD TO HEAR U ARE STILL AROUND, HAVE U QUIT THE SEA? THE IDEA HAS DOMINATED MY MIND FER SUM TIME, BUT IT SEEMS ABT EVERY TIME I MAKE UP MY MIND TO QUIT, I GET A RAISE OR SOMETHING OF THE SORT. SO AM STILL HR—HI.

HV BEEN WORKING FER "ATLANTIC" THE PAST FOUR YEARS, AND IT SEEMS TO BE A VY GOOD OUTFIT.

WAS TORPEDOED THREE TIMES DURING THE WAR, ON THE S/S HAHIRA KUZC, S/S PARISMINA KDAG, (LOST AN OPR. "F. EVANS" ON THAT ONE), AND ON THE S/S ATLANTIC STATES. FIVE OF USE SPENT A COUPLE HECTIC DAYS ABOARD THE "STATES" AFTER SHE WAS HIT, BUT FINALLY WERE TOWED IN TO BOSTON. SHE IS IN SERVICE NOW.

I SPENT ELEVEN ES A HALF MONTHS OF THE PAST YEAR IN THE INDIAN OCEAN, RUNNING OUT OF ABADAN IRAN TO AUSTRALIA, SINGAPORE, INDIA AND ARABIA ABOARD THE S/S PRAIRIE GROVE KYSF; IN PERSIA I RAN INTO A BOY I WENT TO HI SCHOOL WITH AND WHILE IN SYDNEY RAN INTO ONE OF MY OLD BUDDIES FM LAWRENCEBURG TENN. HE MARRIED AN AUSTRALIAN GIRL, THEY BOTH ARE IN THE STATES NW. WE MADE A COMPLETE TRIP AROUND THE WORLD ON THE "PRAIRIE GROVE", WORKED WCC ES WSL EVERY DAY WITH NO TROUBLE ON H/F. RCA 4U UNIT.

THE "ATLANTIC DEALER" IS THE NEWEST SHIP IN THE ATLANTIC FLEET ABT A YEAR OLD. ATLANTIC HAS 15 SHIPS AND OWN THE RADIO EQUIPMENT ON ALL, THE CO, HAS ITS OWN COMMUNICATION DEPT. WE HV A FLEET SCHEDULE ON 425 KC/S AT 8PM CST SHIFTING TO 5530 ABT 8:20 TO 80 PM CST. SO IF U HV A H/F RCVR LSN ON 5530 KC/S SUM NITE. WE USUALLY GET ALL THE POSN'S ON 425 BUT SHIFT TO 5530 TO GET ANI TR'S WE MIGHT HV MISSED ES TO SHOOT THE BREEZE. THE CALL HR IS "AOFK" YOU WILL ALSO HEAR AOFJ, AOAW, AOAV, AOMP ES AOAT ON 5530 THEY ALL SEEM TO MANAGE TO GET UP THERE EVERY NITE.

I BOUGHT AN ARMY SIGN. CORPS RCVR A BC-224-H NEW "SURPLUS" WHICH SEEMS TO BE THE NUTS. RANGE FM 200 KC TO 18 MC AND I THINK IT IS ABT THE BEST RCVR I HV EVER USED. REMEMBER THE HALLICRAFTER S-20R I BOUGHT FM "TROY" AT G.I.? HAD IT SHIPPED UP AND SOLD IT TO THE PUMPMAN ON BOARD FER $60. SUM DEAL—HI!

I HAVEN'T SEEN OR HRD FM ANI ONE IN OUR CLASS AT G.I. FER ABT THREE YEARS. I THINK THE LAST FELLOW I SAW WAS BILL KELLEY IN N.Y. HE WAS IN AFTER BEING TORPEDOED TWICE ABOARD LIBERTY. SED HE MARRIED A GIRL BACK HOME IN GA. THAT WAS ABT 3 YRS AGO. HO ABT YOU? STILL SINGLE, OR MARRIED YET? AS FER ME, "SINGLE."

WE GET 64 DAYS PAID VACATION A YEAR, BUT HAVE TO TAKE IT A MONTH AT A WHACK, EVERY 5 MONTHS,. I'M DUE ONE NW BUT THINK I'LL HOLD OUT TIL ABT JUNE—WOULD RATHER HV IT WHEN IT'S WARM.

THIS IS A GASOLINE WAGON AND WE RUN A CLOCKWORK SCHEDULE EQUAL TO ANI EXPRESS TRAIN. COASTWISE, NEW HAVE, CN., PROVIDENCE, RI, NEW YORK, PHILADELPHIA, WILMINGTON, NC, JACKSONVILLE, FL. ES PORT EVERGLADES, FL. OUT OF PORT ARTHUR, TX HOME OFFICE IS IN PHILADELPHIA. THERE ARE TWO OTHER OF THESE GASOLINE SHIPS IN THE FLEET, THE OTHERS CARRY CRUDE AND FUEL OIL.

WELL DON, THIS IS ABT ALL I CAN THINK OF NW SO HW ABT DROPPING ME A LINE ES LETTING ME KNO WHAT U ARE DOING, ECT.
73'S,
"Pete"

"LOOKING FOR..."

(Editor's Note: This article, submitted for publication to the Spark Gap, was written by Art Bradford as told to him by Percy Reid.)

Seems that Percy's ship, drifting helplessly, was in trouble in the harbor of Anconna, Italy. The Old Man had the anchor dropped, and when it came time to pull the anchor, it wouldn't come.

They sent a diver down, and he found they had dropped the hook right into a sunken Italian cruiser. After cutting the chain, they went out in deep water and anchored near another ship. Percy and another guy swam over to the other ship to visit. It stank with 369 horses in cages aboard. The next night he found out that that ship had sailed and hit a mine and sunk within 15 minutes in the Adriatic Sea. All but nine horses perished.

Please contact Percy if you know who the radio man was on the sunken ship. (Percy currently resides in Jacksonville, FL.)

LOCAL BOY SURVIVES BURNING SHIP AND FLAMING SEA OF GASOLINE

Ensign Robert L. McClendon, radio officer, U.S. Merchant Marines, feels very fortunate in surviving a serious disaster at sea which endangered the lives of two merchant ship crews and threatened destruction of both his ship and its valuable cargo.

The accident occurred when his ship, the *SS Howard L. Gibson*, collided with a British tanker carrying a full load of 100 octane aviation gasoline. The *Gibson* struck the tanker while trying to avoid a similar collision with a heavily ladened ammunition ship which was out of control due to a faulty steering gear. The resulting explosion caused flames to shoot approximately 250 feet into the evening sky, according to Naval observers. The wooden-encased deck cargo of the *Gibson* and the surrounding sea was immediately a blazing inferno due to spraying gasoline and flames.

Most of the *Gibson's* crew assembled at the stern of the vessel to avoid being trapped by the raging flames. In a short time the ship turned into the wind and flames began to leap over the stern, trapping the entire crew. The first engineer, who stood by in the engine room, soon had the situation under control when he backed the ship out of the flaming sea. All hands immediately began fighting the fire, however, only 30 minutes later the ship had to be abandoned.

Survivors were picked up shortly by a Naval escort vessel. But soon it was evident that the *Gibson* was not sinking and the crew volunteered to return and save her. Braving exploding ammunition, the crew fought 30 hours to bring all flames on deck under control. Two members of the Naval Armed Guard aboard the *Gibson* were lost during the disaster and several of the crew received severe burns.

The *Gibson* then set out for port, some 750 miles away. During the four days that the heavily damaged ship was at sea the crew slept very little. Only a single life raft remained undamaged to accommodate the 68 men on board. Upon arrival in port the cargo in the forward holds was still burning and it took the local fire department two days to completely extinguish the fire.

The *Gibson* safely returned to the States after spending 64 days in a foreign port unloading and receiving emergency repairs. During the return voyage the battered ship survived 14 days of severe storms and a two hour running battle with a German submarine.

The captain of the vessel sent the following statement to the Secretary of the Navy:

> *Once again the American seamen's grit in time of fire and stress at sea has proven itself the very best. The SS Howard L. Gibson collided with a tanker carrying 100 octane gasoline on October 14, 1944. Although the vessel was abandoned for two hours because it was feared the forward ammunition magazine was going to explode, the crew willingly and without hesitation returned as volunteers to save her.*

Ensign McClendon, who serves as radio officer aboard merchant ships, has been in the Merchant Marines since October 29, 1942. He has made many voyages carrying war supplies to various theaters of operation. He is now serving in the Pacific on the *SS Enos A. Mills*, Alaska Steamship Company, San Francisco, California.

(Editor's Note: This article, taken from an unknown newspaper account printed in 1944-45, appeared in an issue of "The Anchor Line" newsletter, which was mailed to the Gallups Island Radio Association. According to the newsletter, postmarked November 14, 1991, McClendon was then a member of the SS Stephen Hopkins Chapter, American Merchant Marine Veterans, Dallas-Fort Worth. As a note to the article, "The Anchor Line" referred to page 177 of <u>*The Liberty Ships*</u>*, Second Edition, written by L.A. Sawyer and W.H. Mitchell. According to the book, the SS Howard L. Gibson was scrapped in Philadelphia in October, 1945.)*

LIBERTY SHIPS— A BRIEF TRIBUTE

by William B. Ishee, R-0C1

It may be that old soldiers never die, but old ships certainly do, though memories of them live on.

During World War I, U.S. shipyards built over 2,700 Liberty ships to carry the sinews of war to the world's war plants, staging areas and battlefronts. These war emergency vessels were sometimes called the "expendables" and many of them proved to be just that. More often they were referred to, half disparagingly and half affectionately, as the "Ugly Ducklings."

Regardless of their acclaim or lack of such, they did indeed deliver the goods—a whole multitude of items in quantities that staggered the imaginations. It was said that a Liberty could carry cargo equivalent to that carried by 300 railroad freight cars, or 2,840 jeeps, or 440 light tanks, or 230 million rounds of rifle ammunition, or 3,440,000 C-rations, or maybe enough ducks and Higgins landing craft to storm an enemy beach.

Besides all of this good variety of cargo; solid, liquid, and inanimate; the Liberties also carried an assortment of personnel along with their various and sundry accouterments. From my personal experience, I can testify that they carried troops, troop dependents, prisoners of war, seamen survivors of U-boat and air attacks, and displaced persons (DP's). The first voyage on which I found myself steaming with females on board, was a Liberty bringing war brides out of the Mediterranean to Baltimore.

It is said that they Navy's Liberty troopship skippered by the renowned "Mr. Roberts" was fitted to carry a crew of 200 and 1,000 troops. This sounds like an awful lot of people on a Liberty. On the merchant Liberties, we expected to carry no more than about 50 officers and crew, and about 24 Naval armed guard.

Over 150 of the wartime Liberties were lost to enemy action, and many others were lost in marine disasters of other kinds. It appears that only two of the whole lot have survived to this day. Both are tied to the dock, restored as ship museum—the *SS Jeremiah O'Brien* at San Francisco, and the *SS John W. Brown* at Baltimore—two "heroes" of the biggest military logistical effort ever mounted.

In my years at sea, I sailed on three different Liberty ships which took me to foreign seaports in Cuba, Scotland, England, Germany, Netherlands, Portugal, Italy, and Algeria, with stopovers at Gibraltar. These ships had all been built at the Delta yard in New Orleans.

All of my Liberties were lucky enough to come through the wars. Each lived on to past twenty years of age, but all have long since gone on to "where old ships go to die." The *SS William H. Kendrick* broke up and sank in Mid-Pacific during January 1965, on a voyage from San Francisco to Formosa. The other two went to the shipbreakers to be recycled to scrap metal, the *SS Jellico Seam* at Hamburg in April 1967, and the *SS Groves Connor* at Hong Kong in October 1969.

QUICK TRIMS

by Lorin D. Kinsel, R-21

I learned to barber on Gallups out of necessity. You may remember the barber chair. We were losing liberties because of the number of men in line and only one barber. You sat down with your white hat on and the barber did a 180 degree with his chair and off came the towel. 30 seconds. Out the door. But getting to that 30 seconds was nearly impossible.

My first trip ashore, I bought a barber comb, scissors and squeeze clippers. Before I managed to get a liberty, we had been cutting hair with electric razors and some of that didn't get by on Saturday morning. Which was the reason for my purchase and experiments.

The barber equipment proved successful and I had a job from then on. I cut hair on every ship, and in 1952-53, I was doctor and barber as well. Even for passengers.

ARTIFACT TO PLAQUE

Harry McDade found this artifact on Gallops some time back; it's a rusty I-bolt that was used during our training days and in 1983, Harry presented this I-bolt to Art Bradford of R-16 because of his outstanding performance as a promoter of the Gallops Island Alumni. And in 1986, the plaque was designed and fabricated by our friend Ed Kuler of R-3, who's a very talented craftsman and he incorporated some suggestions made by Art.

Beginning in October of 86 at a GIRA meeting in Duxbury, MA, it was decreed that the plaque would be awarded to an individual in recognition of outstanding efforts in promoting GIRA. Recipients to date: Ralph Albers, R-9; Jim Kehoe, R-103; and Arthur Dean.

THANKSGIVING AT GALLUPS ISLAND, 1991

by Carl Davis, R-8

Thanksgiving, November, 1941, was awaited with great expectations by the members of Radio Group A-2, a name later to be changed to R-8. We had started our seagoing career in August by learning to tie knots and other seafaring chores, and were getting into code and theory. It was a difficult course, with many dropping by the wayside every week. One grade below 70 and you were on probation and one more and you packed your bags. So tensions built up rapidly.

Only a few sailors were going home for Thanksgiving, since most of us were from distant places, so the cooks and bakers prepared a gargantuan feast for us. Remember, this was still peacetime with war far, far away. Several members of the group went on two-day fasts to make room for the anticipated goodies. Two I remember doing the starvation routine were Morris(?) Blum and Charlie Signorelli, both of whom were later killed, early in the war.

The feast itself was something

special. I couldn't believe. There was, of course, turkey and chicken, and pumpkin pie and cranberry sauce and all the other items in unlimited quantities. We just literally stuffed ourselves. The "piece-de-resistance" was the plum pudding. It had been made earlier in the year, put into a tinned container, along with brandy or rum, and sealed so the cake could absorb the pungent liquid. I didn't know such food even existed, coming from a teetotaller family.

Then, just a few weeks later, on a Sunday afternoon some guys listening to the radio said that Japan had attacked Pearl Harbor. I immediately put it down as a hoax, knowing such a thing to be impossible—like the Orson Welles Mars broadcast. But it was true and the world around us changed, and we changed with it.

Edward Rowe Snow, in one of his stories about Gallop's Island, spoke about throwing coins at the base of the Sunken Ship Monument as good luck as they left the island to go to the sea.

THE BERMUDA TRIANGLE

by P. J. Doyle, R-79

The encounter occurred on the darkened seas, far off the eastern coast of the United States, during the last months of World War II. It was an encounter between a T-2 type tanker, the *Esso New Haven*, and some type of a floating flame.

At that time the dull grey mass of the 7240-ton tanker was proceeding north, well within the now-infamous "Bermuda Triangle." The *New Haven* was traveling at approximately fifteen knots, some 1000 miles off the Carolina's, with all lights extinguished—as required by wartime conditions. She was fully loaded with volatile oil which had been pumped aboard in a port in the Gulf of Maracaibo in Venezuela. The *New Haven* had already safely traversed the treacherous Mona Pass area—a small passage between islands of the Caribbean Sea—where Nazi submarines were still thought to lie-in-wait in hopes of destroying any and all Allied shipping required to use those narrow straits.

At this point in her voyage the trip should have been somewhat routine—as it had been on her many previous runs between Venezuela and Bayonee, NJ. But this time, as the ship's scuttlebutt had it, the New Haven would unload at the Imperial Oil Refinery in Halifax, Nova Scotia—substituting for the tanker that usually made that run, but which had been torpedoed a short time before.

It was late at night, around 2400-0100, when the mate at the helm trained a binoculared eye on a distant, faint glow coming from off the starboard bow; a curious sight the likes of which, in all of his seafaring years, he'd never seen before. A light, he mused, that may have been similar to the one which "spooked" the crew of the *Santa Maria* with Christopher Columbus aboard—'. . . a strange light that appeared in the distance late one night.'

Unable to identify the object or to read its signals, the mate quickly summoned the ship's captain, Neil Borgenson, to the bridge. The captain, despite straining through high-powered binoculars for some time, was likewise unable to recognize the object. He surmised, however, that it may have been some seamen, cast adrift in a raft or boat, who were trying to attract the attention of this fast-approaching tanker. Perhaps, he thought, it could even be some of the crew of the torpedoed ship for which the *New Haven* was now making the substitute run.

Captain Borgenson ordered all three of the merchant radio officers and the Navy's Armed Guard signalman to the bridge. He asked each to decipher the signals while ordering the ships' course be set to intercept while gradually slowing speed.

Little could the captain or the crew of the *New Haven* then know the ship was in the general area where, less than twenty years later, that very vessel would disappear from the face of the earth—vanish into a watery grave in the Gulf Stream within the Bermuda Triangle.

Despite concerted and individual efforts, none of the communications personnel could read an intelligible message from the now approaching light. They would call out some letters . . . an "A," or an "I", or and "E," but there were too many "misses" to translate into an understandable message.

The object, whatever it was, came closer and closer to the ship's starboard side until it drifted alongside its 523-foot length. As it passed the now slow-moving ship its light cast an eerie, ghost-like glow along the entire side of the ship—contorting the shadows of the ship's structure so as to make the whole scene appear somewhat unreal.

This "thing" was a low-lying structure with an open configuration consisting of a square base of railroad-like ties, supporting a triangular-shaped erection on top of which was a flame. As the moderate sea swells went over the object its light was extinguished—only to be rekindled when again exposed to the light breezes. But there was no sight or sound of survivors, no boat, no floating debris.

All watched the object as if both hypnotized and puzzled. Aside from the muted slapping of waves against the ship's hull, a grim stillness pervaded the whole scene. A stillness shattered most suddenly when the captain—with an unmistakable sense of urgency in his voice—cried aloud for a hard turn to port to be executed at full speed!

Only then did the ten or twelve shadowy figures huddled on the bridge seem to realize the situation's urgency. The light may very well have been a decoy—a lure set in the seas by a prowling enemy submarine—to attract, slow down and then easily destroy a vessel intent on some believed humanitarian rescue mission.

Struggling to attain maximum speed, the ship shuddered. Steel plates creaked underfoot as it began to put distance between itself and the erractically-glimmering contraption. It seemed a gut-wrenching period of time before the *New Haven's* iridescent wake was leaving its winking attractor safely astern in that inky sea.

Days later, upon arriving at the intended port of destination in Canada, a meeting between the ship's officers and the Port Authority was quickly convened. All attendees were sequestered in an obscure but well-guarded room. There the details of the episode were explored in considerable detail. But none, not one, could be sure of what the object was doing at sea; what it was; where it came from; why it was there.

These maritime and naval authorities had never heard of such an object. Their ultimate decision was to relay all of the information on to some higher command. Meantime, in order to come to some conclusion, however tentative, it was the consensus that the *New Haven* may have encountered some type of night-time target; perhaps a target used by Naval airmen off the coast of Florida which had been carried far out to sea by the Gulf Stream.

As of this date if there is any more definite conclusion or identification of the object, it is not known—at least to this percipient witness-writer.

A DREAM COME TRUE

by Nelson M. Caley, W8EAR SOWP 4504-V, Hoffman Island, 37-R

In 1978 a group in New York City was organized to obtain and preserve a WWII Liberty ship as a memorial to all who had sailed and built over 2,700 of these gallant ships. The Liberty carried troops and equipment that enabled the Allies to defeat the Axis powers of WWII.

A likely candidate was the *SS John W. Brown*, which completed 13 successful voyages to Europe with troops and supplies. In December, 1946 the *Brown* was turned over to NYC, where, for 36 years, she was used by the maritime departments as a vocational high school.

In 1983 she was returned to the government and anchored in the James River in Virginia as part of the National Defense Reserve Fleet. In August, 1988, "Project Liberty Ship," was able to obtain the *Brown* and bring her to Baltimore. On Labor Day, September 5, 1988, she was rededicated and since then over 90,000 hours of volunteer labor have restored her to an active and alive ship. Over $1,000,000 has been contributed by businesses, individuals, and the state of Maryland. Fund raising continues to be an ongoing project.

The long awaited day finally arrived. At 0400 on September 21, 1991, the engineers lit the boilers in preparation for the matron voyage of the *SS John W. Brown.* The one-day Saturday cruise was a sell out and a second cruise was scheduled for Sunday. The ship was dressed from bow to stern with signal flags, and dawn arrived with a beautiful crisp autumn day. By 0800 the passengers began arriving—the thrill of again seeing a real live WWII Liberty ship with a fresh gray paint job! Signal flags flapped in the wind and smoke faintly flowed from the stack, bringing back many memories to the men who brought their wives, children and grandchildren to see a ship like he had sailed many years ago.

Nearly 600 people arrived for the Saturday cruise and another near 600 came for the Sunday cruise. Most were men who had sailed on or served on as Navy Armed Guard or had helped to construct them.; they were there to relive their memories and to show their families what "daddy did during the war."

There was an instant recall—nobody needed directions to find his old haunt. Even the engine room was open for tours. With its volunteer crew eager to explain the mechanics of that huge triple expansion steam engine. Oilers explained how they oiled and wipers how they wiped. There was obvious pride and, yes, love for this old ship that represented all the Liberty ships these men had ever sailed—some for the"duration" and some for many years.

The old RCA Marine unit in the radio room had all the front panels removed and stripped to the bare metal and was given a fresh coat of gray crinkle enamel. The company that originally supplied the paint for these units was located and mixed up a special batch of the original formula in order to restore the unit to its original appearance.

The Naval Armed Guard restored their quarters on the stern, complete with uniforms. photos and even a Betty Grable pin-up. All the guns were obtained from storage and reinstalled in their gun tubs.

The live band set up on the #2 hatch was dressed in WWII khaki for the occasion and a vocal trio impersonating the Andrews Sisters entertained for most of the six-hour cruise with music of the '40s and some military and patriotic tunes.

As we stopped and made our 180 degree turn to return to the dock in Baltimore, a memorial service was held with a floral anchor dropped into the bay in memory of those who were lost. The bugler played Taps and the whole ship was silent. Tears flowed. Soon afterward we were buzzed by four restored AT-6s. The time tunnel was complete.

With so much going on and the weather so perfect the special event amateur station on board did not remain on the air as much as planned. But the equipment will now be available for future use, thanks to the Canton, OH Amateur Radio Club who donated a Tentec Omni receiver and auxiliary equipment to the ship in memory of George Mambourg, WB8VXF.

SS SEA FLASHER
RADIO STATION FILES
BAMS ALL AREAS DEFERRED
FROM CNO CANDR CK235

ALL U.S. MERCHANT SHIPS X DURING THE PAST FIVE YEARS CHRISTMAS HAS COME TO MEN OF THE MERCHANT MARINE AS THEY HUDDLED BEHIND BLACKOUT SCREENS RIDING LOADS OF AMMUNITION OIL AND SUPPLIES FOR WAR X NOW WITH THE AID OF DIVINE PROVIDENCE OUR MERCHANT MARINE CAN SAIL THE SEAS WITH LIGHTS ON X WHEN THE JOB OF BRINGING THE TROOPS BACK HOME IS FINISHED THE MERCHANT MARINE WILL MOVE THE GOODS OF PEACE X DURING THE BLACK YEARS OF WAR THE MEN OF THE MERCHANT MARINE DID THEIR JOB WITH BOLDNESS AND DARING X SIX THOUSAND WERE KILLED OR MISSING IN CARRYING OUT THEIR DUTIES X IN MEMORY OF THOSE MEN AND IN THE INTEREST OF OUR NATION THE UNITED STATES MUST CARRY OUT THE BOLD AND DARING PLAN OF FRANKLIN D ROOSEVELT FOR A MERCHANT MARINE OF THE BEST DESIGNED AND EQUIPPED PASSENGER AND CARGO SHIPS MANNED BY THE BEST TRAINED MEN IN THE WORLD X IN VIEW OF THE MAGNIFICENT JOB THE MERCHANT MARINE HAS DONE THE REAFFIRMATION OF THAT PLEDGE IS THE BEST CHRISTMAS MESSAGE WE CAN GIVE X SIGNED—HARRY S TRUMAN X THE PRESIDENT HAS ASKED THAT HIS CHRISTMAS MESSAGE AND PLEDGE TO THE POSTWAR MERCHANT MARINE BE FORWARDED TO YOU X THERE IS LITTLE I CAN ADD EXCEPT TO BLESS YOU FOR THE JOB YOU HAVE DONE FOR THIS COUNTRY AND FOR HUMANITY X SIGNED— E S LAND

251526

BREWER'S RECOLLECTIONS

1. A class mate, named Digesu being injured by a Beach Cart shell as he lay in that restricted area reading a comic book. It scared us all, but the instructor who discharged the 'gun' was really distressed until it became apparent that Digesu would survive.

2. General Detail. I don't remember the frequency, but it was the week when we were assigned the maintenance of the facilities. For me, it was a good time; even though there was no liberty for that weekend, it was still a time away from the books.

3. We surely were not a bunch of spoiled brats, but not many of us had done our own washing as we were to learn to do. I remember putting our denims in a bucket and pouring full strength bleach directly upon them, resulting in those famous spots.

4. Sailor's Haven. The location was the corner of Milk and Batterymarch in Boston. It was a clean place to spend the night, and the doughnuts were good too. I recall the night of December 7, 1941 when we talked practically all night about Pearl Harbor and what it might mean to us.

5. Movies on the Island. I remember one in particular, "Belle Starr" with Gene Tierney. As I remember, we got it by mistake before any of the movie houses in Boston.

6. I can't remember his name, but we had a class mate who played "Boogie Woogie" on the piano before I ever heard it from Tommy Dorsey.

7. I cannot watch "Spencer for Hire" these days without recognizing the Esplanade, Boston Common and the Public Garden, where so many of us went to meet some or any of the fairer sex from the area. I remember names like Rose, Lorraine, June and Terri, and wonder if they remember me.

8. What few times we were able to eat away from the Island, we usually went to Marshalls Grille on Tremont Street I think. It may have been Washington. There was a waitress there who called me her "little rebel". She was still there in 1945.

9. I have been a total abstainer since 1946, but during those youthful years, I can recall the New Ritz, Ort's Grill and The Silver Dollar when we were out on the town.

10. Fenway Park and Braves Field where we saw the likes of Williams, Feller, Marion and Slaughter.

11. Fire Watch on the island, punching the time clock to prove we were alert.

12. Dock Watch. I was on "watch" when the Yeston arrived one morning. As I slipped the hawser over the cleat on the dock, my hands were caught between the two. Nothing was broken, but badly bruised.

13. The first snow I had ever seen occurred in early January of 1942.

14. The only instructor I recall by name is Summerfield, who used to be in the Marine Corps and was proud of it. In my opinion, he was our best mentor. He taught us more than radio, urged us not to drink any of that "Panther ——" served in Boston, and educated us about the "Man in the Boat".

15. There was a huge fire in Charlestown I think; so large that they called us over to help. Some of our class were overcome by smoke.

16. A group of us was very fortunate to have an acquaintance from Florida who lived at 179 Beacon Street in the very elite neighborhood back then. Everything was fine until a former State official objected to the presence of sailors staying there. This was before Pearl Harbor. When I returned in 1945, we were more than welcome. The owner of the apartment house was named Whitehead.

17. Occasionally we would go to Revere Beach for the same reason we went to the parks mentioned. It seemed to be a long ferry ride. I remember the "Tilt A Whirl" and a girl named Lou. "Hut Sut Ralston",

Prisoners being barged, boarding and using latrine on SS William D. Moseley.

SS Crow Wing, Atlantic, March 1945, Deck cargo, P-51 and P-61.

"Moonlight Cocktail" and "Chattanooga Choo Choo" played on the juke box.

18. I was clowning around and lost part of a tooth in Haymarket Square. We had a thing about stomping each other's white hats, and it got out of hand.

19. The Mafia was active back then too, there were always bodies being found in the bay.

20. We had a rotund swimming instructor who gave lessons down at one end of the island. He always appeared out of shape, but he sure could swim.

21. At first, we met for reception up near our barracks, but later, the recreation hall was set up on what I would call the north end of the island.

22. I personally thought the food served to us was excellent. Certainly better than anything I had been accustomed to.

23. Studying was hard, much was learned by rote. We helped each other with theory and code.

24. The inoculations we received when we first arrived. Many of us had never had shots before, and the small pox injections in particular took its toll.

25. C-1, B-1, B-2 and others had boat crews. We got to be pretty good too, and it was an experience which served me well later on.

26. I don't remember if they had an advantage over the rest of us, but it seems that the students in A-1 and A-2 were superior to most of us. I can recall they were used as tutors for us.

27. We played poker in the head when we were on General Detail. We were caught once, but talked our way out of any penalty.

28. The FCC test for Second Class Radiotelegraph was passed by all but two. Graduation was joy and sadness.

29. Names to remember are: H.C. Newton, Norwood Kornegay, Sam Notaro, Coleman Barber, Oscar Pelligrin, Art Sheddan, Barris Ishee, Red Sadler, C.W. Rampy, James Smart, Howard Orlopp, Leon Aboud, Jack and Jim Dowdy (the buglers), Don Fryklund, Hannah and Morgan.

30. A tribute to the class occured just prior to one of our finals. Someone had written "The shadow knows" on the black board. Summerfield erased it and wrote "C-1 knows." We did.

AMERICAN WAR VETERANS INSULTED . . . *NOT WORTH A $20.00 SERVICE MEDAL*

by Ian A. Millar, Trident Archives Curator

In 1988 when the American merchant seamen of World War II were granted veteran's status, Congress charged the Maritime Administration with having medals designed and struck to go with the service ribbons already awarded for WWII service. To those of us who had family in the Merchant Marine during those trying years, this appeared as a possible tribute and belated thanks for a job well done.

For years we have witnessed the insults and shabby treatment accorded the American merchant seaman by his country. It truly seemed that at last a small ray of sunshine was to penetrate the long-standing national effort to castigate the merchant seaman for serving his country. But we should have realized that such was not possible, such a tribute to brave men and women could not be done when there still existed one last opportunity to kick these veterans in the teeth.

Oh, yes, the medals are a reality; but in that reality lies what may be one of the greatest insults ever hurled at American war veterans. The medals must be purchased at the cost of $20 each! No other veteran had to purchase their original award of war medals. Thousands of medals were struck and awarded for Desert Storm veterans, and none of those veterans had to "buy" their medal.

The callousness of our leaders in Washington to permit such a travesty on those WWII merchant seamen further fortifies the fact that the current leadership has lost all touch with reality. One can only guess the outcry that would have taken place had the government imposed this fee on medals awarded to all veterans. An interesting scenario is that of the bereaved widow who has just received her late husband's Purple Heart and with it a bill for $20. Now, that would cause an outcry. But we find it's OK to treat the merchant seaman in this manner. Why?

Dennis A. Roland was a 2nd officer aboard the *MS Sawokla* during WWII, when the ship was attacked by the German raider, *HSK Michel*. Roland was wounded, captured and in due course, handed over to the Japanese. Like hundreds of others, he slaved on the "Death Railroad," suffered from disease and mistreatment by his captors. Roland gave his all for his country and survived. Some years ago he passed away: if his family would like the medals he earned during the war, they may have them for $20 each.

During World War II, another friend of mine, Winthrop Haskell, served aboard the *SS Alcoa Cadet* in convoy PQ-15 to North Russia. His ship was later mined and lost in Kola Inlet, but not before he and hundreds of other seamen were bombed, torpedoed and dumped into the cold Arctic Ocean. Haskell was among the lucky ones, he made it back. Many did not.

In the same convoy were men of the U.S. Navy who served in the Naval Armed Guard. They were a truly heroic group, and they were accorded the honors of a grateful nation and awarded medals for war service, free of charge. Now over 50 years later, Haskell may at last obtain his medals, but they will cost him $20 each. Certainly what Haskell and his fellow merchant seamen went through to qualify for their medals is more than worth the medal's price tag.

Chief Engineer Rudolph A. Rutz was serving aboard the new Liberty ship *SS Stephen Hopkins* when it came upon the German raider *HSK Stier* in the South Atlantic. In the battle that followed the *Hopkins* was reduced to a mass of burning tangled steel, yet she damaged the German warship to the point that it had to be scuttled. In that battle, Rutz gave his life for his country. He was last seen amidst the hail of enemy shellfire helping put lifejackets on the wounded. The Navy gunners and merchant seamen fought side by side until there was nothing left. The last five rounds that were fired from the sinking *Hopkins* were fired by Edwin O'Hara, an engineering cadet from Kings Point. The survivors of the Navy gun crew and the other's next of kin received the Navy campaign medals due them at no charge. If the next of kin of those lost in the merchant crew would like medals paid for in their relatives blood long ago, they may have them—by paying $20 for each one.

It is pathetic that we have such callous government leaders who stand for this insult to brave Americans. We have leaders who worry about problems in Russia, problems in South America and problems in the Third World. Yet those same leaders ignore the merchant seamen who gave their all and made it possible for them to live in freedom. Over 6,000 merchant seamen were killed in action in WWII. As a group they suffered a higher percentage of casualties than the Army or the Navy. When a veteran's life and service to his country is not worth a $20 medal, then perhaps it's time for us, as a nation, to re-think our priorities and values.

THE WAR YEARS: FIFTY YEARS AGO

by Theodore Bakula, R-1

And it came to pass in the year 1939 Anno Domine, that I read in the *Army Times* applications were being taken for the U.S. Maritime Service. I was about 18 years of age in those days and a mess sergeant in the CCC Company 1754 in Iowa. I was accepted by the Maritime Service and reported to Fort Des Moines Army Base and given free passage to Fort Dix, NJ.

In that group were Dave Calderwood, R-1, myself and a couple of others. We were billeted and milled around a few days somewhere in that huge base when finally we were mustered before some Coast Guard chiefs, tagged, and sent to the medics and told to pee in a bottle, etc.

Most of us as yet had no idea of what was in store for us, so Dave Calderwood told me to sign up for radio or I would wind up in the ship's galley peeling spuds. So I signed up for radio, envisioning radio as one of those Atwater Kent receivers with the big tin speaker so well-known for the decal showing the dog listening to his master's voice.

Then one day after a big meal of Army beans and hot dogs, the Army politely put the whole outfit on Army trucks and transported us to the Philadelphia Army Pier. We boarded the beautiful, white passenger liner, the *SS. American Seaman.* I had dreamed about my sailing on a voyage like this, having read travel magazines when growing up in North Buena Vista, Iowa, near Dubuque. Dave looked a little scared, but I anticipated a great adventure as we scrambled up the gangplank. Everyone mustered on the main deck to receive assignments. I knew enough not to volunteer as a typist because you would be peeling potatoes down in the galley until the ship arrived in Boston. I quickly found a bucket and a mop and if I saw a chief or a boatswain, I was swabbing the deck—avoiding many menial shipboard tasks. Being on deck gave one a good opportunity to see the ocean and also to be in a strategic location when your stomach began churning.

The *SS American Seaman* sailed into Boston Harbor right past an island that was once used as an immigration station. If I had visualized for a moment that this was to be home for a year, I would have been the first man down the gangplank heading for Iowa. No chance, however. Everyone marched aboard a waiting Coast Guard cutter called the *Pequoit.* Then out to Gallups Island to study for a year.

Not too bad. Bed and board for $21.00 a month and learn to be a communications officer. In the meantime we were just apprentice seamen. Shore leave on weekends beginning at noon was quite freely given if you enjoyed sleeping on a park bench in Scolly Square. Finally we discovered a place called Sailors Haven in Charleston where you could bunk for 25 cents. Witnessing twenty or thirty World War I destroyers sailing out of the harbor gave us an idea of the storm clouds to come. These ships were President Roosevelt's gift to England prior to Pearl Harbor.

Yes, we studied hard daytime and evenings. Chief Summerfield scared me half to death. He was a Marine and had transferred to the Coast Guard as a chief—and that didn't mellow him one bit. Dave Calderwood, James Chambers, and Hugh Jones could all recall other great instructors whom I have forgotten. Incidentally, I met these classmates at the Long Beach reunion in 1992, fifty years later.

Back to 1941. Bakula, Devon and, I believe, Maynard went off to Philadelphia to the Atlantic Refining Company. Maynard and Devon were torpedoed but survived. For better or worse, I drew the *SS Albert Hill* under Panamanian registry. That was to comply with the Neutrality Act. If a Panamanian ship was sunk there wasn't much fuss made about it to cause America to get into the war.

While the *SS Albert Hill* was being outfitted in Philadelphia, I visited the old rust bucket and had a strange feeling. Concrete a foot thick covered the whole bridge and radio room. I thought this was strange just to go to Houston for a cargo of oil for Philadelphia.

Difficulties signing on a crew seemed to cause no end of delay. My friend Ray Wilbeck, chief operator, surmised that the crew organizer rounded up enough seamen by being very generous with vodka in the bars near Fort Mifflin terminal. Then signed them on and our ship quickly pulled away from the dock, way out into the harbor, with no means of shore leave.

"Destined for ports unknown," the ship's articles read. Off to the South American bulge to Recife, then to Capetown. The U-boats were active along the U.S. coast, the Gulf and the Caribbean, and raiders prowled the high seas.

Pearl Harbor occurred while we were passing Madagascar along the east coast of Africa. After making several ports we finally arrived at Abadan, Iran, in the Persian Gulf. That was the only source of oil in the Middle East at the time, although some oil trickled out of Bahrain. The British, who were occupying Iran, had exiled the old Shah who was supposedly pro-German. As we docked alongside the oil facility called the Anglo-Iranian Oil Refinery, British military came aboard and forcibly removed our Japanese cook. I looked on sympathetically as he was taken away protesting he was an American citizen. What became of our cook no one seems to know.

Strangely, even after Pearl Harbor, we sailed along the east coast of Africa with lights on until an American ship with a cargo of steel was torpedoed not twenty miles behind us. After that we sailed black out.

By now Singapore was falling to Japan. The Italians were being driven out of Ethiopia and Eritria and General Rommel was marching toward Alexandria and Cairo. Our assignment was to haul oil from the Persian Gulf around the Oman Emirates and up the Red Sea to Suez. The fall of Singapore and much of Burma left the Indian Ocean and India very vulnerable to the Japanese.

On a voyage to Bombay, then around the continent of India to Ceylon, and north to Calcutta, we found then-British India nervously expecting Japanese invasion. Lord Louis Mountbatten, the British Viceroy, was inspecting the non-existent defenses from Calcutta to Ceylon. We felt somewhat safe in Calcutta as General Chenault and his Flying Tigers were located there.

All hell seemed to be breaking loose and our disgruntled American crew was repatriated to the U.S. I will not discuss this, but the result was that the British Ministry of War Transport supplied us with a Lascar crew consisting of: Christian cooks from Portuguese GOA, Hindus for deck hands, and Mohammedan engine crew and one untouchable of lower caste. I believe it was about this time that the British battleship *Prince of Wales* or the British aircraft carrier *Illustrious* was sunk leaving the Indian Ocean open to whoever could control it. It was a devastating loss to the British. I recall Churchill announcing it over BBC.

About this time Captain Guerin, who nicknamed me "Trash," promoted me from junior radio officer to third mate and then later to second mate. I had to acquire my navigating skills overnight. By this time we had acquired several anti-aircraft guns, which we all learned to handle quite well, and a British Navy gun crew to handle the big gun on the stern

Due to my skillful navigating in and out of the Persian Gulf, Red Sea, Bay of Bengal, and Indian Ocean, the *SS Albert Hill* finally circumnavigated the globe to Freemantle, Sidney, Adelaide, Australia and then to Pearl Harbor and Long Beach, California. Arriving in Long Beach, I signed up for the draft as I was not eligible prior to leaving the U.S.

Then I sailed to England across the North Atlantic in convoys which the Navy did a fine job escorting. As we approached England within German bombing distance, up went our inflated air balloons, tethered by piano wire, that kept the stukas from diving in. About this time the deadly V-2 rockets were being released on London from Peenemunde.

During one crossing aboard the *SS La Mesa* there seemed to be four hundred ships methodically spaced in columns and rows and by number, so that even in fog or darkness the Navy, which by now had radar, could let each ship know if it was endangering the other ships by getting out of line—even if it was only by half a ship length. This information was conveyed to each ship by the Navy radio operators with us who had some kind of secret frequencies to communicate on.

Somewhere in the middle of the Atlantic, where the water was real deep and cold, too (I suspect), this huge convoy split. Bad guys to port . . . good guys to starboard. I understand that those that split to port went to Murmansk. In that disastrous group about twenty, out of some thirty, ships were lost. Some information has it that in the North Sea, the British escorts, fearing that the German battleship *Bismarck,* and perhaps the *Tirpitz,* were on the prowl, left the convoy unescorted and the U-boats and bombers attacked. I had the pleasure of meeting one of the survivors of that convoy at the meeting at Treasure Island in St. Petersburg, Florida in 1992.

We good guys went to Casablanca where the British had just disabled the Vichy French battleship *Jean Bart.* We then went on to Gibraltar where depth bombs kept just about anything from sneaking into the Mediterranean. Bari, Italy, on the Adriatic, was our next call after leaving Taranto. The Luftwaffe had just bombed an ammunition ship, some of which crashed through the Church of St. Nicholas uptown. No damage was done to the tomb of St. Nicholas in his beautiful silver sarcophagus, which was located in the basement of the church.

You would think it was time to go home, but things were going on in the Pacific. So, it was back to Panama Canal and a cargo of oil for Quajalin and Eniwetok Islands, too close to Japan for comfort. Finally, back to Corpus Christi in April 1945, or thereabouts. Apparently, the Army was looking for a few good men and the draft board was bugging me after one week at home. I can hear them saying, "Gotcha." So, up to Ft. Snelling to the induction center for 1-A classification The doctors asked me if I was OK. I said, "yes, but I don't feel very good." "What's wrong?" I said I had convoy fatigue. The doctor said he never heard of that. "Move down the line and pee in the bottle."

They never did get around to putting me in the Army; although, in retrospect, I wish I had served because it took fifty years for merchant seamen to be recognized as veterans with benefits. But I certainly have never begrudged anyone being in any branch of service, for I survived six years at sea and I am thankful for that.

My last voyage was out of Galveston sometime in 1945, on the *SS Chester*, in the service of the Ford Motor Company. It was used in the Erie barge canal to haul ore to Edgewater, NJ. At the shipping office I was inquiring about a ship when I met Captain Genussa. He was a friendly fellow and invited me to come with him as his ship was sailing soon. He picked up my luggage and paid for the taxi to the dock. The *Chester* was a flat bottom cracker box. I felt sick. I envisioned a nice liberty ship or T-2 oil tanker, for an officer of my caliber. It was too late to back out. I signed on as my finances were depleted. Off to Venezuela for a cargo of dusty aluminum ore. Up the Orinoco River so far snakes dropped off the trees onto the deck. We spent nearly a year shuttling the bauxite ore to Trinidad where large ALCOA ships reloaded it for Mobile. During this period, of course, Ford forgot to raise our pay.

A near fatal experience in Venezuela might be worthy of mention. I went ashore with the chief mate for Coca-Cola. This guy was a big, strong, 250 pound Swede. We went into this beverage emporium about a quarter mile from the dock and, of course, lady waitresses waited on us. Being of sound mind and body and being and officer and a gentleman, I made the mistake of offering to buy one a coke. They become your private hostess. Not being forewarned and not being accustomed to such places that only U.S. Navy sailors were known to frequent, I was in a predicament. When the Chief and I got ready to leave I was surrounded by a loud commotion and couldn't move. I was supposed to pay for the evening. Fortunately, the Chief sensed danger as I was not about to part with my money. He threw a bunch of pesos out over the floor, hoisted me bodily over his shoulder and ran out the door while the natives were fighting over the booty.

I left the sea in July of 1946 never to return. Only my curiosity that someone out of Class R-1 might still be alive led me to attend the fifty-two year reunion at Long Beach. It was a great event. There stood Dave Calderwood whom I recognized immediately. Then a lady was talking to this old man who was looking for Ted Bakula. She called me over and introduced me to James Chambers. James was a most gracious host. He opened his home to me and Lola which was near by—free bed and board, free transportation. He was really a delightful person.

Before leaving, I thanked James and told him he was really a great guy. He responded, "That's not what you said fifty years ago," whereupon he pulled his class book showing all the names addresses and remarks of Class R-1. On line 1 was Ted Bakula, the date, and in my own handwriting was my remark, "Ya Bum."

Many other things not accounted for herein occurred. Only a few are recorded by me so that you may believe.

MY FIRST VOYAGE

by Raymond C. Anderson, R-12

After leaving Gallups Island with the R-12 group, I was sent to New York to await a ship. From there I was sent to Baltimore, and after a short wait, was assigned to the *SS Felix Taussig*, A World War I vintage vessel bound for Rio de Janeiro via Trinidad with a load consisting mostly of iron rails. All went well for the first several days until we approached Cape Hatteras and we began to encounter choppy seas. They were too much for this small town boy from Minnesota who had never seen a ship. The seas got the better of me and I was a victim of that old malady "seasickness." For three or four days I suffered, standing watch with a bucket within easy reach and not being able to keep a thing in my stomach. The officers, noticing my absence from the dining room, each tried to give me their own recipes for curing this ailment, all to no avail. It was the third mate who suggested an apple and soda crackers eaten simultaneously. After trying everything else, one more "cure" could not hurt. It did the trick, since it was the first thing I was able to hold down in days, and the sickness never returned in the years to come.

After a lay-over in Trinidad, we finally set sail for our destination in convoy TB-1. It was my practice to visit the third mate on his watch after securing the radio room after my final watch of the day. This particular night I mentioned that the "Jerrys" must not have returned from their Christmas holiday since things to that point had been rather uneventful. After chatting for a while, I retired to my room, opened the port to get a little air and retired. It was about 3:30 GCT when I was awakened by a tremendous noise. I opened my eyes and the room was bright as day. In the short time it took me to jump out of bed and go the port, the ship aport of ours was burning from stem to stern and one could see the members of the crew running on the deck. I quickly closed the port and assumed radio watch. In a relatively short time, three ships had been picked off by the submarines about 100 miles north of Paramaribo, Dutch Guiana.

Several days later we broke away from the remaining ships who were headed for Africa and sailed toward Rio on our own. (The three ships sunk had also been bound for Rio.)

About the time we reached the mouth of the Amazon River, word came that we were to sail upstream and anchor, apparently because of sub activity in the area. To be at anchor with the trees overhanging the river banks, the extreme humidity and the torrential rains is an experience that will long be remembered. After a few days at anchor, word came to move out, so we continued to Rio with no other problems. Being ashore in Rio for several weeks while the cargo was discharged and reloaded with coffee beans was a great experience for a small town boy.

GROWING UP ON THE ISLAND

by Raymond C. Anderson, R-12

It was on February 9, 1943 that I arrived at Gallups Island from a small central Minnesota town. In my 18 years, this was the farthest from home I had ever been. Before long my classmates in R-12 were settled in our daily routine of classes in code, theory and seamanship (as well as KP). During our "bull sessions, " I could not help but notice how many of the fellows enjoyed their cigarettes. At the first opportunity for shore leave, my decision was to purchase a pipe so that I too could enjoy the sessions more fully. Purchasing that first pipe was quite a decision, but I finally settled on one that had a silver stem and even an adjustment for air. It was quite a purchase, taking more than half a month's salary. On Sunday night we returned to the island in rather choppy water. It was time for chow which was sauerkraut and wieners. After eating, it was time to break in the pipe, so I loaded it with the best tobacco money could buy, went out in front of the dorm and lit the tobacco, adjusting the air for the best draft and trying to keep it going. It was not long that my head began to twirl and a dizzy feeling came over me. Before long, it was down to the end of the dock where I gave the pipe the "deep six." To this day, I am not sure whether my problems came from the rough boat ride back to the island, the sauerkraut or the pipe. Regardless, today I enjoy an occasional pipe but have never acquired a taste for wieners and sauerkraut.

HIGH SEAS IN THE NORTH ATLANTIC

by Raymond C. Anderson, R-12

For days our convoy, bound for Great Britain, had been fighting the heavy seas in the North Atlantic. One afternoon, about a half hour before chow, the *SS Joseph Hooker*, a Liberty, was buffeted by the heaviest seas I have ever seen. When it was over, the life boats on the port side had been thrown aboard, twisting the davits like hairpins. The solid oak door on the port side boat deck was torn from its frame, permitting the seas to cover the quarters on that deck with several inches of water. Since the scuppers were apparently plugged, the skipper turned out all hands to remedy the situation. The mushroom ventilator in the radio room had been opened only about one inch. Despite that fact, the deck in the radio room was covered with about a quarter-inch of water. When everything was secure, chow was served about two hours late. All was forgotten when the skipper broke out some "refreshments" for both the crew and officer's mess.

BOOKS OF INTEREST

Raiders of the Deep—Lowell Thomas
Memories—Karl Doetz
The Sea Wolves—Wolfgang Frank
We Fight the Merchant Ship—Michael Gannin
Looking for a Ship—John McPhee
History of the Armed Guard
The Captain—Dutch Tug Boat
American Mariner—Hubert Paul Hann
Serpentine—Nova Scotia Seagoing Tug
The Liberty Ships—Sawyer and Mitchell
A Careless Word—A Needless Sinking—Capt. Arthur Moore
The Battle of the North Atlantic—Terry Hughes and John Costello
Convoy North—Philip McCutchan
Steaming to Bambola—Christopher Buckley
Portland Ships Are Good Ships—Herbert Jones
Gray Seas Under—Farlew Mawat
Sea Wolves—Edwin Hoyte
Bloody Winter—John Waters
Convoy—Martin Middlebrook
A Cold Corner of Hell—Robert Carse
The Battle of the Atlantic—Samual Morrison
U Boats Off Shore—Edwin Hoyt
Cruise of the Kron Prinz Wilhelm—Count Alfred Von Niezychowski (Lawrence Davis-K8NUN)

Books by Robert Mirvish Class R-9:
A House Her Own—Sloane
The Eternal Voyages—Sloane
Texana—Sloane
The Long Watch—Sloane
Red Sky at Midnight—Sloane
Woman in a Room—Alvin Redman
Dust in the Streets—Alvin Redman
Point of Impact—Sloane
Cleared Narvic—Alvin Redman
Business Is People—McGraw
The Last Capitalist—Sloane
There You Are—But Where Were You—Dutton
Holy Loch—Sloane

Bob Mirvish made three trips to Murmansk and participated in the beach-head in Italy

Ships of the Esso Fleet—Standard Oil Co.
The Russian Convoys—B.B. Schofield
American Diaries of World War II—Donald Vining
The Destruction of Convoy PQ 17 (Simon & Schuster)
Large, Slow Target—Melvin Barger (Taylor Publishing)
P.T. Boat—Knights of the Sea (Taylor Publishing)
U-505—Danial Gallery
Merchantman or Ship at War—Charles Dana Gibson (Ensign Press, PO Box 638, Camden, ME 04843)
The Ugly Ducklings of WW2—John Gorley Bunker
Gunners Get Glory—Lt. Bob Berry & Lloyd Wendt
The Battle of the Atlantic—John Costello
A Northern Saga—Steve Lawrence

NEW ENGLANDER TAKES ON TANKER TO TEXAS—THROUGH BLIZZARD

by John T. McNulty, R-114

The Winter edition of the *Gallups Islander* just arrived, containing an amusing view on the historic December 8, 1941 Roosevelt speech. The issue also contained the observation "GIRA stirs many memories." It sure did stir mine.

As I scan the names, I can't help noting membership seems to be made up of older classes. I rarely see mention of a class newer than Ray King, which I believe was R-103. What has happened to those latter day Gallups Island graduates? Here's what happened to at least one of them.

My class, R-114, was shuttling to Boston, taking the FCC exams the week of V-J Day. That same week, they let R-115 members jump ahead and take their exams. R-114 was the last class to graduate in a traditional fashion. Later classes were accelerated through the exams, until the final shutdown of the school.

We came out into our requested assignments around the country. I chose San Francisco, never having been farther from Andover, MA than Sheepshead Bay, NY. I took the train across the country, landing in the Bay area early September, 1945. Registering with the War Shipping Administration, I learned it was really doubtful we would ever get out on a vessel as ROs, since peacetime regulations would soon prevail, in which ships would have single ops, who had to have six months experience to serve. Catch 22 for us, since the only seatime we had was on the *Islander* back and forth from the Island to India Wharf.

The WSA paid us off in November and sent us home, via first class train accommodations. I registered at the WSA in Boston, without much hope of getting a ship. December 15, 1945 they called to ask if I would take a T-2 tanker, *SS Marne*, but as purser-pharmacist. Hell, yes! We left in a blizzard December 17, bound for Port Arthur, TX. I was now officially a tanker stiff, and I was soon initiated into the amusements (?) for tankermen offered by Gracie Woodyard, Black Susie and the Ace Hotel in Port Arthur! What a revelation for a 19-year-old, from the up-tight provinces of New England.

Up and down the east coast from Texas ports until April, 1946, when the vessel was laid up in Mobile, AL, along with countless other vessels—tankers, baby carriers, freighters, troopships, etc. I relieved the boredom of boneyard time, and being broke, by running the ship's lifeboat as coxswain, ferrying crew to and from the Mobile docks. One night, in a local saloon, I saw a guy I knew from Gallups, who told me the Army Transportation Corps in New Orleans was hiring crews to go to the Philippines for a year, including Radio Officers, and they did not require six months service! I took off for NOLA, signing up to serve on LSTs the Army would take from the Navy in Subic Bay, PI. A gang of Gallups and Hoffman Islanders were in that cadre, as we moved finally by Army troop train across the country to the Presidio, Ft. Mason, boarding a C4 transport, *SS Marine Cardinal*, sailing on May 7, 1946 for Manila.

In Manila, we were billeted in a camp, a real dump, with sweltering lean-to tin shacks where we languished until July 5, eating powdered everything, or Australian mutton, or Australian canned butter that never melted. Through the local USO, I found another high school classmate who was a Navy radio operator at NPM, operating a high speed CW circuit. I moved out of the camp, borrowed some Navy enlisted men's dungarees and swabby clothes and sneaked into the Navy barracks, where the food was superb! I got to stand a few watches at NPM with my Navy friends, went ashore with them to fascinating places they had found in the months they had been in Manila.

Some of us stood in sweltering heat, near the statue of Rizal, watching Gen. McArthur declare the independence of the Philippines, July 4, 1946. Little did we know at that moment, that a year later the treasury would be empty, looted by the opportunists serving in that first legislature.

A lot of us went ashore to the bars of the 7th street in the port area, celebrating our own independence, since we had been given orders to assemble in the port the next morning, to be loaded into an LCI for a trip up the coast to Subic Bay, where we would join our ships. July 5 was a hell day for most of us, as we rolled in that LCI, like a pencil in a bathtub, up the Luzon shore. We all got seasick, suffered bad sunburns since we had no shelter. Exhausted, we got to Subic and were taken to our various LST assignments on the boneyard, at night, with no power on the ship. They had assigned three radio officers to each LST, and we shared a cabin as big as a large broom closet, and later, a radio shack so small we had to step outside to change our mind! In *LST 1048*, George Reiland of R-110 (?) and a civilian RO from New Orleans, Frank Morton, joined me as the radio gang. We dashed through the boneyard fleet in the ship's LOVP over the next few days, on midnight requisitions, stocking our ship with parts, silverware, linens, working LOVPs, V discs, phonographs, speakers, etc., making life more tolerable, as we got ready to sail on inter-island troop and equipment delivery missions. Finally, after interminable delays from diesel fuel injector repairs, we sailed to Manila harbor in late July, putting the ship ashore on an LST beach constructed in the North Harbor area, walked down the bow ramps to the lean-to beer shacks built by refugees.

Routine in port was one RO covering weather and other nets over 24 hours, while the other two went ashore. What a time to be young and fearless. Included in that bunch of young swingers was Tommy Chadwick of GIRA and your Region 1 leader, Ray King. I enclose a picture of Ray and me with two young ladies at the Malayan Grill, at my usual table. (Note the reservation sign on the table!) My lady was Miss Philippines, 1946. A case of Canadian Club cost $25, at the Army officers PX, which we could use, as we had Simulated Army Rank ID cards as 2nd Lieutenants. The young ROs were the favorites of the ladies working the Manila Gardens, the Santa Anna Cabaret, the Malayan Grill, since we had plenty of time and money, and we were innocent to the ways of the world. The blonds among us, mostly surfers from California, were in big demand from the bar girls, who wanted to have Mestizo babies with blue eyes and sandy hair.

The next year was spent in and out of little out-of-the-way places, anywhere and LST could beach in the islands, delivering Philippine soldiers, construction equipment, portable saw mill equipment and miscellaneous supplies. I can remember it was a time of fun and games for us, with the exception of a few bad storms at sea. The ship conveniently spent a lot of time broken down in Manila Harbor, but we did get to see Samar, Leyte, Cebu, Mindanao and a few other islands, along with a lot of saloon time in between. It all counted toward our six months endorsement for the FCC.

Finally, at the end of the contract, I sailed on the *USAT General Wm. F. Hase*, another C4, to San Francisco, where I was discharged and paid off in May 1947. Back on a train to my home in Andover. Shortly after, I shipped in and out of Boston on a Hog islander, *SS Chipana*, in the sugar trade to Cuban ports. In that fall of 1947, I signed to attend the USMS radar training course at Sheepshead Bay, NY, where a few more Gallups and Hoffman Islanders were encountered. This time, we were treated like officers, not trainees, and we had a great time in New York City.

I continued sailing, mostly tankers until 1954, accumulating some terrific sea stories and experiences. I hammed Maritime Mobil from 1949 through 1951, call W1QOV/MM. I always said I went to the Salt Water University. I piled off the *SS Barbara Lykes*, a C2, in Houston and went to work for the Bell system, as a sales engineer, thence to Honeywell in the same capacity. In 1955, I started a 35-year career in international steel trading, retiring in 1989. I am now in the residential real estate business in Dallas.

In all that time, I still miss the sights and smells of the sea, the ships, the ports. And I miss the friendships formed in those ships with so many great guys, all of them loaded with sea stories. I was sorely tempted this last August to get my license again and sail in the Desert Shield supply fleet, but my wife pointed out I am no longer a kid having an adventure, but an old man of 64 looking for his youth! That was the end of that.

PRELUDE TO: A SHIP WITH A JINX

by Robert F. Clough, R-7

Class A-1, or R-7 as it was known then, graduated on May 12, 1942. Most of us were sent to New York to pick up our first ships. Walter "Tex" Henderson and I checked in at the mid-town YMCA in New York. John De Vleeschower lived at home in New York. The three of us reported to the ACA officers for assignment.

About the second day there were three ships on the board: two tankers and a freighter. Tex and John both wanted the freighter, so I took my choice of the tankers, the *SS Pan Carolina.* The other two flipped a coin to see who would take the freighter—John won the toss. He was torpedoed and lost at sea within 24 hours of signing on. Tex and I both went through the war years and met again in the Flying Tiger Airline in the early 1950s. Tex became a Silent Key in 1952 when the small aircraft he was helping ferry to Japan exploded out of Wake Island and crashed at sea.

I signed articles on the *Pan Carolina*, KOKM, when she came out of the shipyards after the installation of armor plate and defensive armament. A 3-inch 50 forward, a 4-inch 50 aft, and I believe there were 4 tubs of dual 20 mm cannon, plus 2 portable 50 caliber machine guns. One of the 50 caliber guns was placed on the starboard side just outside the radio room door. The radio room was a horrible mess, cluttered with all the chips, scraps and junk left by the shipyard workers. Plus all the windows had been enclosed with armor plate and there was no ventilation until I pried them open.

That evening, I was in the Captain's cabin getting the paperwork that needed to be filled out before we sailed. The screen door banged open and in came a black member of the crew, right behind him was an irate "Southern redneck" with a bowie knife out ready to '"carve that !!#$ *!@% up.'" (Remember this was May 1942, prior to equal opportunities.) Next thing I knew the Captain had "decked" the redneck and disarmed him. The troublemaker was ordered off the ship. At some time during the night the black crew member "disappeared" and we sailed without him. This was my introduction to "life in the Merchant Marine!"

The *Pan Carolina* was a converted Hog Islander with the bridge and officers quarters midships. To protect the vital parts of the ship like the radio room and bridge, there was 3/4-inch steel plate welded around the midship house and the crew quarters aft. The first several days were busy cleaning out the radio room and getting the RMCA equipment to work. There was a motor generator for the main rig. Trouble was, the starting battery was dead. Southbound, we followed the inter-coastal waterway, then slipped around the tip of Florida and across the Gulf to Port Arthur, TX.

At dinner one evening the crew was talking about phosphorescence in the water and how it glowed as the dolphins played around the ship. I commented this was my first trip and I had never seen this. That evening just as I was getting well asleep, the voice tube from the bridge squealed. Piling out of bed in the dark, I headed for the shack only to crash into the 3/4-inch steel splinter shield that protected the doorway to the radio room. When I answered the voice tube, the Captain informed me the phosphorescence was very visible this evening, and invited me to join them on the bridge to watch it. When I got to the bridge, I was still wiping blood out of my eye from a gash on my forehead.

In Port Arthur, I had the radio battery replaced and checked out the transmitter before it was sealed to prevent transmission. Later, the Coast Guard came aboard and took off the Third Mate who was a Nazi sympathizer. The rest of the trip back to New York was uneventful. We anchored overnight "in the stream" and left the next morning for "overseas" which turned out to be Woods Hole, MA. There we had to wait for the convoy to form.

While in New York, the ship was restricted with no liberty. The Captain and I went to the shipping office of National Bulk Carriers in Rockefeller Center. We picked up the ship's mail and several thousand dollars cash for a crew advance. The Captain had other business to attend to, so I took the mail and money back to the ship. Upon boarding the bum boat at Stern Island, I told the driver I wanted to go directly to the ship as I had important papers and funds with me for the ship. My request was understood as being that I wanted to leave the dock right away, not that I wanted to go directly to the ship. The boat swung by customs where the customs inspector found the uncensored mail and a large amount of cash. It took some fancy talking to get the mail passed through. From then on, all mail out of the port of NYK was censored before being allowed on board ship.

We were issued rubberized survival suits while at anchor at Woods Hole. This was over the 4th of July weekend and no one, but no one, would come anywhere near us or let us off the ship. So we "tested" the 20 mm guns and got a warning from the military. The crew tested the survival suits by jumping over the side for a swim. Only trouble was, no one told us to bend over to remove the air before going into the water. Most of the first jumpers came up feet first with air trapped in the legs of their suits. A line thrown over the side hoisted them back aboard.

The trip to Europe was rather pleasant, although we were in fog most of the way across. We had a stop at Halifax both ways. Instead of going on to Russia, we were diverted into Scotland to off load the avgas we had on board. Several days ashore in Glasgow were welcome.

Back on the western side of the Atlantic, we went directly to Aruba, then on to Curacao for our next load of avgas. After loading, we proceeded to Guantanamo Bay, Cuba to form up the convoy. Coming around the east end of Cuba we had a submarine alert. The British whaler to port of us failed to turn fast enough so we gave way and in doing so rammed a freighter loaded with bauxite on our starboard side. At the time, I was copying the usual AMS/BMS broadcast and before I knew it I was on the deck with the typewriter on top of me. It didn't take long to throw the code books into the weighted perforated steel box, jump into my survival suit and head for the rail with the code box in tow. About that time a bellow came from the Captain on the bridge, "NOT YET, SPARKY!" I never did get rid of that darned set of code books. The bow of the *Pan Carolina* was ripped open with the degausing gear running. It's a wonder we didn't go sky high with the load of avgas we were carrying.

On the way back to New York, the gunnery officer requested to exercise his crew. We received permission to drop out of position in the convoy. The crew tossed a 50-gallon barrel overboard and the 4-inch 50 on the stern took a couple of shots at it. We don't know if they hit the barrel, but they did knock all the ceiling loose in the crew quarters and broke most of the dishes in the galley.

Just as the aft gun crew finished firing, a PBY patrol aircraft that was in the area came overhead to see what was happening. This was just as the 20 mm tubs were fired, narrowly missing the PBY. Not having much else to do, the Radio Operator was firing the 20 mm above the starboard wing of the bridge. When firing ceased, there was one loud roar, 'Sparky, get your @#$ down here and shut off this alarm!" The auto alarm had gone off from the firing of the guns and the bridge bell was banging away.

After we put into the shipyard in NYK, I was relieved by a Gallups graduate from Brooklyn. But, as Al Hadad says, I do not remember his name. He was probably from R-10 or R-11. Al says the ship was "jinxed." However, I found it to be an interesting experience. In fact, I was riding the New York City bus from La Guardia airport into Jackson Heights in the summer of 1945 when I heard a group of seamen on the bus talking about having just signed off the *Pan Carolina* in Havana. The crew were on their way back to the shipping office to collect their pay. So, Al, the *Pan Carolina* went through the war in one piece.

A SHIP WITH A JINX OR MY SERVICE ABOARD THE *SS PAN CAROLINA*

by Al Hadad, R-13

My second ship following graduation from Gallups Island, R-13, was a WWI vintage, converted Hog Islander tanker, the *SS Pan Carolina*. She flew the house flag of the National Bulk Carriers. I relieved a Gallups Islander, but time has faded his name from memory. The story is more interesting to me today because Bob Clough, GIRA director of Region 9, just recently informed me that he was relieved by the above unknown individual sometime in September, 1942.

I received my assignment from the ACA hall in Baltimore (Joe Rubin) and proceeded to Newport News, VA via the overnight passenger packet. When I saw the ship, I was really disappointed, as I could foresee many hardships especially in the living quarters and radio shack.

Well, about this time, the City of New York was experiencing a fuel oil shortage. The reasons are well known, as in 1942 the Nazi navy was playing havoc with US shipping, particularly against oil tankers off the east coast. So, we were assigned to make two trips to Texas refineries, load fuel oil and run up to New York and Bayonne for discharge.

This ship was jinxed. The crew had many weird stories to tell. During the first trip up to New York, on a moonless night, we found ourselves running alongside another tanker so close that one could step across easily. The watch hadn't detected the other vessel. Just by chance, the mate happened to walk out on the port wing of the bridge and FELT the presence of the other ship. Things started happening. Both Captains issued commands to the men at the wheel, shouted at each other and made sure that small changes in course would be made to gradually increase separation. From then on, a sharper lookout was kept.

Our next skipper was a retired Swedish gentleman. He was about 80 years old. He loved the bottle and would sit in his room with the door open so that he could get some company to join in with him and hear some of his stories. He felt that Hitler knew his whereabouts and was out to get him! I used to sneak by his quarters to get to the john and if he saw me going by he'd shout, "Hey, Sparks, come in for a drink. I want to tell you about my friend, the Swedish match king." Is that where matches were first invented?

Our third trip was to be interesting. Orders were for us to load avgas at Baton Rouge, LA and proceed to the United Kingdom. We were ordered to proceed from Baton Rouge, down the Mississippi River, across the Gulf of Mexico, up the east coast and join up with a convoy at Halifax. Could we make it? Well, we tried.

After dropping the pilot at the mouth of the Mississippi, we began steaming east, without escort. No sooner were we on our way, when we saw a vessel chasing us. She turned out to be a cabin cruiser type auxiliary, manned by the Coast Guard. We were asked to heave to for orders. She came alongside and we were advised to return to Pilottown, LA and anchor overnight. A convoy was forming as submarines had been reported in the Gulf. So, we reversed course and arrived at Pilottown in the afternoon.

The weather was brilliantly clear and cool. The water, flowing by at 5-6 knots was very cold. The air temperature being quite warm, was causing fog to appear just on the surface, about 6 feet thick.

After chow, I took a stroll on deck and watched the action. Another tanker was anchoring just aft of our position. Both anchors out, by the way, because of the stiff current. She was a Gulf Oil tanker, the *Gulf Pride*.

Coming down the river we could see that we were in for a treat. A baby flattop! As she got closer, we noted that she was flying the British Union Jack, and as she steamed by, her name board proclaimed her to be the *HMS Hunter*. Scuttlebutt had it that she had just been converted and transferred to the Royal Navy.

We were quite surprised when *Hunter* turned around after passing us and headed upstream. We could clearly hear the pilot's orders being given on the p.a. system and were aware that the *Hunter's* intention was to anchor with the rest of us until morning.

When she was just abreast of our port side and about 100 feet away, we heard the command, "Let go port and starboard anchors!" We watched in amazement as only the starboard anchor plunged into the river. Then the command, "Let go the port anchor!" No rattling of the port chains was heard. To our horror, *Hunter's* bow began to veer in our direction. Her port bow anchor was stuck in the hawse pipe and pandemonium reigned aboard the carrier. As I remember, the pumpman and I were on the starboard side of the *Pan Carolina*, midships, watching the action and he shouted towards the carrier, "Stay clear, we are loaded with avgas!"

Without a doubt, those words increased activity along the catwalks on the carrier's starboard side. The p.a. blared orders, "Secure all hatches on the starboard side." We had hoped that just before colliding with us, those of us close to the collision point would be able to step on the catwalk and abandon our own ship. No such luck. The navy ratings were slamming hatches shut in quick time.

With her engines in full ahead and rudder hard over, *Hunter's* bow began to swing toward us. She still had no weigh on her. Nothing was going to prevent a collision. What to do next? Well, I headed for the starboard rail and was preparing to jump overboard when Bill, the pumpman, grabbed me by the arm and said, "For God's sakes, Sparks, stay with the ship!" Little did I realize until later, that Bill had saved my life.

Just then, the carrier collided with us and the air was rent with the screeching of metal on metal and while we watched, the overhanging flight deck struck one of the 20 mm gun tubs, tore it loose from its welded position on a tall tubular pedestal and fell to the deck amidst a shower of sparks and loaded canisters of ammo. Bill and I ran aft not wanting to see any further destruction. We were waiting for the big blast and ammunition exploding, but luckily it never came.

Reaching the fantail, we observed some of the crew jumping overboard into the river. As we looked down at them, they were immediately engulfed in the surface fog on the swift running river and were swept out of sight. Bill and I, almost simultaneously, headed for the life rafts mounted nearby, struck the release mechanism and watched her dive into the water. The raft disappeared from view almost immediately. We hoped that the men would be able to locate the lifesaving raft and climb aboard.

By now, *Hunter* had pulled away, much to our relief, headed upstream, and anchored. I began to calm down and feeling sore in control of my faculties, began a search for the skipper.

I found the Swede surveying damage from the bridge. And what a mess it was. The port lifeboats, both forward and aft, were crushed flat and hanging from damaged davits. The midship gun tub lay on the deck with ammo strewn around. The second 20 mm gun was off its mount in the tub. But the greatest damage was to the hull. Gasoline was leaking from the split seams on the port side, forward of the bridge. We would find out that evening that the *Pan Carolina* was considered a hazard and not seaworthy. The voyage to the UK was aborted and we would receive orders to return to Baton Rogue for unloading.

I told the captain we should send a message to report the accident, and that there were four men overboard who may not last long in the cold water. The captain was cool, maybe even a little high. "Sure, Sparks, go ahead and send whatever you think is necessary." As I crossed over to the starboard side and looked up at the main antenna, I noticed that it had severed and fallen to the deck, no doubt due to the whipping around of the masts, although the emergency antenna appeared intact.

At the radio position, I began following normal startup procedures for the RMCA gear aboard; switched the main transmitter to the emergency antenna and pressed the key to tune up on 500 kHz. NO OUTPUT! I checked all the meter readings and determined that the keying circuit was not functioning. "Oh well," I thought, "I still have the 50 watt emergency transmitter." So, I fired her up and lo and behold, RF current indication!

Pilottown is just down river from New Orleans, so there is normally no problem contacting NOLA-WNU. But this time, things were abnormal. We were in the early evening when the F layers were just merging and signals had severe echoes. I called WNU several times with no response. I called NERK and NCU without success. Then I waited for the silent period.

At 0015GMT, I sent the following message:

XXX XXX XXX de KOKM KOKM KOKM AR K

WNU respond immediately.

WNU de KOKM BT

HAD COLLISION WITH AIRCRAFT CARRIER WHILE AT ANCHOR OFF PILOTTOWN BT SEVERE DAMAGE BT 4 MEN OVERBOARD HEADED DOWNRIVER AND OUT OF SIGHT BT ADVISE COAST GUARD BT

MASTER AR K

KOKM de WNU R QSL URGENT MSG K

WNU de KOKM R TU AR SK

That was a relief. I was just about to leave the shack when I heard a very loud signal in the headphones. The *Gulf Pride*, anchored astern of us was sending a message to WNU advising that a life raft with some men aboard had been sighted as it drifted by and they were in the process of launching their motor lifeboat in an attempt to rescue them. The number of men aboard was not given.

About an hour and a half later, we were hailed by the *Gulf Pride* lifeboat and she came alongside. Only three of our crew climbed aboard. The fourth man, a member of our Armed Guard and on his first trip, was never found. He was just 19 years old. The three survivors were thankful for the pumpman's quick action in releasing the liferaft, otherwise they too would have perished. You can bet your life I felt the same.

The following day, we proceeded up the Mississippi River to Baton Rouge for unloading. Our problems were not over. Just as we sighted the refinery docks, our steering apparatus failed and we ended up ramming the west dike. Some kids coming home from school, and using the roadway on the dike kept shouting, "Whatcha doin', mister?" We invited them aboard, but they ran off, laughing. I don't blame them.

So, the old *Pan Carolina* unloaded the avgas and proceeded to New Orleans for drydocking. As for me, well, you guessed it. . . . Sparks is first off and last on. . . . I signed off and went up to Chicago for some R&R. A couple of weeks later I signed on my first Liberty, the *SS Caleb Strong*, in New York. I will always wonder if the *Pan Carolina* made it through the war. Wouldn't you?

THE LIBERTY SHIP

Laden deep from lands afar
The Liberty stands at the Pilot's bar
A rusty tramp that has euchered fate
Heaves in the swell off the Golden Gate.

Yes she's a Liberty built in twenty days
One of the hundreds that slipped the ways
A champagne bottle, an unknown name
And no designs on future fame.

She was laden deep with tools of war
Below her marks both aft and fore
And as if to tempt the hand of fate
Her decks were piled with plane and crate.

Each time she dives she shudders free
And pounds her way through the angry sea
There are those who would sneer at these
But they have never sailed the seven seas.

The Liberty tramp, the butt of jests
Can-opener ship, and all the rest
An ugly duckling, a rust pot
Waiting for the wrecker's lot.

So now dear lass don't mind the name
Your endless miles have won you fame
From tropic heat to Arctic flow
There's not a sea lane you don't know.

The Murmansk run, and Attu's shore
Salerno beach-head and many more
After the Naval gunfire shook
You were the first to drop your hook.

Why there's not a cargo great or small
But a Liberty tramp has carried them all
High test gas and TNT
from Guadalcanal to Tripoli.

You are the prey of the U-Boat pack
That lurks on the ocean's merchant track
Your hulls lay blasted on the ocean's bed
While your sister ships sail overhead.

We've trod your decks, we're proud to tell
You've sailed us in and out of hell
We've cursed you and we've praised you too
As sailor men so often do.

We'll drink a glass at the Mariner's bar
To you, good ship, so there you are
A merchant ship, a Liberty
Another tramp upon the sea.

But hark, that's the stand-by bell
The Pilot's boarded, all is well
Another trip come to its end
We say farewell to you, my friend.

TESTING AT GALLUPS ISLAND

Anonymous tape recording

". . . The only thing I remember well about Gallups Island was the final test we took. There were three of us in a room taking our final exam, and the instructor fired off a smoke bomb in the room and ran around yelling, 'The ship is sinking!' We had to put the transmitter back on the air in the dark, find the emergency light and see what units had to be moved. If we got rattled and lost our heads and couldn't get that radio transmitter on the air, we would have been thrown out of the school. The final test was to see if we could handle the pressure of an explosion on a ship."

GHOST SHIPS

by Stan Jennings, R-58

The other day as we got out of our Japanese car, carrying our German camera, our neighbor John greeted Jane, my wife, then asked, "Stan, have you got your draft card ready?"

"John," came my reply, "I was in the Merchant Marine in WWII, and I have a saline solution."

Somehow this put me in mind of so many briny buddies I don't see no more. Closest to my heart—or perhaps stomach—was the French chief cook on the *MV Aldebaran* who, with a crust of bread and detailed instructions, cured me forever of my seasickness. He got off at our first port of call: Charleston. I was so sorry to see him go. He watched me struggle for adequate words of farewell, then put out his hand and smiled, *"Bonne chance!"*

Charleston. That's where Joe Barker, the electrician who would be my friend, got on. Joe was from Cleveland, TN, where, by his own admission, the governing body passed an ordinance requiring trains to stop. Joe admitted his electrical training and papers were not full-fledged, but our ship flew a Panamanian flag and needed a crew. Our ship was also from Denmark, so all the electrical/mechanical instructions were forged in iron, in a foreign language. Joe got his electrical education the hard way.

Joe had a double-breasted suit he was proud of. It was the cobalt blue color the ocean took on occasionally under cloudy skies. In Calcutta, an enterprising Indian came aboard offering to dry clean clothing. Joe cautiously obtained his promise to treat the suit with great care. A week later as the *Aldebaran* was about to leave, Joe cursed because his suit was about to stay in India. But the native came back at the last moment, collected, left his package and ran down the about-to be hoisted gangway. Joe unwrapped a suit of a different color, mouse gray, and shrunken to fit a small boy. It had been washed in the Ganges.

Sonny Howcroft and John Patton, of Cincinnati, wiper and oiler, respectively. Sonny never tired of praising the great Scotch he had tasted as a bellboy in a famous Cincinnati hotel. John told us of what was happening right then. He would come sleepily off the 4 to 8, pour the Australian corn flakes we had taken aboard into his bowl. A few roaches would run. A big man, and hungry, John would wave his spoon at them a couple of times. "I gave them their chance," he said, and poured the milk.

In Colombo, Ceylon, we tied up for awhile. Watched movies at night, went swimming amid the spray-splashed rocks at Mt. Lavinia by day. A friend in the Armed Guard and I took a day-trip on the narrow gauge railroad over the mountains to Kandy. He was a nice enough guy who had ambitions to return to Kansas and become a mortician. That, I found difficulty comprehending. We were in a train compartment with two Britishers, looking down, way down, at the tea plantations below. The two English kept discussing Wales. I finally asked one of them if he was Welsh. "No, thank the Lord," he answered, thus giving me a tiny insight to one of the myriad racial attitudes of our planet. In Kandy, high in the mountains, we circled the beautiful lake, cool and serene, shaded by palm trees. The Temple of the Tooth, a Buddhist shrine with closed doors, was one of the buildings off the footpath around the lake. We were told that the tooth in the temple was one of Buddha's. Could be. Other religions have beliefs as unlikely and difficult to prove.

Weeks later, homeward bound, we dropped anchor one afternoon off Perth, Australia where we had previously been ashore all too briefly. This time the word was no shore leave. Our motley crew of Danes, French and Americans was most frustrated. Back aft there was much palavering in three tongues about going to the Danish Captain to protest. Being the only one knowing a few words in two languages, I was elected spokesman.

This surly crew surged toward the boat deck, pushing me in front. My American friends all said they would be right behind me, and at that moment they were. The Captain was big, mean and a Dane, and he was the Captain. Whole thing smacked too much of mutiny. As we neared midships, Arne, a Danish AB called out to me. I stopped to see what he wanted. "Be nice," said Arne. Best advice I ever had.

I knocked on the Captain's door, then turned to look behind me. All the American faces had changed to French.

Seated, the Captain asked us in. There, politely, I told him how we had liked Australia on our one-day earlier visit and how much we wanted to go ashore again. He said that he understood and that he, too, wanted to go with us. But, he said, our orders were to sail at dawn and since it was now dusk, nothing could be done. "If anything changes, I will demand that you get leave." He couldn't have been nicer. We did sail at dawn.

Back in the other ocean, I found myself in the radio shack of one tanker after another. The crews changed, but the succession of characters never ceased.

There was the chief engineer, George O. Speth who told us all at dinner that Franklin Roosevelt had saved our country and our people and that no one had better ever say a word against him, "while I'm around."

On that ship, the *SS Esso Hartford*, Howard Jackson and I were second and third radio operators. If you were newly aboard, you might sit down at the officers' mess table, look over the menu and tell tall, balding, white-haired, hard-bitten Messman Mack that you wanted the eggs but not the bacon. That would be the last time you did so. Because Mack would thunder, "Tell me what you want, not what you don't want!"

When Howard and I went aboard, a quiet man from Marblehead greeted us, carried our bags to our new quarters and made us welcome. I thought, perhaps he was some messman. Soon learned he was a very young and very tough chief mate and my neighbor. Second day at sea, the messman who did call me for the 4 to 8 watch asked if I would mind calling Chief Mate Persons next door before I went up, just to make sure he was up. I did so. Mr. Persons was still asleep. He did not respond to a couple of calls, so I leaned over to give him a gentle shake. Still asleep, he came out of the sack throwing a right hook that would have done Jack Dempsey credit. A fast retreat saved me, but I felt the breeze. Never again, I told the messman.

Then there was the second engineer, nice, chubby chap with a down east accent, fond of looking at world maps in his off hours. He would point at Mexico and explain that when he became President, that would be the first country he would annex and make part of the U.S. Then, gazing north, he said he would invade Canada, since it was contiguous both to the 48 states and the (then) territory of Alaska. He took us ashore in Curacao, to the bar by the bridge that opens for ships. We couldn't understand why the thumping headache after a very few Dutch beers. "Quinine," he said, "They put quinine in the beer."

And the second mate. A dark, lean Bostonian, hawk-faced with no teeth. Once when I was on watch he visited me and chatted some time. Then he asked what was wrong with the flashlight. "Nothing," I said, but he handed it to me and it wouldn't work. He left and I worked on the flashlight some minutes before discovering that the mate had folded up a small piece of paper and inserted it between the batteries.

The first engineer loaned me a wrench one day. I thanked him and headed to do the job I had in mind. "Hey, Sparks," he called, "If you drop that wrench overboard...." "Yes?" I asked. "... hold on to it."

Captain Larsen. He worried that First Mate Persons was such a hard man. "And so young. I hate to think what he'll be when he's my age, and Captain."

Uncle Harvey. Harvey W. Long. Chief operator. From Cape Girardeau, Missouri. Spoke to us two young operators as if were were his idiot sons. He was the only man aboard who took his life jacket wherever he went. No criticism there. Everyone called him "Uncle Harvey," which he enjoyed. Until the first mate started calling him "Uncle Dudley". That he didn't like.

Harvey on watch with glasses, green eye shade and cigar, loved to talk, and might keep you listening long after your own watch was ended. He told us how Babe Ruth on the floor above in the same New York hotel where Esso sailors stayed, used to keep Harvey awake with nocturnal cavortings, noises and feminine accompaniment.

One day we asked Harvey why he had returned to sea. He answered that he and his wife had rented a room in their home to a young mother. One morning Harvey came downstairs early to find her nursing the baby. He commented, "Mmm, that looks good." She answered, "Why, haven't you ever tasted mother's milk?" If he had, said Harvey, he couldn't remember. The lady offered Harvey a taste, and he tried it and found it good. About that time Mrs. Long came down the stairs. She never bought his explanation. "And that," concluded Harvey "is why I'm back at sea."

One other person, never a shipmate, comes to mind. Admiral Emory S. Land, wartime chief of the Maritime Service. Years later, I met Admiral Land at National Geographic, where I worked, and he was a member of the Board of Trustees. Geographic photographer Bob Oakes had to take the Admiral's picture. Bob told me that he could not get him to take the pipe out of his mouth long enough to get a portrait. He would pose the Admiral carefully with the pipe in his hand. By the time Bob got behind his camera, bang! the pipe was once more clenched in the Admiral's teeth.

Bob introduced me to Admiral Land and I told him I had once worked for him—in the U.S. Maritime Service. "Well," said the Admiral, "Your profession and mine has gone to Hell."

It was long ago. If these ships are gone—ghost ships, manned by ghost crews, why then, I must be a ghost writer.

ON TO GALLUPS ISLAND, VIA INDIA

by Stan Jennings, R-58

"This is a day which will live in infamy," he said.

He was speaking to the world from the House of Representatives. Looking down from the gallery, I had little time to listen to what President Roosevelt said on December 8, 1941, because I was too busy running errands, carrying broadcast equipment, and getting a copy of the text for the radio commentator I was working for, Fulton Lewis, Jr. But I do remember how tired and ashen Franklin Roosevelt looked that day, that day which was only the beginning of the war for all of us.

Although it was not, it did at that time seem to me perfectly natural to be there, at the very heart of the free world capital. For I had grown up in Washington, from kindergarten to college, with my first jobs—of which this was one—at a newspaper and a radio station.

On the day of the infamy, only a few hours earlier, Sunday evening, radio station program director, Madeline Ensign, had given me a news bulletin to take to Fulton Lewis, broadcasting Pearl Harbor day happenings live from the White House, two blocks away.

I ran all the way. Through the OPEN iron west gate up to the busy policeman a few feet inside. Showing him the piece of paper, I told him it was a radio bulletin for Mr. Lewis. "Well, I can't let you go in," he said, then paused. "Ask that policemen up ahead." He pointed to a cop twenty feet up the walk, surrounded in the night like himself, by people having business at the White House.

I repeated my little speech to the second officer.

"Did he let you through?" he asked, pointing toward the first.

"Yes."

"Okay, go ahead."

Inside the White House newsroom, the reporters had just been given a briefing and were shouting their stories into phones. Fulton Lewis had his copy ready to read along with the item I gave him. An engineer manned the remote equipment. The *NY Times* reporter yelling into the phone at the desk beside Lewis was easily the loudest voice in the room. A worried Fulton asked me if I could quiet him down. Foolishly, I tapped him on the shoulder. He turned, interrupted and angry, as I explained that Fulton Lewis, Jr. was about to broadcast over the Mutual network and could he please lower his voice a little. Just as the engineer's on-the-air red light flashed, so did the erudite *Times* reporter, with a simple declarative sentence, "Fuck Fulton Lewis, Jr.!"

And these were only the two first days of the war. As further days proved less eventful, I started to think of faraway places. My buddy, Bill Treynor, and I journeyed to faraway (40 miles) Baltimore to investigate the Merchant Marine. As I recall, we learned there was a ship about to sail and all we had to do was sign on. That was a bit more preship-itate than we had planned.

So we returned to D.C. and asked questions of the Army Air Corps. (Billy later did become a Marine aviator in the Pacific.) The enlisting officer I talked to looked into my eyes—searchingly. He found them not 20/20. I thought they were, but evidently the 20 had only to do with my age. I was told that flying was a deadly serious business and that my inadequacy might lead to a dogfight with a Jap (popular term then) in which I would be the underdog. "Forget your desire to fly," the kindly man advised.

So I did. I went down to the US Maritime Service, then under the command of Admiral Emory S. Land, with ship's company at WWI tempos located at 13th and Pennsylvania Avenue. The USMS looked into my eyes, too, didn't like what they saw. Rejected once again, I went back to the radio station, not in a blue jacket, but in a blue funk. I moped about for a while, then had an idea which would change my life. I went in to see Fulton Lewis and told him what happened, asking if there was anything he could do to help me get into the Merchant Marine. He lifted the phone and said to his secretary, Jean Hatton, "Get me Admiral Land."

I was accepted in the USMS the next day.

But there was one little surprise ceremony before I could leave town and start swabbing decks. Day of departure someone called me from my errands to go up to Studio B. When I opened the door, the entire station staff stood there, smiling benignly. This was more terrifying than the North Atlantic. Russ Hodges was in the middle of his afternoon sports broadcast. Russ would later gain national fame by describing, without detailing, a season-ending homerun, thus: "The Giants win the pennant! The Giants win the pennant!" But that afternoon, he simply opened up the microphone, and presented me a watch to wear away to sea. Near half a century later, it is still ticking away on my wrist.

South to St. Petersburg. That's where the USMSTS turned landlubbers into sealubbers. "Off your ass and on your feet! Out of the shade and into the heat! Rise and shine for the Maritime! "Three months of marching, standing watch on a docked sailing ship, the *Tusitalla*, sailing Tampa Bay on a Hog Islander, the *Empire State*, launching lifeboats, more marching, eating poultry so large we were sure they were pelican, more marching. Then, armed with lifeboat tickets and seaman's papers, we went to New Orleans to wait for our ship to come in. There we spent our waiting time rowing lifeboats on Lake Pontchartrain, cleaning barracks, but most of the waiting time was spent on Bourbon Street. December, '42. Christmastime. Nice.

The ship, *MV Aldebaran*, came in, and went out with me aboard.

She was a Danish ship that had been ordered back to port when the Germans took over. The captain, knowing where the orders were coming from, ignored them. Instead, he sailed to New York and donated the *Aldebaran* to the Allies.

When we signed on in New Orleans bound for Charleston, the crew was about a third Danish, about a third French and about a third first-trip Americans from USMSTS, St. Petersburg. And the flag! The flag was Panamanian.

In Charleston, while we new sailors got over our seasickness, the *Aldebaran* took on cargo—ammunition, explosives, bombs, guns—a 3-inch 50, four 20 mm, and a Naval Armed Guard crew.

Where were we going? No clues, but as soon as we were at sea, the word spread from the bow to the fantail: India!

We looped around Cuba to lie at anchor for a week in Gauntanamo Bay. No shore leave. On to Trinidad, Port of Spain, which we observed by night. No shore leave. Through the marvelous Panama Canal, watching the walls of three locks descend before our eyes as we rose 85 feet to the level of Gatun Lake. On to the Pacific, zig-zagging past the Galapagos Islands as the sun set, we would not see land again for forty days.

Our ship, the *Aldebaran*, named for a star, had the Southern Cross for company at night. By day sportive dolphins played back and forth across the bow. We leaned over the rail to watch them having no trouble staying with our 12 knots. When we got far enough south we were joined by albatross, large birds with long, narrow wings, faithful to us not-so-ancient mariners. We would reward them at sundown, dumping the garbage off the stern while they brought their webbed feet down for a meal upon the waves.

That was all the company we had, and all we wanted. We were alone, zigzagging and getting rid of the garbage in the evening to leave no evidence of our passing for enemy subs to sight.

Alone? Well, not quite. Danes. French. Americans. We had each other. What an opportunity to see up close the vagaries of varied cultures.

Arne Bleurger, a Danish AB, and I studied a global map showing just how much real estate the Axis had then expropriated and how far we were from retrieving it. "What a mess the world is in!" he said. Then his eye fell on Majorca and he told how much he loved it and wanted to return. "How long are you going to go to sea, Arne?" I asked. He smiled, but his blue eyes flashed, "Till I get the hot blood out of my veins."

The French were out of sorts about something. Something to do with the food, and I was asked to find out what. My schoollearning was stretched to the limit, but I reported back that the French did not want the fish heads to be thrown overboard with the garbage; they wanted the fish served with the heads on. They were accommodated. Then we learned why. They loved the fish eyes. Popped the mouth of the fish heads and into their Gallic mouths. D'accord.

The French chief engineer rewarded me with some French lessons. What is this? He pointed to the deck. I didn't know. "le pont." I thought "pont" meant bridge. Well, it also means deck. Very logical. Both have to do with getting over water, don't they?

We Americans used to loll about the aft hatch in leisure hours, recalling former days. Captain Dick Fine, the Army cargo officer, had been a traffic cop on the corner of 13th and F in Washington, D.C.; Wiper Sonny Howcroft, a bell boy in Cincinnati; Gunner Marty Salona worked on the lake front in Cleveland; Electrician Joe Barker grew up in another Cleveland. Cleveland, TN, a town where the trains did not stop. The town officials passed an ordinance requiring them to stop. After he told us, we never let Barker forget that. And the galley's two Hargroves, cousins from Kentucky. As conversation waned on an endless ocean, one would tell the other, "I'm going to the house."

Observing every job on the ship, I noted that the best by far seemed to be that of a young fellow from South Carolina by way of Boston, the radio operator.

South of New Zealand the skies turned to lead and the waves into mountains with troughs deeper than valleys. We sometimes could see a liberty ship far behind, rolling to such angles that we doubted she could survive.

Going north, the seas grew calmer till finally we docked in Freemantle. Judging from the ships nearby, others were traveling the route we had come. How strange not to be on an ocean and to ride a narrow gauge railway to Perth. A whole new and fascinating world down under where one Aussie groused about Churchill making a speech while it was left to the lowly to fight the war. Amazing. Not all Brits loved their leader. But then we amazed him by disclosing that in America not everyone revered Roosevelt.

We loved Australia, but it was given us but for one day. The next, we were bound across the Indian Ocean for Karachi. There we saw part of our volatile cargo unloaded. One "bomb," or somesuch, was deposited too roughly on the dock, steamed and emitted what we wondered might be mustard gas. Terror took over. While many thought we were all about to be blown up, a tall, commanding Colonel took over, calmed everyone and got the sputtering threat defused.

Amidst poverty such as we had never seen, one of many Indians asked for baksheesh. We told him we were not so well-to-do ourselves, that we could not afford to give. "Oh, no, sahib. You are a rich man." How often since have I thought, he was right. And memory brings back curious things . . . riding a bus in Calcutta, cowering in a deluge with other humans, all dark, thinking that this surely must be the end of the earth.

It wasn't. Back to Ceylon, Vizagapatam, Perth once more—no shore leave—then the whole journey in reverse.

Back in the other ocean, off Yucatan, we came upon, or he came upon us, a surfaced sub in the night. General alarm. Bodies crashing against each other, each going for his life jacket, always at the other end of the ship when needed. We got away.

Then on to tankers, as explosive as ammo ships, but nice clean cargo!

Many, many months later, in search of that best job on the ship, I managed with great difficulty to get to Gallups Island, and, with more difficulty, to graduate and get the radio license.

WINDING DOWN THE WAR

by Stan Jennings, R-58

On June 6, 1944, the *Esso Hartford* had just unloaded its cargo of 100 octane gasoline in England along with the P-38s lashed to the deck and was one day out bound for New York. This was the second of seven such trips of the *Hartford* that I know about.

As I came sleepily to the 4 to 8 a.m. watch, a wide-eyed and wide-awake Howard Jackson whom I was relieving told me there were ships, troops and planes swarming into France. Then General Eisenhower told us both—and a few million others—"Allied naval forces supported by strong air forces began landing Allied armies this morning on the northern coast of France."

On April 12, 1945, when I came up to take over for Howard, he told me, "President Roosevelt has died."

A few weeks later, May 7 at 4 a.m., Howard shouted to me, "Vee Eee Day!"

"Wait a minute, Jackson," I came back, "Has the whole damn war been fought on the 12 to 4 watch?"

We were only a day from New York and looking forward to the great celebrations we were hearing about. But I copied the message which diverted us south, through Mona Passage to Aruba, and from there to West Africa.

And I, for one, could not get used to the lights being on at night and the port holes open. How we shone. Suppose some sub doesn't get the word, and bam! Sure enough, early one morning in the Caribbean, on 500 KC, it came loud and clear, "Dit-dit-dit, dit-dit-dit, dit-dit-dit, dit-dit-dit." Submarine! I copied the code furiously, then figured the stricken vessel's location. Just over the isthmus in the Pacific. The dawn skip beat had brought him in. Of course the Japanese were still at work. War work.

As the idea of peace-time began to set in, we began bringing cameras aboard. The second mate and I invested in some chemicals and ambitiously decided to develop our own film at sea. The guest room, which had been used either by the cargo officer or the armed guard lieutenant, was vacant. So one afternoon when neither of us was on watch, we went in to process the film. We were in the middle of the procedure when someone, we didn't know who, came banging on the door. We shouted for him to go away.

Later, the Chief Mate told us that the old man had come up on the bridge to tell him, "The Second Mate and the radio operator are in the guest room. They've got the port holes closed, and the lights out and the door locked. And," he continued ominously, "they won't let me in."

This got me to thinking about Chief Snodgrass, the old salt who taught us nautical newcomers at St.Petersburg USMSTS how to tie a bowline. "You make a bight in the line, then pass the bitter end up through the bight, around the line. . . . Then you bring the bitter end back through your asshole. . . . Well," Chief Snodgrass had said, "If you sail on a ship three times, with three good crews, you'll be out there the rest of your life." I mulled this over. I'd had many more than three trips and seen many more than three good crews.

When the *Esso Hartford* sailed up the Cape Fear River in March of '46, I got off the ship at Wilmington. I returned to the world of graphics, pictures and typewriters that had never been to sea. Now, having seen both ships and newspapers go down, I'm not sure which is the sadder spectacle. Both bring a salty tear to the eye.

SOS ON 500!! OR WHO DUNIT?

by Al Hadad, R-13

Things really got boring during the latter part of October 1942 at the USMS Resident Radio School at Gallups Island. Attendance at formal classes was no longer obligatory because most of our classmates had already passed the FCC tests and we

Life on Gallups Island

In formation.

Boots lined up.

W.F. Mansfield, R.W. Dixon, and J.V. Scott along with 62 R-17 buddies were feted after graduation.

Catching a few rays.

Barracks, 1941.

Sea Day inspection.

Resuscitation.

J. Kittle CBM firing Lyle gun.

MS Sailing boats.

Lifeboat racing.

were savoring the exultation of letting everyone within hearing distance know of our accomplishment. "Hey, boot, we've got our ticket and it's just a matter of time before we ship out and get on with the job. And the exam was EASY!" (Oh yeah).

So, to pass the time, we were directed to put in some time at the lab or watchstanding room and gain experience operating the "Coke Machine" and monitoring and logging signals heard on 500 kHz. The watchstanding was fun, but there was some work involved as well. We had to buff the floor to a high gloss and monitor and log whatever we heard on 500 kHz. We worked a deal; one guy would copy everything he heard while the rest of us polished or just plain goofed off. Each hour we would fill our logs from our monitor's copy. We did rotate so that everyone had the chance to experience copying live signals. And we heard plenty . . . SOS's, SSS's, BAMS schedules. Remember copying the WSL press broadcasts on 119 kHz? What terrific QRN! No HF in those days.

Over to the lab. The Mackay "Coke Machine" got lots of exercise. We were given hands-on instruction on tuning procedures, auto-alarm operation and battery charging, to name a few items. We were also informed that the main and emergency transmitters had dummy loads attached and that we could operate them without problems. So, someone began simulating distress traffic on 500! We didn't know any better. To top it off, an SOS was transmitted using a fictional ship's call on 500 and during the SILENT PERIOD! Well, the ship's company was advised the following morning about generating simulated distress traffic in no uncertain terms. It seems that a local monitoring station copied what seemed to be a real distress transmission. Once it was discovered that the ship was fictitious, it didn't take too long for the source to be identified and the authorities notified. The moral of this story is WIRE WOUND RESISTORS DO RADIATE!

SHEEPSHEAD BAY

by R.A. Sutcliffe, R-48

After all these years, the memories of "boot camp" at Sheepshead Bay in Brooklyn, NY in the USMS (United States Maritime Service), tend to get somewhat vague. I do remember what started it all. I was working in a machine shop when a previous employee came back for a visit. He was in uniform with red sparks on his sleeve and was at Gallups Island Radio School. The scenario which he outlined, and which he had gone through, was as follows:

1. Join the USMS, go to boot camp at Sheepshead Bay.

2. While at boot camp, take the examination for radio school and if accepted, go to basic radio school at Huntington, Long Island.

3. After eight weeks at Huntington, if you didn't flunk out, off you went to advanced radio school at Gallups Island in Boston Harbor. After eight months you went to the FCC and got your 2nd Class Radiotelegraph License.

4. You could then go to sea, either through the WSA (War Shipping Administration) or join one of the two unions at that time, ROU (AFL) OR ACA (CIO).

5. The last option, if you desired, was to spend two weeks on the USMS training ship, the *SS American Seamen*, KEVF.

Boot camp at Sheepshead Bay was like the USCG boot camp next door. I remember getting on a train at South Station with a lot of other guys, and eventually winding up at Sheepshead Bay. We were getting a quick physical at 2:00 AM, standing naked, including my first "short arm" inspection; finally getting to bed for a few hours; up with bugle calls; more processing; shots in both arms simultaneously as you walked through a door. Later on there was marching, marching, marching. The MAs (Master at Arms) ran things with a vengeance and there was a Lieutenant in charge of the barracks.

I learned a variation on a word there. If you became disenchanted with the boot camp routine you could leave, or disenroll. Of course, you would then be subject to the draft. The word disenroll was modified to DISENGOD———ENROLL. You fill in the spaces. The other phrase that echoed through the place when a new group came in was, "You'll be sorry!"

We learned to tie knots, row a whale boat, swim, hold the top of your life jacket when you jumped in the water so it wouldn't come up and knock you silly, etc.

I still remember jumping into the swimming pool from the high tower wearing the rubber immersion suit which was a very awkward thing. I swore I would never use one, until I got out in the North Atlantic in the winter, that is. Never had to use it, fortunately.

One night we had a new group come in our section of the barracks, and after things had settled down, one of the "veterans" (been there five weeks) turned on the lights, mustered them out of the sack, and gave them all a short arm inspection. The MAs heard the commotion, came in and caught us and we spent a couple of hours outside standing at attention at 2:00 AM and were restricted to the base that weekend.

Any old time merchant mariners are probably wondering what all this nonsense was about, but merchant seamen were needed in large quantities and when the USMS got rolling, it had decided that a military type boot camp and a little discipline would not hurt. The range of ages at Sheepshead Bay was wide, however, unlike the teens and twenties for military basic training.

THE SS *AMERICAN SEAMAN*

by R. A. Sutcliffe, R-48

Everyone who graduated from Gallups Island had the option of spending two weeks on the training ship, the *SS American Seaman*, KEVF. I decided it was probably worth the time, standing watches, getting accustomed to shipboard routine, etc.

The training ship was primarily for the training of ordinary seamen, oilers, wipers, firemen, and of course radio operators. The radio operators were excluded from ship details like KP, "Captain of the Head," etc., though it didn't always work out that way.

The *Seaman* did its sailing in Long Island Sound safely away from submarines. My memories of the routine on the ship are not very good except for the first time we went to sea. We got out in the sound where there was a very small swell and the ship was rolling maybe two degrees.

For all the trainees from the Midwest, it was their first experience on the ocean and that small amount of ship movement made them all sick. Anyone who sailed on the training ship will remember the long rows of sinks in the head to accommodate the large numbers of people who would be using them normally. They were now being used to throw up in and most of them got plugged. We had all had a hearty breakfast.

Anyone aboard, and that included radio operators like me, that wasn't sick, got drafted to clean up the mess and I will say trying to unblock sinks plugged with vomit was enough to try anyone's stomach.

I didn't learn much on the training ship, but it was an experience and the first endorsement on my nice new 2nd Class license.

THE LIBERTY AS A TRANSPORT

by R. A. Sutcliffe, R-48

Most people who sailed on Liberty ships did not think of them as appropriate for transporting large numbers of people, but the *SS Charles H. Herty*, KUIG, served this purpose once in its career. It carried 1500 people on a several thousand mile trip.

Our lengthy voyage (eight months), started in New York, took us to Italy with general cargo, then to North Africa to pick up a load of military trucks including a full deck cargo of same in "ready to go" condition when unloaded. From there it was the long beat to the Panama Canal (last fling in Cristobal) before sailing alone, heading for Manila in the Philippines. It seemed likely we would be staged there for the invasion of Japan.

Two weeks out into the Pacific, the war with Japan ended. I stood diligent radio watches hoping for a message turning us around. Unfortunately, none came and we wound up some weeks later in Manila. Manila harbor was a mess and getting to an anchorage required a circuitous path around sunken ships.

The military finally unloaded our cargo of trucks and sent us to the Lingayen Gulf area where we swung on the hook for a week. Finally we were sent to Davao on the island of Mindanao.

There, latrine facilities were built outboard of the ship on both sides, fore and aft, sort of a multi-outhouse with many holes and no surrounding structure. No privacy. This was at the main deck level.

Our passengers were then brought aboard: 1500 Japanese, mostly civilians who had been sent to the Philippines to work on the plantations. Bunks were not supplied for these people. They slept on Tatami mats on the main deck, between decks and lower hold areas. Many were in poor health and several were buried each day off the fantail, with appropriate Shinto ceremonies.

While in the Okinawa area, we passed through the aftermath of a typhoon. The large swells were coming in on the starboard bow, resulting in that sickening corkscrew motion of the ship and with most of the passengers seasick, the odor was something.

We finally arrived in the Yokohama area and as we approached, there it was, looming over the low clouds like something out of a travelogue movie, the snow covered peak of Fujiyama.

We went to a dock in Yokohama to unload the passengers but were not given shore leave. The Army was not sure there weren't fanatics who refused to accept the Japanese surrender. Most of the crew, including me, did run down the gangway so we could at least say we had been ashore in Japan in 1945.

We left Yokohama and went back to the States via the Panama Canal, arriving in Galveston, TX a few days before Christmas.

I came to the conclusion that the *SS Charles H. Herty* did not know how to make a short trip, said sayonara to it, and went home for awhile.

SEA ASSIGNMENTS

by Roscoe C. Maricle, R-0B2

I was given my release and furnished a train ticket to New York in February, 1942. I arrived the next morning at the R.O.U. Manhattan office, met Fred Howe, and by mid-day boarded the *SS West Kyska.* The *Kyska* was a Hog Islander cargo vessel, for the Waterman SS Line sailing under Captain V. T. Wood.

We sailed that evening to Philadelphia and spent a week loading aviation gas, Mack truck chassis'and flour. We then sailed for points east. I experienced two-days seasickness along the Cape Hatteras coast swells. We passed near the Miami Keys, where we sighted a sub just before windward passage. He dived; we ran (13 knots TOPS!). We proceeded to Trinidad where we off-loaded the remains of a Philly longshoreman who had been killed and dumped into our chain locker.

We proceeded to Capetown, South Africa. A German small battleship or gunboat was sunk within 100 miles of our position while crossing. Then to the Port of Suez through the Suez Canal and then Port Said. Stayed overnight then departed in a three-ship convoy during the early afternoon. At dusk, Italian planes flew over us to bomb and mine Port Said. We arrived at Alexandria the following day. We were about ten days off-loading. Alexandria was bombed three nights. The city sustained some damage. I would come out in the morning and walk in six inches of shrapnel on the deck. Good old steel and concrete radio shack!

I managed to take a train to Cairo. King Farouk and a portion of his harem entered the train through my coach on their way to their private coach. In Cairo, the day was spent going to and entering the Great Pyramid, visiting the old city (where a wedding was in progress on the main boulevard), and purchasing a few items.

As a lone visitor, I was ushered into the main mosque and given special guest treatment. I was ushered into a huge domed room. The Koran—a huge book on a massive pedestal—was in the center where the King himself read to the faithful. I was taken to the underground area where deceased formal royalty were buried in approximately 5' x 12' white marble coffins sitting on 3' to 4' full size pedestals. The stone caskets were inscribed with one-inch wide and deep, scrollwork filled with silver and gold that covered all four sides and lid. A stone carving artist was working on the reigning-king Farouk's coffin which was about half completed. He informed me that he was well paid and would receive a lifetime pension but would lose both hands on completion of the project. I learned later that those other than the Muslim faithful were allowed to enter only one day out of the year. A very interesting day!

We came close to being assigned to a convoy to Malta where the Germans and Italians had attacked continuously for months. Our flour was transferred to another ship. We gladly sailed from Alexandria returning through the Suez Canal into the Red Sea to Aden and then to Bombay, India. We loaded manganese ore for a week via a continuous human chain—each person having a thick padded platelike headpieces holding the ore. The line circled from port deck to the ship's hatch's opening. Two circles and close to one hundred workers were involved. We partially loaded, then sailed to Calcutta for topping off with hemp and tea. We sailed alone and were the first American ship into Calcutta after a Japanese Cruiser had destroyed a large convoy in the Bay of Bengal that had originated in Australia. The Hoogli River, the muddiest in the world I believe, was a three-day entrance run to the Calcutta seaport. We learned that the Malta convoy from Alexandria had been demolished but that one from Gibraltar at the same time had made it.

The ship was loaded in ten days and sailed to Durban, South Africa. This trip included a week of typhoon weather with days of 50-and 60-foot waves. The antenna I had spliced mid-point between masts on leaving Philadelphia survived the whole trip. We sailed to Capetown, then to Trinidad where we joined a large escorted convoy to New York arriving very late in September, 1943. The East Coast sub battle had peaked and abated somewhat. The captain had lost a brother-in-law; Chief Mate's relative had a ship torpedoed under him and was chased into port several times and was a nervous wreck. I signed out from Waterman SS Company. Right after Christmas, 1942, I sailed on the Pan Maryland under Captain Lund, a Pan American tanker to Aruba, loaded, and returned to New York the first of February, 1942. This was the last of my merchant ship assignments.

I was hired by American Export Airlines based at the New York LaGuardia Airport. It was a Flying Boat company covering the North and South Atlantic carrying passengers and cargo. I was a Flight Radio Officer for two years, still observing a lot of the coasts and the sea. We did get home on the average of every three weeks—a worthwhile consideration at the time.

DECEMBER 7, 1941

By 2nd Lt Harrison W. Moore, Jr.
U.S. Army Signal Corps
Electronics Training Group

The closing credits faded out, superimposed over that familiar silhouette of Alfred Hitchcock. Cary Grant and Joan Fontaine had held us in suspense for well over an hour. The movie was "Suspicion," and it was Hitchcock at his best.

The curtains at stage right parted, revealing the organ console rolling forward into the spotlight. Dick Leibert was playing a scintillating segue into "Nice Gourd," the title of the stage show soon to begin with the full corps of Rockettes doing their famous high kicks.

It was Radio City Music Hall and 2nd Lt. John Graves and yours truly, 2nd Lt. Harrison W. Moore, Jr., both newly commissioned in the U.S. Army Signal Corps, were enjoying Sunday afternoon on leave from Fort Monmouth, NJ in the company of my kid sister, Ethel Many Moore.

As the house lights came up to signal that the next show would begin shortly, an usher reached in from the aisle and tapped me on the shoulder. He asked us to follow him to the lobby, where he informed us of the attack on Pearl Harbor. "All military men are to report back to their bases immediately," he said. When we asked for details he suggested we check the new ticker in the lower lobby. The teletype clattered out sketchy details of the Japanese attack and ended with the command to all military to "Report back to base."

A quick call from the adjacent phone booth reached the orderly room of Electronics Training Group, Fort Monmouth. It was still a quiet Sunday afternoon there. In the background I could hear the sound of a ping pong ball bouncing back and forth over the net on the table in the day room. The non-commissioned officer on duty asked us when we were going to return to base from our one day trip to the big city. We said we planned to be back early in the morning. "Ok," the NCO said, "That will be in plenty of time."

Four days later Graves and Moore were on their way to England. Their orders, classified "C O N F I D E N T I A L" had been cut in October. They were two of a group of signal corps lieutenants in group number five being sent to serve with the British Army and Royal Air Force studying British "radio direction finding." We were to study in their RDF school for three months and then serve with British forces for six months in return. It was a lend-lease of personnel agreement arranged by President Roosevelt and Prime Minister Churchill through the U.S. Office of Scientific Research and Development.

Radio Detection and Ranging (RADAR) was a new and highly classified subject. Our parent unit, Electronics Training Group, was established to provide the U.S. Army Signal Corps, and later the Army Air Corps, with a cadre skilled in the technical and tactical operation of ground and airborne radar equipment. At that time, the British were well ahead of the Americans in both the science and tactical operation radar.

On December 11, 1941, Moore and Graves, with 48 other Group Five lieutenants, left Grand Central Station in New York City, by train to Montreal and thence Nova Scotia to await formation of a small convoy of 11 vessels and two destroyers. After a week of waiting, the convoy set sail for Greenland. Once there, four ships sailed on alone, running a zigzag course for Liverpool. Ours, the *SS Latetia*, a Greek cruise ship making her first trip as a troop transport, was one of them. We were without escort until being picked up by a Sunderland Flying Boat. It provided air cover until we made port in Liverpool on December 25, a most momentous Christmas Day.

Two hundred officers from ETG Group One through Four were already assigned as "civilian observers" to the American Embassy and were attached to various RAF and Royal Artillery units throughout England. We created quite a stir on December 8, 1941, when orders went out for us to put on our American uniforms. Many a comment was made by military and civilian alike: "How did all of these 'Yanks' get over here so fast?" and "Don't they look smashing in their green and grey uniforms with those colored crossed signal flags on their collars!"

So much for December 7, 1941 in the life of Moore, who now, over 50 years later, serves as the president of CONEX, the Council of Northeastern Chapters of the Quarter Century Wireless Association. 2nd Lt. Harrison W. Moore, Jr. retired from the Signal Corps as a Colonel. His last major assignment was as Commandant of The Army Reserve School in Norfolk, VA.

REINSTATED

by Maurice Reller, R-11

My grades in math and theory were low from the very start. I was able to spend all my free time studying my shortcomings as I was the only one of the top three in radio code. We had weekly tests, and finally, I missed my allotted three failures and flunked out. I knew that I could get it together enough to pass the 2nd class FCC exam, if I only had more time.

So I stayed on the Island and did general detail work, much of which consisted of grounds maintenance. After several weeks of studying extra hard and long, I requested and received several days of accumulated leave. I rented a room at the Charleston YMCA, and put in several days of intensive review. I then took my FCC exam about a week before my regular class did. I passed the exam and received my 2nd class Telegraph Certificate.

When I came back to the Island, a friend of mine, Bob Klay, insisted I go with him to the Commander (I don't remember his name) and request to be reinstated. Bob did the talking and I was reinstated.

SEXUAL HARASSMENT

by Maurice Reller, R-11

This is a true story of sexual harassment in 1942.

I caught my first ship in Hoboken, New Jersey. There was no action for almost six weeks at the War Shipping Administration, so I joined the American Communications Association (ACA) and was assigned a ship in a few days.

It was absolutely the worst, stinking ship I had ever heard of. The *SS Chiloil* was a 5,000 ton tanker that, later, either came apart or was torpedoed in the Caribbean. It belonged to National Bulk Carriers, who, I now suspect, was ripping off the War Shipping Administration with this ship.

After about one day, I realized that the Skipper was gay. That's not the word that we used for reference in those days. He would come into my cabin and pretend to stumble and thereby catch his balance by thrusting out his hand and landing right on my crotch. If I came out on the bridge after dark, he would give me a cheek to cheek bear hug. In those days I looked like a clean-cut young boy. I barely even had to shave then and I am sure I had that fresh-from-the-farm look. I had to watch my step for the whole six weeks until I signed off in New York.

In these days, I completely accept anyone's sexual orientation and do not prejudice my thinking. But, I do have a keen sense of empathy when I read about female sexual harassment.

THE *S. S. ROGER WILLIAMS*: FIGHTING SHIP

by Max Stein, R-45

The S.S. Roger Williams *was on a "SHUTTLE RUN" in the Mediterranean Sea for its entire voyage* noted above. We carried ammunition, participated in invasion, and beach-heads and was the ship that carried 7,000 tons of mustard gas, 275,000 175mm shells of mustard gas into Europe (as a preventative). I still do not wish to name the city or country we delivered the cargo to.

The *SS Roger Williams* should go down in history as the fighting ship in the Maritime Service. All the time, ships that were in the service in the Mediterranean would toot their horns at us to acknowledge their respect. All the time in the Mediterranean, we ran around mines like it was a dance. Please note, that on the second day of the invasion of France, one of the engine room boilers blew up and we limped along all the time until we came back to the States. I should note that we were always under submarine attacks in the South Atlantic, North Atlantic, and Mediterranean Sea.

My ship shot down two (2) German planes, confirmed by the Admiral of the convoy, in the Mediterranean Sea, south of the Beleric Islands, (Spain) enroute to Naples. This air attack lasted for one hour and twenty minutes, one of the longest air attacks on record. A total of nine German planes were shot down, seven by the Admiral's cruiser. Air attacks, too numerous to mention, about 50 or more. Submarine attacks in convoy. Going through the Straits of Gibraltar, 19 ships were sunk in the convoy by submarines. We were carrying ammunition.

Captain volunteered for mission to the city of Civitavecchia, Italy. All by ourselves. To bring food, ammunition, and bring back the wounded on the Beach. Secret mission. Attacked for four (4) days and three (3) nights continuously. We were the only ship, or target in the Bay; never anchored, just moved around all the time. Attacked by German "E" boats (P.T. boats) during the evenings; attacked by Italian 2-man submarines; attacked by German scuba divers during the nights. We drove them all off. American wounded troops refused to stay on ship and went back to the Beach where it was safer. We had 6 men disappear and presumed captured. Three men died and were buried at sea. The Chief Steward was taken off in Naples, on return, in a straight jacket, going insane.

POEM OF WAR

'Twas off the shores of Eniwetok
In the South Pacific Isles,
Far away from the the USA
Some fifty one hundred miles.
We were riding to our anchor,
And all was calm and still
When through the quite of early morn
We heard a warning shrill.
Enemy planes approaching,
Gunners! Man your guns.
All grim-faced and fighting mad,
We waited for those sons.
The brilliant beam of a searchlight
Was lancing through the air
Searching for those enemy planes
That were hidden up there.
They soared around above us.
They dropped their guns at will
But a sudden burst from an Ack Ack Gun
sent one of Tojo's boy to hell.

Midst the deafening roar of bursting bombs,
Fires burning everywhere,
The steady pom pom guns
As shrapnel filled the air.
Another plane came in too low, strafing on our side
Our D.E.'s guns spoke through the night
And that was his last ride.
In those worthless isles of shell-torn palms
With their shores of coral sand
"Twas a mighty price we have to pay
In that far-off distant land.
But it is for a worthy cause,
Yes, a worthy cause I must say.
It's to protect our homes, the land we love,
The good old USA.

MY GALLUPS ISLAND EXPERIENCE

By Al Hadad, R-13

I was born on December 9, 1922, and raised in Toronto, Canada. Great Britain and the rest of the Commonwealth declared war against the Axis in September, 1940 when I was 16. Having been a member of the Canadian Sea Cadets for two years, I wanted to become involved in the fight. The Navy rejected me for physical reasons. The next step was to try and join the Merchant Marine. In this I was successful, so on February 24, 1940, I became a member of the crew aboard the Panamanian flag ship, *SS Beaconlight*. The crew was composed of all Canadian citizens, many being too old to be in the military.

I remained aboard the *Beaconlight* for a year as an ordinary seaman. The ship travelled the world over and included one trip to Southampton just at the time of the evacuation of France by the British and their allies. The cargo consisted of 80,000 barrels of avgas. We departed England in June, bound for San Pedro, via the Panama Canal. From that time on the *Beaconlight* steamed across the Pacific several times, visiting the Philippines, French Indo-China, Dutch East Indies, New Zealand and Japan. My last trip was from Japan to San Pedro and I was paid off on February 25, 1941.

While aboard the *Beaconlight*, I was always fascinated with the position of Radio Officer. Having been trained in the Cadets as a signalman, I was familiar with the Morse code. Our "sparks" was a very friendly fellow and permitted me to visit with him in the "Radio Shack." I would monitor 500 kHz and identify the various stations that I heard. I was not very proficient.

When I signed off, I was given a Pullman ticket to Toronto and a big payoff of $85.00. My salary was $45.00/month, of which $30.00 was sent to my mother who was living in Chicago.

On arrival in Chicago, I was met at the station by my father, Lewis. I had not seen him for seven years. He informed me that I was to remain with him, in Chicago. He said, "There's a war going on up there. Anyhow, your mother is living here as well. No need to return to Toronto and your grandmother."

So, at age 18, I moved in with my dad, found a job and was accepted as a student at the American Television Laboratories. My objective then was to acquire my 1st Class Radiotelephone license and work as a TV cameraman. The school was owned and operated by the Sanabria brothers with Dr. Lee de Forest as a figurehead.

This lifestyle was not to last. December 7, 1941 changed it forever. My attempts to join the Navy and Coast Guard on the 8th were rejected because of myopia/astigmatism. By the end of December, I felt, sadly, that the draft was to be my alternative. Fortunately, I happened to see an advertisement in the Chicago Tribune that would keep me out of the Army and set a course for the rest of my life. The ad, placed by the U.S. Maritime Service, called for men to be trained for positions aboard merchant ships of the United States. It was just what I was looking for—given my past sea-going expertise.

The USMS recruiting office was located in the Monadnock Building in downtown Chicago. On entering, I was surprised to find the staff of two holding down a small two room office and me the only recruit! After filling out the required papers, I was sent to a physician for my physical. Naturally, I flunked the eye test. When the USMS officer-in-charge told me that I was being rejected because of my eyesight, I informed him that I had been a seaman for a year and that I managed the job with that handicap without any problems. That did the trick. He gave me a waiver on the spot and informed me that I would be getting a call in a month or so. I would be trained for upgrading to the position of Able-bodied seaman at Hoffman Island, NY. Enrollment date was February 2, 1942.

At Hoffman Island, my classmates were all experienced seamen and expecting to up-grade to AB. There were about six of us, all ranked as Seaman First Class. Our uniforms were similar to the Coast Guard Chiefs' with the USMS badge and no stripes. We were quarantined for several weeks while getting our shots. We sure wanted to get ashore!

In the beginning, we were trained to fire and maintain the Lewis machine gun and load and fire the 4-inch 50 caliber gun. A certificate was issued stating that the holder was qualified to serve as a member of a gun crew for Armed Guard duty.

By the time gunnery school was completed, the island was full of recruits attending seamanship classes, cooks' and bakers' school, loading and unloading simulated cargoes. Instead of training us, we were made acting Team Leaders and our jobs changed from trainee to trainer. March to chow, march to class, march to chow, etc., etc. We were bored.

Then, sometime in March, a notice appeared on the bulletin board asking for volunteers to attend radio school at Gallups Island. . . . if you could pass a test. As I remember it, the test was being able to distinguish the Morse character "A" from the character "N." I signed up immediately and received orders to transfer the first week in April, 1942.

So, a group of us, including Charlie Buckingham, took the "Red Eye" to Boston and early in the morning found ourselves aboard the cutter *Yeaton*. We even had breakfast served to us, no doubt because the uniforms we were wearing may have indicated we were Chiefs instead of S 1/C. It wasn't long before we changed into our sailor suits.

I was assigned to platoon R-14 and not long afterwards, to R-13 from which I graduated in November, 1942. Our platoon leader was Lyell Thomas, from Kentucky.

Study was accelerated. Some of the students had a hard time learning code or the mathematics of electronics. We had two or three mates who were licensed amateurs. They seemed to have no difficulties with their studies.

My two closest mates were Charlie Buckingham and Clark Ison. Charlie was from New York City and Clark from Blackfoot, ID. It has taken me over 40 years to find these guys. Charlie, retired, lives in Mashpee, MA in the Cape Cod area and Clark, retired from the Department of Agriculture, lives in Camp Springs, MD. Both are now members of GIRA.

Life at Gallups was interesting, if not difficult for me. The weekend shore liberty was very helpful to continue working for my license. I became acquainted with a young lady and her family from Somerville on my second week end ashore. It became a ritual every week to spend the week end with the family. Brown bread and Boston baked beans are never to be forgotten! The family encouraged me to keep working on the subjects being taught and not give up. The King family gave me the support I needed to reach the goal of my license.

On November 7, 1942, Jean King and I celebrated my passing the FCC 2nd Class Radiotelegraph license exam. The place? None other than the famous Coconut Grove nightclub. Lucky for us, because two or three weeks later the club had a disastrous fire with many lives lost.

Having completed training and passing the FCC test, life became boring the week of November 14. One night, a few of the R-13 gang decided to initiate a new platoon of boots on the first deck of our barracks. After lights out, we removed the fire hose outside their dorm, silently entered their dorm, then turned on the water, anointing everyone. We ran like hell back to our dorm and not long afterwards the Master-at-Arms appeared, and put all of us on report. Even after we all complained that we were fast asleep!

November 13, 1942 was a red letter day for some of the R-13 gang. Lt. Casilli informed us at Captain's Mast that we would never become Radio Officers and told us to pack up because by 5 p.m. we would be shipped off the island. True to his word, that evening we were on the "RedEye" to New York arriving there the following morning. So long Gallups Island!

Accompanied by classmate Carl Thenemann, we travelled to New York for assignment to our first ship. I never saw Carl again. About four years ago, I heard that Carl lost his life in a boating accident in the Gulf of Mexico while on vacation at the time.

From the time I went aboard my first ship, I have felt that my future was directed by first, the urging of my friend, "Sparks," aboard the *SS Beaconlight*, to return to school and better myself and for the fine training I received from the USMS staff on Gallups Island. They created the path that led to a satisfying career, 43 years long, in communications.

During this time period, I was employed as follows: three years sailing as Radio Officer; thirteen years with TWA as Ground Radio Operator, Flight Radio Officer and Staff Technician; twenty-three years with Philco, later Ford Aerospace (FACC) as Field Engineer to Senior Engineering Specialist. I retired from FACC February 1, 1985.

What do I remember most during my eight months on the island? Here is a short list: KP week. Well, almost, because I was able to get that week off to travel back to Chicago to see my family. The reason I used was I had been on duty since first arriving on Hoffman Island in February. Whaleboat training. Great fun and exercise and competitive racing. Weekends ashore. The people of Boston were extremely hospitable. The Boston Common was a great place to catch a little EYE time! The crazy antics of some of my classmates. In particular Leon Holster and Jack Heffernan.

Favorite instructors were: W.O. Summerfield, Chief Sweeney, Chief Lodge, Chief Lane, Mr. McElroy and Mr. Canavan.

Special date: October 7, 1942. I will never forget watching in awe as the *SS Queen Mary* passed by the island on her way to the Boston Navy Yard. Her bow, severely damaged as a result of a collision with one of her cruiser escorts, was repaired in drydock. In the afternoon she transmitted a test on 500 kHz. Why would she do that? Maybe showing off.

I received my Amateur Radio License in 1956 and now that I am retired, I devote more time to the hobby. I make contact with many former Gallups Islanders every week and enjoy the camaraderie it brings including the"Sea Stories." My call sign is AA6SB.

Having maintained my 1st Class Telegraph ticket, I was able to sail once more, sail as R.O. aboard a tanker in 1989. GIRA member "Shanghai" Ray Maurstad had a hand in getting me the job with Texaco aboard the *SS Texaco Massachusetts*. After a 45 year absence, I found I could still do a satisfactory job. Thank you U.S. Maritime Service.

What a shame that by 1999 the job of Radio Officer will have disappeared forever.

As GIRA Region 9's first Director, I have enjoyed my association with our members and working with everyone to make the organization grow. The national and regional reunions have been particularly enjoyable because one meets friends, old and new. My sincerest thanks to our officers who have done a tremendous job in keeping interest at a high level and publishing great stories of the past in both the "Spark Gap" and "Islander" newsletters. Congratulations.

THREE YEARS SHIPBOARD

by Robert Milus Bond, R-16

This is my chronology of voyages completed while serving in Merchant Marine from February 1943 to February 1946. Upon completion of radio school at Gallups Island I was sent to San Francisco and assigned to a new ship, the *SS Dashing Wave*, as chief radio operator. The maiden voyage was with cargo to Pearl Harbor and returned to San Francisco.

While in San Francisco the *Dashing Wave*, a C-2 cargo, was modified to carry troops. First part of our voyage was to Port Hueneme, CA, to take a battalion of Navy SeaBees and their equipment to New Caledonia.

With other cargo aboard, the trip continued to Guadalcanal. Jap planes attacked enroute without success. Bombs were dropped at our bow. Fortunately the bombs were duds and only splashed water on the gun crew. In Guadalcanal, the cargo was unloaded on barges during daylight. Frequent bomber attacks were made at night.

My third voyage took me back to New Caledonia with SeaBees and cargo for the Espiritu Santo Islands. On the return trip the ship carried wounded and other troops stateside. Enroute, traveling with darkened vessel, another ship collided with us at about midnight, striking #4 hold and caving the side in about six inches. Voyage was continued as no leaks occurred. Next port was San Diego and then on to San Francisco.

From there to Port Hueneme and Pearl Harbor where we spent considerable time at anchor. Finally, we joined a convoy bound out to the Atoll of Tarawa, arriving a few days after the main battle. Small arms fire was continuing and there were nightly enemy aerial attacks. Back in Honolulu the news was out that we had been sunk with all hands lost! After regrouping, the ship was routed back to San Francisco to pick up cargo. Just outside the Farallon Isles, we came upon a Navy float plane drifting out to sea. The aircraft engine mounts had broken, disabling the unit. We found them on Christmas night—undoubtably the best gift the two crewmen had ever received.

With cargo a return was made to Hawaii and we then took SeaBees to Eniwetok Atoll. Enemy aerial attacks came nightly. Return trip to San Francisco for more cargo.

We returned to Hawaii and sailed with Army personnel to Eniwetok Atoll for the second time. While cargo was off loaded the Army set up camp on shore. When cargo was cleared out the Army unit came back on board and returned to Hawaii. Our ship went to San Francisco for more cargo.

With cargo on board it was back to Pearl Harbor and more waiting. Shuttled Army troops between islands. In June we joined a convoy which was bound for the Invasion of Saipan. We arrived there shortly after the battle had begun. We endured nightly enemy aerial attacks, presumably from Guam. After a few days of unloading cargo into landing boats all Merchant vessels were ordered to depart in haste, without even closing cargo hatches. A convoy of about a hundred ships formed and sailed east many miles. We sailed in a very large circle for about a week, then returned to Saipan to complete unloading. Nightly bomber attacks continued. With cargo off loaded the *Dashing Wave* returned to San Francisco. The crew was relieved as the Navy took control of the ship.

I have since learned that the reason for the odd departure was the Mariannas Turkey Shoot, carried out by our Navy Carrier Task force occurring at that time.

Being relieved from the *SS Dashing Wave* after 18 months, I took leave to my home in Texas. After leave I went to New York and on to Boston to join the *SS Hubert Bancroft* as chief radio operator.

The *Bancroft*, a Liberty ship, was an ammunitions ship. From Boston it sailed to Searsport Ammo Depot. After loading 7,000 tons of H.E. shells and deck cargo, we returned to Boston. We sailed in convoy to Halifax, Nova Scotia and on to Belfast, North Ireland then to Le Havre, France to unload. Returned to New York and Philadelphia for a second load. From the Philadelphia Ammo Depot we sailed in convoy first to Wales then on to Cherbourg, France to unload and return to New York.

New York to Houston and Galveston, TX where we took on 5,000 tons of sulphur and food stores plus deck cargo. Sailed back to Norfolk, VA where a convoy was forming for the Mediterranean. Stop over at Marseilles, France were deck cargo was removed overnight. Next day we sailed on to Toulon, France where all cargo except the sulphur was off loaded. Cargo of sulphur carried to Sete, France. Sulphur was considered dangerous cargo, similar to ammunition. War in Europe ended while in Toulon. Returned to New York.

Took leave home to Texas. Journeyed to New Orleans and joined the crew of the *SS Calvin Austin* as chief radio operator. The *Austin*, another Liberty ship, took cargo (mostly beer) and construction equipment on deck to the Philippines via the Panama Canal. While enroute, the A-bombs were dropped and Japan ended WWII in the Pacific. Returned stateside with some homebound troops through the Canal and on to Baltimore, where I left the *Austin*.

Traveled across country to San Francisco, thence to Portland joining the crew of the *SS Sea Flasher*, a C-3 troop/cargo ship. From Portland we sailed direct to Manila, where we embarked approximately 2,000 homebound troops. Delivery was to San Francisco. In February 1946, I ended service and returned to my home in Texas.

A SHIP OF MANY NAMES

by John J. Ward, R-19

Scuttled by the Italians at the beginning of WWII, the old ship had sat on the muddy river bottom for six weeks and should have been left there. Instead it was raised and towed to Galveston for refitting. Built in Scotland in 1908, it already had been called by several names, flying the Union Jack, Italian, and possibly other flags. The Americans named it after the Greek, Alcibiades, put it under the Panamanian flag, and made Cities Service Oil Company its operator.

After the raw, cold winter on Gallups Island and a severe attack of tonsillitis near graduation time, I opted to go south to New Orleans and await the War Shipping Board's wishes. A fortnight later they sent several of us to Houston. Still later, I went on to Galveston. A naive hillbilly with no idea of how the system worked, I didn't know the better assignments came through the union halls. After still more waiting I was assigned to the *SS Alcibiades.* At this point I would have accepted a dugout canoe.

The deck house and living quarters had been rebuilt, but basically the *Alcibiades* was a 1908 ship and, in comparison, a Liberty ship seemed like a cruise liner. The crew was a polyglot of losers: a few Americans, one or two Canadians, but mostly Europeans. The Captain was a Norwegian with an interrelated woman and drinking problem. He promptly came around with the cash advance sheet and said, "Sign here. I've just borrowed fifty dollars each from you, the purser, and the third mate." Several others got to me in the following weeks before pulling the disappearing acts.

After more delay we finally sailed one evening in a small convoy and into the middle of a weather system. By morning there wasn't another ship in sight, and the Skipper was passed out drunk, a condition he managed to retain until we limped into New York harbor days behind schedule. I'm convinced any lurking subs didn't think we were worth a torpedo.

The Skipper got fired when caught trying to sneak the ship's sextant ashore to pawn for booze and was replaced by a Canadian. It required several days for the recalcitrant pumps to discharge our bunker C cargo. We then went into the Perth Amboy shipyard for more repairs. The purser got fired for being AWOL and borrowed fifty dollars from me to get back home in Houston. The only one to repay my loans, he later went to a lot of trouble to find out my whereabouts.

The old ship had a dry cargo hold back of the midships house that was never used except for off duty crewmen to sit upon. The natural draft smoke stack thrust skyward. An ancient steam whistle, activated by a mechanical wire, blew condensed water all over everything within 20 feet. Long seconds passed, as everybody on deck scrambled for cover, before it was clear of water and any sound actually emitted. Engine problems continued and leaking steam pipers were jerryrigged with rags tied around them. The next few months saw our time split about 50/50 between various shipyards and actual sailing. At 4 a.m. in the Gulf of Mexico, an old Canadian Second Assistant Engineer came out of the stifling hot engine room, walked over to the rail for a breath of air and plunged overboard to eternity. The builders had failed to extend the railing underneath a life raft. This defect was promptly remedied but too late to help the hapless engineer, a non-swimmer.

When I approached or passed by the off-duty crew members loafing on the cargo hatch, typically someone would begin, "There he is. He don't smoke, he don't drink, and he don't go to whorehouses." Then a companion would exclaim, "What in the hell does he live for?" Everybody would laugh uproariously.

In early December we again went into the Galveston shipyard to replace our guns, as ancient as the old ship itself, with a three-incher forward, a five-incher aft, and in place of the old machine guns, we got 20 millimeters. Rumors that our destination was the South Pacific sent most of the crew scurrying ashore. Just before Christmas, the *Alcibiades* sailed for Australia. We made our disorganized way toward the Panama Canal joining a convoy to Guantanamo Bay where the steward took on fresh supplies which proved a tragic mistake. Subs were about in the Caribbean and our convoy commander ordered all radios off including receivers. Depth charges kept us awake at night then on the second day our engine faltered, and we became a straggle. A PC dropped back and circled us until we could rejoin the 8-knot convoy.

Once we were through we breathed easily again. For the next 45 days we never spotted another ship and only one tiny island looking like a flyspeck on the endless ocean. Every day was like the one before bringing to mind a stanza from the Rime of the Ancient Mariner: "...no wind nor any motion, like a painted ship upon a painted ocean." We had no short wave radio equipment aboard and, consequently, no news. About three weeks out, on BAMS, I copied an encoded message, QSA 2 on the very low frequency band, warning us that we might pass a ship during the hours of darkness. It proved to be a worthless piece of information. A week out of Australia we got a second message diverting us from Townsville to Brisbane. Three days from Brisbane, the ocean that had been like glass for six weeks became violent, whipped by gale force winds. The *Alcibiades'* arrival in Australia was received calmly.

Our mission of delivering the ship to Australia where, we were told, a Navy crew was supposed to be awaiting was completed, in theory. However, we topped off—half our cargo of bunker C had been burned on the trip from Balboa—and were ordered to deliver it to Milne Bay in New Guinea. Two naval officers joined us for familiarization training. There were more lights in Milne Bay than Boston Harbor. A few days later we pressed on to Buna and anchored about 200 meters from the end of Buna's airstrip as sort of a filling station to refuel the 7th Fleet, cynically called MacArthur's navy. Japanese planes attacking the airstrip made their runs within a few hundred feet of our mast. Our gunners were ordered not to fire on them after some wild shooting that did more damage to Air Corps facilities than the enemy. Conversely, the attackers showed no interest in the old *Alcibiades.*

When it was time to move farther up the coast our windless couldn't pull up the anchor. We were delayed helplessly for days until a barge with a crane came to help us get the anchor up.

With the stores in Guantanamo, a cadre of cockroaches came aboard and proliferated in the hot, humid tropical atmosphere. They were in everything including the awful bread our disinterested cooks baked. Nobody had to worry about weight problems.

Then, as now, the country had a plague of lawyers, and large numbers of them ended up in the Navy and were assigned to merchantmen as gunnery or communication officers. Most suffered from ego trips, but the one exception was Stanley Raymer, a former prosecuting attorney from Indiana who joined us in Texas and became a lifelong friend. The old ship would swing on the hook for days then a few warships would come alongside for fuel—we could do two cruisers at a time or four destroyers. Finally empty and eager to sail for Australia, we were disappointed when a T-2 tanker came alongside to refill us. More boring days. The understudy naval officers were replaced by two others. The second cook went over the edge and was hospitalized. We lost the pumpman to scurvy. Fights broke out. Empty again, our spirits plunged when another T-2 came along side to refill us. Our former Captain was the First Mate. He said he was making plenty of money now, took my address, and promised his wife would send the borrowed fifty dollars. Of course, I never heard from him again.

Somebody threw our eating utensils (not silverware) over the side and for months we managed with makeshift wooden forks and personal pocket knives. A CB came aboard to help with minor repairs and was good enough to promote some replacements. After several months in Hollandia, (New Guinea), we moved farther north to a tranquil bay. Fresh water was a perpetual problem, always rationed. Salt water showers made everything worse. B. O. was universal. The engine gang suffered most. After days of solitude and without warning, the quiet would be broken by frantic firing and explosions. Then tranquility returned as if we'd only dreamed of an attack.

The idle 20-mm gun crew (38 merchantmen) began getting into mischief such as fights, stealing food. Half a dozen drank quantities of Aqua Velva and orange juice, went blind temporarily and nearly died. This caused increasing strain between the Skipper and Lt. Raymer, who had the shape of a marsupial when he came aboard but now resembled Mahatma Gandhi's twin. One of the gunners kept busy cutting our hair with only a comb and paper scissors.

Empty for the dozenth time, were elated to receive orders for Australia. But in the Coral Sea our spirits again plunged when radioed instructions came to reload in Port Morsbey, deserted except for its now diminished military garrison. Ashore to clear ship, the Skipper, Raymer, and I stopped at Australian Officer's Club and were asked to leave by a spit and polish steward because of our casual dress (khakis). We discharged the cargo on Thursday Island, also deserted except for small military detachment and a few "abos." The island's main industry was mother of pearl brought to the surface by big-footed natives.

Our skin amber-toned from taking atebrine (anti-malaria pills), we finally got to Brisbane in September where the Navy did commission the ship. The new Skipper, who had been with us for several months, was a former merchantmen. They manned it with a crew of 142 and so much equipment, of dubious worth, that their cargo capacity was reduced to barely more than half our former capacity with the 38-member crew. The *Alcibiades* got still another name when the Navy christened it the *USS Andrea Doria.* I said goodbye to Lt. Raymer and spent a fortnight in a tent city amid a muddy and cold race track before boarding the Dutch troop ship *Klipfontain* for San Francisco. I then took a wartime cross-country train with all the amenities of a troop ship. My home stay was cut short when the local draft board presented me with a 1A and threatened to have me arrested.

In New York I was prompted assigned to the *SS Red Canyon,* a T2 tanker, and within days was back in the Pacific in time for the invasion of Leyte. During 1944-45, the *Red Canyon* called at most of the important Pacific islands. One night we fired on a surfaced submarine that protested it was one of ours as we turned tail and fled. Near the war's end, between Hawaii and California, we were diverted on an unlikely dog leg to the south. After we thought they were no longer a threat, a Japanese sub had sunk an American ship on our doorstep.

On the return leg of the next voyage, we lost a propeller blade and came into San Pedro with the remaining three, vibrating fiercely. Got a new one and was back in the Philippines when a B-29 dropped the big one on Hiroshima. As we entered Manila, a ship's blinker inquired if there was a John Ward aboard. It was my old friend Lt. Raymer on the *USS Andrea Doria* (old *Alcibiades*) and we spent VJ Day together in Manila where a few fires still smoldered. Stanley confided that except for my fast friendship and understanding during the *Alcibiades* duty that he would have certainly gone mad.

I never saw Stanley Raymer again, but we corresponded over the years, remaining lifelong friends. He returned to his northern Indiana law practice and became a judge. Stanley named his first son after me, the ultimate compliment.

Upon returning to Los Angeles I bid the *Red Canyon* adieu. It has been a two-ship war for me and, in a sense, a draw. I killed as many of them as they did me.

Having experienced the cross-country train trip and no desire for an encore, I signed on a Liberty ship sailing through the Canal for New York and the boneyard. The following winter I spent on a T-2 (*Wahoo Swamp*) carrying crude oil from the north coast of South America to England and then on coastwise or nearby foreign trade.

The following October I went to Lima, Peru, to fly for Pan American-Grace Airline. That southern continent was then a paradise compared to its beleaguered conditioned today. From South America I went to college, a rough transition.

I went back for the Korean War making two trips around the world, one in either direction. On one of many voyages to the arctic (Tulie, Greenland) we hit a huge iceberg head-on that came closer to sinking us than any German submarine or Japanese aircraft had.

After that (second) war I flew the Atlantic for TWA and worked briefly for Braniff Airways then signed on with the FAA's air traffic division for the duration. Using six weeks of accumulated leave from the FAA, I made one final voyage on a C-2 freighter to Viet Nam in that unfortunate war (number 3). Unbelievably disorganized, it took more than a month to unload and my six-week trip became ten. My FAA employer was less than delighted.

My FAA career began in Honolulu, and I moved five times in the first five years then said "no more." Resisting the constant pressure of their mobility program, I spent 22 years at Phoenix's Sky Harbor Airport, a record at the time. When they moved the facility to Prescott, I retired to the quiet and relatively smog-free atmosphere of the Foothills region north of Phoenix to garden, hike and write. We did the 23-mile trek across the Grand Canyon in one day, spent two nights in the lodge, and hiked back. A long trail for a sexagenarian.

This autumn we're making another voyage, a 7-day cruise on the Caribbean, on the brand new *MS Ecstasy*. It's as different from the old *Alcibiades* as the Boeing 747 is from the open cockpit Stearman.

Bon voyage.

CODE INSTRUCTOR

by Walter C. Warman

Every classroom had something like 72 positions—36 positions on each side of the room with an aisle going up the center. At each position there was a typewriter, a key, a set of headphones and a switchbox that could be switched to any one of eight settings.

There was a raised area and podium where the instructor would teach from. There was also an oscillator, an HRO receiver, and patch panel. We could connect to adjoining code rooms of patch into the receiver. On a table were two tape machines that would feed the oscillator. You could send on two speeds with those machines, or if you wanted to send by hand you could send from the instructor's desk.

Also at the instructor's desk were nine switch boxes that monitored nine tables in the room. Each student would take a table, and the instructor could monitor the students sending and receiving messages to one another.

One day, while we were having sending practice, I was listening around to see how they were doing. One fellow was going like a bat out of hell with no spaces in between words, and not too much in between characters. I listened to him for a little bit and decided to record what he was sending to play back to him later. I told this student to put a piece of paper in his typewriter and to copy this. I played back what he had sent, and he sat there, and he looked around, looked at the typewriter, then back to me, then back to the typewriter. I stopped the machine and asked him what he got, and he said, "I didn't get anything." I said, "Well, you just sent it, you ought to be able to read it. I think it would be a good idea if you start practicing putting spaces in there." I sent a little to him to demonstrate.

It was real interesting reaction to watch him try to read what he had just sent.

"ROOSEVELT MORT"

by Joseph Harold, R-22

. . . Marseilles, France. This city had only recently been re-taken from the Germans. For about two blocks up from the harbor the buildings had been bombed so badly that it was just debris and devastation. Saturation bombing is about as effective as atomic bombs, just takes a little longer.

One afternoon sitting at a sidewalk cafe in downtown Marseilles, a woman came by selling newspapers. She was crying, tears streaming down her cheeks, and the headline in the newspaper read "ROOSEVELT MORT." It surprised me that a resident of a foreign country would take the loss of our U.S. President so hard. . . .

THE "PONDEROSA LEMON"

by Ken Slabotsky, R-47

One evening in November 1943 while chewing the fat in the barracks, the members of R-47 and R-48 got into a discussion about the relative merits of citrus fruit grown in California and Florida, as some of the fellows who had come to Gallups by way of Catalina thought the fruit from California was the best and the group who had been in St. Petersburg thought Florida citrus was the better fruit. The discussion became a little heated and I decided to put in my two-cents worth as a "disinterested" middle American from Kansas.

I told them that they were both wrong as the best citrus by far was grown in Texas! Nobody had any idea that Texas had a citrus crop and right away they belittled my statement. But I stayed in the discussion and told them that Texas grew a lemon that was as big as either Florida or California grapefruit. Well, nobody would buy that and gave me the horse laugh.

What they didn't know was that I had an uncle who grew citrus in the Rio Grande Valley and I'd seen the "Ponderosa Lemon" which is as big as a grapefruit and really is a lemon in all respects.

The upshot of that evening's go-around was that I wrote to my uncle in McAllen, TX and explained the situation and asked him to send me a Ponderosa lemon. About a week before Christmas I received a half-bushel of hand picked fruit from McAllen with labels on each piece which told what it was and how it was grown, etc. And right in the middle of this beautiful package was a big Ponderosa lemon! The others were very surprised to say the least and had to admit I'd been right all along.

Our barracks was connected to the Mess hall and the cooks took that lemon and made two pies out of it which were enjoyed at Christmas dinner.

It is only one of the memories I have of the weeks spent on Gallups Island and I wanted to share it with everyone and I wonder if anyone who was there remembers it.

POST GALLUPS ISLAND EXPERIENCE

by Samuel T. Hucke, Jr. R-15

On my first assignment *(SS Baldhill)* I got seasick from the gentle land swell off the West Coast. Coming back to the docked tanker in Seattle, I stood on the dock and looked at my ship and got seasick before ever going back aboard—talk about suggestion.

Much later, when I was assigned to the *SS Thomas Kearns,* and while we were returning from Europe after Germany's surrender, we had such heavy seas in a North Atlantic storm that we headed south for two days to ride out the storm. The seas were so high that we couldn't prepare food and we ate sandwiches. Some of our crew members even tied themselves to their bunks when they laid down.

Later, after we had headed back westward, a Navy aircraft carrier gradually overtook us. When it was abeam I called them on the blinker light and asked, "How many troops do you have aboard?" The answer: "1800." I then asked, "How many seasick aboard?" The answer: "1800." That must have been a mess, but I was an old salt by then and didn't get sick. Later, when I was back in the Pacific still on the *SS Thomas Kearns,* we had been lying in Okinawa harbor for several days. When we were sent out to sea in a convoy to ride out a typhoon, I got seasick again.

A BANANA BOAT GOES TO WAR

AND TAKES A COTTON PICKER ALONG

by William B. Ishee, R-6

March 29, 1942

Boarded train at Boston's South Station on this Sunday morning, traveling south with three other Gallups graduates. I am almost certain that one other man in the group was Dempsey Breeland of Tangipahoa, LA.

We arrived at New York's Penn Station late in the afternoon, in a light snow flurry. Taxied to YMCA's William Sloane House on 34th Street and took lodging for the night. We, or at least some of us, attended the Sunday evening performance of the Rockettes at Radio City Music Hall. The long program at Music Hall that evening included some colorful Easter skits; a patriotic re-enactment of Fort McHenry and the composition of our National Anthem; the screening of Cecil B. DeMille's "Reap The Wild Wind," starring Ray Milland and John Wayne; and last but not least scary, about a half hour of newsreel, mostly holocaustic scenes of torpedoed burning tankers off our East Coast.

March 30, 1942

Taxied to Maritime Service placement office at 45 Broadway in Lower Manhatten with orders to report to Captain Conway at 9:00 a.m.. We were promptly steered to a placement officer who informed us that he had four ships to be manned with radio operators; two American flag tankers, one American flag freighter and one British flag freighter. Since the tanker assignments were considered to be more nearly suicidal, he declined to make assignments, choosing to avoid favoritism by having each man draw for the vessel that he would take. I drew the British freighter and was, at least momentarily, somewhat relieved.

In short order, the four of us said our goodbyes and departed from 45 Broadway to find our ships. I was picked up by the Vacarro Line (Standard Fruit & Steamship Co.) port purser and driven to their offices on Pier 19, East River. This location was off South Street, not too far from the Brooklyn Bridge and hard by the Fulton Fish Docks, the world's largest (and smelliest) fish markets.

Several hours were spent there with briefing, paper work, and trying to arrange transportation to Norfolk, VA, where I would join the banana boat, *SS St. Mary* . During briefing, I learned that the sistership to St. Mary, the *SS Miraflores,* was then overdue in New York since February 20th, having departed Port au Prince, Haiti on February 14th and disappeared without trace. (Post-war search of the German naval archives revealed the Miraflores had been torpedoed and sunk on Feb. 19th by *U-432*, Schultze commanding.)

Knowing the fate of the sistership, and considering the calamitous state of affairs along our East Coast—the whole picture was not very reassuring to this landlubber about to embark on his first ocean voyage. There were no seats available on plane flights south, so I took a pullman out of Penn Station late in the afternoon.

April 1, 1942

At about daybreak on April Fool's Day, the *SS. St. Mary* pulled anchor, and headed out to sea, turning southward for a voyage to Santiago, Cuba. All ships sailed alone—no convoy—no escort. We were told to steam in daylight hours, and pull into safe anchorage at nightfall.

This marked the beginning of my "special assignment" to the British Merchant Navy. For my first day at sea, I was kept busy copying distress traffic (SOS), as the U-boats continued to pick them off the Virginia-Carolina coast. To make my day, I was seasick. The Jamaican cook did curried lamb for lunch, and the galley ventilator poured out the fumes which then poured into the radio room porthole. That just about did me in.

We were fortunate enough to escape attack and after a week of our "hop, skip, and jump" voyage, we arrived at Santiago on April 7th.

The *St. Mary,* its owners and its crew, were a most unusual encounter about which a book has been written, and a few more could be written. The vessel was one year older than I, having been built in Sunderland, England in 1919. I was the only U.S. citizen aboard. The London-born captain and the Dublin-born chief engineer were naturalized citizens. In a surprising arrangement, these two gentlemen, both veterans of the British Royal Navy, World War I, shared the captain's table with me, a veteran of the Mississippi cotton fields turned "wireless operator." All other officers and crew were citizens of "somewhere else."

The ship's owners were the Vacarro brothers, Sicily born Italian fruit peddlers, transplanted to the New Orleans French Market. In the late 19th century, they were said to have started in business selling in the streets from push carts. By the 1930's they owned a fleet of refrigerated banana boats, "half of Honduras," "half of New Orleans," including the Roosevelt Hotel, and a piece of a few other banana republics. Their ships were all built in England and all but two flew the Honduran flag, with none under the U.S. flag.

So far as I could tell, they employed no Italians on their ships. All officers came from the U.K., and port captains as well. For crew, they seem to have recruited "escapees" from every war and revolution lost in this century. Ashore as well as afloat, they had some odd ones in their employ. To my friends, I have jokingly cited three examples; first, of course, myself; secondly, Seymour Weiss, a famous and/or infamous personality in Louisiana politics and business; and lastly, Jose Sandino, the founder of that bothersome bunch called the Sandinista.

HORD, GEORGE S., R-17

Killed in action, August 19, 1943.

This reminds me of the World War II song, "I Came Here to Talk For Joe" because obviously George can't write his story. George Hord and I (Ray Doerhoff) were the only two Missourians in R-17, and we left Missouri for Gallups Island in June of 1942. I first met George when we arrived on the island and we became very good friends with very much in common. He was known as "Little Missouri" and I was known as "Big Missouri" because of the difference in height . He was the pitcher on our platoon softball team and I was the catcher. We both came from small towns about 50 miles apart in Central Missouri, he from Tebbets and I from St. Elizabeth, each with a population of less than 200. We both left girl friends at home that we intended to marry.

Most of the information in this article came from George's widow, the former Vanita Zeugin, who still lives in Tebbets, MO. and is now married to Bernard Boffa, and from my experiences and conversation with George while on Gallups Island.

George learned about the United States Maritime Training Station for radio operators either through his draft board or from his father, who was head of the county A.S.C. office at that time. George decided he would rather serve his country as a radio operator in the merchant marine rather than take a random assignment to military duty in some other branch of service. This was at a time when the U.S. merchant marine was suffering heavy losses of men and ships, especially in the North Atlantic, due to German U-boats and warships. The U.S. Maritime Service was trying to train radio operators as fast as possible to replace those lost in action and to fill the ever increasing number of new Liberty ships that were being built. With this in mind the draft board agreed that service in the Merchant marine was just as important as service in any branch of the "armed" services.

I will always remember one of our first conversations while we were waiting to be assigned to a barracks on Gallups. George wanted to know where I had taken my Maritime Service physical and I told him, St. Louis. He said, "Did they make you have a circumcision before they passed you?" I told him this was not one of the requirements. He said they had required it of him and as a result he had to spend a few very painful days in the hospital before heading for Gallups Island. Neither of us could figure out how this operation was going to make him become a better radio operator. George made friends easily and soon was on a first name basis with more than 50 people in platoon R-17. He often talked of his girlfriend Vanita, and when he returned from Christmas leave in December of 1942 he proudly announced that he and Vanita were married while he was on leave.

George graduated with R-17 in February, passed his FCC exam, and chose New Orleans for his first ship assignment. He was assigned to the *SS Sandwich*, owned by United Fruit Lines. He boarded the ship on March 27, 1943 for a run to Trinidad. They returned to the U.S. on May 20, 1943. He then returned home to his wife for about a month because United Fruit Lines was waiting for a new Liberty ship to come from the shipyards.

George was then assigned to the *SSJ Pinckney Henderson*, a new Liberty ship from the Todd Houston Shipbuilding Corp. He left Port Arthur, TX on this ship on August 1, 1943. The ship made a stop in the New York area to wither form a convoy or take on additional cargo. According to the publication "The Liberty Ships" published by Lloyd's of London Press, the Henderson was on its maiden voyage from New York to Liverpool carrying a cargo of magnesium, glycerin resin, oil and wax. The next communication Mrs. Hord had regarding her husband was the following telegram from United Fruit Company dated September 4, 1943.

WE DEEPLY REGRET TO INFORM YOU ON ADVICE OF THE WAR SHIPPING ADMINISTRATION THAT YOUR HUSBAND GEORGE S. HORD HAS BEEN REPORTED MISSING AT SEA SINCE NINETEENTH AUGUST 1943.

Mrs. Hord then contacted United Fruit Company for more information and in a letter dated September 13, 1943 received the following information:

THE INFORMATION WE HAVE IS TO THE EFFECT THAT THE S.S. J. PINCKNEY HENDERSON WAS TOTALLY DESTROYED AS A RESULT OF A COLLISION WITH AN OIL TANKER; THAT THE ENTIRE CREW LEAPED OVERBOARD BUT ONLY THREE SURVIVORS WERE ACTUALLY PICKED UP; UNFORTUNATELY, MR. HORD WAS NOT ONE OF THEM.

WE TAKE THIS OCCASION TO EXPRESS TO YOU OUR DEEPEST SYMPATHY IN YOUR BEREAVEMENT.

The information was probably the best available at that time due to war time communication, etc. the following additional information is taken from the publication, "The Liberty Ships."

J. PINCKNEY HENDERSON—AUGUST 18, 1943—IN COLLISION ON MAIDEN VOYAGE OFF NEWFOUNDLAND 44.12 NORTH 53.58 WEST WITH TANKER J.H. SENIOR WHICH WAS LADEN WITH HIGH OCTANE AVIATION SPIRIT. BOTH VESSELS DRENCHED IN AVIATION FUEL AND BECAME RAGING INFERNOS, THE FLAMES SPREADING SO QUICKLY THAT ONLY NINE PERSONS FROM BOTH CREWS SURVIVED. AUGUST 31, 1943 ARRIVED SYDNEY, NOVA SCOTIA IN TOW. BEACHED. FOR MORE THAN THREE WEEKS FLAMES AND EXPLOSIONS RACKED THE SHIP AND WHEN FINALLY EXTINGUISHED THE VESSEL WAS COMPLETELY GUTTED. LATER REFLOATED, TOWED HALIFAX. JANUARY 14, 1944 TOWED TO NEW YORK, JULY, 1944 SCRAPPED IN PHILADELPHIA.

In a letter dated September 25, 1943 from United Fruit Co. Mrs. Hord received the following information:

WE DO KNOW THAT THE SHIP WAS IN COLLISION. AT NIGHT, IN A DENSE FOG AND WE ALSO KNOW THAT THERE WERE BUT THREE SURVIVORS. WE HAVE BEEN ADVISED THAT THE DIVISION OF WARTIME INSURANCE OF THE WAR SHIPPING INSURANCE ADMINISTRATION AND THE MARITIME WAR EMERGENCY BOARD AFTER A VERY THOROUGH INVESTIGATION HAVE CONCLUDED THAT YOUR HUSBAND WAS LOST AND THAT A CERTIFICATE OF PRESUMPTIVE DEATH WILL BE ISSUED. WE ARE ABLE NOW TO ADVISE THAT A MEMORIAL SERVICE WAS HELD AT A CANADIAN PORT AND APPROPRIATE SERVICES WERE CONDUCTED BY THE REVEREND FATHER RONALD MACLEAN AND REVEREND C. K. WHALLEY AND REVEREND T. S. SHERREN OF THE CHURCH OF ENGLAND.

Mrs. Hord made an effort to contact the three survivors from the Henderson and eventually received letters from all three of the survivors. The following are excerpts from those letters:

YOU SEE WE DID NOT HAVE A CHANCE AND IT WAS AT NIGHT. THE SHIPS WERE AFIRE ALL OVER THE WATER. I AM STILL A LITTLE SICK AND WILL WRITE MORE NEXT TIME. (This survivor had been badly burned.)

THE ONLY HONEST FACTS I KNOW ARE THESE: IT HAPPENED AT NIGHT, ABOUT 9:35 DURING AN EXTREMELY DENSE FOG. WE WERE ALMOST IMMEDIATELY AFLAME FROM STEM TO STERN AND CONSEQUENTLY THERE WAS PANIC—DISASTrOUS PANIC. HOW THE THREE OF US LIVED WILL REMAIN A MYSTERY TO ME AS LONG AS I LIVE.

In considering how such a collision could happen we must remember what war time conditions at sea were like. No lights on any of the ships, radar was not yet available, no radio communications permitted, and on top of all this, a very dense fog.

On March 4, 1944 Mrs. Hord gave birth to a son, appropriately named George, in honor of his late father. Because of the tragedy the previous August, father and son were denied the joy of growing up together. Many of us more fortunate now realize how precious a gift this really is. George Hord Jr. now lives in Plano, TX in case any Gallups

Islanders who knew his father would like to contact him.

In a letter dated February 18, 1949 Mrs. Hord received the following information from the War Department in Washington, D.C.

THE DEPARTMENT OF THE ARMY DESIRES THAT YOU BE GIVEN THE LATEST INFORMATION REGARDING THE DISPOSITION OF THE REMAINS OF YOUR HUSBAND, THE LATE GEORGE S. HORD.

IN VIEW OF THE CIRCUMSTANCES OF HIS DEATH, WHICH ALSO CAUSED THE DEATH OF HIS OTHER SHIPMATES FOLLWING THE DISASTER TO THE S.S. J. PINCKNEY HENDERSON ON 19 AUGUST 1944, IT WAS NOT POSSIBLE TO ESTABLISH INDIVIDUAL IDENTIFICATION WHEN THE REMAINS WERE RECOVERED AND BURIAL MADE IN HARDWOOD HILL CEMETERY, NOVA SCOTIA. SUBSEQUENTLY THE REMAINS WERE TRANSFERRED TO NEW YORK AND REPROCESSED BY A SPECIALLY TRAINED IDENTIFICATION UNIT WHICH WAS FURTHER UNABLE TO ESTABLISH INDIVIDUAL IDENTIFICATION OF HIS REMAINS. THE REMAINS REPRESENTED IN THIS GROUP ARE SCHEDULED FOR FINAL BURIAL IN JEFFERSON BARRACKS NATIONAL CEMETERY, LOCATED AT SAINT LOUIS, MISSOURI.

In a letter dated November 13, 1951 Mrs., Hord received the following information from the Department of the Army in Washington, D.C.

REFERENCE IS MADE TO THE INTERMENT OF YOUR HUSBAND, THE LATE GEORGE S. HORD, AND HIS COMRADES, WHICH WAS MADE IN GRAVE 8, SECTION 82, JEFFERSON BARRACKS NATIONAL CEMETERY, ST. LOUIS, MISSOURI. IT IS REGRETTED THAT BECAUSE OF THE FACT IT WAS IMPOSSIBLE TO IDENTIFY INDIVIDUALLY THE REMAINS OF YOUR HUSBAND, YOU WERE DEPRIVED OF THE COMFORT AND CONSOLATION WHICH YOU MIGHT HAVE BEEN AFFORDED BY INTERRING HIS REMAINS AT HOME. IT IS FELT THAT YOU MIGHT LIKE TO HAVE THE ENCLOSED PHOTOGRAPH OF THE STONE WHICH HAS BEEN PLACED AT THE GRAVE. YOU ARE ASSURED THAT THE GRAVE WILL ALWAYS BE CARED FOR IN A MANNER FULLY COMMENSURATE WITH THE SACRIFICE YOUR HUSBAND MADE. ANY DESIRED INFORMATION CONCERNING THE GRAVE OR THE CEMETERY WILL BE FURNISHED UPON REQUEST.

George Hord, our fellow Gallups Islander, chose the merchant marine as the best way to serve his country at a time when the merchant marine was suffering many casualties. Those of us who are writing experiences for this book are the fortunate ones who survived. As I stated at the beginning, this reminds me of the World War II song, "I Came Here to Talk for Joe," so I came here to write for George because he gave everything he had to help assure that the rest of us could continue to live in freedom.

HIGHEST W. W. II CASUALTIES

Members of the U. S. Merchant Marine serving aboard U. S. cargo ships and tankers during W. W. II suffered a greater percentage of war related deaths than did the nations regular armed forces combined.

Branch of Service	Total Number Who Served	War Deaths	Percent of Members	Odds of Being Killed
ARMY	11,268,000	234,874	2.88%	1 in 48
NAVY	4,183,466	36,958	.88%	1 in 114
MARINES	669,108	14,733	2.94%	1 in 34
COAST GUARD	241,093	574	.24%	1 in 417
TOTAL:	16,361,667	287,139	1.76%	1 in 57
MERCHANT MARINES	215,000	6,795	3.16%	1 in 32

Parades and Celebrations

Buddies' Day Parade, Boston, 1945.

Maritime Day, 1945.

Buddies' Day, 1945, two Platoon, 110 and ?.

Mickey Rooney

Greer Garson

Dorothy Merritt and Platoon leader Len Buresh

Betty Hutton

Fred Astaire

Bond rally, Boston, R-35.

Harpo Marx.

Judy Garland.

4th of July Parade, Quincy, MA.

James Cagney

R-35 Color Guard.

Gallups Island and Merchant marine memorials

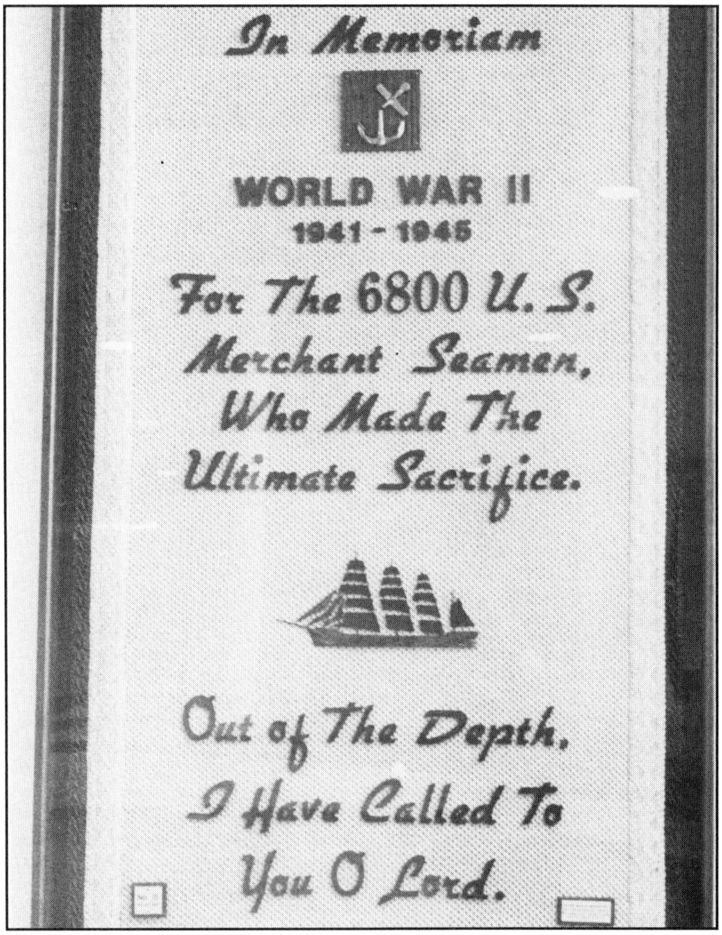
In Memoriam
WORLD WAR II
1941 - 1945
For The 6800 U. S.
Merchant Seamen,
Who Made The
Ultimate Sacrifice.
Out of The Depth,
I Have Called To
You O Lord.

By Their Deeds
Measure Yours
U.S. MARITIME SERVICE
RADIO TRAINING STATION
SUPERINTENDENTS
APRIL 1942 – SEPT. 1942

Nix's Mate Island, in 1700.
What was once a 12 acre island is now marked
by a navigational beacon called "Nix's Mate."
In colonial times the island was used as a sheep pasture.
Captain John Gallop
The island is named for its first owner, Captain John Gallop. A native of Dorset, England, Gallop achieved renown in the colonies for his skill and courage in piloting ships through the treacherous waters of Boston Harbor. At his death in 1649, Gallop owned Nix's Mate Island, a Long Island pasture, and a farm on this island. Successive owners of Gallop's Island included farmers who produced vegetables and milk for ships anchored in the harbor and Margaret Newcomb who established a guest house in 1833.
Lovell's Island and the Brewsters, from Gallop's Island.
King's Handbook of Boston Harbor, M.F. Sweetser, 1885.
Boston Harbor Islands State Park
The preparation of this sign was funded in part through a grant from the
National Endowment for the Humanities.

Aerial drawing of the island.
Photograph over tops of buildings to pier.
Photograph of seamanship class.
Radio communications class.
United States Maritime
Service Radio School
In 1941 the United States Maritime Commission established a radio school on Gallops Island to train merchant marine radio operators. Each year as many as 1,000 students attended classes. Many of the students became radio communicators aboard World War II naval ships. Operation of the school ceased in 1945 and the buildings were subsequently dismantled and removed.
Photograph of island from harbor.
Courtesy of Edward Rowe Snow.
Boston Harbor Islands State Park
The preparation of this sign was funded in part through a grant from the
National Endowment for the Humanities.

Gallups Island Radio Association Roster

ORVAL AAKHUS
LEON ABOUD
JACK ADAMS
RALPH ALBERS
JOSEPH ALEKSUN
CARROLL ALLEN
REX ALLENDER
LESLIE ALT
GEORGE ANDERHEGGEN
OWEN ANDERSON
GREG ABEYTA
AL ADAMS
CECIL ADLER
AMORY L ALLEN
FREDERIC C ALLEN
ROBERT ALLGAIER
FRANCIS ANDERSON
DONALD ALTHOLZ
RAYMOND ANDERSON
WILLIAM ANDERSON
SAMUEL ANDREWS
HERMAN AROND
LEON ASHTON
JAMES AULD
ROBERT AULIK
EDWARD BACKOWSKI
DON BAKER
R LEE BALZER
ROBERT ANDERSON
RICHARD ANDREN
WILLIAM ARCHAMBEAU
DALE D ASHLEY
FRANK ATWOOD
LEONARD AULIK
ALEXANDER AVERA
DAVID M BAKER
JESSE BAKER
JACK BANDAZIAN
COLEMAN BARBER
WALLACE BARNEKE
BENJAMIN BARTOLINI
ANTHONY BASTONE
C WALTER BATY
JACK BAXTER
WILFRED BEAULIEU
OTIS BELL
WILLIAM BENEDICT-PATTERSON
WILLIAM BENNETT
WILLIAM BARKER
CHARLES BARTLETT
WILLIAM BASTIEN JR
ANTHONY BATTISTELLA
STEPHEN BAUMAN
WILLIAM BEADING
KENNETH O BECK
HOWARD BELLMAN
GEORGE BENKO
ROBERT BENCHING
WILLIAM BENTON
JOHN BERENS
CHARLES K BERGEY
ROBERT BERGMAN
NORBERT BERINGER
GORDON O C BESCH
STANTON M BIER
VINCENT BIRGE
JOHN BLACK
ROBERT R BLACK
GEROGE BERCOS
ROBERT BERGER
ROBERT BERGGREN
DR FRED BERGMANN
JOHN BERST
ROBERT BETHEL
CLARENCE BIRELINE
H A BISHOFF
PETER BLACK
ROBERT BLAIR
ZERNIA BLALOCK
DONALD BOGART
GLEN BOHNER
ROBERT BOMAN
JACK BOOKMAN
ARTHUR BOOTH
ROBERT BOUCHARD
DONALD BOYLE
ARTHUR BRADFORD
ROBERT BRAINARD
BASIL BLETHEN
EDWARD BOGDA
JOSEPH BOISSONNEAU
ROBERT BOND
LYNN COOREN
CORNELL BOSTWICK
GERALD BOYD
ROBERT BRACH
MICHAEL BRADLEY
HENRY H BREAZEALE JR
CARL BRENDEL
W L BREWER
ANTHONY BRIGGS
WILLIAM BRIGLIA
DONALD BRONSON

HOWARD BROWN
MARWIN BROWN
JOHN BRUCKER
CHARLES BRUSS
VICTOR BRZOZOWSKI
JOHN BRENNAN
WILLIAM BRIDGES
SALVATORE BRIGLIA
ROBERT BRISSETT
GILBERT BROWN
JOHN G BROWN
MAX BROWNSTEIN
GEORGE BRUSH
HORACE BRYSON
C NELSON BUCKLES
DAVID D BULKLEY
EDWIN BUNDAY
LEONARD BURESH
CHARLES BURKINS
JAMES BURNETT
ROBERT BUSHER
WILLIAM BUTLER
KENNETH BUXTON
CHARLES CADA
ISAAC CALEF
LAWRENCE BULTEMA
GEORGE BURBRINK
EUGENE BURKE
RAYMOND BURLETON
JOHN BURNS
WILLIAM BUSICK
HARVEY BUTT SR
ROBERT BYRON
DAVID L CALDERWOOD
NELSON M CALEY
WILLIAM CALLAHAN
ROBERT CALVELAGE
CHARLES CAMERON JR
LOUIS L CANADAY
HAROLD CANTOR
WILLIAM CANTRELL
HARRY CARLSON
ARCHIE CARPENTER
EARL CARTER
WILLIAM CARTER
RICHARD CALLIHAN
DELMAR CALWHITE
FEDELE CAMPAGNOLA
MAX CANNON
NED CANTRELL
FRANKLIN CARLSEN
CARMINE CARPENELLI
EDWARD CARSTARPHEN
JACK CARTER
ROBERT CARTIER
JOSEPH CARUSO
F R CATELLIER
JAMES CHAMBERS
DR ROBERT CHEATMAN
ROY D CHRISTENSON
MARCEL CLAING
G PRESTON CLANCEY
HENRY CLARK
ROY CLARK
OTTO CLAUS
RAYMOND CASH
BARTHOLOMEW CAWLEY
HERBERT CHARROW
J LAWRENCE CHOINIERE
RALPH CHRISTMAN
ARNOLD CLAMAN
F HOWARD CLARK
LESTER A CLARK
ADRIAN E CLARK JR
JAMES CLAUSEN
JOHN CLEARY
ERNEST CLIFFORD
DAVID COATS
WALTER COLE
RAYMOND COLLINS
EDWARD COLOMBI
GARNIE D CONE
CHARLES CONNER
CLIFFORD CONRAD
WILLIAM COOKE
LEO CLEM
ROBERT CLOUGH
ELMER COCKE
LEO COLEMAN
WILLIAM COLLINS
RICHARD COMBS
WILLIAM CONKLE
PAUL CONNER
FRANK CONWELL
ROBERT COOLEY
MARION COOPER
DESMOND CORCORAN
MELVIN CORDIE
DAVID CORY
WILLIAM COWGER
RICHARD COX
JAMES CRENCA
FRANCIS CROMPTON
GOLDEN CROSS

L TOM CRUSE JR
NATHAN COPELAND
WILLIAM CORCORAN
JERRY CORETTI
FLOYD COVEL
HERBERT COX
BILLY CRAWFORD
CARL CRESSEY
RICHARD H CROSBY
STANLEY CROSS
EDWARD J CUNNINGHAM
DONALD CURCIE
PAHILIP CURTIN
GEORGE CUSHMAN
WALTER DAHLBERG
FRANK DANIK
DALLAS DARLAND
GEORGE DAVIE
CARL DAVIS
JOHN W DAVIS
JACKSON DAVIS
BESSIE CURRY
ROY A CURTIS
WALTER CZESNOWSKI
DORMAN DAHLEN
ALBERT DANKER
WILLIAM DAVENPORT
DAVID DAVIES
DONALD F DAVIS
THEODORE C DAVIS
DAVID DE BREE
JOSEPH DEGEORGE
HERBERT DELANEY
LESLIE DEAN
ANGELO DEMATTEI
WILLIAM DENNEY
WILLIAM DEVOE
RAY DIAMOND
JIMMY DICUS
CHARLES DIPPOLD
VIRGIL DOBSON
STAN DEGIOVANNI
VICTOR DELUCIE
ALBERT DELANO
NORBERT DEMPSEY
FRANCIS DERWIN
JAMES DEZAUCHE
J O DICKINSON
CHARLES DINSMORE
WOODWARD DIXON
ROBERT W DOCTOR
RAY DOERHOSS
JOHN DOLAN
WILLARD DONOVAN
ALBERT DORVAL
WILLIAM DOUGHTY
KEN DOWLING
JOHN DOWNING
PAUL DOYLE
DAL DREHER
SIDNEY DRUCKER
KENNETH DOERING
CHARLES DONLY
ARTHUR DORSEY
PHILIP DOUD
JAMES DOWDY
E F DOWNEY
HENRY DOXSEE
JOHN DRAPER
PAUL DRESSEN
RAYMOND DUBOIS
CHARLES DUARTE
PATRICK DUNNE
MARTIN DURCIANSKY
BERTRAM DYKES
HARRY EDGREN
RUSSELL D EDWARDS
JOHN ELIASON
KIRBY ELTON
PETER ENGEL
MAURICE ENRIGHT
RICHARD DUNHAM
HAROLD DUPUY
LEWIS DYER
C JAY EARL
EDWARD EDWARDS
JOHN EHRLER
EDGAR ELLIS
NAT ENGEL
GORDON ENGLISH
LESLIE WNZ
DONALD EPPLER
MAC ERPS
HENRY ESSIG
ROBERT FALARDEAU
VINCENT FARENGA
WESLEY FARNUM
JOHN FARRINGTON
IRVING FAY
JHAMES FERGUSON
JOSEPH FERLAZZO
FRANCIS ERDLE
WILLIAM ESSAY
KJOHN FAIRFIELD

JOSEPH FARACA
DEL FARISS
JAMES FARRELL
GERARD FASANO
WILLIAM FELLS
WENDALL FERGUSON
ROBERT FIELD
ARTHUR FILLEBROWN
GEORGE FISHER
JOHN FISHER
JAMES FISHED
ROBERT FITZPATRICK
BERNARD FLATOW
EDWIN FLEMING
JOHN FORSBERG
BYRON FORTNER
HENRY FOUNDAS
REINHOLD FISCHER
HOWARD FISHER
SAUL FISHER
PAUL FITZGERALD
JOHN FITZSIMMONS
JAMES FLAVIN
WILLIAM FOGLEMAN
ARTHUR FORSYTH
ARNOLD FOSTER
WESLEY FOWLER
JEROME FOX
ARTHUR FREER
JONATHA FULTON
MERLE FUNDERBURK
FRANCIS GALASSO
STEVEN CARGIA
JACK GATTENIO
WILLIAM GAY
DONALD GEE
WILLIAM GEISEL
EDWARD FRANK
RICHARD FRICK
RICHARD FULTON
DONALD GAGNE
WILLIAM GALVEZ
SHELDON GATES
CHARLES GAVIN
JOHN GEGEON
ARTHUR GEHRMAN
PATRICK GEISELMAN
JOSEPH GELLER
ANTHONY GERCHMAN
CHARLES GIBSON
JOHN GIBSON
ROBERT GILL
ROBERT GISSLOW
MELVIN GLZER
SOL GNATT
DARRELL GODWIN
BERT GOLDRATH
HARRY GERALDI
CHAUNCEY GERCKEN
HOMER GIBSON
THOMAS GIBSON
JOE GILMAKER
RAY GLASCOCK
KENNETH GLESZER
ROBERT GODOWN
JOHN GODZYK
JOSEPH GOMBATTO
VINCENT GONZALEZ
BERTRSAM GOODMAN
JAMES GOODWIN
ROBERT GRANGER
CHARLES GREGORY
LORENZO GRIFFIN
ROBERT GROFF
MAX GROSSMAN
REMINGTON92045
GEORGE GUNDLACH
SAMUEL GOODLEY
DANHY GOODMAN
JOSEPH GRABER
FRED GRANT
JACK GREY
MARLIN GRINNA
BARNEY GROSSMAN
MICHAEL GROSSO
DOUGLAS GULLICKSON
URBAN GUNTNER
GLENN GUTHRIE
RICHARD HACKENBERGER
QLEXANDER HADAD
JOHN HALLORAN
ROBERT HALVORSON
EUGENE HAMMES
MARVIN HAND
CULLEN HANSEN
HAROLD HANSON
RICHARD HARM
LAWRENCE HACAY
W GLENN HACKNEY
DAYLE HALL
RALPH HALTON
ROBERT HAMMAR
CARL HAMMOND
KEITH HANDSAKER

VINCENT HANSEN
RAYMOND HANSON
JOSEPH HAROLD
NAZARET HARONIAN
JAMES HARRELL
JOHN HART
FREDERICK HARTMANN
THOMAS HARZULA
CHARLES HAVENS
FRED HAYES
JOHN HEATON
VERNE HEGGE
ALBERT HEIMBACH
CLAUD HARPER
J MICHAEL HARRIS
WILLIAM HART
RAYMOND HARVEY
WILLIAM HATTWICK
EDWARD HAYDEN
RONALD HAZEN
H JOHN HEFFERNAN
JOSEPH HEIDT
JOHN HEINSTAND
HARVEY HELM
FRANCIS HENRY
PAUL HERMENAU
NEAL HESSELTON
DONALD HEWITT
CHARLES HICKEN
FLOYD HILL
ANGUS HINES
DELBERT HISH
FRANK HOBZA
VINCE HENCIER
WILLIAM HENRY
BERNARD HERR
JIM HESTER
ROSS HICKS
JAMES HIKDRETH
GEORGE HILTON
HERMAN HINZ JR
JOSEPH HOBBS
JAMES HJOCHTRITT
GEORGE HOGAN
EDWARD HOLLAND
RAYMOND HOLLAND
PHILLIP HOLMES
LEON HOLSTER
GUY HORNER
WILLIAM HOUSTON
SAMUEL HUCKE
J CLIFTON HULBERT
STANLEY HOLDEN
KENNETH HOLLAND
EDWARD HOLLEY
H CHARLES HOLLISIAN
RON HOLMQUIST
NICHOLAS HORANSKY
THOMAS HOULDENB
FRANK HUBBARD
JAMES HUETT
JOSEPH HULL
WILLIAM HUME
WARREN HYSON
WILLIAM ISHEE
CLARK ISON
CHARLES JACKSON
RALPH JACKSON
ARNOLD JANGORD
JAMES JARRARD
FRANK JENNE
NORMAN JENNINGS
GEORGE HUTRCHINS
LEONARD ISBISTER
BOSTON HARBOR ISL STATE PARK
SAMUEL ISSOKSON
HOWARD JACKSON
HROLD JAMES
FLAVIUS JANKAUSKAS
CLARENCE JEFFERS
HOWARD JENNINGS
STANLEY JENNINGS
DENNIS JENSEN
A KENNETH JOHNSON
ELMER JOHNSON
HENRY JOHNSON
JAY L JOHNSON
WALTER JOHNSON
CLARK JONES
FRANK JONES
JAMES JONES
SHERMAN JONES
MARCUS JINKS
EDGAR JOHNSON
GOLDEN JOHNSON
JAMES JOHNSON
M HERBERT JOHNSON
JAMES JOLLY
DONALD JONES
HUGH T JONES
ROBERT JONES
LEONARD JORDAN
RAY JORGENSON

JOHN KATTAK
JOHN KAY
J ORVIS KEENE
CHARLES KELLEY
LAWRENCE KELSEY
JOHN KESLER
JACK KILGORE
JAMES KING
GORDON JORGENSEN
CHARLES KALSON
EUGENE KAUDER
WILLIAM KEE
JOHN KEIGLEY II
WILLIAM KELLY
CLYDE KERNS
ROY KIDMAN
JOHN KIMBLE
RAYMOND E KING
JAMES KINKEL
WILLIAM KIRES
CHESTER KLINGENSMITH
MYRLEN KNUTESON
ROY KOLEHMAINEN
WALTER KONOPKA
HENRY KOWALSKI
EUGENE KRENT
ALFRED KRON
ALBERT KUBANY
LORIN KINSEL
WALLACE KLANDRUD
WALTER KLIS
DR DEL KOHLER
W DURWARD KOLL
EDWARD KOST
EUGENE KRAWCZAK
ORLANDO KROLL
RAYMOND KRUEGER
NORBERT KUCALA
CHARLES KUHRY
KENNETH KUNZE
JAMES KYLER
JACK LACEY
JOSEPH LANG
ROBERT LANPHERE
ELDON LARSON
NORMAN LASSETTER
ROBERT LESAGE
WENDELL LEAVITT
LOUIS KULMA
HARRY KUZYK
EDWARD LAROCQUE
L J LANCASTER
CHARLES LANTZ
LESLIE LARSON
PHILIP LAYEUX
RUFUS LEA
ROBERT H LEE
ROBERT LEFERVE
ARMAND LEMMA
EDWARD LEWIS
ROBERT LEWIS
ALBERT LINDEN
MEL LINDSEY
JOSEPH LODATO
JAMES LOGSDON
THEODORE LONG
MAYNARD LONIS
LESLIE LEIGH
SCHLEY LEMMA
ELBERT LEWIS
MICHAEL LIASKOS
GARETH LINDER
ROBERT LOCKE
BENJAMIN LODWICK
KENNETH LONG
JAMES LONG
FRANK LOPATA
WILLIAM LORRAINE
PHILIPPE LOZEAU
ED LUTZ
JULIAN LYNCH
GEORGE MACZALI
WILLIAM MADSON
MELVIN MAHER
JOHN MALLORY
GEORGE MANN
LYLE MAPLES
JOHN LOWE
ROBERT LUKENBILL
JOHN LYNCH
WILFRED MACLEOD
BILL MADDOX
LLOYD MAFFETT
FRANK MALINZAK
BRUNO MANCUSO
LEO MANUS
SAM MARCHINSKY
ROSCOE MARICLE
JOHN MARION
CLYDE MARRS
HAROLD MARSCH
PALMER MARTINO
BERNARD MASTERS
RAYMOND MAURSTAD
KENNETH MAYNARD
EDWARD MCCARTHY
GERALD MCCLISH
SALVATORE MARINO
MARTIN MARKOWITZ
LARRY MARSALA
GEORGE MARSHALL
CHARLES MASON
GLENN MAUER
ROBERT MAYHEW
HOMER MCANINCH
JOSEPH MCCARTHY
REUBEN MCCONNEY
RICHARD MCCORD
LOUIS MC COURT
JAMES MCCULLOUGH
TERANCE MC DONALD
DANIEL MCELHINNEY
RUSSEL MCILHATTAN
CLAUDE MCKEE
NORMAN MCKELVEY
JAMES MCKINLEY
ARTHUR MCMULLEN
JOHN MCCORMICK
EUGENE MCCOY
CHARLES MCDONALD
JOHN MCDONNELL
EUGENE MC HALE
WARREN MCKAY
JACK MCKEEHAN
PETER MCKEON
JOHN MCKINNON
MAX MCMULLEN
JAMES MCNALLY
HAROLD MCPHEE
IRVING MELTZER
ALBERT MERCURIO
GEORGE MERRIMAN
GEORGE MEUSE
JOHN MEZEY
DARWIN MIKESELL
EDWARD MILLER
JEFFERSON MILLER
JOHN MCNULTY
BRUCE MEADOWCROFT
JOHN MENSING
EDWARD MERRILL
THOMAS MESTAYER
WILLIAM MEYER
JULIAN MIKELL
BILLY MILLER
JAMES MILLER
JOHN D MILLER
JOSEPH MILLER
RADFORD MILLER
RALPH MILLOY
CARL MINTZLAFF
ROBERT MIRVISH
BRUCE MITCHELL
JAMES MITCHELL
ROBERT MITCHELL
RUDOLPH MITCJKO
LEONARD MOFFITT
RICHARD MOODY
LOWELL MILLER
ROBERT MILLER
PATRICK MINEO
PHILIP MIONE
STAN MISORSKI
DOANLD MOFFENBEIER
DONALD MOODY
MURDOCK MOORE
ROBERT MOORE
HARRISON MOORE
JOHN MORAN
MARION MORAN
WARREN MORGAN
GERALD MORRELL
WILLIAM MORRIS
ARTHUR MORSE
MELVIN MORSE
HUGO MOSSER
CHARLES MUNYAN
ROY MOORE
ROBERT MORGENBERGER
RONALD MORRIS
WILLIAM MORRISON
CURTIS MORSE
EWALTER MOSKWA
ALVIN MULLEN
JOHN MURPHY
JACK MURRAY
LEON MYERS
ARTHUR NATVIG
JACK NEESON
MELVIN NELSON
NELS NELSON
ALOYSIUS NEPPL
ALEXANDER NEWBOLD
STANLEY NEWMAN
HUBERT NEWTON
JOHN MYERS
CONRAD NAGLE
PAUL NEEDLER
ARVID NELSON
MICK NELSON
PAUL NELSON
EDGAR NESMAN
CHARLES NEWBOLD
FRED NEWSOM
PAUL NICHOLSON
HAROLD NIELSEN
ROBERT NORDBY
HAROLD NOWLAND
JOHN O'HARE
CHARLES O'LEARY
EARLE OLSON
WESLEY OLUND
PAUL ORLOVE
JOHN OSIKOWICZ
SALVATORE NOCELLA
LOREN NORBERG
CHARLES NORVIL
JOHN O'DONNELL
JOSEPH O'HEARN
DON OLSON
HUBERT OLSON
WILLIAM OPALKA
JOHN ORR
JOHN OST
JOSEPH OSTROFF
LAWRENCE OWENS
JAMES PAFUMI
DOMINIC PALLAZOLA
KENNETH PALMER
CLAUDE PALUSKI
FRANK PAPP
CLIFFORD PAQUIN
CHARLES PARISH
GEORGE PARKER
JAMES OWENS
PAUL IOZBUN
JAMES PAINE
JOHN PALMER
RICK PALMER
JOSEPH PANAK
PETER PAPPAS JR
JAMES PARCE
CHARLES PARKER
RICHARD PARKS
DON PAULSEN
SAM PEARSALL
MARVIN PEASE
CHARLES PEDERSON
FRED PEPER
JOHN PERKINS
GEORGEW PETERSON
ALBERT PETTUS
ROBERT PHELPS
OSWALD PAXTON
EVERETT PEARSON
THOMAS PEDDICORD
ROBERT PEEBLER
CHESTER PEREZ
FRANCIS PERRY
JOHN PETERSEN
ANDREW PETRICK
JOSEPH PHELPS
THEODOERE PHELPS
PASQUALE PIACENTE
EDWARD PLEULER
HOWARD POOLE
BOBBY PIUNCEY
ALEXANDER PRASINOS
WARREN PRATT
FRANK PRITCHETT
CONRAD PROFT
JOSEPH PUGLIA
JOHN PINO
ROBERT POLLITT
WILLIAM PTEET
CARROLL POUNCEY
EDWARD PRATHER
THOMAS PRIOR
RALPH PROCK
JOSEPH PRUSAS
ROSARIO PULEO
TONY RADSPIELER
C W RAMPY
EDWIN RASANEN
EVREMONDE RAYNOR
GEORGE REILAND
ALEXANDER REIS
GROVER RENICK
BARNIE REYNOLDS
FRANK RICKER
DONALD RIDER
EWUGENE RAETZ
MORTON RAPHELSON
LESTER RAUBER11765
PERCY REID
ALBVERT REINHARDT
FRANK REISDORF
SELIG RESNICK
ROBERT RICHELSON
GERALD RICKSHAM
WILLIAM RIEDEL
DANIEL RIEU
FRANK RINES JR
FRED ROBERTS
EARL ROBINSON
JOHN RODGERS
MARTIN ROEMER
PETER ROGERS
ORVILLE ROONES
EUGENE ROSENFIELD
HENRY ROWLAND
CARL RILEY
EDGAR RIST
JAMES ROBERTS
WILLIAM ROBINSON
FRED ROEBUCK
BEN ROESHMAN
JAMES ROHAM
CLAIR ROSENBERGER
PAUL ROSTAK
S JEROME RUBIN
WALTER RUDAT
DAVID RULE
WILDURR RUSSELL
NORMAN RYAN
FRANK SABOURIN
JAMES SADLER
MAYNARD SANDBERG
CARROLL SAUNDERS
WILLIAM SCHAFER
RICHARD SCHIFF
ROBERT RUEDISUELI
DON RUNMARK
VERNON RUTTER
OAPL RYSTAD
CLARENCE SACKETT
CLARENCE SACKETT
EDWARD SAMARDAK
MELVIN SANDERSON
ROBERT SCHACHT
RICHARD SCHIFANI
WILLIAM SCHRADE
FRANK SCHROEDER
ROBERT SCHULTZ
PAUL SCHWAB
JAMES SCOTT
ANDREW SCRUDATO
HARRY SEAMAN
LEE SEESE
TRUMAN SEILER
MARVYN SEXAUER
RICHARD SHARKEY
LELAND SCHULTZ
CHARLES SHUUR
ROY SCHWARTZ
STEPHEN SCOTT
JACK SEALE
GEORGE SEATON
WILLIAM SEIGLER
GEORGE SENSIBAR
JOHN SHAKESPEARE
WILLIAM SHARPE
CLARENCE SHAW
ARTHUR SHEDDAN
FRANCIS SHELGREN
DANIEL SHEPPARD
BOB SHETTERLY
JOSEPH SIFFERLIN
BILL SIMMS
GEORGE SIMPSON
NORBERT SINAMARK
FRED SIWIEC
MELVIN SHEARER
THOMAS SHELDON
CHARLES SHELLENBERGER
ROBERT SHERMAN
LEON SIENKIEWICZ
JOSEPH SILVA
JOHN SIMON
THOMAS SIMPSON
LAWRENCE SINTAY
MARVIN SKYDELL
KENT SLABOTSKY
JOHN SLOAN
ARTHUR SMITH
CARL SMITH
EDWARD SMITH
JAMES SMITH
JARED SMITH
JOSEPH SMITH
ELMR SMITLEY
G DAVID SPAINHOWER
LLOYD SLATER
ARNOLD SMITH
BILL SMITH
DR DONALD SMITH
I BRUCE SMITH
JAMES SMITH
JOSEPH SMITH
SAMUEL SMITH
THOMAS SOFRANKO
CHARLES SPELLINGS
EDGAR TIMM
KLIMENT TISTAN
HAROLD TOMPKINS32904
GERALDS TOZER
JOHN TRAVERS
GLEN TRIMBLE
ROBERT TROTTER
JUNN TROY
JAY TURNER
THOMAS UNRUH
LEONARD TIRNAUER
EDWIN TJARDES
GENE TONTONI
GEORGE TRANSOM
WILLIAM TRICE
RAYMOND TRIPLETT
GORDON TROUT
ALFRED TSCHAEKOFSKE
GEORGE UHRIG
JOHN VALESKA
JACK VALLEY
JOHN VENTERS
WILLIAM VIENT
ROBERT VINSON
JAMES VON TUNGELN
WALDEN WADSWORTH
CAREY WALDEN
CHARLES WALKER
VERNON WALKER
BILLY WALLER
DOANLD VARNUM
WILLIAM VERNON
THEODORE VINCENT
RICHARD VOLLSTEDT
WILLIAM VOSBURGH
DR RICHARD WAECHTER
DONALD WALDROP
ROY WALKER
KEITH WALLACE
ANDERSON WARD
JOHN WARD
WALTER WARMAN
WILLIAM WARNER
CARL WATSON
KENNETH WATSON
ROBERT WATTS
DAN WEBER
JOHN WEHRY
DANIEL WEISSER
JOHN WELCH
FRANK WARMACK
OTHO WARNER
CLARENCE WARREN
JAMES WATSON
RICHARD WATSON
J RUSSELL WEBB
MORRIS WEEKS
WALTER WEISS
MORRIS WEISSMAN
WALTER WELLINGTON
JOHN WENTZ
ALBERT WESLEY
MERRILL WEST
JEFF WHISENHUNT
JOHN WHITAKER
CECIL WHITE
STANLEY WHITEMAN
HARRY WICKHAM
EDWARD WILDER
CHARLES WILLIAMS
MAYNARD WENTZEL
HAROLD WEST
RICHARD WESTLEY
WILLIAM WHISLER
ARTHUR WHITE
HOWARD WHITEHOUSE
STANLEY WHITNEY
ANTHONY WIEJEK
JULIAN WILDER
ARCHIE WILLIS
HERBERT WILLIS
EDMUND WILLRICH
EDWARD WILSON
ROBERT WILSON
ROSS WINEBARGER
RICHARD WISE
CHARLES WOEBER
WARREN WOOD
CARL WORTMAN
WYIT WRIGHT
MONROE WILLNER
DELOY WILSON
MASSILON WILSON
SAWYER WIMBERLY
CXLYDE WINSTEAD
WILLIAM WITTKOWSKI
WILLIAM WOGLOM
LORING WORDELL
STEPHEN WOWCZUK
NICHOLAS WYNNICK
MARTIN YANUZZI
SIDNEY YOUNG
JOHN YURICEK
SAMUEL ZIEGLER
EDDIE ZINK
JOHN YORK
WILLIAM YOUNT
FRED ZAHRN
HARRISON ZIESKE
HAROLD ZISKIN

INDEX

OF SURNAMES, SHIPS AND KEYWORDS

ALL CAPS =Submitted Biography; Ships listed by name only; Platoon = Beginning of that group's biographies.

F

G

H

I

J

K

L

S

T

U

V

W

Y

Z

U.S. MARITIME SERV
RESIDENT RADIO SCHOOL
GALLUPS ISLAND
BOSTON MASS.
ADMINISTERED FOR THE U.S. MARITIME COMMISSION
BY THE
U.S. COAST GUARD